# Lasers and Dynamic of Systems

# Lasers and Dynamic of Systems

Editors

**Vicente Aboites**
**Juan Hugo García López**
**Rider Jaimes-Reategui**

MDPI

Basel • Beijing • Wuhan • Barcelona • Belgrade • Novi Sad • Cluj • Manchester

*Editors*
Vicente Aboites
Laboratorio de Laseres
Centro de Investigaciones
en Óptica
León de los Aldamas
Mexico

Juan Hugo García López
Centro Universitario
de los Lagos
Universidad de Guadalajara
Lagos de Moreno
Mexico

Rider Jaimes-Reategui
Centro Universitario
de los Lagos
Universidad de Guadalajara
Lagos de Moreno
Mexico

*Editorial Office*
MDPI
St. Alban-Anlage 66
4052 Basel, Switzerland

This is a reprint of articles from the Special Issue published online in the open access journal *Photonics* (ISSN 2304-6732) (available at: www.mdpi.com/journal/photonics/special_issues/Y716EFPT9R).

For citation purposes, cite each article independently as indicated on the article page online and as indicated below:

| Lastname, A.A.; Lastname, B.B. Article Title. *Journal Name* **Year**, *Volume Number*, Page Range. |
|---|

**ISBN 978-3-7258-0918-9 (Hbk)**
**ISBN 978-3-7258-0917-2 (PDF)**
**doi.org/10.3390/books978-3-7258-0917-2**

# Contents

# About the Editors

**Vicente Aboites**

Vicente Aboites is a Fellow of the British Institute of Physics; he graduated as a physicist from the Universidad Autónoma Metropolitana in Mexico and as a philosopher from Birkbeck College of the University of London in Great Britain. He carried out doctoral research at the Rutherford Laboratory in Oxfordshire and presented his thesis at the University of Essex in England. He completed specialty studies at the University of Paris and a postdoctorate at the University of Berlin. He was a Marie Curie Fellow visiting researcher at Imperial College London, among other institutions. In 1986, he founded the Laser Laboratory of the Optical Research Center (CIO), where he is a professor and senior researcher. He has also been a visiting researcher and teacher at the Mathematics Research Center (CIMAT) and the Faculties of Philosophy, Physics, and Mathematics of the University of Guanajuato, among other national and foreign institutions. He has published twenty books and around two hundred international articles with strict review. He has supervised several dozen PhD, Master's, and Bachelor's students. He is a member of the British Society for the Philosophy of Science, the National System of Researchers, the International Society of Philosophy, and the Mexican Academy of Sciences.

**Juan Hugo García López**

Juan Hugo Garcia Lopez is Senior Researcher Professor Titular C of the Department of Science and Technology of the University of Guadalajara, Campus CULagos, from Jalisco, Mexico. J.H. Garcia Lopez studied for his Ph.D. at the Center for Optical Research of the University of Guanajuato in 2003; he conducted his doctoral research work at the Blackett Laboratory of the Imperial College of the University of London with the thesis "Solid-state laser pumping by high power laser diode" and obtained the mention of "Honored work,". He obtained his Master's degree in Electric Engineering (specialty: Instrumentation and Digital Systems) from the University of Guanajuato (1999) and his Bachelor's degree in electrical engineering from the Technological Institute of Tuxtla Gutierrez (1994). He is a member of the National Researcher System (SNI I) and a member of the Mexican Academy of Physics (SMF). He was Head of the Department of Science and Technology and Academic Secretary of CULagos of the University of Guadalajara, an expert in the design and modeling of laser systems, non-linear and quantum optics, and laser dynamics, and the author of many scientific papers.

**Rider Jaimes-Reategui**

Rider Jaimes-Reátegui received his Ph.D. from the A. C. CIO Optics Research Centre in Mexico. He conducted his doctoral research at the University of Los Angeles, California, USA, on the Dynamics of Complex Systems with Parametric Modulation: Duffing Oscillators and a Fiber Laser (2004). Currently, he is a member of the National System of Researchers (S.N.I.) Level II Mexico, with recognition as Professor with Desirable Profile by the Secretariat of Public Education (SEP) Mexico; since 2004, R. Jaimes-Reategui has been Senior Research Professor C of the Centro Universitario de los Lagos of the University of Guadalajara, in Lagos de Moreno, Jalisco, Mexico. In addition, he is a member of the Consolidated Academic Board: Optics, Complex Systems, and Innovation of the University of Guadalajara and a member and founder of the Laboratory of Dynamic Systems and Lasers, Centro Universitario de los Lagos, University of Guadalajara. His skills and experience are in nonlinear optics, nonlinear fiber optics, laser dynamics, erbium-doped fibers, multistability fiber lasers, laser diodes, optoelectronics, fiber optics technology, nonlinear dynamics synchronization, chaotic dynamics, mathematical modeling, and dynamical systems.

# Preface

The ongoing advancements in lasers and their diverse applications, coupled with their interaction with dynamic systems, emphasize the importance of studying them. These advancements have led to numerous applications across various fields. Recently, there has been a surge in research interest surrounding disruptive technologies, and this reprint, generated from the Special Issue "Laser and Dynamics of Systems", aims to consolidate new ideas and present groundbreaking discoveries that could pave the way for future applications.

Laser technology, which has been extensively explored and studied, has significantly influenced the progress of other disciplines. This reprint seeks to present interesting research, showcasing highlights from original investigations at the forefront of laser technology and system dynamics.

**Vicente Aboites, Juan Hugo García López, and Rider Jaimes-Reategui**
*Editors*

photonics

MDPI

icle

# oute to Chaos in a Unidirectional Ring of Three Diffusively oupled Erbium-Doped Fiber Lasers

é Octavio Esqueda de la Torre [1], Juan Hugo García-López [1,*], Rider Jaimes-Reátegui [1,*], illermo Huerta-Cuellar [1], Vicente Aboites [2] and Alexander N. Pisarchik [3]

[1] Optics, Complex Systems and Innovation Laboratory, Centro Universitario de los Lagos, Universidad de Guadalajara, Enrique Díaz de León 1144, Paseos de la Montaña, Lagos de Moreno 47460, Mexico; jose.edelatorre@alumnos.udg.mx (J.O.E.d.l.T.); guillermo.huerta@academicos.udg.mx (G.H.-C.)

[2] Lasers Laboratory, Optical Research Center, Loma del Bosque 115, Col. Lomas del Campestre, Leon 37150, Mexico; aboites@cio.mx

[3] Center for Biomedical Technology, Universidad Politécnica de Madrid, Campus Montegancedo, 28223 Madrid, Spain; alexander.pisarchik@ctb.upm.es

* Correspondence: jhugo.garcia@academicos.udg.mx (J.H.G.-L.); rider.jaimes@academicos.udg.mx (R.J.-R.)

**Abstract:** We numerically investigate the dynamics of a ring consisting of three unidirectionally coupled Erbium-Doped Fiber Lasers (EDLFs) without external pump modulation. The study focuses on the system behavior as the coupling strength is varied, employing a six-dimensional mathematical model that includes three variables for laser intensities and three variables for population inversions of all lasers. Our primary objective is to understand the system evolution towards chaos from a stable equilibrium in the ring, considering the impact of increasing coupling strength. To analyze the system's behavior, we employ various techniques such as time series analysis, power spectra, Poincaré sections, bifurcation diagrams, and Lyapunov exponents. During the transition to chaos, the system undergoes a Hopf bifurcation and a series of torus bifurcations. An essential aspect of this study is the exploration of a rotating wave propagating along the ring, where the wave nature (periodic, quasiperiodic, or chaotic) depends on the coupling strength. Additionally, we observe the coexistence of periodic and chaotic orbits within a specific range of the coupling strength. However, for very strong coupling, this bistability disappears, resulting in a monostable system with a single limit cycle. This regime exhibits potential for applications that demand short laser pulses with a substantial increase in peak power, reaching nearly 20 times higher levels compared to the continuous mode when the lasers are uncoupled. This discovery holds particular importance for optical communication systems, especially considering the attenuation optical signals experience when transmitted over long distances.

**Keywords:** laser; network; ring; dynamics; coupling; synchronization; multistability

check for updates

**ion:** Esqueda de la Torre, J.O.; ía-López, J.H.; Jaimes-Reátegui, Iuerta-Cuellar, G.; Aboites, V.; rchik, A.N. Route to Chaos in a lirectional Ring of Three ısively Coupled Erbium-Doped : Lasers. *Photonics* **2023**, *10*, 813. ;://doi.org/10.3390/ onics10070813

ived: 25 May 2023
sed: 6 July 2023
pted: 9 July 2023
ished: 12 July 2023



## 1. Introduction

In recent decades, there have been remarkable advancements in the research and commercialization of fiber lasers. These lasers have undergone revolutionary progress, fueled by their widespread utilization in various fields such as optical communications, optical sensing, laser surgery, nonlinear optics, and optical materials [1–6]. Additionally, fiber amplifier technology has emerged as an exceptionally practical platform for industrial applications, primarily due to its compactness, robustness, reliability, and high efficiency. The alignment-free structure and spatial beam profile further enhance its appeal in industrial settings.

Among various types of fiber lasers, erbium-doped fiber lasers (EDFLs) offer several advantages that make them highly suitable for optical communications [7]. The EDFL active medium comprises an optical fiber doped with erbium ions. When the diode-pumped laser light interacts with these erbium ions, it results in high gain and supports a single transverse mode, provided that the fiber parameters are appropriately chosen. First, ED can easily be integrated into optical communication networks due to the compact s of their optical components. Second, the laser wavelength, particularly around 1550 is widely utilized in optical communication systems due to its minimal losses in opti fibers [8]. Third, EDFLs exhibit a diverse range of dynamical behaviors, including per doubling, chaos, and multistability [9,10]. Their dynamics can be controlled and levera not only for chaotic communication [2,11] but also for numerous other applications, such spectral interferometry [12], optical coherence tomography [13], optical sensing [14], opt metrology [15], industrial micromachining [16], LIDAR systems [17], and medicine [18

Nonlinear effects in fiber lasers are typically considered undesirable due to th potential to disrupt stable laser operation, introduce deviations from a Gaussian pu shape, and hinder the achievement of a diffraction limit [19]. However, recent studies h revealed that nonlinear effects can be harnessed for certain applications. Conseque there has been significant research conducted on nonlinear phenomena in EDFLs [19– with a specific focus on investigating multistability in these laser systems [9,10,22–26].

Multistability characterized by the coexistence of multiple stable states or attract under a specific parameter set, is a fascinating phenomenon observed in various fie of science, including physics, engineering, chemistry, biology, and medicine (see [27] a references therein). The ability to control multistability enables the selection of desi attractors or the elimination of undesirable ones [28]. In the context, EDFLs subjected periodic modulation of the laser pump current or cavity losses exhibit the coexistence of to four periodic attractors. These attractors have different periodicity, with longer peri corresponding to higher pulse energies. For instance, the laser pulse amplitude in period-5 regime is approximately 50 times higher than that in the period-1 regime [ High pulse power fiber lasers have found numerous applications, including in cutt welding, and surgery, and particularly in optical communications [7], where they enable long-distance transmission of optical signals without the need for frequent amplificat Recently, by employing selective control of multistability, giant pulses were successfu generated in an array of EDFLs [30].

While the dynamics of a solitary EDFL have been extensively studied [9,10,22– there is still a lack of comprehensive investigation into the dynamics of coupled ED in different coupling configurations. The study of the dynamics of networks of coup oscillators has garnered significant attention across various scientific disciplines. W multiple oscillators are coupled, the range of potential behaviors becomes more comp making the analysis of governing equations challenging. Each oscillator may only connected to a few immediate neighbors [31].

One of the simplest network configurations is a cycle ring of coupled oscillators [ The ring dynamics can be highly complex, particularly when the oscillators are coup unidirectionally, even if the individual uncoupled units exhibit stable equilibrium. Previ works have explored interesting dynamics of ring-coupled oscillators, including cha synchronization in three coupled electrical circuits [33], chimera states in nonlocally coup oscillators [34], and the generation of delays in coupled CMOS inverters [35], among oth

In addition, the ring configuration is particularly attractive because it allows rotat phase waves to propagate along the coupled nodes [36–39]. Such waves were first foun a ring reactor of reaction–diffusion systems [40,41]. Rotating waves arise when a homo neous state becomes linearly unstable due to a Hopf bifurcation [42]. Later, Nekorkin a colleagues [43] discovered traveling waves propagating in a ring of coupled bistable ph oscillators with sinusoidal nonlinearity. It should be noted that unidirectional couplin of particular interest because a signal is transmitted from one subsystem to another with receiving feedback. Moreover, unidirectional rings were explored in coupled electr circuits based on Chua [44], Lorenz [45,46], and Duffing [47,48] models where rotat waves were also discovered. Transitions from a stable equilibrium through quasiper icity to chaos and hyperchaos with respect to the coupling strength were observed in rings of unidirectionally coupled Lorenz [45,49], Duffing [50], and Rulkov [51] oscillat

The mechanism leading to such transitions was studied in detail in coupled autonomous Duffing oscillators [47,52–54].

When studying dynamics in ring-coupled lasers, it is important to acknowledge that previous investigations have predominantly focused on semiconductor lasers [55–57]. These studies have demonstrated that when operating in an uncoupled state, semiconductor lasers operate in a continuous-wave regime. However, once the coupling strength surpasses a certain threshold, the lasers transform into an oscillatory state. As the coupling strength further increases, the dynamics follow a route to chaos through a series of Hopf bifurcations, leading to periodic, quasiperiodic, and chaotic oscillations. Within the chaotic range, various synchronization states, ranging from asynchronous behavior to phase synchronization, have been observed.

In this paper, we study, for the first time to our knowledge, the behavior of a ring of EDFLs as the coupling strength is increased, aiming to address the following key questions: What is the route to chaos observed in the EDFL ring? How do the lasers achieve synchronization along this route? By answering these questions, we can gain valuable insights into the collective behavior of coupled EDFLs and make informed decisions regarding optimal laser parameters and configurations for network performance. Since fiber and semiconductor lasers belong to the same class-B lasers [58], we anticipate observing a similar route to chaos in the ring of coupled EDFLs. Through the application of time series analysis, bifurcation diagrams, Poincaré sections, power spectra, and Lyapunov exponents, we demonstrate that while the dynamics of this ring share similarities with other coupled oscillators [50], it also possesses distinctive characteristics inherent to EDFLs.

Long ago, Landau and Hopf discovered a transition to turbulence via a sequence of successive Hopf bifurcations [59,60]. Later, Newhouse, Ruelle, and Takens (NRT) [61] proved that the 3D torus decays into a strange chaotic attractor immediately after the third successive Hopf bifurcation, due to the effect of an arbitrarily small perturbation of the so-called NRT scenario. Although several validations of this effect were verified in a large family of frameworks, little attention was paid to a study of the NRT scenario in optical systems such as lasers, and especially in fiber lasers. Therefore, one of the aims of this work is to analyze this scenario in a cyclic ring of three diffusively coupled EDFLs.

This paper is structured as follows. In Section 2, we present the model of a single EDFL with pump modulation. In Section 3, we introduce the model of a ring of three coupled EDFLs. In Section 4, we investigate the dynamics of the system through the analyses of time series, bifurcation diagrams, Poincaré sections, power spectra, and Lyapunov exponents. Finally, in Section 5, we summarize the main conclusions drawn from our study.

## 2. Laser Model

To describe the dynamics of diode-pumped EDFLs, we employ the power-balance approach, which accounts for excited state absorption (ESA) in erbium at the 1.5 μm wavelength. This approach also considers the averaging of population inversion along the pumped active fiber. By incorporating these factors, the model captures the underlying mechanisms that give rise to intrinsic oscillations in the laser, even in the absence of external modulation, as observed experimentally [10,23].

The balance equations for the intracavity laser power, denoted as $P$ and expressed in units of $s^{-1}$ (representing the sum of powers of the contra-propagating waves inside the cavity), and the averaged population $y$ of the upper level (referred to as "2" and ranging from 0 to 1 as a dimensionless variable), are derived as follows:

$$\dot{P} = \frac{2L}{T_r} P\{r_w \alpha_0 (y[\xi - \eta] - 1) - \alpha_{th}\} + P_{sp} \tag{1}$$

$$\dot{y} = -\frac{\sigma_{12} r_w P}{\pi r_0^2} (\xi y - 1) - \frac{y}{\tau} + P_{pump}, \tag{2}$$

where $\sigma_{12}$ is the cross-section of the absorption transition from the ground state "1" to the upper state "2". We suppose that the cross-section of the return stimulated transition $\sigma_{12}$ is practically the same in magnitude that gives $\xi = (\sigma_{12} + \sigma_{21})/\sigma_{12} = 2$, $\eta = \sigma_{23}/\sigma_{12}$ be the coefficient that stands for the ratio between ESA $\sigma_{23}$ and ground-state absorption cr sections at the laser wavelength. $T_r = 2n_0(L + l_0)/c$ is the lifetime of a photon in the ca ($l_0$ being the intra-cavity tails of FBG-couplers), $\alpha_0 = N_0\sigma_{12}$ is the small-signal absorpt of the erbium fiber at the laser wavelength, $N_0 = N_1 + N_2$ is the total concentratior erbium ions in the active fiber), $\alpha_{th} = \gamma_0 + nL(1/R)/(2L)$ is the intra-cavity losses on threshold ($\gamma_0$ being the nonresonant fiber loss and $R$ the total reflection coefficient of FBG-couplers), $\tau$ is the lifetime of erbium ions in the excited state "2", $r_0$ is the fiber c radius, $w_0$ is the radius of the fundamental fiber mode, and $r_w = 1 + \exp[2(r_0/w_0)^2]$ is factor addressing a match between the laser fundamental mode and erbium-doped c volumes inside the active fiber.

The population of the upper laser level "2" is given as

$$y = \frac{1}{n_0 L}\int_0^L N_2(z)dz,$$

where $N_2$ is the population of the upper laser level "2", $n_0$ is the refractive index of a "c erbium-doped fiber core, and $L$ is the active fiber length),

$$P_{sp} = \frac{10^{-3}y}{\tau T_r}\frac{\lambda_g}{w_0}\frac{r_0^2\alpha_0 L}{4\pi^2\sigma_{12}}$$

is the spontaneous emission into the fundamental laser mode, and the pump power is

$$P_{pump} = P_p\frac{1 - \exp[-\beta\alpha_0 L(1-y)]}{n_0\pi r_0^2 L},$$

where $P_p$ is the pump power at the fiber entrance and $\beta = \alpha_p/\alpha_0$ is the ratio of absorpt coefficients of the erbium fiber at pump wavelength $\lambda_p$ and laser wavelength $\lambda_g$. We assu that the laser spectrum width is $10^{-3}$ of the erbium luminescence spectral bandwidth. N that Equations (1) and (2) describe the laser dynamics without external modulation.

The parameters used in our simulations correspond to the real EDFL with an act erbium-doped fiber of $L = 70$ cm. Other parameters are $n_0 = 1.45$, $l_0 = 20$ cm, $T_r = 8.7$ $r_0 = 1.5$ cm, and $w_0 = 3.5 \times 10^{-4}$ cm. The last value was measured experimentally a it was a bit higher than $2.5 \times 10^{-4}$ cm given by the formula for a step-index single-m fiber $w_0 = r_0(0.65 + 1.619/V^{1.5} + 2.879/V^6)$, where the parameter $V$ relates to numer aperture $NA$ and $r_0$ as $V = 2\pi r_0 NA/\lambda_g$, while the values $r_0$ and $w_0$ result in $r_w = 0.3$

The coefficients characterizing resonant-absorption properties of the erbium-do fiber at lasing and pumping wavelengths are $\alpha_0 = 0.4$ cm$^{-1}$ and $\beta = 0.5$, respectiv and correspond to direct measurements for heavily doped fiber with erbium concen tion of 2300 ppm, $\sigma_{12} = \sigma_{21} = 3 \times 10^{-21}$ cm$^2$, $\sigma_{23} = 0.6 \times 10^{-21}$ cm$^2$, $\xi = 2$, $\eta =$ $\tau = 10^{-2}$ s [10], $\gamma_0 = 0.038$, and $R = 0.8$ that yields $\alpha_{th} = 3.92 \times 10^{-2}$. Finally, the gen tion wavelength $\lambda_g = 1.56 \times 10^{-4}$ cm ($h\nu = 1.274 \times 10^{-19}$ J) is measured experiment while the maximum reflection coefficients of both FBGs are centered on this wavelen The pump parameters are the excess over the laser threshold $\varepsilon$ defined as $P_p = \varepsilon P_{th}$, wh the threshold pump power

$$P_{th} = \frac{y_{th}}{\tau}\frac{n_0 L\pi w_p^2}{1 - \exp[-\alpha_0 L\beta(1 - y_{th})]}$$

and the threshold population of the level "2"

$$y_{th} = \frac{1}{\xi}\left(1 + \frac{\alpha_{th}}{r_w\alpha_0}\right)$$

with the pump beam radius taken, for simplicity, to be the same as that for generation ($\omega_p = \omega_0$).

### 3. Normalized Equations

To simplify the laser model and generalize it in a dimensionless form, we transform Equations (1) and (2) into the simple form (all parameters and other details can be found in [62]):

$$\frac{dx_1}{d\theta} = ax_1y_1 - bx_1 + c(y_1 + r_w), \quad (8)$$

$$\frac{dy_1}{d\theta} = -dx_1y_1 - (y_1 + r_w) + e\left\{1 - \exp\left[-\beta\alpha_0 L\left(1 - \frac{y_1 + r_w}{\xi_2 r_w}\right)\right]\right\}, \quad (9)$$

where $x_1$ is the laser intensity and $y_1$ is the population inversion with parameters

$$a = 2L\left(\frac{\tau_{sp}}{T_r}\right)\left(\frac{\xi_1}{\xi_2}\right)\alpha_0 = 6.620 \times 10^7,$$

$$b = \left\{2L\left(\frac{\tau_{sp}}{T_r}\right)\left[\alpha_{th} - \frac{\alpha_0(\xi_1 - \xi_2)}{\xi_2} r_w\right]\right\} = 7.4151 \times 10^6,$$

$$c = \left(\frac{\tau_{sp}}{\xi_2 r_w}\right) = 0.0163,$$

$$d = \left(\frac{\tau_{sp}\xi_2 r_w \sigma_{12}}{\pi r_0^2}\gamma\right) = 4.0763 \times 10^3,$$

$$e = P_p \frac{\tau_{sp}\xi_2 r_w}{N_0 \pi r_0^2 L} = 506.$$

By numerically solving the system of Equations (8) and (9) using the fourth-order Runge–Kutta method, we generate time series. To simulate the laser dynamics, we utilize parameters that are close to those employed in the experimental study [22]. Specifically, we set the initial pump power $P_p^0$ to $7.4 \times 10^{19}$ s$^{-1}$, resulting in the laser relaxation oscillation frequency $f_0$, equal to $f_r = 28.724$ kHz. This frequency is determined from the power spectrum of the damped oscillations presented in Figure 1a. It is evident that solving the laser Equations (8) and (9) leads to a stable fixed point. Although the period of these oscillations is not constant, we determine the dominant frequency $f_r$ in the spectrum shown in Figure 1b, which corresponds to the average frequency of relaxation oscillations. This frequency closely aligns with the laser's fundamental frequency, and we adopt this value for our simulations.

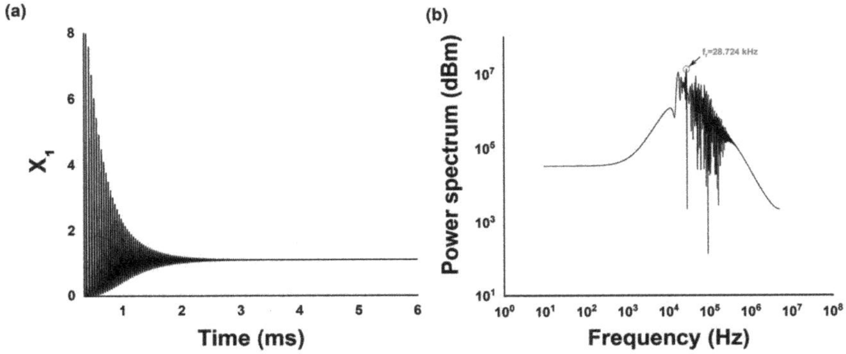


**Figure 1.** (**a**) Time series with relaxation oscillations and (**b**) power spectrum with relaxation oscillation frequency $f_r$ of EDFL given by Equations (8) and (9).

## 4. Dynamics of the Ring of Three Unidirectionally Coupled EDFLs

Ring-coupled oscillators can be conceptualized as a recurrent cycle of interactions [
Even with just three oscillators, this simple network motif can exhibit thirteen disti
configuration patterns [64], distinguished by various combinations of unidirectional a
bidirectional couplings. In this study, our primary focus is on the configuration w
unidirectional coupling, as depicted in Figure 2. In this configuration, each laser acts
both a slave and a master oscillator simultaneously. The dynamics of this ring can
mathematically described by two differential equations representing the laser intensity
$(j = 1, 2, 3)$ and the population inversion $y_j$, given as follows:

$$\frac{dx_j}{d\theta} = ax_jy_j - bx_j + c(y_j + 0.3075),$$

$$\frac{dy_j}{d\theta} = dx_jy_j - (y_j + 0.3075) + P_{pmod_j}\left\{1 - \exp\left[-18\left(1 - \frac{1 - (y_j + 0.3075)}{0.6150}\right)\right]\right\}$$

with pumping

$$P_{pmod_j} = 506\left[1 + k(x_{j-1} - x_j)\right],$$

where $k$ is the coupling coefficient.

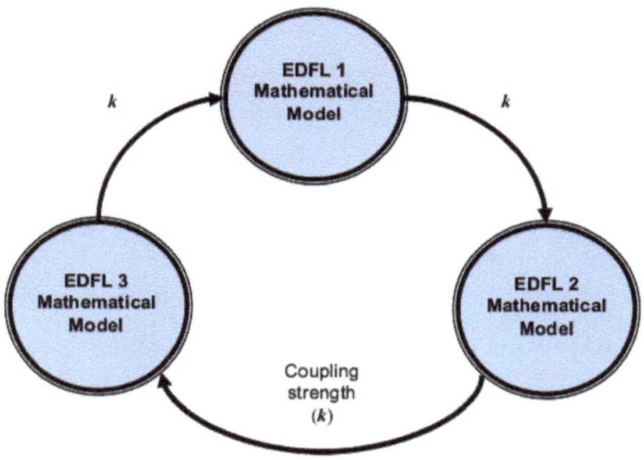


**Figure 2.** Schematic diagram of a unidirectional ring of three diffusively coupled EDFLs.

With the ring configuration's inherent symmetry, the dynamics of each laser wit
the ring are identical. In Figure 3, we present the bifurcation diagram illustrating the p
amplitude of a single laser ($x_1$) and the largest Lyapunov exponent $\lambda$ as functions of
coupling strength $k$. The bifurcation scenario approaches the Landau route, characteri
by the transition from a stable equilibrium to chaos through quasiperiodicity via success
Hopf bifurcations [59,60]. However, another scenario was described by Newhouse, Rue
and Takens (the so-called NRT scenario [61]), who found that just after the third H
bifurcation, a chaotic attractor appears in the form of a three-dimensional torus. Remarka
our model described by Equations (10) and (11) exhibits a similar scenario, display
a transition to hyperchaos as the coupling strength $k$ is increased.

The term $k(x_{j-1} - x_j)$ represents a diffusive coupling, which refers to the differe
in intensities between neighboring lasers. Experimentally, the coupling between adjac
lasers can be realized as follows. First, the laser intensities are measured with photode
tors. Then, the difference between these intensities is multiplied by the desired coupl
strength $k$, and finally the light with the resulting intensity is injected into the neighb
ing laser.

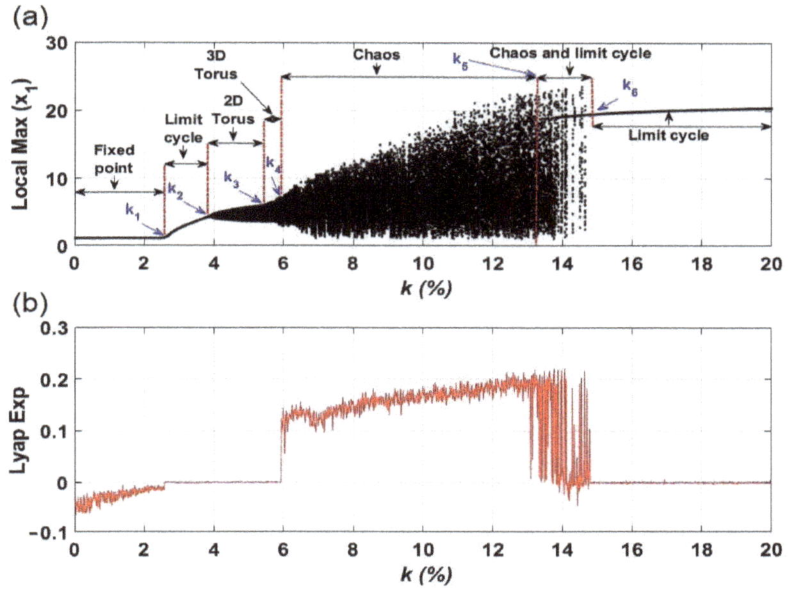


**Figure 3.** (**a**) Bifurcation diagram of the peak intensity (in arbitrary units) and (**b**) largest Lyapunov exponent as a function of $k$.

The time series and Poincaré sections depicted in Figure 4 provide a detailed visualization of the dynamic regimes observed on the route from a stable fixed point to chaos. As the coupling strength is increased from $k = 0$ to approximately $k_1 \approx 2.58$, the system undergoes a Hopf bifurcation, transforming the equilibrium state (Figure 1) into a limit cycle (Figure 4a). During this transition, the largest Lyapunov exponent approaches zero, indicating the onset of a periodic behavior. This periodic regime persists within a relatively small range of $2.58 < k < 3.81$.

Subsequently, at $k_2 = 3.82$, the limit cycle undergoes a transition to a quasiperiodic regime characterized by a two-dimensional (2D) torus, as depicted in Figure 4b. This transition is observed when the second largest Lyapunov exponent approaches zero. It should be noted that within the 2D torus region, the Lyapunov exponent occasionally takes positive values. This occurrence can be attributed to the disparity in step sizes employed by Wolf's algorithm [65]. However, these positive values are transient, as the Lyapunov exponent returns to zero in subsequent steps.

As the coupling strength further increases, a three-dimensional (3D) torus emerges at $k_3 = 5.49$ (Figure 4c), and the third largest Lyapunov exponent approaches zero. Thus, within the region $5.49 < k < 5.83$, the system exhibits a quasiperiodic regime.

At the critical value $k_4 = 5.84$, the system enters a chaotic regime (Figure 4d), characterized by a positive largest Lyapunov exponent. Another critical point occurs at $k = 13.2$, where an inverse crisis bifurcation emerges. In this scenario, a stable periodic orbit arises from the initial chaotic regime, resulting in a bistable behavior for the particular value of $k$, featuring both periodic oscillations and chaos. As the coupling strength continues to increase, a stable limit cycle is once again established at $k_6 = 14.81$ (Figure 4e), as indicated by the return of the largest Lyapunov exponent to zero.

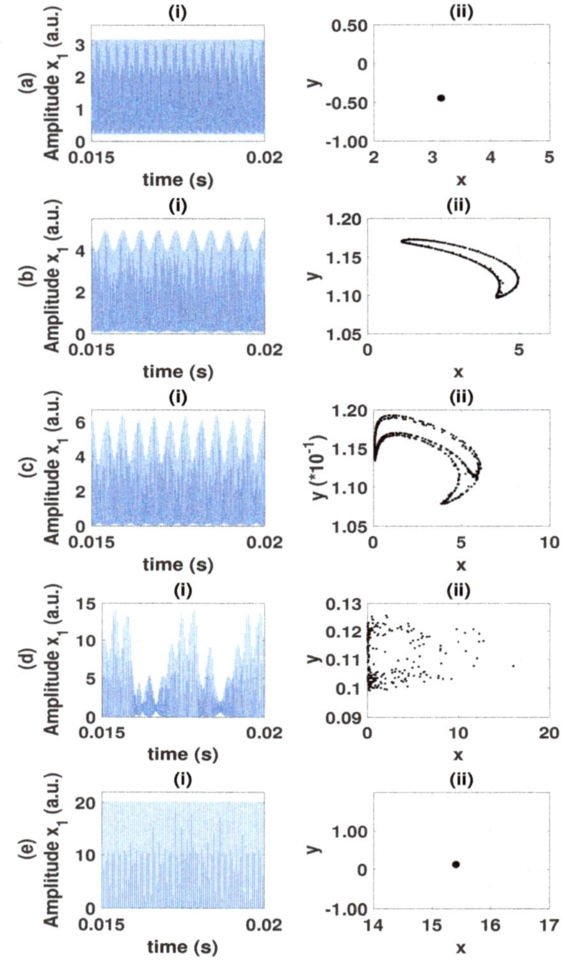


**Figure 4.** (**i**) Time series and (**ii**) Poincaré sections at (**a**) $k = 2.58$ (periodic orbit), (**b**) $k = 3.82$ torus), (**c**) $k = 5.49$ (3D torus), (**d**) $k = 5.84$ (chaos), and (**e**) $k = 14.81$ (periodic orbit).

### *4.1. Rotating Wave*

An intriguing phenomenon known as rotating wave occurs at certain values of coupling strength. This rotating wave arises from the phase difference between nei boring lasers and propagates along the ring structure. In the time series depicted Figure 4b–d, the rotating wave is evident as a slowly evolving envelope, showcasing p odic, quasiperiodic, or chaotic behavior. It serves as a distinctive dynamical feature of ring-coupled oscillators.

The existence of a periodic rotating wave was initially discovered in a ring of coup Chua oscillators [66,67], and later in the rings of coupled Lorenz [49,68] and Duffing osc tors [47,54,69]. The rotating wave operates in a similar manner to external modulation, induc low-frequency oscillations. When the periodic rotating wave interacts with the local limit cy of each oscillator, a local 2D torus is formed, resulting in quasiperiodic dynamics (Figure 4c

As the coupling strength is further increased, the local 2D torus interacts with quasiperiodic rotating wave, leading to the emergence of a local 3D torus (Figure With continued increases in the coupling strength, the rotating wave interacts with local 3D torus and undergoes a transition to chaotic behavior (Figure 4e). Finally, for very strong coupling, the interaction between the rotating wave and the chaotic orbit results in the stabilization of a limit cycle.

The oscillatory regimes are depicted through time series plots in Figure 5 for four distinct coupling strengths: $k = 2.58$, $k = 3.82$, $k = 5.49$, and $k = 5.84$. In these plots, the only variation among the oscillators lies in their initial phase, resulting in phase shifts at each successive node. Consequently, this leads to the formation of a rotating wave within the cyclic ring. The left column of the figure showcases the two-dimensional time series patterns of the three oscillators, clearly revealing oblique stripes that represent the existence of rotating waves. The inclination of these stripes indicates the propagation of phase waves along the ring of oscillators. The right column of the figure illustrates the phase portraits of the corresponding attractors, which are identical for all oscillators due to the system's symmetry. Notably, as the coupling strength is increased, the size of the attractor expands. In the right column, the one-dimensional time series of the three lasers are presented, providing further evidence of the rotating wave by depicting the phase difference between the lasers.

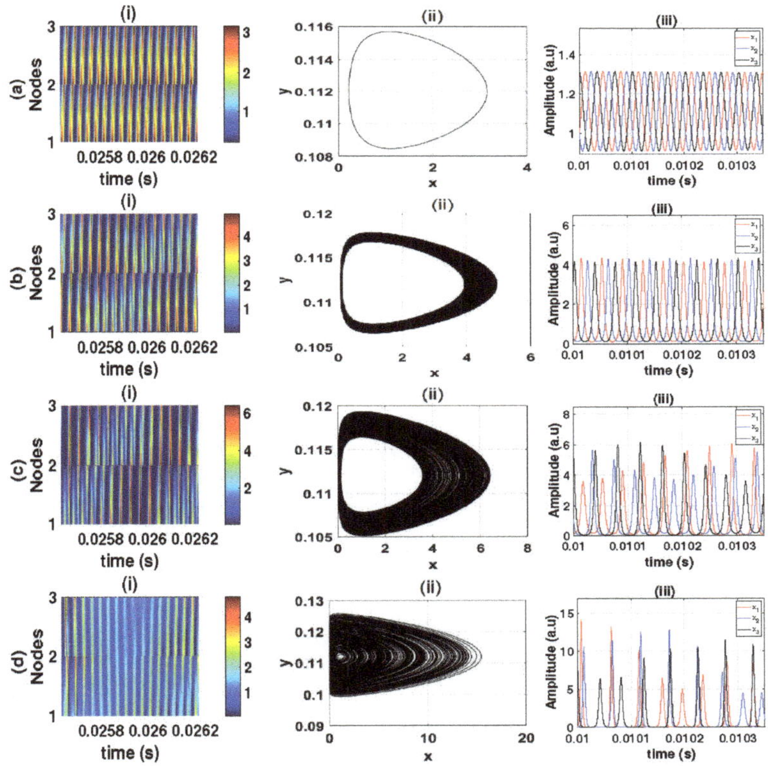


**Figure 5.** (i) Rotating wave **(ii)** phase portraits and **(iii)** time series for **(a)** $k = 2.58$, **(b)** $k = 3.82$, **(c)** $k = 5.49$, and **(d)** $k = 5.84$. The color in the left column indicates the laser intensity.

*4.2. Power Spectrum Analysis on the Route to Chaos*

Power spectrum analysis utilizing the fast Fourier transform (FFT) is a valuable t in various scientific and engineering disciplines for investigating system dynamics [70, This analysis complements traditional qualitative and quantitative research technique the study of dynamical systems, including Poincaré maps, bifurcation diagrams, and L punov exponents.

In Figure 6, we present a bifurcation diagram showcasing the dominant frequency the power spectra of $x_1$ as a function of the coupling strength $k$. Furthermore, in Figur we provide illustrative power spectra for fixed values of $k$. Upon inspecting the bifurcat diagram depicted in Figure 6, it becomes evident that the first Hopf bifurcation occur $k_1$, signifying the transition from a stationary to a periodic solution. In the power sp trum displayed in Figure 7a, a solitary peak corresponding to the fundamental oscillat frequency $\Omega_0$ emerges.

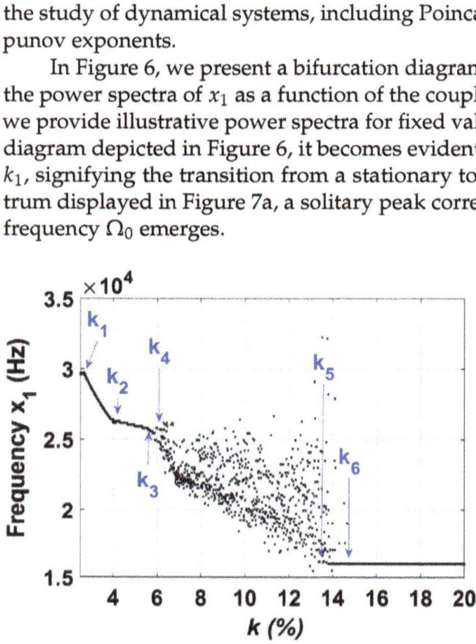


**Figure 6.** Bifurcation diagram of the dominant frequency in the power spectra of $x_1$ as a functio the coupling strength $k$.

As the coupling strength is further increased, we encounter the second Hopf bi cation at $k_2$, leading to the transformation of the limit cycle into a quasiperiodic solut characterized by two incommensurate frequencies, namely $\Omega_0$ and $\Omega_1$ (refer to Figure The existence of a 2D torus is observed until the subsequent Hopf bifurcation at $k_3$, wh gives rise to a transition towards a quasiperiodic solution with three frequencies, result in a 3D torus. In the power spectrum displayed in Figure 7d, the emergence of a th independent frequency $\Omega_2$ becomes prominent. The dominance of the 3D torus is obser within the interval $k_3 < k < k_4$. While the power spectrum in Figure 7c appears bro band and indicative of chaotic behavior, a comparison between the Poincaré section Figure 4c(ii),d(ii) clearly reveals the distinction between the 2D and 3D tori.

A continued increase in the coupling strength $k$ eventually leads to the destruct of the 3D torus, resulting in a direct transition to chaos within the range of $k_4 < k <$ The chaotic response is evident in the power spectrum, displaying numerous freque peaks that are randomly distributed with varying amplitudes. This type of behavior been observed in other dynamical systems as well. For instance, Sánchez et al. [49] stud a ring of unidirectionally coupled Lorenz oscillators and observed the transition fr a periodic rotating wave to a chaotic rotating wave via quasiperiodicity. Additiona Borkowski et al. [53] reported the observation of a rotating wave in a ring of seven directionally coupled Duffing oscillators using spectral and bifurcation analyses. Th examples highlight the presence of similar phenomena in different dynamical systems

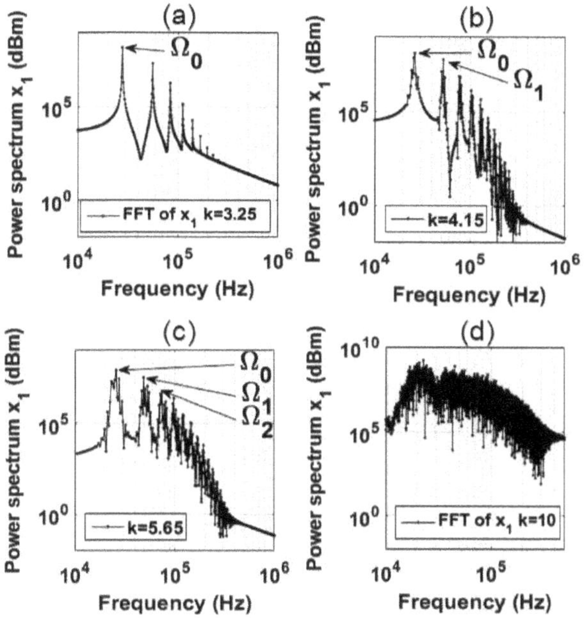


**Figure 7.** Power spectra for (**a**) $k = 3.25$, $\Omega_0 = 28$ kHz (periodic orbit), (**b**) $k = 4.15$, $\Omega_0 = 52.32$ kHz, $\Omega_1 = 25.83$ kHz (2D torus), (**c**) $k = 5.65$, $\Omega_0 = 25.83$ kHz, $\Omega_1 = 49.01$ kHz, $\Omega_2 = 74.83$ kHz (3D torus), and (**d**) $k = 10$ (chaos).

### 4.3. *Coexistence of Attractors*

In the course of the transition from a stable fixed point to chaos, a region of bistability emerges within a narrow range of the coupling strength, specifically $k_5 < k < k_6$. Within this range, two distinct attractors coexist. Figure 8 showcases the phase portraits and corresponding power spectra of these coexisting periodic and chaotic regimes for a specific value of $k = 14.08$. Recently, a similar phenomenon of multistability was observed in the dynamics of a ring of three fractional-order double-well Duffing oscillators [72]. In that study, the interplay between the fractional order index and coupling strength led to the coexistence of stable fixed points, limit cycles, 2D and 3D tori, as well as chaotic behavior, with the specific outcome depending on the initial conditions.

In our system, for coupling strengths in the range of $13 < k_6 < 14.81$, a stable limit cycle coexists with chaos, as depicted in the bifurcation diagram shown in Figure 3. The chaotic and periodic orbits interact with the rotating wave, resulting in the emergence of a monostable limit cycle. This observation bears a resemblance to findings reported in other studies (e.g., [73,74]), where multistability was controlled through a secondary sinusoidal perturbation. In our case, the rotating wave functions analogously to the secondary sinusoidal perturbation, effectively amplifying the laser pulse power. As a result, the lasers operate in a pulsed regime, emitting extremely short pulses with substantial amplitudes. It is important to note that this regime differs fundamentally from traditional Q-switching, which typically requires cavity quality losses, such as an intra-cavity saturable absorber [75,76]. In our case, the pulsed regime arises solely from the increasing coupling strength.

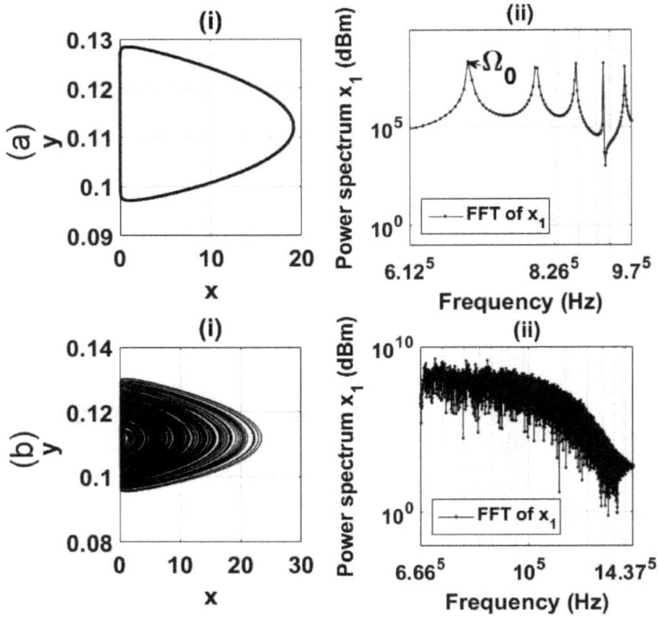


**Figure 8.** (**i**) Phase portraits and (**ii**) power spectra of coexisting (**a**) stable limit cycle and (**b**) chao $k = 14.08$.

### 4.4. Synchronization

The rotating waves in the ring of coupled oscillators can be analyzed from the p spective of synchronization. Specifically, phase synchronization between a pair of EDFL and $j$) can be quantitatively characterized by the difference between their instantane phases [64]:

$$\Delta\phi_{i,j} = \phi_i - \phi_j,$$

$$\phi_{i,j} = \arctan\left(\frac{y_{i,j}}{x_{i,j}}\right).$$

At the same time, identical synchronization between a pair of EDFLs can be determi by the synchronization error

$$\epsilon_{ij} = \sqrt{(x_i - x_j)^2 + (y_i - y_j)^2}.$$

Figure 9 illustrates the synchronization scenarios described by Equations (13) and ( The plot demonstrates the dependence of time-averaged phase synchronization and aver synchronization error on the coupling strength $k$. It reveals the synchronization patt transitioning from a stable equilibrium to chaos.

Within the interval $k_1 < k < k_2$, a stable limit cycle is observed, characterized by ph locking near zero or perfect phase synchronization between the $\Delta\phi_1$ and $\Delta\phi_2$ phases o and $x_2$, respectively. As the coupling strength is increased, the system enters the range $k_2 < k < k_3$ and $k_3 < k < k_4$, where phase locking is lost, and two additional frequen $\Omega_1$ and $\Omega_2$ emerge, leading to the formation of 2D and 3D tori, respectively. Moving i the range of $k_4 < k < k_6$, phase synchronization is completely lost, resulting in cha behavior. Finally, for $k > k_6$, the phase locking regime reappears.

Thus, the phase difference between the coupled EDFLs gives rise to the phenomenon of the rotating wave, which propagates along the ring in a periodic, quasiperiodic, or chaotic manner. Consequently, achieving identical synchronization becomes unattainable, as evidenced by the nonzero average synchronization error depicted in Figure 9, down.

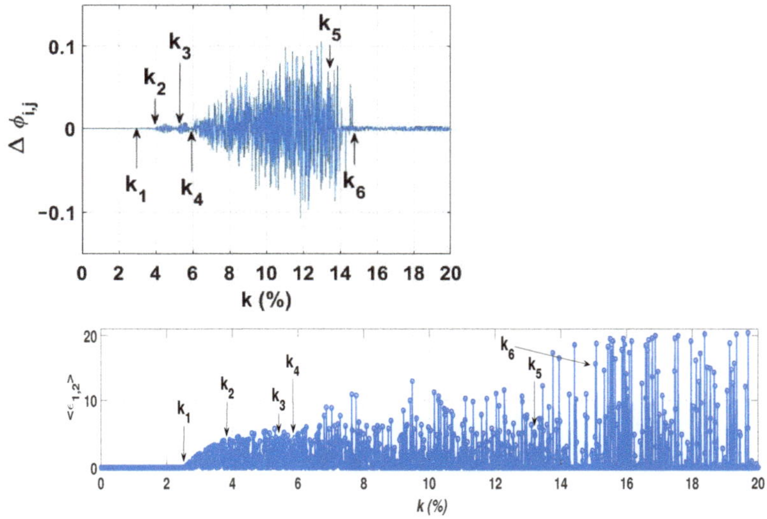


**Figure 9.** (**Up**) Averaged phase synchronization and (**down**) synchronization error versus coupling strength $k$.

## 5. Conclusions

In this study, we conducted a numerical investigation into the dynamics of a system consisting of three unidirectionally ring-coupled EDFLs as a function of the coupling strength. By employing a six-dimensional mathematical model that accounts for laser intensities and population inversions of all lasers, we extensively examined the system's behavior on the route from a stable equilibrium to chaos in the ring. Our analysis encompassed various tools including time series analysis, bifurcation diagrams, power spectra, Poincaré sections, and Lyapunov exponents. A key focus of our study was the exploration of the rotating wave propagating along the ring, which exhibited periodic, quasiperiodic, or chaotic behavior depending on the coupling strength.

Furthermore, we made a noteworthy observation regarding the coexistence of periodic and chaotic rotating waves within a specific range of coupling strength. As the coupling strength increased, this bistability ceased to exist, and the system became monostable, characterized by a single limit cycle. We explained this stabilization mechanism as the interaction between the chaotic and periodic orbits with the rotating wave, where the rotating wave acted as a secondary sinusoidal perturbation, leading to the annihilation of the chaotic attractor.

Of particular interest was the significant increase in the peak power of laser pulses due to phase locking under strong coupling conditions. In fact, for coupling strengths exceeding $k_6$, all EDFLs operated in the pulsed mode, producing very short high-amplitude pulses. This finding has promising implications for applications requiring high-power laser pulses. We achieved a nearly 20-fold increase in peak power compared to the continuous mode observed in the absence of coupling. This outcome holds considerable significance in optical communication, where optical signals experience significant attenuation during propagation along optical fibers. The utilization of optical amplifiers based on the nonlinear properties of EDFLs allows for the generation of sufficiently high power in transmitted optical signals. The coexistence of different pulsed regimes with varying pulse amplitudes, along with the proper control of bistability, offers valuable opportunities for obtain high-power laser pulses.

It is important to acknowledge the limitations of this study. Our investigation prima concentrated on the simplest configuration of a three-laser ring, and as such, the con sions drawn may not directly extrapolate to larger laser networks. However, we h the belief that certain dynamical characteristics exhibited by our system could po tially extend to larger laser networks, thereby warranting further investigation in fut research endeavors.

**Author Contributions:** J.O.E.d.l.T.: writing—original draft, writing—review and editing, metho ogy, software, validation, and visualization. J.H.G.-L.: writing—review and editing, resources, project administration. R.J.-R.: writing—original draft, supervision, funding acquisition, writing—rev and editing, and resources. V.A.: writing—review and editing, and resources. G.H writing—review and editing, validation, and methodology. A.N.P.: writing—review and edit visualization, conceptualization, and data curation. All authors have read and agreed to the publis version of the manuscript.

**Funding:** This project was supported by: Programa Presupuestario F003 CONACYT–MEXI Convocatoria "Ciencia Básica y/o Ciencia de Frontera. Modalidad: Paradigmas y Controversia la Ciencia 2022", under project number: 320597.

**Institutional Review Board Statement:** Not applicable.

**Informed Consent Statement:** Not applicable.

**Data Availability Statement:** Not applicable.

**Acknowledgments:** J.O.E.d.l.T. thanks CONACYT for financial support (CVU-854990). R.J.-R. tha CONACYT for financial support, project No. 320597.

**Conflicts of Interest:** The authors declare no conflict of interest.

## References

1. Digonnet, M.J. *Rare-Earth-Doped Fiber Lasers and Amplifiers, Revised and Expanded*; CRC Press: Boca Raton, FL, USA, 2001.
2. Luo, L.; Chu, P. Optical secure communications with chaotic erbium-doped fiber lasers. *J. Opt. Soc. Am. B* **1998**, *15*, 2524–2 [CrossRef]
3. Shay, T.; Duarte, F. Tunable Fiber Lasers. In *Tunable Laser Applications*; CRC Press: Boca Raton, FL, USA, 2009; pp. 179–196.
4. Pisarchik, A.N.; Jaimes-Reátegui, R.; Sevilla-Escoboza, R.; García-López, J.H.; Kazantsev, V.B. Optical fiber synaptic sensor. *Lasers Eng.* **2011**, *49*, 736–742. [CrossRef]
5. Mary, R.; Choudhury, D.; Kar, A.K. Applications of fiber lasers for the development of compact photonic devices. *IEEE J. Sel. Quantum Electron.* **2014**, *20*, 72–84. [CrossRef]
6. Zhao, L.; Li, D.; Li, L.; Wang, X.; Geng, Y.; Shen, D.; Su, L. Route to larger pulse energy in ultrafast fiber lasers. *IEEE J. Sel. Quantum Electron.* **2017**, *24*, 1–9. [CrossRef]
7. Zervas, M.N.; Codemard, C.A. High power fiber lasers: A review. *IEEE J. Sel. Top. Quantum Electron.* **2014**, *20*, 219–241. [Cross
8. Castillo-Guzmán, A.; Anzueto-Sánchez, G.; Selvas-Aguilar, R.; Estudillo-Ayala, J.; Rojas-Laguna, R.; May-Arrioja, D.; Martí Ríos, A. Erbium-doped tunable fiber laser. In Proceedings of the Laser Beam Shaping IX, International Society for Optics Photonics, San Diego, CA, USA, 11–12 August 2008; Volume 7062, p. 70620Y.
9. Saucedo-Solorio, J.M.; Pisarchik, A.N.; Kir'yanov, A.V.; Aboites, V. Generalized multistability in a fiber laser with modula losses. *J. Opt. Soc. Am. B* **2003**, *20*, 490–496. [CrossRef]
10. Reategui, R.; Kir'yanov, A.V.; Pisarchik, A.N.; Barmenkov, Y.O.; Il'ychev, N.N. Experimental study and modeling of coexist attractors and bifurcations in an erbium-doped fiber laser with diode-pump modulation. *Laser Phys.* **2004**, *14*, 1277–1281.
11. Ke, J.; Yi, L.; Xia, G.; Hu, W. Chaotic optical communications over 100-km fiber transmission at 30-Gb/s bit rate. *Opt. Lett.* **2** *43*, 1323–1326. [CrossRef]
12. Keren, S.; Horowitz, M. Interrogation of fiber gratings by use of low-coherence spectral interferometry of noiselike pulses. *Lett.* **2001**, *26*, 328–330. [CrossRef]
13. Lim, H.; Jiang, Y.; Wang, Y.; Huang, Y.C.; Chen, Z.; Wise, F.W. Ultrahigh-resolution optical coherence tomography with a fi laser source at 1 μm. *Opt. Lett.* **2005**, *30*, 1171–1173. [CrossRef]
14. Wu, Q.; Okabe, Y.; Sun, J. Investigation of dynamic properties of erbium fiber laser for ultrasonic sensing. *Opt. Express* **2** *22*, 8405–8419. [CrossRef]
15. Droste, S.; Ycas, G.; Washburn, B.R.; Coddington, I.; Newbury, N.R. Optical frequency comb generation based on erbium fi lasers. *Nanophotonics* **2016**, *5*, 196–213. [CrossRef]

Kraus, M.; Ahmed, M.A.; Michalowski, A.; Voss, A.; Weber, R.; Graf, T. Microdrilling in steel using ultrashort pulsed laser beams with radial and azimuthal polarization. *Opt. Express* **2010**, *18*, 22305–22313. [CrossRef]
Philippov, V.; Codemard, C.; Jeong, Y.; Alegria, C.; Sahu, J.K.; Nilsson, J.; Pearson, G.N. High-energy in-fiber pulse amplification for coherent lidar applications. *Opt. Lett.* **2004**, *29*, 2590–2592. [CrossRef]
Morin, F.; Druon, F.; Hanna, M.; Georges, P. Microjoule femtosecond fiber laser at 1.6 μm for corneal surgery applications. *Opt. Lett.* **2009**, *34*, 1991–1993. [CrossRef]
Sanchez, F.; Le Boudec, P.; François, P.L.; Stephan, G. Effects of ion pairs on the dynamics of erbium-doped fiber lasers. *Phys. Rev. A* **1993**, *48*, 2220. [CrossRef]
Colin, S.; Contesse, E.; Le Boudec, P.; Stephan, G.; Sanchez, F. Evidence of a saturable-absorption effect in heavily erbium-doped fibers. *Opt. Lett.* **1996**, *21*, 1987–1989. [CrossRef]
Rangel-Rojo, R.; Mohebi, M. Study of the onset of self-pulsing behaviour in an Er-doped fibre laser. *Opt. Commun.* **1997**, *137*, 98–102. [CrossRef]
Pisarchik, A.N.; Barmenkov, Y.O.; Kir'yanov, A.V. Experimental characterization of the bifurcation structure in an erbium-doped fiber laser with pump modulation. *IEEE J. Quantum Electron.* **2003**, *39*, 1567–1571. [CrossRef]
Pisarchik, A.N.; Kir'yanov, A.V.; Barmenkov, Y.O.; Jaimes-Reátegui, R. Dynamics of an erbium-doped fiber laser with pump modulation: Theory and experiment. *J. Opt. Soc. Am. B* **2005**, *22*, 2107–2114. [CrossRef]
Pisarchik, A.N.; Barmenkov, Y.O. Locking of self-oscillation frequency by pump modulation in an erbium-doped fiber laser. *Opt. Commun.* **2005**, *254*, 128–137. [CrossRef]
Huerta-Cuellar, G.; Pisarchik, A.N.; Barmenkov, Y.O. Experimental characterization of hopping dynamics in a multistable fiber laser. *Phys. Rev. E* **2008**, *78*, 035202. [CrossRef]
Pisarchik, A.N.; Jaimes-Reátegui, R.; Sevilla-Escoboza, R.; Huerta-Cuellar, G.; Taki, M. Rogue waves in a multistable system. *Phys. Rev. Lett.* **2011**, *107*, 274101. [CrossRef]
Pisarchik, A.N.; Hramov, A.E. *Multistability in Physical and Living Systems: Characterization and Applications*; Springer: Berlin/Heidelberg, Germany, 2022.
Pisarchik, A.N.; Feudel, U. Control of multistability. *Phys. Rep.* **2014**, *540*, 167–218. [CrossRef]
Huerta-Cuellar, G.; Pisarchik, A.N.; Kir'yanov, A.V.; Barmenkov, Y.O.; del Valle Hernández, J. Prebifurcation noise amplification in a fiber laser. *Phys. Rev. E* **2009**, *79*, 036204. [CrossRef] [PubMed]
Jaimes-Reátegui, R.; Esqueda de la Torre, J.O.; García-López, J.H.; Huerta-Cuellar, G.; Aboites, V.; Pisarchik, A.N. Generation of giant periodic pulses in the array of erbium-doped fiber lasers by controlling multistability. *Opt. Commun.* **2020**, *477*, 126355. [CrossRef]
Strogatz, S.H.; Stewart, I. Coupled oscillators and biological synchronization. *Sci. Am.* **1993**, *269*, 102–109. [CrossRef]
Boccaletti, S.; Latora, V.; Moreno, Y.; Chavez, M.; Hwang, D.U. Complex networks: Structure and dynamics. *Phys. Rep.* **2006**, *424*, 175–308. [CrossRef]
Kyprianidis, I.; Stouboulos, I. Chaotic synchronization of three coupled oscillators with ring connection. *Chaos Solitons Fractals* **2003**, *17*, 327–336. [CrossRef]
Abrams, D.M.; Strogatz, S.H. Chimera states in a ring of nonlocally coupled oscillators. *Int. J. Bifurcat. Chaos* **2006**, *16*, 21–37. [CrossRef]
Maneatis, J.G.; Horowitz, M.A. Precise delay generation using coupled oscillators. *IEEE J. Solid-State Circ.* **1993**, *28*, 1273–1282. [CrossRef]
Ermentrout, G. The behavior of rings of coupled oscillators. *J. Math. Biol.* **1985**, *23*, 55–74. [CrossRef] [PubMed]
Keener, J.P. Propagation and its failure in coupled systems of discrete excitable cells. *SIAM J. Appl. Math.* **1987**, *47*, 556–572. [CrossRef]
Yamauchi, M.; Wada, M.; Nishio, Y.; Ushida, A. Wave propagation phenomena of phase states in oscillators coupled by inductors as a ladder. *IEICE Trans. Fundam. Electron. Comput. Sci.* **1999**, *82*, 2592–2598.
Van der Sande, G.; Soriano, M.C.; Fischer, I.; Mirasso, C.R. Dynamics, correlation scaling, and synchronization behavior in rings of delay-coupled oscillators. *Phys. Rev. E* **2008**, *77*, 055202. [CrossRef]
Cohen, D.S.; Neu, J.C.; Rosales, R.R. Rotating spiral wave solutions of reaction-diffusion equations. *SIAM J. Appl. Math.* **1978**, *35*, 536–547. [CrossRef]
Noszticzius, Z.; Horsthemke, W.; McCormick, W.; Swinney, H.L.; Tam, W. Sustained chemical waves in an annular gel reactor: A chemical pinwheel. *Nature* **1987**, *329*, 619–620. [CrossRef]
Alexander, J. Patterns at primary Hopf bifurcations of a plexus of identical oscillators. *SIAM J. Appl. Math.* **1986**, *46*, 199–221. [CrossRef]
Nekorkin, V.I.; Makarov, V.A.; Velarde, M.G. Spatial disorder and waves in a ring chain of bistable oscillators. *Int. J. Bifurcat. Chaos* **1996**, *6*, 1845–1858. [CrossRef]
Matias, M.; Pérez-Muñuzuri, V.; Lorenzo, M.; Marino, I.; Pérez-Villar, V. Observation of a fast rotating wave in rings of coupled chaotic oscillators. *Phys. Rev. Lett.* **1997**, *78*, 219. [CrossRef]

45. Sánchez, E.; Matías, M.A. Transition to chaotic rotating waves in arrays of coupled Lorenz oscillators. *Int. J. Bifurcat. Chaos* 1
9, 2335–2343. [CrossRef]
46. Horikawa, Y. Metastable and chaotic transient rotating waves in a ring of unidirectionally coupled bistable Lorenz syste
*Physica D* **2013**, *261*, 8–18. [CrossRef]
47. Perlikowski, P.; Yanchuk, S.; Wolfrum, M.; Stefanski, A.; Mosiolek, P.; Kapitaniak, T. Routes to complex dynamics in a rin
unidirectionally coupled systems. *Chaos* **2010**, *20*, 013111. [CrossRef]
48. Jaimes-Reátegui, R. Dynamic of Complex System with Parametric Modulation: Duffing Oscillators. Doctoral Dissertation, Ce
de Investigaciones en Optica Leon, Leon, Mexico, 2004; pp. 112–119.
49. Sánchez, E.; Pazó, D.; Matías, M.A. Experimental study of the transitions between synchronous chaos and a periodic rota
wave. *Chaos* **2006**, *16*, 033122. [CrossRef]
50. Barba-Franco, J.; Gallegos, A.; Jaimes-Reátegui, R.; Gerasimova, S.; Pisarchik, A.N. Dynamics of a ring of three unidirection
coupled Duffing oscillators with time-dependent damping. *Europhys. Lett.* **2021**, *134*, 30005. [CrossRef]
51. Bashkirtseva, I.A.; Ryashko, L.B.; Pisarchik, A.N. Ring of map-based neural oscillators: From order to chaos and back. *C*
*Solitons Fractals* **2020**, *136*, 109830. [CrossRef]
52. Perlikowski, P.; Yanchuk, S.; Wolfrum, M.; Stefanski, A.; Kapitaniak, T. Dynamics of a large ring of unidirectionally coup
Duffing oscillators. In *IUTAM Symposium on Nonlinear Dynamics for Advanced Technologies and Engineering Design: Proceedings o*
*IUTAM Symposium on Nonlinear Dynamics for Advanced Technologies and Engineering Design, Held Aberdeen, UK, 27–30 July 2*
Springer: Berlin/Heidelberg, Germany, 2013; pp. 63–72.
53. Borkowski, L.; Stefanski, A. FFT bifurcation analysis of routes to chaos via quasiperiodic solutions. *Math. Probl. Eng.* 2
*2015*, 367036. [CrossRef]
54. Borkowski, L.; Perlikowski, P.; Kapitaniak, T.; Stefanski, A. Experimental observation of three-frequency quasiperiodic solutio
a ring of unidirectionally coupled oscillators. *Phys. Rev. E* **2015**, *91*, 062906. [CrossRef] [PubMed]
55. Buldú, J.M.; Torrent, M.C.; García-Ojalvo, J.O. Synchronization in semiconductor laser rings. *J. Light. Technol.* **2007**, *25*, 1549–1
[CrossRef]
56. Gao, Z.C.; Wu, Z.M.; Cao, L.P.; Xia, G.Q. Chaos synchronization of optoelectronic coupled semiconductor lasers ring. *Appl. P*
*B* **2009**, *97*, 645–651. [CrossRef]
57. Arroyo-Almanza, D.A.; Pisarchik, A.N.; Ruiz-Oliveras, F.R. Route to chaos in a ring of three unidirectionally coupled semi
ductor lasers. *IEEE Photon. Tech. Lett.* **2012**, *24*, 605–607. [CrossRef]
58. Arecchi, F.T.; Harrison, R.G. *Instabilities and Chaos in Quantum Optics*; Springer Science & Business Media: Berlin, Germany, 2
Volume 34.
59. Landau, L.D. On the problem of turbulence. *Dokl. Akad. Nauk USSR* **1944**, *44*, 311.
60. Hopf, E. A mathematical example displaying features of turbulence. *Commun. Pure Appl. Math.* **1948**, *1*, 303–322. [CrossRef]
61. Newhouse, S.; Ruelle, D.; Takens, F. Occurrence of strange axiom A attractors near quasi periodic flows on $T^m$, $m \geqq 3$. *Comm*
*Math. Phys.* **1978**, *64*, 35–40. [CrossRef]
62. Barba-Franco, J.; Romo-Muñoz, L.; Jaimes-Reátegui, R.; García-López, J.; Huerta-Cuellar, G.; Pisarchik, A.N. Electronic equiva
of a pump-modulated erbium-doped fiber laser. *Integration* **2023**, *89*, 106–113. [CrossRef]
63. Alon, U. Network motifs: Theory and experimental approaches. *Nat. Rev. Genet.* **2007**, *8*, 450–461. [CrossRef] [PubMed]
64. Boccaletti, S.; Pisarchik, A.N.; Del Genio, C.I.; Amann, A. *Synchronization: From Coupled Systems to Complex Networks*; Cambri
University Press: Cambridge, UK, 2018.
65. Wolf, A.; Swift, J.B.; Swinney, H.L.; Vastano, J.A. Determining Lyapunov exponents from a time series *Physica D* **1985**, *16*, 285–
[CrossRef]
66. Matias, M.; Güémez, J.; Pérez-Munuzuri, V.; Marino, I.; Lorenzo, M.; Pérez-Villar, V. Size instabilities in rings of cha
synchronized systems. *Europhys. Lett.* **1997**, *37*, 379. [CrossRef]
67. Marino, I.; Pérez-Muñuzuri, V.; Pérez-Villar, V.; Sánchez, E.; Matıas, M. Interaction of chaotic rotating waves in coupled ring
chaotic cells. *Physica D* **1999**, *128*, 224–235. [CrossRef]
68. Matías, M.; Güémez, J. Transient periodic rotating waves and fast propagation of synchronization in linear arrays of cha
systems. *Phys. Rev. Lett.* **1998**, *81*, 4124. [CrossRef]
69. Borkowski, L.; Stefanski, A. Stability of the 3-torus solution in a ring of coupled Duffing oscillators. *Eur. Phys. J. Spec. Top.* 2
*229*, 2249–2259. [CrossRef]
70. Krysko, A.; Awrejcewicz, J.; Papkova, I.; Krysko, V. Routes to chaos in continuous mechanical systems: Part 2. Model
transitions from regular to chaotic dynamics. *Chaos Solitons Fractals* **2012**, *45*, 709–720. [CrossRef]
71. Awrejcewicz, J.; Krysko, A.; Papkova, I.; Krysko, V. Routes to chaos in continuous mechanical systems. Part 3: The Lyapu
exponents, hyper, hyper-hyper and spatial–temporal chaos. *Chaos Solitons Fractals* **2012**, *45*, 721–736. [CrossRef]
72. Barba-Franco, J.; Gallegos, A.; Jaimes-Reátegui, R.; Pisarchik, A.N. Dynamics of a ring of three fractional-order Duffing oscilla
*Chaos Solitons Fractals* **2022**, *155*, 111747. [CrossRef]
73. Pisarchik, A.N.; Jaimes-Reategui, R. Control of basins of attraction in a multistable fiber laser. *Phys. Lett. A* **2009**, *374*, 228–
[CrossRef]
74. Meucci, R.; Marc Ginoux, J.; Mehrabbeik, M.; Jafari, S.; Clinton Sprott, J. Generalized multistability and its control in a laser. *C*
**2022**, *32*, 083111. [CrossRef]

Pando, C.L.; Meucci, R.; Ciofini, M.; Arecchi, F.T. $CO_2$ laser with modulated losses: Theoretical models and experiments in the chaotic regime. *Chaos* **1993**, *3*, 279–285. [CrossRef] [PubMed]
Doedel, E.J.; Pando, C.L. Multiparameter bifurcations and mixed-mode oscillations in Q-switched $CO_2$ lasers. *Phys. Rev. E* **2014**, *89*, 052904. [CrossRef] [PubMed]

photonics

Article

# Implementation of Logic Gates in an Erbium-Doped Fiber Las (EDFL): Numerical and Experimental Analysis

**Samuel Mardoqueo Afanador Delgado [1], José Luis Echenausía Monroy [1,2], Guillermo Huerta Cuellar [1], Juan Hugo García López [1] and Rider Jaimes Reátegui [1,*]**

1 Dynamical Systems Laboratory, CULagos, Universidad de Guadalajara, Centro Universitario de los Lag Enrique Díaz de León 1144, Paseos de la Montaña, Lagos de Moreno 47460, Mexico
2 Applied Physics Division, Center for Scientific Research and Higher Education at Ensenada (CICESE), Carr. Ensenada-Tijuana 3918, Zona Playitas, Ensenada 22860, Mexico
* Correspondence: rider.jaimes@academicos.udg.mx

**Abstract:** At a time when miniaturization and optimization of resources are in the foregrou the development of devices that can perform various functions is a primary goal of technolog development. In this work, the use of an Erbium-Doped Fiber Laser (EDFL) is proposed as a b system for the generation of an optical logic gate. Taking advantage of the dynamic richness of type of laser and its use in telecommunication systems, the dynamic response is analyzed when system is perturbed by a digital signal. The emission response of the system is controlled by intensity of the digital signal, so that it is possible to obtain different logic operations. The numer results are in good agreement with the experimental observations. The presented work raises r aspects in the use of chaotic systems as a means of obtaining optical logic gates.

**Keywords:** Erbium-Doped Fiber Laser (EDFL); logic gate; logic operation; chaos computing



**Citation:** Afanador Delgado, S.M.; Echenausía Monroy, J.L.; Huerta Cuellar, G.; García López, J.H.; Jaimes Reátegui, R. Implementation of Logic Gates in an Erbium-Doped Fiber Laser (EDFL): Numerical and Experimental Analysis. *Photonics* **2022**, *9*, 977. https://doi.org/10.3390/photonics9120977

Received: 16 November 2022
Accepted: 6 December 2022
Published: 12 December 2022



## 1. Introduction

In optical fibers doped with rare earths (also called laser fibers), absorption proces take place, both spontaneous and stimulated emission, when excited with certain er gies [1–5]. These characteristics give this type of fibers the necessary properties for th use as optical amplifiers, lasers, sensors, to name a few. The ability of the fibers to gener light in different wavelengths by doping with rare earths has been used in fields such telecommunications, signal processing, medicine, study of meteorological phenomena a modeling of real multistable systems, all this with the help of various pumping meth and laser resonators.

The research and commercialization of Erbium-Doped Fiber Lasers has recently in sified. Reasons include their long pumping interaction with active ions, high gain mar transverse mode of operation, tiny size equivalent to the width of a hair, high threshold damage, excellent heat dissipation, and the fact that they have no bandwidth limitati leading to better efficiency in laser pumping applications [6–8]. Although their proper are diverse and rich, this type of laser is also relevant from the perspective of nonlin systems, as they are very sensitive to external disturbances that lead to destabilizat of the laser dynamics, causing them to oscillate in a nonlinear regime, leading to cha behavior [9,10].

Taking advantage of the dynamic properties of chaotic systems, the so-called "ch computing" has become popular. It is essentially based on the use of an analog cha oscillator that emulates the behavior of various logic gates, allowing it to switch betw different modes of operation and thus the resolution of various arithmetic operations [ In this sense, many proposals on chaotic logic gates have been made, such as in [12], wh a SR (Set-Reset) flip-flop has been developed based on NOR gates implemented by a C circuit. In [13] it is described how to implement fundamental logic gates based on a logistic map. In [14] the fundamental NOR gate is implemented based on threshold control of chaotic systems. The use of a nonlinear system with noise is discussed in [15], where logic functions are obtained by changing the nonlinearity of the model, and in [16] a circuit with dynamic logic architecture is presented, showing NOR, NAND and XOR logic gates. Potential applications of chaos computing can be approached by changing the pump inputs in time in a $CO_2$ laser with modulated losses, where changing the modulation depth of the periodic losses leads to different chaotic and periodic attractors [17,18].

Most of the logic gates described above are based on the use of chaotic systems with two attractors located at different equilibrium points. This allows for a dynamic discretization that takes into account the response of the system to a bias and the change in behavior due to the position of the attractor [19–21]. Considering the possible application of a chaotic system that can function as a logic gate based on a telecommunication system, and the few works that deal with this topic from the optical systems point of view [22,23], the present work deals with the generation of AND and NOR logic gates using an Erbium-Doped Fiber Laser as a dynamic system. By perturbing the system with a bias signal, it is possible to change the operating regime of the laser, obtaining a completely different logic state that is stable in time. The numerically obtained results are experimentally validated and largely agree.

The remainder of the paper deals with the following topics: Section 2 describes the EDFL mathematical system. The implemented experimental setup and the natural dynamics of the system are also described in this section. The obtained results are described in detail in Section 3. The discussion of the obtained results is given in Section 4. Section 5 describes the conclusions, implications, and future work.

## 2. Materials and Methods

### 2.1. Numerical Model

The dynamics of an EDFL pump diode are described by Equation (1), and better described in [24–27].

$$\begin{aligned} \frac{dP}{dt} &= \frac{2L}{T_r} P\{r_w \alpha_0 [N(\xi_1 - \xi_2) - 1] - \alpha_{th}\} + P_{sp}, \\ \frac{dN}{dt} &= \frac{\sigma_{12} r_w P}{\pi r_0^2}(N\xi_1 - 1) - \frac{N}{\tau} + P_{pump}, \end{aligned} \tag{1}$$

where $P$ is the power of the laser in the cavity, $N = (1/n_0\ L)\int_0^L N_2(z)dz$ is the average of the upper level of the laser potential, where $N_2$ is the upper plane in the $z$-coordinate, (propagation direction of the laser emission), $n_0$ is the refractive index of the cold core of the erbium-doped fiber, $\xi_1$ and $\xi_2$ define the relationship between the cross sections and the ground absorption state. The variable $T_r$ is the complete travel time of a photon in the cavity, $\alpha_0$ is the absorption coefficient of the erbium-doped fiber, $\alpha_{th}$ is the threshold loss counter of the laser cavity, where $\tau$ is the lifetime of an excited erbium ion, and $r_w$ is the factor comprising the coincidence between the fundamental mode of the laser and the volume of the erbium-doped core.

The spontaneous emission within the fundamental mode of the laser is defined as:

$$P_{sp} = N\frac{10^{-3}}{T_r\tau}\left(\frac{\lambda_g}{\omega_0}\right)^2 \frac{r_0^2\alpha_0 L}{4\pi^2\sigma_{12}}, \tag{2}$$

where $\lambda_g$ is the wavelength of the laser, 1.5 µm. The pump power is equivalent to:

$$\begin{aligned} P_{pump} &= P_p \frac{1 - \exp[-\alpha_0\beta L(1-N)]}{N_0 \pi r_0^2 L}, \\ P_p &= p[1 - m\sin(2\pi f_m t)]. \end{aligned} \tag{3}$$

In Equation (3), $P_p$ is the pump power at the fiber input and $\beta$ is a dimensionl coefficient. When the input current of the laser diode is modulated with a harmonic sig the EDFL describes the behavior given in Equation (1), which represents the coexiste of up to four attractors defined as periods one, three, four, and five ($P_1$, $P_3$, $P_4$, and $P_5$) shown by the red dashed line in Figure 1a.

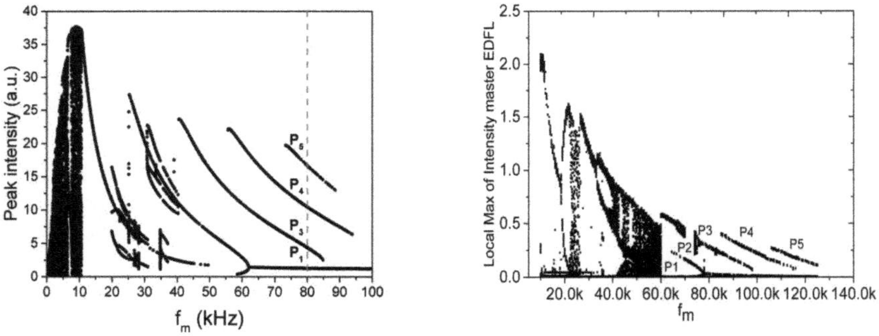


(a) (b)

**Figure 1.** EDFL Bifurcation diagram. (**a**) Numerically calculated, (**b**) experimentally determined

To understand the dynamics of the EDFL, the bifurcation diagram of the local maxi of the laser intensity as a function of the pump modulation frequency $f_m$ is calculated. perform numerical simulation, the equations described in Equation (1) are normalizec obtain Equation (4). The normalization procedure and further details can be found in appendix of reference [5,28], and the system coefficients are described in Table 1.

$$\begin{aligned} \dot{x} &= axy - bx + c(y + r_w), \\ \dot{y} &= dxy - (y + r_w) + e\left(1 - \exp\left[-\beta\alpha_0 L(1 - \frac{y + r_w}{\xi_2 r_w})\right]\right). \end{aligned}$$

**Table 1.** Parameters of the normalized system of equations for the EDFL, Equation (4).

| Parameter | Value |
|---|---|
| $a$ | $6.6207 \times 10^7$ |
| $b$ | $7.4151 \times 10^6$ |
| $c$ | 0.0163 |
| $d$ | $4.0763 \times 10^3$ |
| $e$ | 506 |
| $r_w$ | 0.3075 |
| $\xi_2 r_w$ | 0.6150 |
| $-\beta\alpha_0 L$ | −18 |

At a modulation amplitude $m = 1$, the laser system exhibits dynamic richness varying the modulation frequency, showing a bifurcation diagram as in Figure 1a. T sequence of periodic orbits (P3, P4, P5, etc.) in the periodically modulated EDFL Figure 1 is known as Period-Adding Cascades (PAC) and typically arises as a saddle-n bifurcation in which there is a homoclinic to a periodic orbit. In the context of infrared las PAC has also been observed in simulations, and by numerical continuation in bifurcat analysis in the $CO_2$ laser with a saturable absorber in [29,30].

### 2.2. Experimental Set-Up

The experimental setup used in this work, shown in Figure 2, consists of a 1560 nm EDFL pumped by a 977 nm Laser Diode (LD). The laser cavity consists of an erbium-doped fiber of approximately 88 cm long with a core diameter of 2.7 µm and two fiber Bragg gratings (FBG1 and FBG2) with reflectivities of 100% and 95.88%, respectively. All components are interconnected with a monomode fiber. The laser diode is controlled by the pump controller (LDC). For all experiments performed, the pump current is set to 145.5 mA (20 nW). The choice of the pump current is to ensure a relaxation frequency of the laser close to $F_r = 30$ kHz. To control the dynamics of the laser and to achieve the different logic gates, both the harmonic modulation and the bias signal are injected into the pump diode via a function generator (WFG) and a digital interface, respectively.

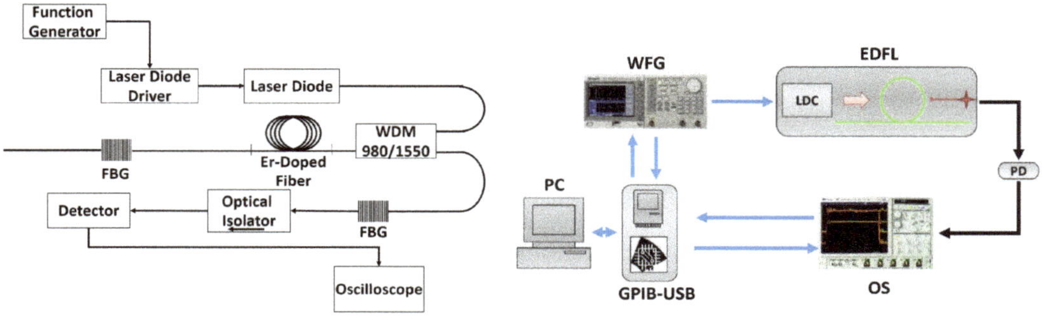


(a) (b)

**Figure 2.** Experimental setup. (**a**) Optical elements required to put the Erbium-Doped Fiber Laser into operation (**b**) Elements required to implement the logic gate based on the operation of the EDFL. *(WFG-Wave Form Generator) (LDC-Laser Diode Controler), (PD-Photo-Detector), (WDM-Wavelength Division Multiplexing), (OS- Osciloscope), (FBG-Fiber Bragg Grating).*

Figure 1b shows the bifurcation diagram of the experimental system, which agrees with the numerical estimate.

### 2.3. Methods

For the implementation of the logic gate, a modulation frequency range $40 \leq f_m \leq 60$ kHz was defined, and the modulation amplitude remains fixed $A_m = 1$. The frequency selection results in the laser response being at $P_2$, as shown in Figure 3. In this case, the bias signal is added to the modulation signal and causes a shift in the laser response that allows switching between different periodic regimes of the system.

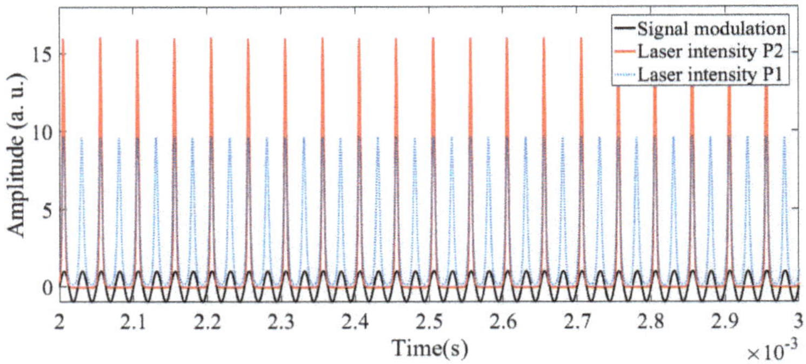


**Figure 3.** Dynamic response of the EDFL in a regimen of $P_2$ to a modulation of $f_m = 40$ kHz (
line). The blue line denotes the behavior of the EDFL in period one. The modulation signal is sho
in black.

**Remark 1.** *In this work, the dynamic response of the laser is referred to as "period N (P*
*i.e., N* $\in \mathbb{R}^+$. *The EDFL system is modulated with a periodic sinusoidal signal that determi the laser emission. A* $P_1$ *type output response indicates that the laser emission occurs sequenti with respect to the modulation period. A type* $P_3$ *output response, in turn, indicates that th periods of the periodic signal were injected for each pulse of the laser response. Figure 2 illustra the relationship with the superposition of the laser modulation (black line), the laser response in (red line) and the type* $P_1$ *emission (blue line).*

To characterize the response of the EDFL to logic gate generation, we need to defir reference signal $I_d$ that represents the bias voltage that is injected into the pumping proc as shown in Equation (5). Consider an input signal $I_d$ defined as the sum of two dig decorrelated signals $I_d = I_1 + I_2$. Since $I_i$ are digital signals, i.e., can only take the val 0 and 1, the combination of these signals can only result in four possible logical inp $(I_1, I_2) : (0,0),\ (0,1),\ (1,0),\ (1,1)$.

$$P_p = A_m \sin(2\pi f_m t) + A_d I_d.$$

**Remark 2.** *Note that the logical input combinations* $(0,1)$ *and* $(1,0)$ *result in the same signa so combining the logical signals* $(I_1, I_2)$ *yields only three possible combinations. Thus, the sur two decorrelated logic signals results in an aperiodic signal with three states, as shown in Figu For a set of inputs* $(0,0)$*, there is a value of* $I_d = 0$*. For a combination* $(0,1)/(1,0)$ *the valu* $I_d = 1$*, and finally for* $(1,1)$ *we have* $I_d = 2$*. In our implementation, we propose a digital gain ( to control the amplification of the generated signal.*

**Remark 3.** *The approach considered in this paper is based on the shift between* $P_2$ *and* $P_1$*. P considered as a logical 1, while the shift to* $P_1$ *is a logical 0.*

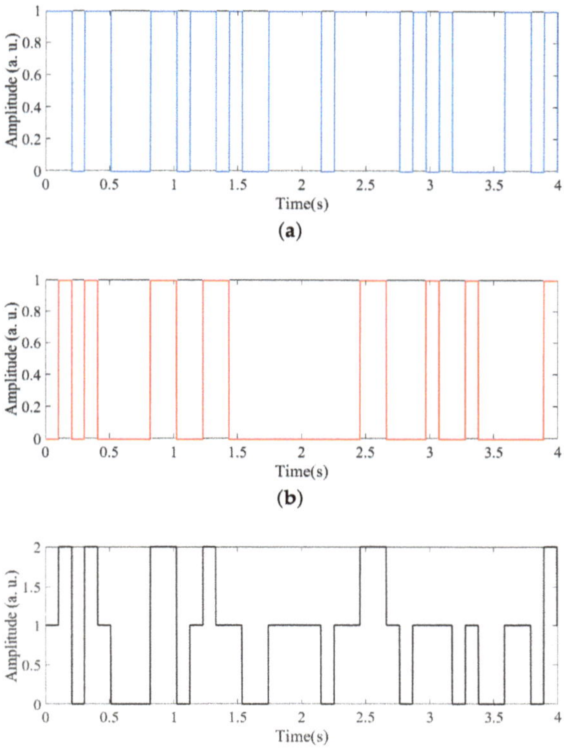


**Figure 4.** Structure of the digital signal $I_d$ used as bias and fed to the laser pump. (**a**) Temporal behavior of $I_1$, while $I_2$ is shown in (**b**), being $I_d = I_1 + I_2$ shown in (**c**).

## 3. Results

The following section describes the results obtained, both for the numerical simulation and for the experimental implementation.

### 3.1. Numerical Results

Any dynamic system changes its behavior when it is perturbed with the necessary energy. In this case, we assume that the laser changes its behavior when a bias signal is added to the modulation. In our implementation, a square wave signal $(I_d)$ was added to the laser modulation.

Figure 5a–e shows the effects of the bias signal on the dynamic response of the laser when a fixed modulation frequency is used and the gain of the digital signal is varied. If we use the signal shown in Figure 5e as example, in Figure 5f again shows the digital input signal $I_d$, in which we can observe the three states given by the arithmetic sum of the two decorrelated digital inputs. In Figure 5e we see the behavior of the laser changing its behavior into different periods due to the bias voltage, and in Figure 5g we can see the output of the logic gate, where the states are defined by the change of the period, as mentioned above. Thus, $P_2$ is assigned a high value and a $P_1$ is assigned a low value.

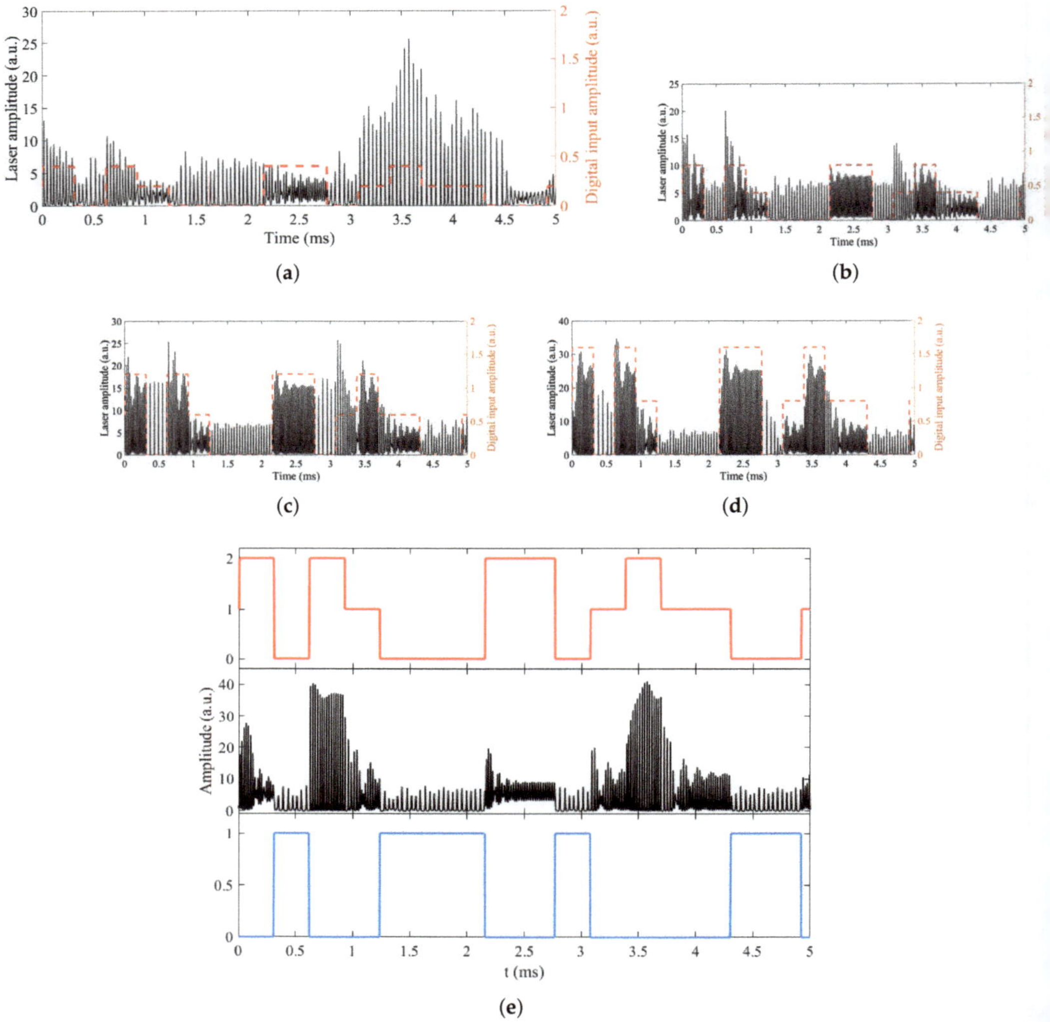


**Figure 5.** (**a–d**) Behavior of laser intensity (solid black line) when adding a digital signal (red line amplitude (**a**) 0.2 V, (**b**) 0.4 V, (**c**) 0.6 V, (**d**) 0.8 V. (**e**) (black line) EDFL behavior for a digital amplit of 1 V. (red line) Digital input signal $I_d$, (blue line) output of logic gate determined when $P_2$ is ta as logical 1 and response in $P_1$ is taken as a logical 0.

Having confirmed the original hypothesis that the system changes its dynamics injecting a digital signal into the laser pump, a characterization of the laser dynamics w performed using a frequency sweep in the range defined by $40 \leq f_m \leq 60$ kHz, which w varied in steps of 1 kHz, and the amplitude of the digital signal $0.2 \leq A_d \leq 1$ was chang in steps of $\Delta_{A_d} = 0.1$ (a.u.). For each set of parameters $A_d, fm$, a time series analysis w performed to discretize the laser dynamics taking into account the period changes, result in the behavior shown in Figure 6.

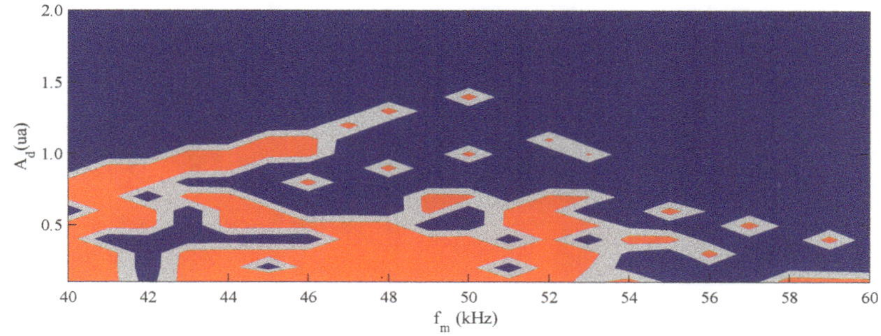


**Figure 6.** Color map of the different emission regimes of the laser, based on the influence of the bias signal. The red area corresponds to $P_2$, the blue area corresponds to $P_1$, and the gray area corresponds to an undefined state.

On the color map shown in Figure 6, it is possible to distinguish different color regions corresponding to the emission regime of the laser. The gray region has no defined period, the red region corresponds to $P_2$, and the blue region corresponds to $P_1$. Thus, if we are at a frequency of $f_m = 44$ kHz and the amplitude of the signal is $A_d = 0.2$ (a.u.), the laser will respond with $P_2$, while at $A_d = 0.8$ (a.u.) it will respond with $P_1$. In this way we can build the logic gates.

### 3.2. *Experimental Results*

To experimentally verify the results of the numerical EDFL model, the following parameters are used: the modulation amplitude remains fixed $A_m = 1$ V, the modulation frequency was varied from $40 \leq f_m \leq 60$ kHz, the amplitude of the bias $0.2 \leq A_d \leq 1$ V. The digital inputs $I_d = I_1 + I_2$ were generated using a data acquisition board (DAQ), which was later added to the modulation.

Figure 7 shows the change in laser intensity when the digital signal $I_d$ is added to the modulation. The modulation frequency is fixed $f_m = 40$ kHz and the amplitude of the digital signal $0.4 \leq A_d \leq 0.8$ V. Here (Figure 7a–e), the solid black line represents the laser intensity response and the red line represents the bias signal used. As in the previous section, a change in amplitude $A_d$ results in a change in the emission response of the laser. For an amplitude $A_d > 0.6$ V (Figure 7c,d) three periods are shown, one for each level of the signal $I_d$, but for an amplitude $A_d = 0.4$ V the signal oscillates between $P_1$ and $P_2$ (Figure 7a).

In Figure 7f,g, the amplitude $A_d = 0.4$ V and the modulation frequency $f_m = 40$ kHz remain fixed. This choice is due to the fact that the laser shows a response between P1 and P2. Figure 7f shows the bias signal $I_d$, Figure 7e shows the laser intensity, and Figure 7g shows the discretized gate output of the laser response. At a value of $I_d = 0.0$ V, the (0,0) state, the laser response is $P_2$, while at $A_d = 0.4$ V and $A_d = 0.8$ V, the (0,1)/(1,0) and (1,1) states, respectively, the laser response corresponds to $P_1$. If, as mentioned in the methodology section, you define $P_2$ as logic 1 and the period change to $P_1$ as logic 0, you obtain a logic gate with the behavior NOR.

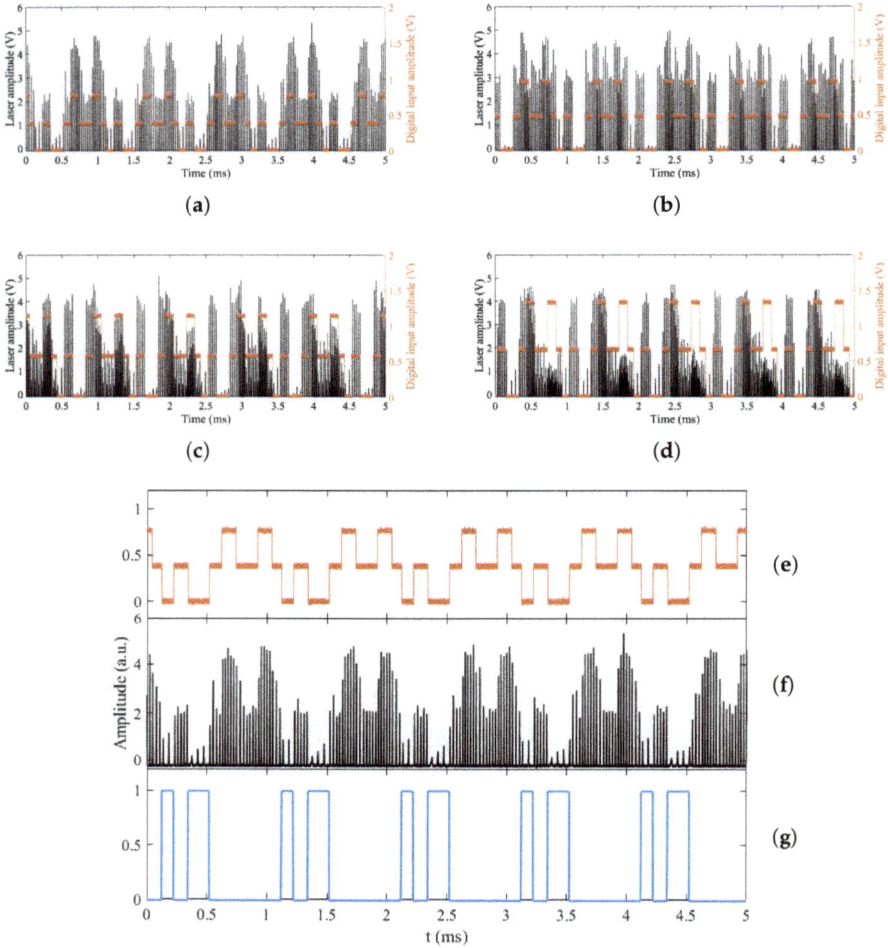


**Figure 7.** (**a–e**) Behavior of the laser intensity (solid black line) at $f_m = 40$ kHz when bias $I_d$ (ora line) is added to the laser modulation. The amplitude of the digital inputs is $A_d =$ (**a**) 0.4 V, (**b**) 0. (**c**) 0.6 V, (**d**) 0.7 V, and (**e**) Digital input signal $I_d$ for $A_d = 0.8$ V, while (**f**) shows the EDFL beha and (**g**) represents the output of the logic gate, which is determined when $P_2$ is taken as logic 1 the response in $P_1$ is taken as logic 0.

Similar to the numerically obtained color map (Figure 6), a corresponding swee the modulation frequency and amplitude of the digital signal was performed for the ex imental scheme. Once the laser signals are acquired, a time series analysis is performe determine the period exhibited by the laser for each combination of parameters ($A_d$, The result is the color map shown in Figure 8. In contrast to the numerical model which only $P_1$ and $P_2$ are represented. In the experimental implementation, we can four regions corresponding to $P_4$ (red), $P_3$ (yellow), $P_2$ (light blue), and $P_1$ (dark blue we define the modulation frequency $f_m = 44$ kHz and the amplitude of the digital sig $A_d = 0.4$ V, we have P1 for one state (0,0) and $P_2$ for the other states.

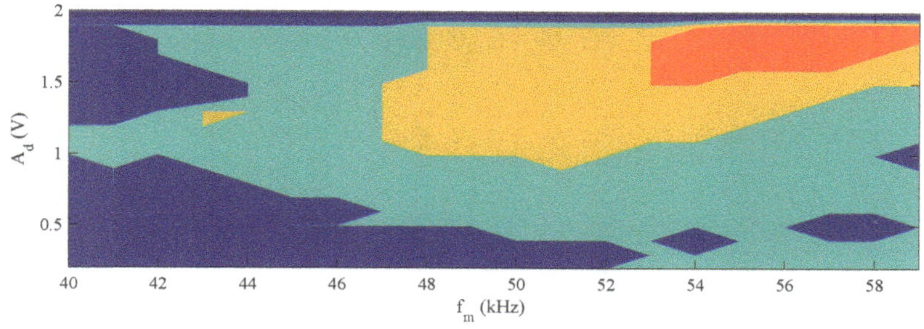


**Figure 8.** Map resulting from the discretization of the periods of the experimental scheme. The red region corresponds to $P_4$, the yellow region to $P_3$, the light blue region to $P_2$, and the dark blue region to $P_1$.

## 4. Discussion

Since the EDFL is modulated by the current, the system has a defined response when a fixed modulation frequency ($f_m$) is added and is insensitive to initial conditions. However, by adding a signal to this modulation, it is possible to vary the current in the system, resulting in a change in the dynamic response of the laser. This phenomenon can be observed both numerically and experimentally in Figure 5 and Figure 7, respectively.

In Figures 5 and 7, it is possible to see that the laser changes its emission period when the amplitude of the bias voltage is increased. This is because the modulation signal oscillates between a positive and a negative voltage, Equations (3) and (5), where the voltage limits are given by $A_m$ and the frequency between the positive and/or negative states is given by the ratio $f_m/2$ and has the same duration. When a constant signal of amplitude $A_d$ is added to the laser modulation, the amplitude of the valley $A_v = A_v - A_d$ decreases and the amplitude of the peak $A_p = A_p + A_d$ increases. As a result, the time when the modulation signal has a negative value decreases and the positive value increases. This phenomenon is shown in Figure 9.

Considering the type of laser used in the study, the laser shows no emission when the modulation signal has no positive component, while the laser is able to emit when the modulation signal is positive. In this way, the addition of the bias signal increases the probability of laser emission, resulting in a dynamic change from $P_2$ to $P_1$, which is reflected in an increase in the amplitude of laser emission, as shown in Figures 5 and 7.

The color maps in which the dynamics of the laser are discretized (Figures 6 and 8 in numerical and experimental ways, respectively) provide a broader perspective on what happens to the laser when both the modulation frequency and the amplitude of the digital signal are changed. With this discretization, it is possible to create the logic gates by defining a high logic output state when the laser is emitting at $P_2$ and a logic zero when the laser emission is on $P_1$.

Analyzing the results of the numerical implementation (Figure 6) with a frequency of $f_m = 44$ kHz and $A_d = 0.2$ (a.u.), we obtain a NAND logic gate as shown in Table 2. If the parameters are changed, it is possible to obtain a range of gates, i.e., for a bias signal amplitude $A_d = 0.4$ (a.u.) and $f_m = 42$ kHz, the obtained gate will be an XOR. Table 3 shows other gates resulting from changes in the parameter space ($f_m, A_d$)

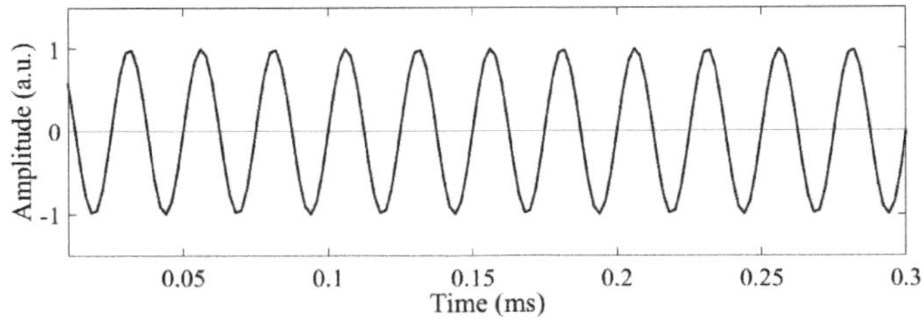

(**a**)

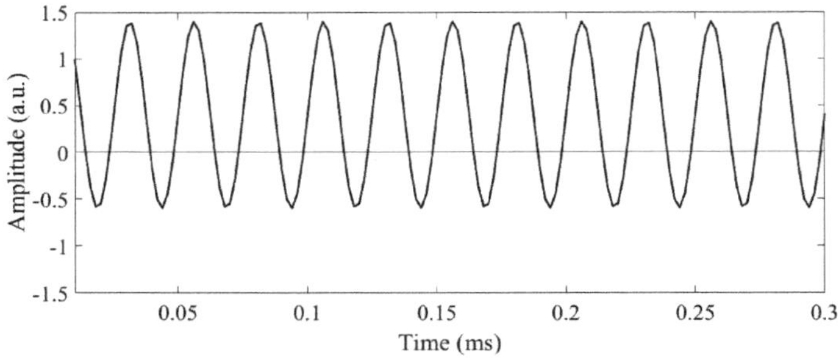

(**b**)

**Figure 9.** EDFL response to a modulation signal (**a**) without bias, (**b**) with a bias of 0.4 V. The addi of a positive bias increases the amplitude of the peak and decreases the amplitude of the valley, v respect to zero. It also shortens the time the signal is in a negative state and lengthens the time wi positive value.

**Table 2.** NAND logic gate constructed from the numerical laser response defining $P_2$ as logic 1 $P_1$ as logic 0. The modulation frequency is fixed at 44 kHz.

| **Numeric NAND Logic Gate** | | | |
|---|---|---|---|
| $(I_1, I_2)$ | **Inputs** $A_d - I_d$ | **Period** | **Logic Output** |
| $(0,0)$ | 0 | $\mathbf{P_2}$ | **1** |
| $(0,1)/(1,0)$ | 0.2 | $\mathbf{P_2}$ | **1** |
| $(1,1)$ | 0.4 | $\mathbf{P_1}$ | **0** |

**Table 3.** With parameters $f_m = 40$ kHz and $A_d = 0.5$ V, a NAND gate can be obtained by numerical model. On the other hand, for $f_m = 42$ kHz and $A_d = 0.4$ V, an XNOR gate is genera For $f_m = 46$ kHz and $A_d = 0.7$ V, a NOR gate is obtained.

| **Numeric Logic Gates for Different Input Values** | | | |
|---|---|---|---|
| $(I_1, I_2)$ | $A_d = 0.5$ **V**, $f_m = 40$ **kHz** **NAND** | $A_d = 0.4$ **V**, $f_m = 42$ **kHz** **XNOR** | $A_d = 0.7$ **V**, $f_m = 46$ **kHz** **NOR** |
| $(0,0)$ | 1 | 1 | 1 |
| $(0,1)/(1,0)$ | 1 | 0 | 0 |
| $(1,1)$ | 0 | 1 | 0 |

Similarly, it is possible to determine different logic gates from the dynamic characterization of the experimental system, see Figure 8. Consider $A_d = 0.5$ V and $f_m = 44$ kHz, with this combination of values it is possible to build an XNOR gate using the EDFL. On the other hand, if $A_d = 0.2$ V and a modulation frequency of 42 kHz is used, a NAND gate is obtained. The different logic gates and their corresponding experimental values are described in Table 4.

**Table 4.** With the parameters $f_m = 44$ kHz and $A_d = 0.5$ V an XNOR gate is obtained, while for $f_m = 46$ kHz and $A_d = 0.2$ V a NAND gate is generated.

| Experimental Logic Gates for Different Input Values | | |
|---|---|---|
| $(I_1, I_2)$ | $A_d = 0.5$ **V**, $f_m = 44$ **kHz** **XNOR** | $A_d = 0.2$ **V**, $f_m = 46$ **kHz** **NAND** |
| (0,0) | 1 | 1 |
| (0,1)/(1,0) | 1 | 0 |
| (1,1) | 0 | 1 |

It is worth noting that the ranges of values and the behavior between the numerically and experimentally determined values are not identical. This is because the numerical model did not take into account or model the noise of the instruments used to generate the signal, the resolution and the noise of the electronic components used to connect the instruments to the laser. However, it is confirmed that it is possible to build logic gates using this method and to change the nature of the gate by changing the parameter space $(f_m, A_d)$.

## 5. Conclusions

In the present work, it was possible to develop an optical device capable of solving logical operations and enabling the implementation of logic gates based on the dynamic richness of an EDFL. The operation of the EDFL as a logic gate generator is based on the dynamic richness of the system by injecting a periodic signal into the laser pump.

The obtained results are based on the perturbation of the modulation of an Erbium-Doped Fiber Laser. The system exhibits a dynamical richness ranging from the presence of periodic states to chaos and coexistent states; all of which are caused by the modulation of the system. Starting from this dynamic richness, the behavior of the laser emission is studied when a digital signal with random frequency is added to the laser modulation.

As a result, the EDFL is able to respond constantly to the perturbations induced by the digital signal, which allows the dynamic response to be discretized so that a logic gate generator system can be obtained by optical means.

The numerically obtained results are validated by the characterization of the experimental system. In a space defined by the modulation frequency and the amplitude of the digital signal ($f_m$, $A_d$), the system is able to give a consistent response with up to four different logic gates.

**Author Contributions:** S.M.A.D.: Writing—Original Draft, Writing—Review & Editing, Methodology, Software, Validation, Data Curation, Visualization. J.L.E.M.: Writing—Original Draft, Writing—Review & Editing, Visualization. G.H.C.: Formal analysis, Review & Editing. J.H.G.L.: Writing—Review & Editing, Resources, Project administration. R.J.R.: Supervision, Funding acquisition, Writing—Review & Editing, Conceptualization, Resources, Project administration. All authors have read and agreed to the published version of the manuscript.

**Funding:** This project was supported by CONACYT under project number 320597.

**Institutional Review Board Statement:** Not applicable.

**Informed Consent Statement:** Not applicable.

**Data Availability Statement:** The data is available on article submitted to Data journal, from MI (ID:2075448) under the title: Numerical and Experimental Data of the Implementation of Logic Gate an Erbium-Doped Fiber Laser (EDFL), and available at the repository with DOI: 10.17632/ctnhmgbr

**Acknowledgments:** J.L.E.M. thanks CONACYT for financial support (CVU-706850, project: A 26123, and project: 320597). R.J.R. thanks CONACYT for financial support, project No. 320597.

**Conflicts of Interest:** The authors certify that they have NO affiliations with or involvement in organization or entity with any financial interest (such as honoraria; educational grants; participat in speakers' bureaus; membership, employment, consultancies, stock ownership, or other eq interest; and expert testimony or patent-licensing arrangements), or non-financial interest (such personal or professional relationships, affiliations, knowledge or beliefs) in the subject matte materials discussed in this manuscript.

## References

1. Digonnet, M.J.; Gaeta, C. Theoretical analysis of optical fiber laser amplifiers and oscillators. *Appl. Opt.* **1985**, *24*, 333– [CrossRef] [PubMed]
2. Digonnet, M. Theory of operation of three-and four-level fiber amplifiers and sources. In *Proceedings of the Fiber Laser Sources Amplifiers*; SPIE: Cardiff, Wales, 1990; Volume 1171, pp. 8–26.
3. Sola, I.; Martín, J.; Ávarez, J.; Jarabo, S. Erbium doped fibre characterisation by laser transient behaviour analysis. *Opt. Comm* **2001**, *193*, 133–140. [CrossRef]
4. Tang, C.L.; Statz, H.; deMars, G. Spectral output and spiking behavior of solid-state lasers. *J. Appl. Phys.* **1963**, *34*, 2289–2 [CrossRef]
5. Magallón, D.A.; Jaimes-Reátegui, R.; García-López, J.H.; Huerta-Cuellar, G.; López-Mancilla, D.; Pisarchik, A.N. Contro Multistability in an Erbium-Doped Fiber Laser by an Artificial Neural Network: A Numerical Approach. *Mathematics* **2** *10*, 3140. [CrossRef]
6. Arecchi, F.T.; Harrison, R.G. *Instabilities and Chaos in Quantum Optics*; Springer Science & Business Media: Berlin/Heidelb Germany, 2012; Volume 34.
7. Lacot, E.; Stoeckel, F.; Chenevier, M. Dynamics of an erbium-doped fiber laser. *Phys. Rev. A* **1994**, *49*, 3997. [CrossRef]
8. Saucedo-Solorio, J.M.; Pisarchik, A.N.; Kir'yanov, A.V.; Aboites, V. Generalized multistability in a fiber laser with modula losses. *JOSA B* **2003**, *20*, 490–496. [CrossRef]
9. Pisarchik, A.N.; Barmenkov, Y.O.; Kir'yanov, A.V. Experimental characterization of the bifurcation structure in an erbium-do fiber laser with pump modulation. *IEEE J. Quantum Electron.* **2003**, *39*, 1567–1571. [CrossRef]
10. Meucci, R.; Marc Ginoux, J.; Mehrabbeik, M.; Jafari, S.; Clinton Sprott, J. Generalized multistability and its control in a la *Chaos Interdiscip. J. Nonlinear Sci.* **2022**, *32*, 083111. [CrossRef]
11. Sinha, S.; Ditto, W.L. Dynamics based computation. *Phys. Rev. Lett.* **1998**, *81*, 2156. [CrossRef]
12. Cafagna, D.; Grassi, G. Chaos-based SR flip–flop via chua's circuit. *Int. J. Bifurc. Chaos* **2006**, *16*, 1521–1526. [CrossRef]
13. Munakata, T.; Sinha, S.; Ditto, W.L. Chaos computing: Implementation of fundamental logical gates by chaotic elements. *II Trans. Circuits Syst. I Fundam. Theory Appl.* **2002**, *49*, 1629–1633. [CrossRef]
14. Murali, K.; Sinha, S.; Mohamed, I.R. Chaos computing: Experimental realization of NOR gate using a simple chaotic circuit. *P Lett. A* **2005**, *339*, 39–44. [CrossRef]
15. Murali, K.; Sinha, S.; Ditto, W.L.; Bulsara, A.R. Reliable logic circuit elements that exploit nonlinearity in the presence of a n floor. *Phys. Rev. Lett.* **2009**, *102*, 104101. [CrossRef]
16. Campos-Cantón, I.; Pecina-Sánchez, J.; Campos-Cantón, E.; Rosu, H.C. A simple circuit with dynamic logic architecture of b logic gates. *Int. J. Bifurc. Chaos* **2010**, *20*, 2547–2551. [CrossRef]
17. Pando L, C.; Meucci, R.; Ciofini, M.; Arecchi, F. $CO_2$ laser with modulated losses: Theoretical models and experiments in chaotic regime. *Chaos Interdiscip. J. Nonlinear Sci.* **1993**, *3*, 279–285. [CrossRef]
18. Pando, C.; Acosta, G.L.; Meucci, R.; Ciofini, M. Highly dissipative Hénon map behavior in the four-level model of the $CO_2$ la with modulated losses. *Phys. Lett. A* **1995**, *199*, 191–198. [CrossRef]
19. Singh, K.P.; Sinha, S. Enhancement of "logical" responses by noise in a bistable optical system. *Phys. Rev. E* **2011**, *83*, 046 [CrossRef]
20. Murali, K.; Sinha, S.; Kohar, V.; Kia, B.; Ditto, W.L. Chaotic attractor hopping yields logic operations. *PLoS ONE* **2018**, *13*, e0209 [CrossRef]
21. Murali, K.; Sinha, S.; Kohar, V.; Ditto, W.L. Harnessing tipping points for logic operations. *Eur. Phys. J. Spec. Top.* **2** *230*, 3403–3409. [CrossRef]
22. Jaimes-Reátegui, R.; Afanador-Delgado, S.; Sevilla-Escoboza, R.; Huerta-Cuellar, G.; García-López, J.H.; López-Mancilla, Castañeda-Hernández; Pisarchik, A.N. Optoelectronic flexible logic gate based on a fiber laser. *Eur. Phys. J. Spec. Top.* **2** *223*, 2837–2846. [CrossRef]

García-López, J.H.; Jaimes-Reátegui, R.; Afanador-Delgado, S.M.; Sevilla-Escoboza, R.; Huerta-Cuéllar, G.; López-Mancilla, D.; Chiu-Zarate, R.; Castañeda-Hernández, C.E.; Pisarchik, A.N. Experimental and Numerical Study of an Optoelectronics Flexible Logic Gate Using a Chaotic Doped Fiber Laser. In *Recent Development in Optoelectronic Devices*; IntechOpen: London, UK, 2018; pp. 97–114.
Reategui, R.; Kir'yanov, A.; Pisarchik, A.; Barmenkov, Y.O.; Il'ichev, N. Experimental study and modeling of coexisting attractors and bifurcations in an erbium-doped fiber laser with diode-pump modulation. *Laser Phys.* **2004**, *14*, 1277–1281.
Pisarchik, A.N.; Kir'yanov, A.V.; Barmenkov, Y.O.; Jaimes-Reátegui, R. Dynamics of an erbium-doped fiber laser with pump modulation: Theory and experiment. *JOSA B* **2005**, *22*, 2107–2114. [CrossRef]
Pisarchik, A.N.; Jaimes-Reátegui, R.; Sevilla-Escoboza, R.; Huerta-Cuellar, G.; Taki, M. Rogue waves in a multistable system. *Phys. Rev. Lett.* **2011**, *107*, 274101. [CrossRef] [PubMed]
Barba-Franco, J.; Romo-Muñoz, L.; Jaimes-Reátegui, R.; García-López, J.; Cuellar, G.H.; Pisarchik, A. Electronic equivalent of a pump-modulated erbium-doped fiber laser. *Integration* 2022, *in press*. [CrossRef]
Reategui, R.J. Dynamic of Complex Systems with Parametric Modulation: Duffing Oscillators and a Fiber Laser. Ph.D. Thesis, Centro de Investigaciones en Optica, Leon, Mexico, 2004.
Doedel, E.J.; Carlos, L.; Pando, L. Isolas of periodic passive Q-switching self-pulsations in the three-level: Two-level model for a laser with a saturable absorber. *Phys. Rev. E* **2011**, *84*, 056207. [CrossRef]
Doedel, E.J.; Carlos, L.; Pando, L. Multiparameter bifurcations and mixed-mode oscillations in Q-switched $CO_2$ lasers. *Phys. Rev. E* **2014**, *89*, 052904. [CrossRef]

*photonics* MDPI

*Article*

# Flexible, Stretchable, Tunable, and Switchable DFB Laser

**Daniele Eugenio Lucchetta [1,2,*], Andrea Di Donato [3], Oriano Francescangeli [2], Cristiano Riminesi [4,5], Gautam Singh [6] and Riccardo Castagna [4,5,*]**

1 Optoacoustic Lab, Dipartimento di Scienze e Ingegneria della Materia dell'Ambiente ed Urbanistica (SIM Università Politecnica delle Marche, Via Brecce Bianche, 60131 Ancona, Italy
2 Dipartimento di Scienze e Ingegneria della Materia dell'Ambiente ed Urbanistica (SIMAU), Università Politecnica delle Marche, Via Brecce Bianche, 60131 Ancona, Italy
3 Dipartimento di Ingegneria Dell'Informazione (DII), Università Politecnica delle Marche, Via Brecce Bian 60131 Ancona, Italy
4 URT-CNR@UNICAM, Università di Camerino (UNICAM), Polo di Chimica, Via Sant'Agostino, 1, 62032 Camerino, Italy
5 CNR, Institute of Heritage Science, Via Madonna del Piano, 10, 50019 Sesto Fiorentino, Italy
6 Department of Applied Physics, Amity Institute of Applied Sciences, Amity University, Noida 201313, Ir
* Correspondence: d.e.lucchetta@univpm.it (D.E.L.); riccardo.castagna@cnr.it (R.C.)

**Abstract:** We are reporting on easily-made, flexible, wearable, and stretchable holographic grati for tunable distributed-feedback lasers. The laser device is formed by the integration of a reflec volume phase grating in a flexible substrate. The grating is recorded in a photopolymerizable mix by using optical holography. The photopolymerizing material is a new compound derived fro recently developed mixture which is based largely on haloalkanes and acrylates. An appropr choice of photoinitiators promotes the photoactivation of the monomers at the writing wavelengt 460 nm. The laser device has a low emission threshold due to the high efficiency of the photop merization process at 460 nm. Finally, the amplified spontaneous emission of the device can continuously tuned by simply bending the film. This peculiarity is promising for the manufactur of optical pumped tunable organic lasers and flexible displays.

**Keywords:** distributed-feedback (DFB) laser; holographic reflection gratings; flexible sensors; stretcha displays; free-standing gratings; tunable plastic laser



**Citation:** Lucchetta, D.E.; Di Donato, A.; Francescangeli, O.; Riminesi, C.; Singh, G.; Castagna, R. Flexible, Stretchable, Tunable, and Switchable DFB Laser. *Photonics* **2023**, *10*, 12. https://doi.org/10.3390/photonics10010012

Received: 01 November 2022
Revised: 05 December 2022
Accepted: 22 December 2022
Published: 23 December 2022



## 1. Introduction

The world of interaction between composite polymer materials and light is extrem interesting because it allows for the development of new devices adopting a wide ra of previously unimaginable new technologies [1–12]. At the time of writing, flexi electronics is one of the most intriguing research areas, the development of whicl essentially based on polymeric composite materials. Flexible, stretchable, and weara electronics have become increasingly important in today's society. A significant ni of this field is occupied by opto-electronics and, as a consequence, an important rol played by distributed-feedback lasers (DFB) [13]. In recent years, the evolution of flexi DFB lasers [14] has further amplified the perspectives of this sector. As an exam in [15], conjugated polymers are used to produce flexible DFB lasers emitting at multi wavelengths when appropriately stretched. Recently, we presented a novel polyn composite mixture suitable for use in high-density optical data storage and in DFB las due to its very high transparency, high-diffraction efficiency, and the resolution of recorded volume phase reflection gratings [16,17]. The mixture is based on a combinatior multi-acrylate dipentarythritol-hexa-acrylate (DPHA) and a low refractive index mate usually a haloalkane. The presence of haloalkanes induces a lowering of the aver refractive index of the mixture, an increase in the sample transparency in the visi (Vis)-region, a better phase separation process, and a lowering of the viscosity, wh can be inserted into micrometric-sized devices even at environmental temperature. Here, we intend to further exploit its characteristics and propose a very easy-to-make, flexible, and wearable holographic device that can produce highly efficient and frequency-tunable and/or -switchable DFB lasers. The central principle is the insertion of a polymerized material containing a high-resolution reflection grating in a double-layered scotch tape flexible film. The first layer of scotch tape is used to peel off the DFB grating from the recording glass support, while a second layer covers and seals the entire structure in order to protect it from possible aerobic degeneration. At the same time, this particular configuration makes it more manageable and suitable for tunability in terms of DFB frequencies emission and intensity (tunable and/or switchable DFB laser).

## 2. Materials and Methods

### 2.1. Materials

DPHA, butyl-bromide (BB), hexyl-bromide (BH), 6-Oxocamphor (6OC), and RH6G are purchased from Merck, Darmstadt, Germany.

### 2.2. Methods

#### 2.2.1. Holographic Mixture

The procedure is similar to that detailed in the Methods section of Ref. [13]. In brief, BB (20%, $w/w$), BH (10%), DPHA (c.a. 68%), 1.9% 6OC, and 0.05% RH6G are blended together till a homogeneous pale orange color is obtained. After that, the mixture is stirred at room temperature, in dark aerobic conditions for 6 days and stored for 90 days before being used.

#### 2.2.2. Cell Preparation

The above prepared mixture is then inserted by capillarity between two glass slides separated by a thickness of $\approx$50 μm.

#### 2.2.3. Hologram Recording Set-Up

The recording set-up is shown in Methods of Ref. [13]. In this work, we used a different polymerizing continuous laser source, a Coherent Genesis MX-460 single longitudinal mode (SLM) operating at $\lambda_w$ = 460 nm. The irradiation time is 1.2 s.

#### 2.2.4. Flexible Grating Preparation

After the grating recording, the whole structure (including the material surrounding the grating area) is further polymerized with the help of a high-power UV-A lamp ($\lambda$ = 365 nm; P = 0.5 W) placed directly in contact with the glasses containing the sample for one minute. After post-polymerization, the cell is opened and the polymer film strip—containing the DFB grating—is peeled-off with the help of a transparent scotch tape. Once the polymer grating strip is attached on the scotch tape strip, a further tape layer is used to seal the polymerized structure. The final result is a DFB grating inserted between two transparent scotch tape strips. The thus prepared free-standing film is finally clamped in the optical pumping setup.

#### 2.2.5. Pumping and Detection Set-Up

The pumping set-up is made by a Nd-YAG pulsed laser operating at $\lambda$ = 532 nm $\tau$ = 4 ns. The pulsed light impinges at 45° to the clamped free-standing sample containing the photo active mixture. The light emitted by the device is selected and amplified by the presence of the DFB grating and collected by a thin lens at the entrance of an optical fiber connected to a spectrometer.

## 3. Results and Discussion

Figure 1 shows the transmission spectrum of the recorded DFB structure. The fi position of the reflection peak depends on the recording angle and on the shrinka phenomenon as will be discussed below in the text. The normalization of the transmit signal is made outside of the grating area. This explains a transmission higher than 10 in the left region of the spectrum, since the transparency in the grating area is higher th that in the other regions of the sample. In the inset of the same figure we report, in bla the transmission spectrum of the polymerized free-standing film (see the Section 2) and, comparison, in blue, the transmission spectrum of the flexible polymerized film confir between two transparent scotch tape strips. The spectra slightly differ from one anot being composed of polypropylene and acrylate. Irradiation at the wavelength $\lambda_w$ = 460 for 1.2 s, gives an efficient phase separation process in our mixture allowing the recordi of reflection gratings with a 50% value of diffraction efficiency.

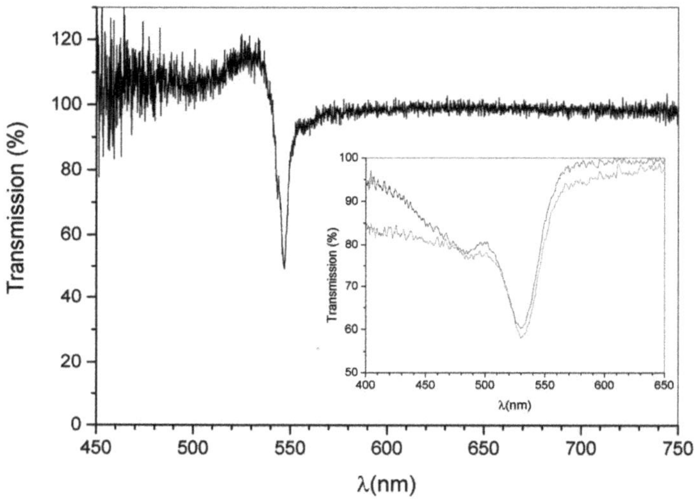


**Figure 1.** Transmission spectrum showing the normalized reflection peak at the end of the record process. The normalization is performed outside the grating area. This explains a transmiss higher than 100% in the first region of the spectrum. In the inset we report, in black, the transmiss spectrum of the polymerized free-standing film and, for comparison, in blue, the transmiss spectrum of the flexible polymerized film confined between the two transparent scotch tape stri

The recording process is however affected by a considerable shrinkage. For this reas it is essential to monitor in real time the growth of the reflection peak and its displacem (blue-shift) in order to stop the irradiation when the peak reaches the maximum value diffraction efficiency. A typical behavior is shown in Figure 2. Reported in this figure is growth of the reflection peak after ≈2 s of irradiation. As it can be seen, the peak grows to a maximum value (after ≈1.2 s) and after that the diffraction efficiency decreases an parasite grating appears [16]. For this reason, we decided to stop the irradiation after 1 For a more complete treatment of the shrinkage phenomenon please see ref [18–21] a references quoted therein.

An amplified spontaneous emission (ASE) effect [22–25] can be easily seen by pumpi the sample outside the grating area, see Figure 3. This effect is typical when a laser dye su as the rhodamine 6G (R6G) is used. R6G, like many other laser dyes, is characterized, un excitation, by a complex energy levels diagram involved in which are transitions betwe singlet and triplet states. The electronic levels are wide in energy due to a continuum vibrational, rotational, and solvent states. The diffracted energy of a grating containi a small amount of R6G (typically in the range $10^{-2}$–$10^{-4}$ mol) depends on the square of the incident energy. For high values of the impinging energy a saturation of the diffracted energy connected with the saturation of the excited state population is observed. IN deep information and energy level diagrams of R6G can be found in Ref. [26].

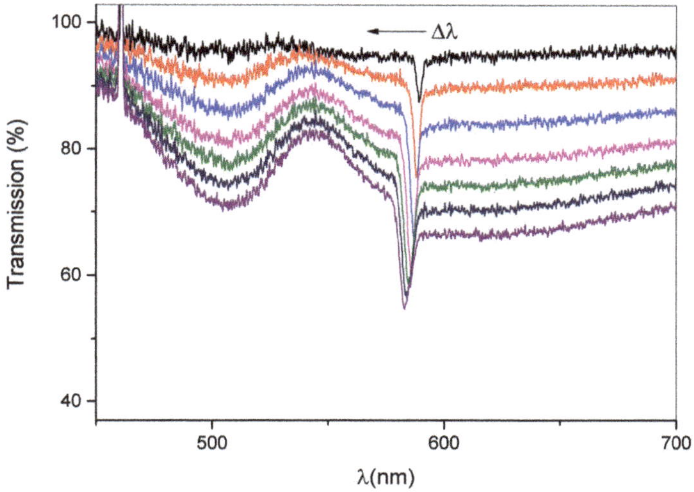


**Figure 2.** Real-time growth of the reflection peak under irradiation at $\lambda$ = 460 nm. The acquisition time is 100 ms. The first peak (black curve) appears at t = 0.15 s. The total irradiation time is ≈2 s and the maximum of diffraction efficiency is obtained after ≈1.2 s. The blue-shift $\Delta\lambda$ of the reflection peak is due to the shrinkage under irradiation.

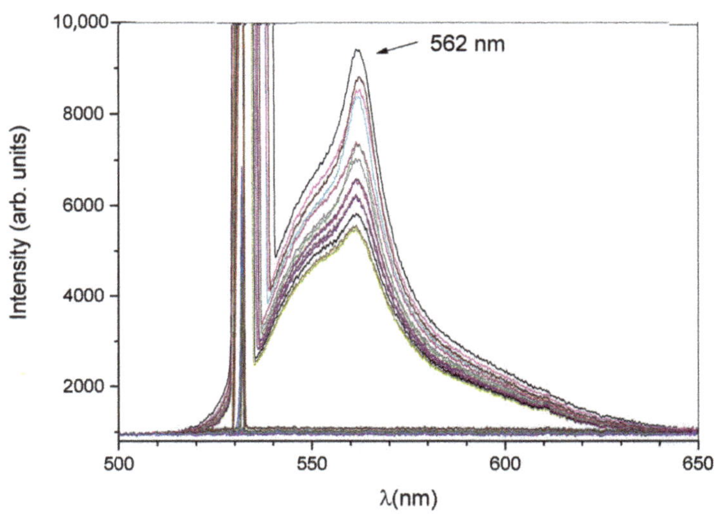


**Figure 3.** The ASE effect is observed by pumping the free-standing film outside the grating area. The peak on the left side is due to the pumping beam. The wide peak on the right side has a full width at half maximum (FWHM) of about 5 nm and is due to the presence of the R6G dye. The range of fluences used is 10–80 mJ/cm$^2$, the ASE threshold is located at ≈15 mJ/cm$^2$. Irradiation wavelength $\lambda$ = 532 nm, pulse duration $\tau$ = 4 ns.

It is important to note that the full width at half maximum (FWHM) of the emitted li remains stable at 5 nm for the whole set of measurements. Figure 4 reports the emiss spectra of our device for values of the irradiation energy ranging from 4 to 20 mJ wh when an irradiated area has a 3 mm radius corresponds to ≈15–70 mJ/cm$^2$. In the inse the same figure, the intensity of each single peak as a function of the pumping energ reported. The shown behavior does not suggest any lasing effect.

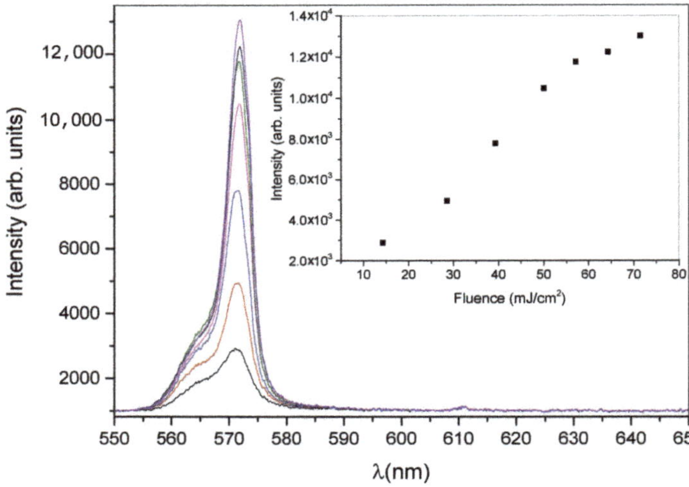


**Figure 4.** The ASE effect is observed by pumping the free-standing film at $\lambda$ = 532 nm and $\tau$ = 4 outside the grating area. In the inset the fluence values used to pump the free-standing structur

By properly setting the writing angle and the exposure time of the recording sourc reflection peak showing a 50% diffraction efficiency, centered around 585 nm, can be recorc The minimum of the peak located at 585 nm is far from the naturally growing peaks of mixture containing the R6G dye. By pumping the system with a pulsed Nd-YAG sourc clear laser effect can be detected at 588 nm as shown in Figure 5. Lasing happens on edge of the photonic bandgap where the density of states (DOS) is higher. The DOS c one-dimensional periodic structure is defined as: $D(\omega) \propto \int dk\delta[\omega - \omega(k)] = \left|\frac{d\omega}{dk(\omega)}\right|^{-1} =$ It shows Van Hove singularities at the two edges of the photonic bandgap [27,28]. T singularity becomes a narrow sharp peak in the DOS that improves ASE bringing to las once a threshold of ≈3.5 mJ/cm$^2$, in our specific case, is reached. From a physical poin view, at the band edges, the photons experience multiple reflections causing them to mc collectively slowly. This behavior allows for large population build-up and an increase in optical path length. The presence of the R6G, ensures light amplification and lasing on one the edges of the photonic stop band. The emitted light satisfies the condition of construct interference, see Refs. [29–31]. Above a threshold value the light emitted from the excited R molecules is amplified in a resonant way and finally laser emission occurs. Narrowing of emission band of the R6G is observed by monitoring the behavior of the emission peak the band as a function of the excitation energy. The results are reported in Figure 6 wh reports the emission spectra of the DFB grating when pumped with energies in the rar 3.5 mJ/cm$^2$–≈70.0 mJ/cm$^2$. The inset of the same figure shows the behavior of the heigh the peaks as a function of the fluence. The analysis of this figure shows a nonlinear behav characterized by an abrupt change of the slope at ≈10 mJ/cm$^2$ directly connected to appearance of the laser emission.

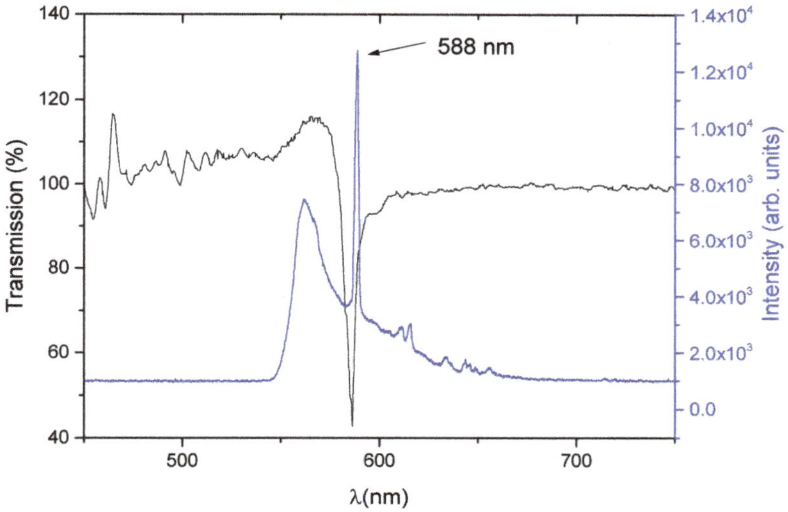


**Figure 5.** Typical lasing effect (in blue) observed by pumping the grating structure with a fluence of $\approx$40 mJ/cm$^2$, $\lambda$ = 532 nm and $\tau$ = 4 ns. In black the reflection peak recorded in the active mixture.

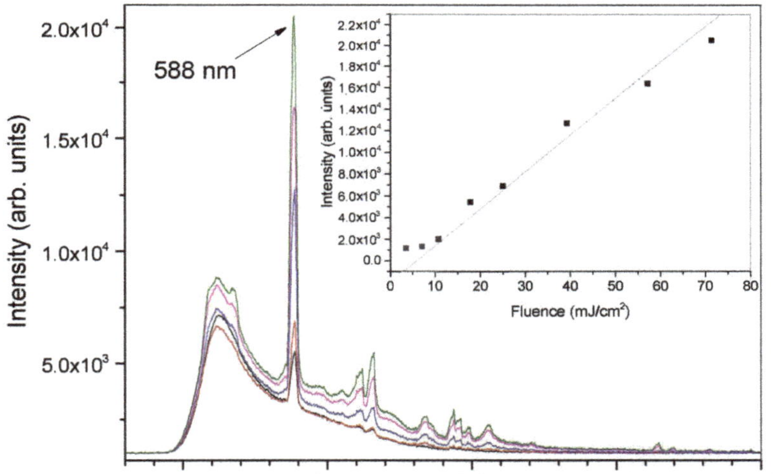


**Figure 6.** Lasing from the free-standing DFB structure for different values of the impinging energy. The energy values are reported in the inset. The laser threshold is around $\approx$3.5 mJ/cm$^2$.

During our experiments, we also observed an interesting phenomenon linked with the bending of our free-standing film. The phenomenon involves the mechanical tuning of the device obtained by simply stretching the film along the directions indicated by the red arrows in the sketch of Figure 7. By starting with the film bent in the position indicated by the number 1, and moving through different bending states along the blue arrow until a straight film is obtained, we noticed that, by pumping outside the grating area, the emission changes from 568 to 572 nm. A similar behavior is obtained by simply rotating the bent sample. We believe that the change in the emitted wavelength is linked to variations in the impact cross section of the pumping source when the bend conditions or the rotatio angle change. The observed variation in the emitted wavelength could also be due t bend-induced change of the grating pitch or simply by a non uniformity of the grati structure among the irradiated area. This situation can only be investigated by an accur SEM analysis of the sample cross section. Connected with the changes in the emitt wavelength we observe also a remarkable growth of the emitted laser intensity which c be switched from low to high values. This behavior is explainable with the bend-induc change of the direction of the grating vector. It is important to underline that these preliminary measurements. In Figure 7, we only reported the experimentally observ phenomena which certainly deserves a future deeper investigation.

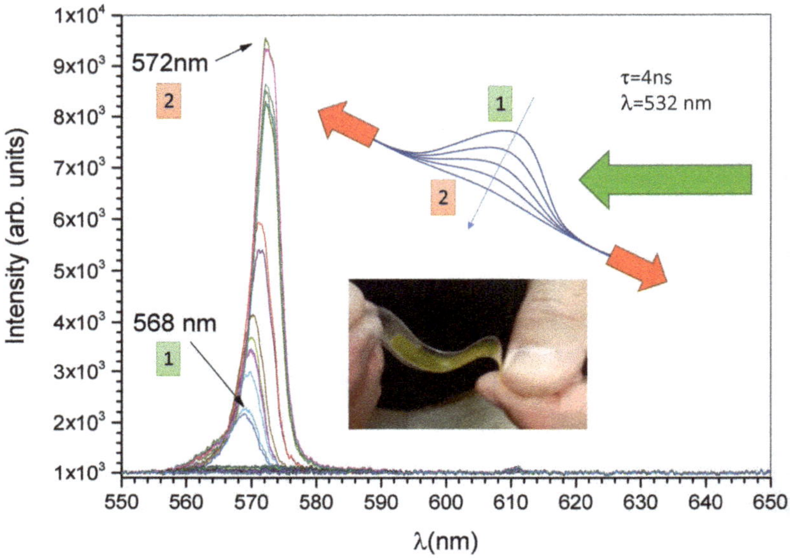


**Figure 7.** Demonstration of the tuning possibilities of the free-standing film. On the right sid sketch of the experimental setup. By stretching the film along the directions indicated by the arrows, bending occurs and the emission can be tuned from 568 to 572 nm. The real device is sho in the inset.

## 4. Conclusions

In conclusion, a new holographic mixture is used to record phase reflection grati showing very good diffraction efficiency values. The mixture containing the laser dye R can be optically pumped obtaining a lasing effect in a free-standing configuration. A effect and lasing are observed outside and inside the grating area. An interesting tunabil of the ASE emission is reported. This flexible, stretchable, tunable, and switchable DFB-la could open up new perspectives in the field of plastic organic laser devices.

**Author Contributions:** Conceptualization, R.C. and D.E.L.; methodology, R.C. and D.E.L.; validati R.C. and D.E.L.; formal analysis, A.D.D.; investigation, R.C. and D.E.L.; writing—original d preparation, R.C. and D.E.L.; writing—review and editing, R.C., D.E.L., A.D.D. and G.S.; supervis C.R., G.S. and O.F. All authors have read and agreed to the published version of the manuscript.

**Funding:** R.C. thanks MARLIC—POR Marche FESR 2014-2020.

**Institutional Review Board Statement:** Not applicable.

**Data Availability Statement:** Data are available from the authors under reasonable request.

**Acknowledgments:** Authors thank Simona Sabbatini for the NIR characterization of the scotch tape. R.C. thanks "Marche Applied Research Laboratory for Innovative Composites" (MARLIC) financed by Regione Marche in the frame of POR Marche FESR 2014-2020.

**Conflicts of Interest:** The authors declare no conflict of interest.

## ferences

He, J.; Zhao, Y.; Zhao, Y. Photoinduced bending of a coumarin-containing supramolecular polymer. *Soft Matter* **2009**, *5*, 308–310. [CrossRef]

Lucchetta, D.; Vita, F.; Francescangeli, D.; Francescangeli, O.; Simoni, F. Optical measurement of flow rate in a microfluidic channel. *Microfluid. Nanofluid.* **2016**, *20*, 9. [CrossRef]

Kondo, M.; Takemoto, M.; Fukae, R.; Kawatsuki, N. Photomobile polymers from commercially available compounds: Photoinduced deformation of side-chain polymers containing hydrogen-bonded photoreactive compounds. *Polym. J.* **2012**, *44*, 410–414. [CrossRef]

Lucchetta, D.; Spegni, P.; Di Donato, A.; Simoni, F.; Castagna, R. Hybrid surface-relief/volume one dimensional holographic gratings. *Opt. Mater.* **2015**, *42*, 366–369. [CrossRef]

Hu, Y.; Wu, G.; Lan, T.; Zhao, J.; Liu, Y.; Chen, W. A Graphene-Based Bimorph Structure for Design of High Performance Photoactuators. *Adv. Mater.* **2015**, *27*, 7867–7873.

Castagna, R.; Lucchetta, D.E.; Rippa, M.; Xu, J.H.; Donato, A.D. Near-frequency photons Y-splitter. *Appl. Mater. Today* **2020**, *19*, 100636. [CrossRef]

Zhang, X.; Yu, Z.; Wang, C.; Zarrouk, D.; Seo, J.W.T.; Cheng, J.C.; Buchan, A.D.; Takei, K.; Zhao, Y.; Ager, J.W.; et al. Photoactuators and motors based on carbon nanotubes with selective chirality distributions. *Nat. Commun.* **2014**, *5*, 2983. [CrossRef]

Lucchetta, D.; Simoni, F.; Hernandez, R.; Mazzulla, A.; Cipparrone, G. Lasing from chiral doped nematic liquid crystal droplets generated in a microfluidic device. *Mol. Cryst. Liq. Cryst.* **2017**, *649*, 11–19. [CrossRef]

Angeloni, A.S.; Caretti, D.; Carlini, C.; Chiellini, E.; Galli, G.; Altomare, A.; Solaro, R.; Laus, M. Photochromic liquid-crystalline polymers. Main chain and side chain polymers containing azobenzene mesogens. *Liq. Cryst.* **1989**, *4*, 513–527.

Kawatsuki, N.; Hasegawa, T.; Ono, H.; Tamoto, T. Formation of Polarization Gratings and Surface Relief Gratings in Photocrosslinkable Polymer Liquid Crystals by Polarization Holography. *Adv. Mater.* **2003**, *15*, 991–994. . adma.200304988. [CrossRef]

Shalit, A.; Lucchetta, D.; Piazza, V.; Simoni, F.; Bizzarri, R.; Castagna, R. Polarization-dependent laser-light structured directionality with polymer composite materials. *Mater. Lett.* **2012**, *81*, 232–234. [CrossRef]

De Sio, L.; Serak, S.; Tabiryan, N.; Ferjani, S.; Veltri, A.; Umeton, C. Composite Holographic Gratings Containing Light-Responsive Liquid Crystals for Visible Bichromatic Switching. *Adv. Mater.* **2010**, *22*, 2316–2319. [CrossRef] [PubMed]

Lucchetta, D.E.; Di Donato, A.; Singh, G.; Castagna, R. Lasing in Haloalkanes-based polymeric mixtures. *Opt. Mater.* **2022**, *131*, 112614. [CrossRef]

Karl, M.; Glackin, J.M.E.; Schubert, M.; Kronenberg, N.M.; Turnbull, G.A.; Samuel, I.D.W.; Gather, M.C. Flexible and Ultra-Lightweight Polymer Membrane Lasers. In Proceedings of the 2019 Conference on Lasers and Electro-Optics Europe & European Quantum Electronics Conference (CLEO/Europe-EQEC), Munich, Germany, 23–27 June 2019. [CrossRef]

Castro-Smirnov, J.; Sousaraei, A.; Osorio, M.; Casado, S.; Hernandez, J.J.; Wu, L.; Zhang, Q.; Xia, R.; Granados, D.; Wannemacher, R.; et al. Flexible distributed feedback lasers based on nanoimprinted cellulose diacetate with efficient multiple wavelength lasing. *NPJ Flex. Electron.* **2019**, *3*, 17. [CrossRef]

Castagna, R.; Lucchetta, D.; Vita, F.; Criante, L.; Greci, L.; Simoni, F. Haloalkane-based polymeric mixtures for high density optical data storage. *Opt. Mater.* **2008**, *30*, 1878–1882. [CrossRef]

Lucchetta, D.E.; Di Donato, A.; Francescangeli, O.; Singh, G.; Castagna, R. Light-Controlled Direction of Distributed Feedback Laser Emission by Photo-Mobile Polymer Films. *Nanomaterials* **2022**, *12*, 2890. [CrossRef]

Castagna, R.; Vita, F.; Lucchetta, D.E.; Criante, L.; Greci, L.; Ferraris, P.; Simoni, F. Nitroxide radicals reduce shrinkage in acrylate-based holographic gratings. *Opt. Mater.* **2007**, *30*, 539–544. [CrossRef]

Waldman, D.; Li, H.; Horner, M. Volume shrinkage in slant fringe gratings of a cationic ring-opening holographic recording material. *J. Imaging Sci. Technol.* **1997**, *41*, 497–514.

Castagna, R.; Davis, P.; Vasu, V.; Soucek, K.; Cross, C.; Greci, L.; Valacchi, G. Nitroxide radical TEMPO reduces ozone-induced chemokine IL-8 production in lung epithelial cells. *Toxicol. Vitr.* **2009**, *23*, 365–370. [CrossRef]

Ramos, G.; Álvarez-Herrero, A.; Belenguer, T.; del Monte, F.; Levy, D. Shrinkage control in a photopolymerizable hybrid solgel material for holographic recording. *Appl. Opt.* **2004**, *43*, 4018–4024. [CrossRef]

Peters, G.I.; Allen, L. Amplified spontaneous emission I. The threshold condition. *J. Phys. A Gen. Phys.* **1971**, *4*, 238. [CrossRef]

McGehee, M.D.; Gupta, R.; Veenstra, S.; Miller, E.K.; Díaz-García, M.A.; Heeger, A.J. Amplified spontaneous emission from photopumped films of a conjugated polymer. *Phys. Rev. B* **1998**, *58*, 7035–7039. [CrossRef]

Malcuit, M.S.; Maki, J.J.; Simkin, D.J.; Boyd.; W., R. Transition from superfluorescence to amplified spontaneous emission. *Phys. Rev. Lett.* **1987**, *59*, 1189–1192. [CrossRef] [PubMed]

25. Ganiel, U.; Hardy, A.; Neumann, G.; Treves, D. Amplified spontaneous emission and signal amplification in dye-laser syste *IEEE J. Quantum Electron.* **1975**, *11*, 881–892. [CrossRef]
26. Eichler, H.; Günter, P.; Pohl, D. *Laser-Induced Dynamic Gratings*; Springer Series in Optical Sciences; Springer: Berlin/Heidelb Germany, 2013.
27. Kneubühl, F. *Theories on Distributed Feedback Lasers*; Handbook of laser science and technology; Harwood Academic Publish Reading, UK, 1993.
28. Criante, L.; Lucchetta, D.; Vita, F.; Castagna, R.; Simoni, F. Distributed feedback all-organic microlaser based on holograp polymer dispersed liquid crystals. *Appl. Phys. Lett.* **2009**, *94*, 111114. [CrossRef]
29. Fu, Y.; Zhai, T. Distributed feedback organic lasing in photonic crystals. *Front. Optoelectron.* **2020**, *13*, 18–34. [CrossRef]
30. Zhai, T.; Ma, X.; Han, L.; Zhang, S.; Ge, K.; Xu, Y.; Xu, Z.; Cui, L. Self-Aligned Emission of Distributed Feedback Lasers on Opt Fiber Sidewall. *Nanomaterials* **2021**, *11*, 2381. [CrossRef]
31. Kok, M.H.; Lu, W.; Lee, J.C.W.; Tam, W.Y.; Wong, G.K.; Chan, C.T. Lasing from dye-doped photonic crystals with graded layer dichromate gelatin emulsions. *Appl. Phys. Lett.* **2008**, *92*, 151108. [CrossRef]

photonics

MDPI

mmunication

# ;eneralized Poincaré Beams in Tight Focus

ctor V. Kotlyar [1,2], Alexey A. Kovalev [1,2,*] and Alexey M. Telegin [2]

1 Image Processing Systems Institute of the RAS—Branch of FSRC "Crystallography & Photonics" of the RAS, 151 Molodogvardeyskaya St., 443001 Samara, Russia
2 Samara National Research University, 34 Moskovskoe Shosse, 443086 Samara, Russia
* Correspondence: alanko@ipsiras.ru

**Abstract:** We study the tight focus of generalized (hybrid) Poincaré beams. A conventional Poincaré beam is a coaxial superposition of two optical vortices, one with left circular polarization and a topological charge (TC) of $m$, while the other has a right circular polarization and a TC of $-m$. The generalized Poincaré beams are also composed of two optical vortices, but their TCs are different, for instance, $p$ and $q$. Here, we theoretically investigate the generalized Poincaré beams with the TCs $p = m + 1$ and $q = -m$ in tight focus. In this case, both transverse components of the strength vector of the initial electric field have a topological charge of $1/2$, and the beam itself is a cylindrical vector beam of fractional order $m + 1/2$. Analytical expressions are derived for the components of the strength vectors of the electric and magnetic field at the focus as well as for the intensity distribution, the longitudinal component of the spin angular momentum (SAM), and for the components of the Poynting vector (energy flow density). We show that the intensity at the focus has $2m - 1$ local maxima residing evenly in a certain circle radius with the center on the optical axis. We also demonstrate that the radial spin and orbital Hall effects occur at the focus, i.e., the longitudinal SAM component has different signs in the circles of different radii, and the azimuthal component of the transverse Poynting vector also has different signs.

**Keywords:** Poincaré beam; tight focus; generalized Poincaré beam; spin angular momentum; Hall effect; Poynting vector

check for updates

**ation:** Kotlyar, V.V.; Kovalev, A.A.; :gin, A.M. Generalized Poincaré ms in Tight Focus. *Photonics* **2023**, 218. https://doi.org/10.3390/ tonics10020218

eived: 27 January 2023
ised: 12 February 2023
epted: 14 February 2023
lished: 16 February 2023



## 1. Introduction

Poincaré beams, whose polarization state is related to the polarization Poincaré sphere [1–3], are actively studied in optics, starting with works [3–6]. In a general case, a Poincaré beam is a superposition of two optical vortices with different topological charges (TC) $p$ and $q$ and with orthogonal polarizations. For the optical vortices, the conventional Laguerre–Gaussian beams of different indices [7–9] can be chosen, or diffraction-free Bessel beams, or the Bessel–Gaussian beams generated by axicons [10–12]. The Poincaré beams can be generated similarly to all the other vector beams, by using liquid-crystal light modulators, half-wave and quarter-wave plates [13–15], or by using lasers and q-plates [16], and metasurfaces [12]. In Ref. [17], the polarization singularity index (Poincaré–Hopf index) of the Poincaré beams was studied. In Ref. [18], the optical Hall effect was theoretically discovered in the tight focus of the Poincaré beams. The optical (or photonic) Hall effect is divided into spin [19,20] and orbital [21,22]. Usually, the Hall effect in optics is observed when a light field is reflected from an interface between media [21,22], or when it passes through multilayered media [23], crystals [24,25], or through a metasurface [26]. There are known works investigating the Hall effect in the tight focus of a laser radiance [27,28] or the vicinity of the focal plane [29]. We note that the abovementioned works do not contain theoretically obtained key characteristics of the generalized Poincaré beams in tight focus using the Richards–Wolf formalism [30]: amplitudes of the electric and magnetic vectors, intensity distribution, distributions of components of the Poynting vector, and the spin angular momentum (SAM) vector.

In this work, adopting the Richards–Wolf approach, we obtain analytical expressic describing key characteristics of the generalized Poincaré beams in a case when the to logical charges (TC) of the two optical vortices with left and right circular polarizati are equal with respect to $p = m + 1$ and $q = -m$. We demonstrate that at the focus of su beams, radial spin and orbital Hall effects take place. We note that in [31], we demonstra the spin Hall effect for fractional-order cylindrical vector beams at the focus plane. the current work, at $p = m + 1$ and $q = -m$, there is also a cylindrical vector beam w a fractional order of $m + 1/2$. Therefore, we can expect that there is also the spin Hall eff at the focus of such a generalized Poincaré beam. The work of [31] does not contain a lytical expressions for electric field components at the focus of fractional-order cylindri vector beams. In the current work, we derive such analytical expressions.

In our previous work [32], we have shown that the orbital Hall effect occurs bef and after the focus of the conventional vectorial cylindrical beams, which are a spec case of Poincaré beams when the optical vortices have the TCs $m$ and $-m$, and that lo areas in the beam cross section, where the transverse energy flow is rotating clockwise counterclockwise, reside in pairs on a certain circle radius with the center on the opti axis. In this work, energy flows, rotating clockwise or counterclockwise, reside on circ with different radii. Therefore, this orbital Hall effect is called radial.

We note that the Richards–Wolf formalism [30] adequately describes the light fi only near the focus. The work of [33] investigates the behavior of light at the focus by usi an exact solution of the Helmholtz equation in the spherical coordinates, which is correct the whole space. However, generating such a light field at the focus requires generating the initial plane all three components of the electric vector. This is a challenging proble In our case, only the transverse components of the electric field should be generated in initial plane, which is easy to implement in practice.

## 2. Vector Field in the Initial Plane

We consider here the following Jones vector of the initial light field:

$$\begin{aligned} E_1 &= a\frac{\exp(ip\varphi)}{\sqrt{2}}\begin{pmatrix} 1 \\ i \end{pmatrix} + b\frac{\exp(iq\varphi)}{\sqrt{2}}\begin{pmatrix} 1 \\ -i \end{pmatrix}, \\ a^2 &+ b^2 = 1, \end{aligned}$$

with $(r, \varphi)$ being the polar coordinates in the initial plane, and $a$, and $b$ being complex c stants. If $p = -n$ and $q = n$, the beam from Equation (1) reduces to a conventional Poinc beam [4–6]. If $a = b = 1/\sqrt{2}$, then the field (1) becomes maximally inseparable [18]:

$$E_2 = \exp\left(\frac{i(q+p)\varphi}{2}\right)\begin{pmatrix} \cos\left(\frac{q-p}{2}\right)\varphi \\ \sin\left(\frac{q-p}{2}\right)\varphi \end{pmatrix}.$$

When $p = q$, the field (1) reduces to a linearly polarized optical vortex with the to logical charge (TC) $q$. When $p = -q$, the field (1) is a cylindrical vector beam of the or $q$ [34]. When $p = -m$ and $q = m + 1$, the field (2) is given by

$$E_2 = \exp\left(\frac{i\varphi}{2}\right)\begin{pmatrix} \cos\left(m+\frac{1}{2}\right)\varphi \\ \sin\left(m+\frac{1}{2}\right)\varphi \end{pmatrix} = \frac{\exp(i(m+1)\varphi)}{2}\begin{pmatrix} 1 \\ -i \end{pmatrix} + \frac{\exp(-im\varphi)}{2}\begin{pmatrix} 1 \\ i \end{pmatrix}.$$

The field (3) is interesting because it is a cylindrical vector beam of a half-integer or In Ref. [35], the beam (3) is not quite correctly called a vector vortex beam with a fractio topological charge. As was already shown in [31], in the tight focus of the fractional-or cylindrical vector beams, subwavelength areas are generated with elliptic polarization different handedness, that is, the polarization vector in these areas is rotating clockwise counterclockwise. We note that the initial light field (3) is linearly polarized at each point its cross section. Therefore, similarly to [31], it should be expected that the focused field should also contain the areas with elliptic polarization of different handedness. It is also seen from Equation (3) that the initial field is a coaxial superposition of two optical vortices with left and right circular polarization and with different topological charges of $m + 1$ and $-m$. Since these topological charges do not compensate for each other, it is reasonable to expect circular energy flow at the focus. This means that at the focus, a nonzero distribution of the axial component of the angular momentum vector should be present. Below, we show that this is indeed so.

## 3. Components of the Strength Vector of the Electric Field at the Focus

The Richards–Wolf formalism [30] allows access to all components of the strength vector of the electric field at the tight focus of the initial field (3):

$$
\begin{aligned}
E_{2x} &= \frac{i^m}{2}\left[e^{i(m+1)\varphi}I_{0,m+1} + e^{i(m-1)\varphi}I_{2,m-1} - ie^{-im\varphi}I_{0,m} - ie^{-i(m-2)\varphi}I_{2,m-2}\right],\\
E_{2y} &= \frac{i^m}{2}\left[-ie^{i(m+1)\varphi}I_{0,m+1} + ie^{i(m-1)\varphi}I_{2,m-1} + e^{-im\varphi}I_{0,m} - e^{-i(m-2)\varphi}I_{2,m-2}\right],\\
E_{2z} &= i^m\left[ie^{im\varphi}I_{1,m} + e^{-i(m-1)\varphi}I_{1,m-1}\right],
\end{aligned} \tag{4}
$$

where

$$
I_{\nu,\mu} = \left(\frac{4\pi f}{\lambda}\right)\int_0^{\theta_0}\sin^{\nu+1}\left(\frac{\theta}{2}\right)\cos^{3-\nu}\left(\frac{\theta}{2}\right)\cos^{1/2}(\theta)A(\theta)e^{ikz\cos\theta}J_\mu(\xi)d\theta, \tag{5}
$$

where $k = 2\pi/\lambda$ is the wavenumber of light with the wavelength $\lambda, f$ is the focal length of an aplanatic system (ideal spherical lens), $\nu$ = 0, 1, and 2, $J_\mu(\xi)$ is the $\mu$th-order Bessel function of the first kind, $\xi = kr\sin\theta$, $\theta$ is the polar angle that defines the tilt of the optical axis of rays converging into the focus, $\theta_0$ is the maximal angle that defines the numerical aperture of the aplanatic system ($NA = \sin\theta_0$), $(r, \varphi, z)$ is the cylindrical coordinate system with the origin at the focus ($z$ = 0 is the focus plane), and $A(\theta)$ is the amplitude of the initial circularly symmetric field (real-valued function).

## 4. Intensity Distribution of the Electric Field at the Focus

From the components of the electric vector (4), we can derive the intensity distribution of the light field at the focus plane ($z$ = 0):

$$
\begin{aligned}
I &= I_\perp + I_z = I_x + I_y + I_z\\
&= \tfrac{1}{2}\Big[\Big(I_{0,m+1}^2 + I_{2,m-2}^2 - 2\sin((2m-1)\varphi)I_{0,m+1}I_{2,m-2}\Big)\\
&+\Big(I_{0,m}^2 + I_{2,m-1}^2 - 2\sin((2m-1)\varphi)I_{0,m}I_{2,m-1}\Big)\\
&+2\Big(I_{1,m}^2 + I_{1,m-1}^2 - 2\sin((2m-1)\varphi)I_{1,m}I_{1,m-1}\Big)\Big],
\end{aligned} \tag{6}
$$

where the first two terms in the round brackets describe the transverse intensity $I_\perp = I_x + I_y$, whereas the third term in the round brackets describes the longitudinal intensity $I_z$. As seen from Equation (6), the intensity is a nonnegative function ($I \geq 0$) since each term in the round brackets in Equation (6) is nonnegative, for the sum of two squared numbers is equal to or greater than their doubled product. Equation (6) also follows that the intensity distribution contains $2m - 1$ local maxima and $2m - 1$ local minima (or intensity nulls) that reside on a certain circle radius with the center on the optical axis. Thus, the number of these intensity maxima and nulls is always odd ($2m - 1$).

## 5. Longitudinal Component of the Spin Angular Momentum Vector at the Focus

Using the components of the electric field vector at the focus (4), we can derive the longitudinal component of the spin angular momentum (SAM) vector of the field (3), since the longitudinal SAM component $S_z$ is equal to the third Stokes parameters $S_3$, whose magnitude indicates the presence of elliptic or circular polarization in the beam cross section. The SAM vector is defined by the following expression [28]:

$$\mathbf{S} = \frac{1}{16\pi\omega}\mathrm{Im}(\mathbf{E}^* \times \mathbf{E}),$$

with $\omega$ being the angular frequency of light. Below, we omit the constant factor $1/(16\pi$ for brevity. It can be seen from Equation (7) that the longitudinal SAM component (with the constant) coincides with the nonnormalized third component of the Stokes vector:

$$S_3 = S_z = 2\mathrm{Im}\left(E_x^* E_y\right).$$

Substitution of expressions (4) for the electric field components into Equation (8) yie

$$S_z = \tfrac{1}{2}\Big[\Big( I_{0,m}^2 + I_{2,m-1}^2 - 2\sin((2m-1)\varphi) I_{0,m} I_{2,m-1}\Big) - \Big( I_{0,m+1}^2 + I_{2,m-2}^2 - 2\sin((2m-1)\varphi) I_{0,m+1} I_{2,m-2}\Big)\Big].$$

A comparison of Equations (6) and (9) reveals that if the transverse intensity is a s of two positive terms $A$ and $B$, then

$$I_\perp = A + B = \tfrac{1}{2}\Big[\Big( I_{0,m+1}^2 + I_{2,m-2}^2 - 2\sin((2m-1)\varphi) I_{0,m+1} I_{2,m-2}\Big) + \Big( I_{0,m}^2 + I_{2,m-1}^2 - 2\sin((2m-1)\varphi) I_{0,m} I_{2,m-1}\Big)\Big], \quad ($$

and the longitudinal SAM component is a difference between these terms:

$$S_z = B - A. \quad ($$

According to Equation (9), similarly to the intensity distribution in Equation (6), SAM distribution also has $2m-1$ local maxima and $2m-1$ local minima. As seen fr Equation (11), if $B > A$, then $S_z > 0$ (polarization vector is rotating counterclockwise), a vice versa, if $B < A$, then $S_z < 0$ (polarization vector is rotating clockwise). In the areas wh $B = A$ ($S_z = 0$), the polarization is linear. The points in the beam cross section at the foc where $S_z = 0$, are called [36] topological spin defects. Thus, it follows from Equations and (11) that there are areas with different spins at the focus: positive ($S_z > 0$) and negat ($S_z < 0$). The spatial separation of areas with left circular and right circular polarizatio called the spin Hall effect [27–29]. In the Simulation section below, these conclusions confirmed by concrete examples.

## 6. Energy Flow Density at the Focus

Here we derive the Poynting vector (energy flow density) at the focus of the field To do this, we should first obtain the components of the strength vector of the magne field at the focus. In the same way, as we obtained the components of the electric vector by using the Richards–Wolf theory [30], we can also obtain the magnetic vector:

$$\begin{aligned} H_{2x} &= \tfrac{i^m}{2}\left[ie^{i(m+1)\varphi} I_{0,m+1} + ie^{i(m-1)\varphi} I_{2,m-1} - e^{-im\varphi} I_{0,m} - e^{-i(m-2)\varphi} I_{2,m-2}\right], \\ H_{2y} &= \tfrac{i^m}{2}\left[e^{i(m+1)\varphi} I_{0,m+1} - e^{i(m-1)\varphi} I_{2,m-1} - ie^{-im\varphi} I_{0,m} + ie^{-i(m-2)\varphi} I_{2,m-2}\right], \quad ( \\ H_{2z} &= -i^m\left[e^{im\varphi} I_{1,m} + ie^{-i(m-1)\varphi} I_{1,m-1}\right]. \end{aligned}$$

The Poynting vector is defined by the well-known formula [30]:

$$\mathbf{P} = [c/(8\pi)]\mathrm{Re}[\mathbf{E} \times \mathbf{H}^*], \quad ($$

where $c$ is the vacuum speed of light, Re is the real part of a complex number, $\mathbf{E} \times \mathbf{H}$ is cross product, and * is the complex conjugation. Below, we omit the constant $c/(8\pi)$

brevity. Substituting components (4) and (12) into Equation (13), we obtain the components of the Poynting vector at the focus of the field (3):

$$\begin{aligned} P_x &= Q(r)\sin\varphi, \\ P_y &= -Q(r)\cos\varphi, \\ P_z &= \frac{1}{2}\left(I_{0,m}^2 + I_{0,m+1}^2 - I_{2,m-1}^2 - I_{2,m-2}^2\right), \\ Q(r) &= I_{1,m-1}(I_{0,m} + I_{2,m-2}) - I_{1,m}(I_{0,m+1} + I_{2,m-1}). \end{aligned} \tag{14}$$

Passing to the polar components $P_r$ and $P_\varphi$ of the transverse Poynting vector, we obtain:

$$\begin{aligned} P_r &= 0, \\ P_\varphi &= -Q(r). \end{aligned} \tag{15}$$

As seen from Equation (14), the longitudinal component of the Poynting vector at the focus has a circularly symmetric distribution and does not depend on the azimuthal angle $\varphi$. It is also seen from Equation (14) that if $m = 1$ or $m = 2$ then there is a reverse energy flow on the optical axis since for $m = 1$ or $m = 2$, we obtain the following on the optical axis:

$$P_z(r=0) = -\frac{1}{2}I_{2,0}^2 < 0.$$

Equation (15) indicates that the transverse energy flow at the focus is rotating along a closed trajectory with the center on the optical axis, clockwise if $Q(r) > 0$, and counterclockwise if $Q(r) < 0$. Since the function $Q(r)$ is of different signs on different radii $r$, it can be stated that the radial orbital Hall effect occurs at the focus of the light field (3). This also follows on from the expression for the longitudinal component of the angular momentum vector $\mathbf{J}$ of field (3) when it is written by definition using the azimuthal component of the energy flow [28]:

$$\begin{aligned} \mathbf{J} &= \mathbf{r} \times \mathbf{P}, \\ J_z &= rP_\varphi = -rQ(r). \end{aligned} \tag{16}$$

The energy flow at the focus is rotating along a spiral around the optical axis since the topological charges of the two optical vortices, which are present in the superposition in the initial field (3), do not compensate for each other as they have different magnitudes: $m + 1$ and $-m$.

## 7. Simulation

Using the Richards–Wolf formalism [30], we computed the distributions of intensity and the longitudinal component of the SAM vector (spin density) at the tight focus of the light field using the initial distribution given by Equation (3). We supposed that the field amplitude in the initial plane was constant, i.e., $A(\theta) = 1$, wavelength $\lambda = 532$ nm, focal length $f = 10$ μm, and the numerical aperture $NA = 0.95$. Figure 1 shows the distributions of the longitudinal component of the spin angular momentum $S_z$ (Figure 1a–d) (red and blue colors denote positive and negative values), the intensity $I$ (Figure 1e–h) (black and yellow colors denote zero and maximal values), and the angular component of the Poynting vector $P_\varphi$ (Figure 1i–l) (red and blue colors denote positive and negative values) of a light beam with polarization (3) of different order $m$ at the tight focus. The beam orders in Figure 1 are $m = 1$ (Figure 1a,e,i), $m = 2$ (Figure 1b,f,j), $m = 3$ (Figure 1c,g,k), and $m = 5$ (Figure 1d,h,l). The arrows in Figure 1i–l illustrate the directions of the angular energy flow. The scale mark in each figure denotes 1 μm.

As seen in Figure 1 (2nd row), the number of local maxima in the intensity distribution at the focus is consistent with the theory [Equation (6)] and is equal to $2m - 1$: 1 (Figure 1e), 3 (Figure 1f), 5 (Figure 1g), and 9 (Figure 1h). It is also seen in Figure 1 (1st row), that, according to Equation (9), the SAM distribution also contains $2m - 1$ local maxima (red color in Figure 1a–d), where $S_z > 0$, which reside on a certain circle with the center on the optical axis. On a circle with a larger radius, (blue color in Figure 1a–d), $S_z < 0$. The black color in Figure 1a–d denotes the areas with zero spin, i.e., where polarization is linear. Since the brightness of the blue color in Figure 1a–d is 2–3 times lower than that of the red color, elliptic polarization in the areas of positive spin is closer to circular polarization, wher the polarization ellipses in the areas of negative spin are more elongated and closer to lin polarization. Nevertheless, the spatial separation of the areas with positive and negat spin at the focus demonstrates the spin Hall effect.

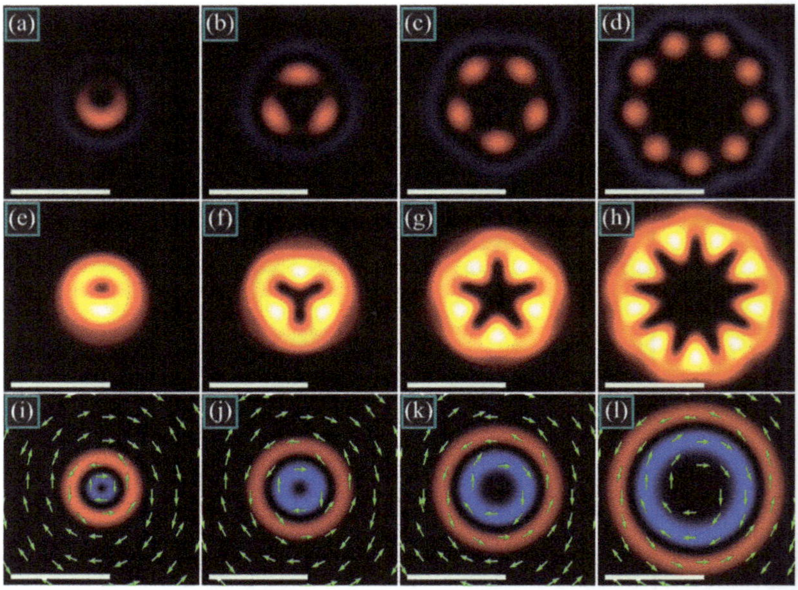


**Figure 1.** Distributions of the longitudinal component of the spin angular momentum (**a–d**) ( and blue color denote, respectively, positive and negative values), intensity (**e–h**) (black means z and yellow means maximum), and the angular component of the Poynting vector (**i–l**) (red me positive and blue means negative values) of a light beam with polarization (3) and with a differ order at the tight focus. Arrows (**i–l**) denote the directions of the angular energy flow. The scale m in each figure denotes 1 μm.

Figure 1i–l (3rd row) confirms theoretical predictions [Equation (15)] and demonstra that the transverse energy flow at the focus plane rotates. On a circle closer to the opti axis (blue color in Figure 1i–l), $P_\varphi < 0$, i.e., the transverse energy flow is rotating clockw On a larger circle radius (red ring in Figure 1i–l), $P_\varphi > 0$, and the energy flow is rotati counterclockwise. The spatial separation of the orbital energy flux in opposite direction a manifestation of the radial orbital Hall effect at the focus.

## 8. Discussion of the Results

Here, we compare the transverse components of the electric field in the initial pla (3) and the focus plane (4). Although in the initial plane, the components $E_x$ and $E_y$ of field (3) have the same phase, and thus the field has inhomogeneous linear polarizati at the focus, the transverse components of field (4) acquire a relative phase delay of $\pi$ or $3\pi/2$. This leads to the areas with elliptic polarization at the focus. On the other ha the longitudinal SAM component (7) in the initial plane is equal to zero $S_z = 0$, wh the energy flow (13) has only one longitudinal component, which is equal to the u $P_z = 1$. At the focus plane, the SAM density is given by Equation (9), but if the function is integrated over the whole focus plane, then it is equal to zero. Thus, the full longitu nal SAM component is conserved and equal to zero. The field (3) in the initial plane h a nonzero density of the longitudinal component of the orbital angular momentum (OAM) vector [37]:

$$L_z = \text{Im}\left( E_x^* \frac{\partial}{\partial \varphi} E_x + E_y^* \frac{\partial}{\partial \varphi} E_y \right) = \frac{1}{2}. \quad (17)$$

If the OAM density (17) is integrated over the angle φ from 0 to $2\pi$, then the full OAM in the initial plane is nonzero and equal to half of the initial beam power $W/2$.

At the focus plane, the longitudinal OAM component can also be obtained:

$$\begin{aligned} L_z &= \text{Im}\left( E_x^* \tfrac{\partial}{\partial \varphi} E_x + E_y^* \tfrac{\partial}{\partial \varphi} E_y + E_z^* \tfrac{\partial}{\partial \varphi} E_z \right) \\ &= \tfrac{1}{2}\Big\{ (m+1) I_{0,m+1}^2 - m I_{0,m}^2 + 2m I_{1,m}^2 + (m-1) I_{2,m-1}^2 - 2(m-1) I_{1,m-1}^2 \\ &-(m-2) I_{2,m-2}^2 + \left[ I_{0,m} I_{2,m-1} - 3 I_{0,m+1} I_{2,m-2} - 2 I_{1,m} I_{1,m-1} \right] \sin(2m-1)\varphi \Big\}. \end{aligned} \quad (18)$$

According to Equation (18), the OAM density depends on the angle as $\sin(2m-1)\varphi$. This means that on a certain circle radius with the center on the optical axis, the OAM has $(2m-1)$ local maxima and minima, similar to the SAM distribution (9). It can be shown that if the OAM density is integrated over the whole focus plane, then this also yields half of the initial beam power $W/2$. Thus, in this case, the full SAM and OAM are conserved separately. Therefore, it can be concluded that the spin Hall effect at the focus of the beam (3) arises due to the conservation of the full longitudinal SAM of the beam. Since the longitudinal SAM of the whole beam is zero, the areas with the spin of a different sign should arise in pairs. In the same way, the radial orbital Hall effect at the focus occurs due to the conservation of the full longitudinal OAM of the beam.

Summing the SAM (9) and OAM (18) densities, we obtain:

$$\begin{aligned} L_z + S_z &= \tfrac{1}{2}\Big\{ m\left( I_{0,m+1}^2 + 2I_{1,m}^2 + I_{2,m-1}^2 \right) - (m-1)\left( I_{0,m}^2 + 2I_{1,m-1}^2 + I_{2,m-2}^2 \right) \\ &- \sin(2m-1)\varphi\left[ I_{0,m} I_{2,m-1} + I_{0,m+1} I_{2,m-2} + 2 I_{1,m} I_{1,m-1} \right] \Big\} \end{aligned} \quad (19)$$

The comparison of the density of the longitudinal component of the angular momentum (AM) vector in Equation (16) with the sum of the longitudinal SAM and OAM components in Equation (19) reveals that they are not equal to each other: $J_z \neq L_z + S_z$. We considered the reason for this inequality earlier in [37].

In concluding this section, we consider the difference between the Hall effect near the tight focus [18,31,32,37] and the Hall effect which occurs when light is reflected off the interface between two media [22,38]. As was shown in [22], when an optical vortex is reflected from a plane glass surface, the annular intensity distribution becomes inhomogeneous. For the optical vortices with the topological charges $m$ and $-m$, the intensity maxima on the ring appear in different places, i.e., shifted relative to each other (orbital Hall effect). In Ref. [38], it was shown that when a $p$-polarized Gaussian beam (polarization vector is in the incidence plane) is reflected from the glass surface under an angle close to the Brewster angle, the spin Hall effect occurs when the reflected light is split into two beams with opposite spins in the direction orthogonal to the incidence plane. In the tight focus [18,31,32,37], spin and orbital Hall effects occur due to the conservation of the full angular momentum. The light with opposite spins and/or with opposite energy rotation at the focus plane is concentrated in different places. A different manifestation of the Hall effect at the focus, as investigated in the different works [18,31,32,37], is explainable since the different types of the initial vector fields were considered. In Ref. [18], the focusing of the conventional Poincaré beams was considered, while [31] dealt with the focusing of fractional-order cylindrical vector beams. In Ref. [32], the Hall effect of the cylindrical vector beams of an integer order arises before and beyond the focus, whereas it is absent at the focus itself. In Ref. [37], the Hall effect was studied at the focus of a circularly polarized optical vortex. In contrast to these works, we here studied the tight focusing of a generalized Poincaré beam, whose topological charge is equal to 1/2 and whose order of inhomogeneous linear polarization is equal to $m + 1/2$. It is impossible to predict in advance, based on the initial light field, whether or not the Hall effect will arise at the tig focus. Thus, each new type of initial vector beam should be considered separately.

## 9. Conclusions

Based on the Debye integrals [30], we have investigated both theoretically and num ically generalized (hybrid) Poincaré beams at a tight focus. A generalized Poincaré be is a coaxial superposition of two optical vortices with left and right circular polarizati and with the TC of $p$ and $q$. For certainty, we studied the case when $p = m + 1$ and $q =$ - [Equation (3)]. Simple analytical expressions have been obtained for the components of electric and magnetic strength vectors at the focus [Equations (4) and (12)], for the intens distribution [Equation (6)], for the longitudinal component of the spin angular moment [Equation (9)], and for the components of the Poynting vector [Equation (14)]. It has be shown that the intensity at the focus has $2m - 1$ local maxima, residing evenly on a cert circle radius with the center on the optical axis. In addition, radial spin and orbital H effects have been demonstrated. This means that the longitudinal SAM component h different signs on circles with different radii in the focal plane, and the azimuthal com nent of the transverse Poynting vector also has different signs (Figure 1). Such beams c be used for the simultaneous trapping of several micro- or nanoparticles (Figure 1h) in the local intensity maxima that should simultaneously rotate around their centers of m (Figure 1d) and move along the ring (Figure 1l). In addition, when moving along the ri the particles will need to overcome the 'breaks' in the intensity distribution (Figure 1h)

**Author Contributions:** Conceptualization, V.V.K.; methodology, V.V.K.; software, A.A.K. and A.M validation, V.V.K.; formal analysis, V.V.K.; investigation, V.V.K. and A.A.K.; resources, V.V.K.; d curation, V.V.K.; writing—original draft preparation, V.V.K.; writing—review and editing, V.V.K.; v alization, A.A.K. and A.M.T.; supervision, V.V.K.; project administration, V.V.K.; funding acquisit V.V.K. and A.A.K. All authors have read and agreed to the published version of the manuscript.

**Funding:** This research (theory) was funded by the Russian Science Foundation, grant number 22-12-00

**Institutional Review Board Statement:** Not applicable.

**Informed Consent Statement:** Not applicable.

**Data Availability Statement:** Not applicable.

**Acknowledgments:** We acknowledge the support of the Ministry of Science and Higher Educat of the Russian Federation within a government project of the FSRC "Crystallography and Photon RAS (simulation).

**Conflicts of Interest:** The authors declare no conflict of interest. The funders had no role in the des of the study; in the collection, analyses, or interpretation of data; in the writing of the manuscript in the decision to publish the results.

## References

1. Holleczek, A.; Aiello, A.; Gabriel, C.; Marquardt, C.; Leuchs, G. Classical and quantum properties of cylindrically polarized sta of light. *Opt. Express* **2011**, *19*, 9714–9736. [CrossRef] [PubMed]
2. Chen, S.; Zhou, X.; Liu, Y.; Ling, X.; Luo, H.; Wen, S. Generation of arbitrary cylindrical vector beams on the higher order Poinc sphere. *Opt. Lett.* **2014**, *39*, 5274–5276. [CrossRef] [PubMed]
3. Yi, X.; Liu, Y.; Ling, X.; Zhou, X.; Ke, Y.; Luo, H.; Wen, S.; Fan, D. Hybrid-order Poincaré sphere. *Phys. Rev. A* **2015**, *91*, 023 [CrossRef]
4. Beckley, A.M.; Brown, T.G.; Alonso, M.A. Full Poincaré beams. *Opt. Express* **2010**, *18*, 10777–10785. [CrossRef]
5. Milione, G.; Sztul, H.I.; Nolan, D.A.; Alfano, R.R. Higher-order Poincaré sphere, Stokes parameters, and the angular moment of light. *Phys. Rev. Lett.* **2011**, *107*, 053601. [CrossRef]
6. Milione, G.; Evans, S.; Nolan, D.A.; Alfano, R.R. Higher Order Pancharatnam-Berry Phase and the Angular Momentum of Li *Phys. Rev. Lett.* **2012**, *108*, 190401. [CrossRef]
7. Galvez, E.J.; Khadka, S. Poincare modes of light. *Proc. SPIE* **2012**, *8274*, 82740Y. [CrossRef]
8. Galvez, E.J.; Khadka, S.; Schubert, W.H.; Nomoto, S. Poincare-beam patterns produced by nonseparable superpositions Laguerre-Gauss and polarization modes of light. *Appl. Opt.* **2012**, *51*, 2925. [CrossRef]
9. Lopez-Mago, D. On the overall polarisation properties of Poincaré beams. *J. Opt.* **2019**, *21*, 115605. [CrossRef]

Boucher, P.; Del Hoyo, J.; Billet, C.; Pinel, O.; Labroille, G.; Courvoisier, F. Generation of high conical angle Bessel–Gauss beams with reflective axicons. *Appl. Opt.* **2018**, *57*, 6725–6728. [CrossRef]
Li, D.; Feng, S.; Nie, S.; Chang, C.; Ma, J.; Yuan, C. Generation of arbitrary perfect Poincaré beams. *J. Appl. Phys.* **2019**, *125*, 073105. [CrossRef]
Liu, M.; Huo, P.; Zhu, W.; Zhang, C.; Zhang, S.; Song, M.; Zhang, S.; Zhou, Q.; Chen, L.; Lezec, H.J.; et al. Broadband generation of perfect Poincaré beams via dielectric spin-multiplexed metasurface. *Nat. Commun.* **2021**, *12*, 2230. [CrossRef]
Gu, Z.; Yin, D.; Gu, F.; Zhang, Y.; Nie, S.; Feng, S.; Ma, J.; Yuan, C. Generation of concentric perfect Poincaré beams. *Sci. Rep.* **2019**, *9*, 15301. [CrossRef]
Alpmann, C.; Schlickriede, C.; Otte, E.; Denz, C. Dynamic modulation of Poincaré beams. *Sci. Rep.* **2017**, *7*, 8076. [CrossRef]
Fu, S.; Gao, C.; Wang, T.; Zhai, Y.; Yin, C. Anisotropic polarization modulation for the production of arbitrary Poincaré beams. *J. Opt. Soc. Am. B* **2018**, *35*, 1–7. [CrossRef]
Naidoo, D.; Roux, F.S.; Dudley, A.; Litvin, I.; Piccirillo, B.; Marrucci, L.; Forbes, A. Controlled generation of higher-order Poincaré sphere beams from a laser. *Nat. Photonics* **2016**, *10*, 327–332. [CrossRef]
Kotlyar, V.V.; Kovalev, A.A.; Stafeev, S.S.; Zaitsev, V.D. Index of the Polarization Singularity of Poincare Beams. *Bull. Russ. Acad. Sci. Phys.* **2022**, *86*, 1158–1163. [CrossRef]
Kotlyar, V.V.; Stafeev, S.S.; Zaitsev, V.D.; Telegin, A.M. Poincaré Beams at the Tight Focus: Inseparability, Radial Spin Hall Effect, and Reverse Energy Flow. *Photonics* **2022**, *9*, 969. [CrossRef]
Ling, X.; Yi, X.; Zhou, X.; Liu, Y.; Shu, W.; Luo, H.; Wen, S. Realization of tunable spin-dependent splitting in intrinsic photonic spin Hall effect. *Appl. Phys. Lett.* **2014**, *105*, 151101. [CrossRef]
Yin, X.; Ye, Z.; Rho, J.; Wang, Y.; Zhang, X. Photonic spin Hall effect at metasurfaces. *Science* **2013**, *339*, 1405–1407. [CrossRef]
Kumar, R.N.; Gupta, S.D.; Ghosh, N.; Banerjee, A. Probing the rotational spin-Hall effect in a structured Gaussian beam. *Phys. Rev. A* **2022**, *105*, 023503. [CrossRef]
Zhang, J.; Zhou, X.X.; Ling, X.H.; Chen, S.Z.; Luo, H.L.; Wen, S.C. Orbit-orbit interaction and photonic orbital Hall effect in reflection of a light beam. *Chin. Phys. B* **2014**, *23*, 064215. [CrossRef]
Kavokin, A.; Malpuech, G.; Glazov, M. Optical spin Hall effect. *Phys. Rev. Lett.* **2005**, *95*, 136601. [CrossRef] [PubMed]
Fadeyeva, T.A.; Rubass, A.F.; Volyar, A.V. Transverse shift of a high-order paraxial vortex-beam induced by a homogeneous anisotropic medium. *Phys. Rev. A* **2009**, *79*, 053815. [CrossRef]
Fu, S.; Guo, C.; Liu, G.; Li, Y.; Yin, H.; Li, Z.; Chen, Z. Spin-orbit optical Hall effect. *Phys. Rev. Lett.* **2019**, *123*, 243904. [CrossRef]
Zhang, F.; Guo, Y.H.; Pu, M.B.; Li, X.; Ma, X.L.; Luo, X.G. Metasurfaces enabled by asymmetric photonic spin-orbit interactions. *Opto-Electr. Eng.* **2020**, *47*, 200366. [CrossRef]
Shu, W.; Lin, C.; Wu, J.; Chen, S.; Ling, X.; Zhou, X.; Luo, H.; Wen, S. Three-dimensional spin Hall effect of light in tight focusing. *Phys. Rev. A* **2020**, *101*, 023819. [CrossRef]
Bliokh, K.Y.; Ostrovskaya, E.A.; Alonso, M.A.; Rodríguez-Herrera, O.G.; Lara, D.; Dainty, C. Spin-to-orbital angular momentum conversion in focusing, scattering, and imaging systems. *Opt. Express* **2011**, *19*, 26132–26149. [CrossRef]
Man, Z.; Xi, Z.; Yuan, X.; Burge, R.E.; Urbach, H.P. Dual coaxial longitudinal polarization vortex structures. *Phys. Rev. Lett.* **2020**, *124*, 103901. [CrossRef]
Richards, B.; Wolf, E. Electromagnetic diffraction in optical systems, II. Structure of the image field in an aplanatic system. *Proc. R. Soc. Lond. A* **1959**, *253*, 358–379. [CrossRef]
Stafeev, S.S.; Nalimov, A.G.; Zaitsev, V.D.; Kotlyar, V.V. Tight focusing cylindrical vector beams with fractional order. *J. Opt. Soc. Am. B* **2021**, *38*, 1090–1096. [CrossRef]
Kotlyar, V.V.; Stafeev, S.S.; Kovalev, A.A.; Zaitsev, V.D. Spin Hall Effect before and after the Focus of a High-Order Cylindrical Vector Beam. *Appl. Sci.* **2022**, *12*, 12218. [CrossRef]
Volyar, A.V.; Shvedov, V.G.; Fadeeva, T.A. Structure of a nonparaxial gaussian beam near the focus: III. Stability, eigenmodes, and vortices. *Opt. Spectrosc.* **2001**, *91*, 235–245. [CrossRef]
Zhan, Q. Cylindrical vector beams: From mathematical concepts to applications. *Adv. Opt. Photon.* **2009**, *1*, 1–57. [CrossRef]
Weng, X.; Miao, Y.; Wang, G.; Zhan, Q.; Dong, X.; Qu, J.; Gao, X.; Zhuang, S. Light beam carrying natural non-integer orbital angular momentum in free space. *arXiv* **2021**, arXiv:2105.11251. [CrossRef]
Wang, H.; Wojcik, C.C.; Fan, S. Topological spin defects of light. *Optica* **2022**, *9*, 1417–1423. [CrossRef]
Kotlyar, V.V.; Kovalev, A.A.; Telegin, A.M. Angular and Orbital Angular Momenta in the Tight Focus of a Circularly Polarized Optical Vortex. *Photonics* **2023**, *10*, 160. [CrossRef]
Li, S.-M.; Chen, J. Spin Hall effect of reflected light from an air-glass interface around the Brewster's angle. *Appl. Phys. Lett.* **2012**, *100*, 071109. [CrossRef]

*photonics*

MDP

*Communication*

# Multi-Bessel Beams Generated by an Axicon and a Spatial Ligl Modulator for Drilling Applications

**Christian Lutz [1,*], Simon Schwarz [1], Jan Marx [2], Cemal Esen [2] and Ralf Hellmann [1]**

1 Applied Laser and Photonics Group, University of Applied Sciences Aschaffenburg, Würzburgerstraße 4 63743 Aschaffenburg, Germany
2 Applied Laser Technologies, Ruhr University Bochum, Universitätsstraße 150, 44801 Bochum, Germany
* Correspondence: christian.lutz@th-ab.de

**Abstract:** We report on an optical setup to generate multi-Bessel beam profiles combining a refract axicon and a spatial light modulator. Based on their particular beam profile, Bessel beams o advantageous properties for micro drilling processes and internal volume processing, especia for transparent materials. In addition, the laser power of industrial, ultrashort pulsed lasers l increased significantly over the last few years, offering the possibility for highly efficient proces using multi-spot profiles. Our optical concept combines the dynamic possibilities of beam splitt using a spatial light modulator with the benefits of Bessel beams, which facilitates multi-Bessel be processing. Beside the simulation and experimental evaluation of the generated multi-Bessel bea we exemplify the applicability of the developed module for the perforation of thin metal foils micro drilling.

**Keywords:** Bessel beam; axicon; optical system; ultrashort pulsed laser; material ablation



**Citation:** Lutz, C.; Schwarz, S.; Marx, J.; Esen, C.; Hellmann, R. Multi-Bessel Beams Generated by an Axicon and a Spatial Light Modulator for Drilling Applications. *Photonics* **2023**, *10*, 413. https://doi.org/10.3390/photonics10040413

Received: 14 March 2023
Revised: 3 April 2023
Accepted: 3 April 2023
Published: 6 April 2023



## 1. Introduction

Diffraction-free Bessel beams were first discovered and described by Durnin et al. in 1987, and they represent the propagation-invariant solution of the Helmholz equati While Durnin et al. [2] showed convergences to the ideal Bessel beams for a finite ran in theory, they are of an infinite transverse extent and they carry infinite energy, so th cannot be generated experimentally [3,4]. To design a so-called quasi-Bessel beam, bas on a Gaussian beam profile, there are several commonly used components, such as ri apertures [2], binary-coded holograms [5], axicon lenses [6,7], converging lenses with sph ical aberrations [8], tunable acoustic gradient lenses [9], phase-shifted Fresnel axicons [1 and spatial light modulators (SLMs) [11,12], respectively, which have been successfu employed. Using a SLM, it is also possible to create an array of Bessel beams with a flexi separation instead of a single Bessel beam [13–15].

Quasi-Bessel beams offer advantages for various micro-machining processes due their particular intensity distribution. The long and high intensity range in the propagati direction of a quasi-Bessel beam makes it an excellent tool, especially for laser drilling ho with a high aspect ratio in transparent materials. As Xie et al. [4] showed, the advantage the higher focal length of Bessel beams leads to a 52 times higher aspect ratio as compared a focused Gaussian beam profile. With a further improvement of the focal position and used pulse energy, the authors achieved an aspect ratio of up to 330:1 in PMMA, for ho with a diameter between 1.5 μm and 2.4 μm. In addition to drilling holes in transpare materials, Bessel beams offer the possibility to induce cracks in glass in a controlled man by means of internal volume modification. For example, Dudutis et al. [16] demonstra the possibility of using a pulse energy of 2 mJ to produce a 2.3 mm-long modification an elliptical shape to create cracks in the axial direction. Finally, it was shown that i possible to control the orientation of the cracks by rotating the polarization of the Bes beam profile. Furthermore, Mitra et al. [17] fabricated nanovoids with a length up to 5 mm in borosilicate glass, corresponding to an aspect ratio of 1200:1.

While the processing of transparent materials is often possible with one or a few pulses [4], the processing of non-transparent materials, such as metals, requires a large number of pulses, due to the significantly smaller optical penetration depth. According to Alexeev et al. [18], the aspect ratio of holes drilled by a Bessel beam in non-transparent materials is significantly lower, as exemplified by this for drilling 100 μm-thick copper sheets with an exit hole diameter of 15 μm. Due to the typical Bessel beam profile, which is also characterized by ring-shaped intensities around the intense maximum, ring-shaped modifications or even ablation can be seen on the processed surface of non-transparent materials [18]. To minimize these effects, He et al. [19] developed a tailored Bessel beam using a binary phase plate in front of the axicon, which enables the drilling of 6 μm holes in a 100 μm silicon wafer.

The most common method of creating a quasi-Bessel beam for drilling applications is to use an axicon in combination with an imaging lens assembly [3,16,20,21]. Although this allows processes with low power losses and high-quality drilled holes, it only generates one single Bessel beam. Additionally, the generation of Bessel beam arrays is a promising method for drilling applications with multi-spot beam profiles [22]. Contrary to conventional axicons, which generate one quasi-Bessel beam behind their apex, axicon arrays can be used to form a multitude of these beams. Schwarz et al. [23] showed a femtosecond laser ablation process with a polishing step using a $CO_2$ laser for the production of micro axicon arrays, including a characterization of the resulting multi-Bessel beams. While axicon arrays have a fixed separation of the Bessel beams due to their fixed geometry, the separation of the Bessel beams can be adapted to the machining process using variable optical elements.

Spatial light modulators are variable optical elements for a lot of different beam-shaping applications, such as spherical aberration correction [24] and multi-spot processing [25]. By means of SLM, it is possible to create multi-Bessel beams by combining axicon phase patterns or beam splitter functions [26–28]. Studies by García-Martínez et al. [26] show the generation of multi-Bessel beams by a combination of phase patterns consisting of linear axicons or binary axicons with a Dammann grating, resulting in a focal depth of more than 125 cm. The limited phase shift of an SLM allows only very small axicon angles compared to commercial axicons made of glass, resulting in very long Bessel beams of several centimeters or even meters in length [26,29].

Reducing the period length of the phase mask imaged on the SLM results in a larger angle, but the diffraction efficiency of the SLM decreases drastically, and efficient operation with this setup is not possible due to high power losses. Based on these correlations, in this contribution we demonstrate an optical system based on a simulation and experimental evaluation that combines the flexible beam-shaping capabilities of an SLM with an axicon-generated Bessel beam profile to create multi-Bessel beams. The SLM enables flexible changes of the Bessel beam separation by the calculated computer-generated hologram without the exchange of optical components and subsequent adjustment work on the optical setup. This approach allows the drilling of holes in non-transparent materials using a multi-Bessel beam, which is exemplified by an experimental study of a multi-Bessel beam.

## 2. Experimental

We used an ultrashort pulsed laser (Light Conversion, Pharos, Vilnius, Lithuania) with a wavelength of 1028 nm, an adjustable repetition rate of up to 610 kHz, and a pulse duration between 230 fs and 15 ps. The linearly polarized and collimated laser beam with a diameter of 3.4 mm (FWHM) is shaped by a Holoeye SLM (Berlin, Germany) with a resolution of 1920 × 1080 pixels with a pixel pitch of 8 μm (Pluto 2, NIR-015-D, Berlin, Germany) in combination with a 170° precision axicon from Eksma Optics (Vilnius, Lithuania) and different lenses to complete the optical setup. After reflecting the laser beam from the SLM, the first lens (f = 100 mm) is placed 200 mm behind the SLM to image its phase pattern onto the following axicon. Behind the axicon, the resulting Bessel bean imaged using a 4f-setup with two lenses with f = 100 mm and f = 50 mm into the fo plane. By superimposing the phase pattern of the SLM and the imaged Bessel bean multi-Bessel beam results in the focus of lens 3. To image the intensity distribution camera (IDS UI-1490SE-M-GL, Farnham, UK) is used, which is replaced by a workpiece order to conduct experimental drilling applications.

The optical setup shown in Figure 1 allows for the adjustment of the separation a multi-spot beam profile using the SLM with different beam splitter patterns and generation of multi-Bessel beams with the included axicon. As stated before, the generat of Bessel beams only using an SLM leads to very long intensity profiles, while this set combines short Bessel beams generated by the used axicon with the flexibility regardi the beam-splitting possibilities of the SLM.

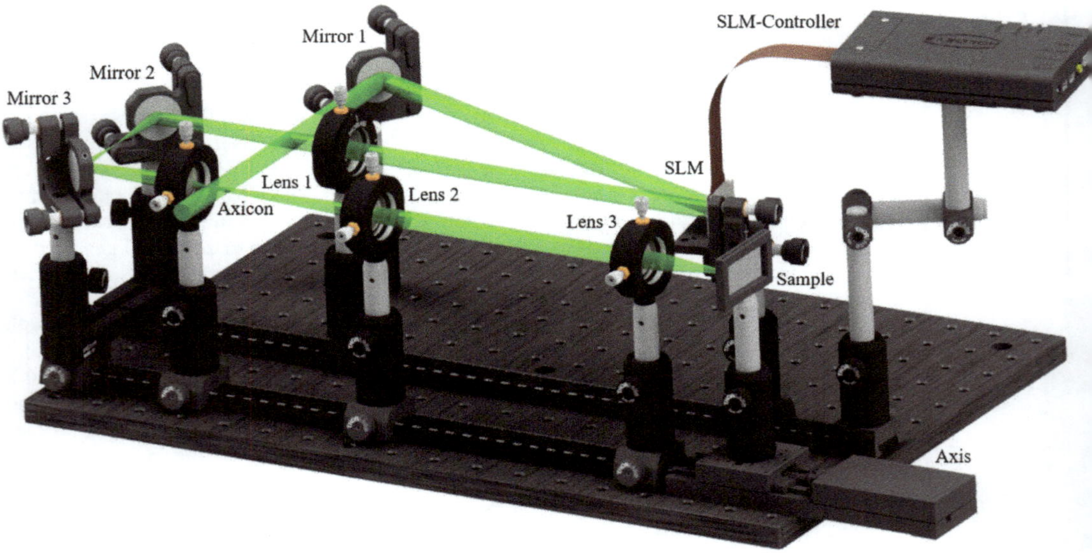


**Figure 1.** Schematic illustration of the optical system for multi-Bessel beam generation using a spa light modulator, a 4f-telescope, an axicon, focusing lenses, and an exemplary sample in form c metal sheet, which can be replaced by a camera for recording the beam profiles.

## 3. Simulation

To simulate parameters such as the spot separation and the axicon angle, Virtual L from LightTrans International GmbH (Jena, Germany) is used. The simulation results Figure 2 show an optical system that generates multi-Bessel beams with the describ combination of a spatial light modulator and an axicon. The input beam is defined 3.4 mm in diameter ($1/e^2$) with a Gaussian shape, which is directed onto the SLM. T phase pattern shown in Figure 2a is a computer-generated hologram (CGH), which calculated with an iterative Fourier transform algorithm with a spot separation of 1 mm a focal length of 100 mm. Thus, at the focal plane between lens 1 and the following axic Figure 2b shows a 2 × 2 regular beam splitter with Gaussian-shaped spots and a separat of 1 mm. As a result of the superposition of the phase pattern, in 30 mm distance fro the axicon, a multiplied Bessel beam with a separation of 200 µm is shown in Figure Through the following two lenses, the beam profile is imaged into the focal plane w a separation of 1 mm, where material processing is possible (Figure 2d). Furtherm the intensity distribution along the propagation direction of the single Bessel beam w simulated. The beam profiles shown in Figure 3 were obtained by detecting the intens in the cross-section at different Z propagations: (a) shows the intensity distribution of the beam directly after the axicon with a length of 68.8 mm (FWHM) for a single Bessel beam, while the SLM is treated as a mirror. After a propagation of z = 21 mm, the Bessel beam reaches its maximum intensity and has a core diameter of 15.3 µm (FWHM). The ring-shaped secondary maxima that are typical for Bessel beams can also be seen, which move away from the center of the beam profile with increasing propagation, due to the divergent input beam of the axicon. Through lenses two and three, the beam is imaged according to the focal lengths of the lenses in the ratio 2 to 1. As can be seen in Figure 3b, the diameter at the point of maximum intensity, which is reached at a propagation length of z = 5.2 mm, is halved to 7.6 µm (FWHM). Due to the special propagation properties of Bessel beams, the imaging ratio of 2:1 has a quadratic effect on the length of the intensity distribution [30]. The length of the Bessel beam is thus 17.5 mm. In addition, the ring-shaped side maxima behave accordingly, whereby a higher number can be seen by imaging in the same cross-sectional area.

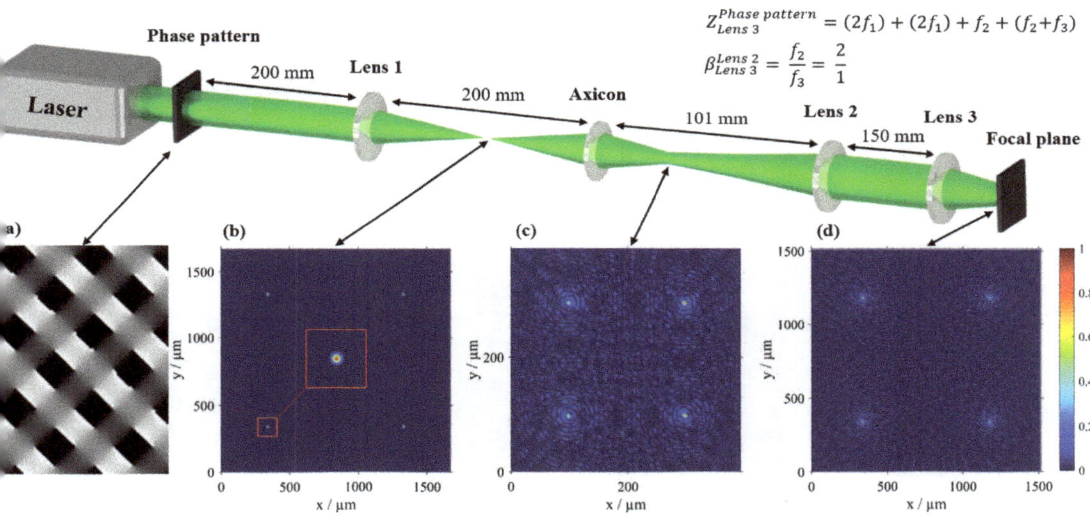


**Figure 2.** Schematic illustration of the simulated beam path including an SLM with a 4f-telescope and an axicon with a following imaging setup. Additionally, beside the shown phase pattern (**a**), the multi-spot beam profile in the intermediate focus (**b**), and the multi-Bessel beam after the axicon (**c**) as well as after the imaging setup (**d**) are shown. While the variable $z_{Lens\ 3}^{Phase\ pattern}$ in the shown formula defines the propagation distance of the beam path between the SLM and lens 3, $\beta_{Lens\ 3}^{Lens\ 2}$ indicates the magnification of lens 2 and 3.

While for the simulation results in Figure 3, the SLM was used only as a mirror, with respect to an exemplifying application in material processing, a binary-blazed grating is now applied to create a 2 × 1 beam splitter on the SLM. The binary-blazed grating has a periodical length of 16 pixels. In Figure 4, beam profiles are shown with the generated multi-Bessel beam at different propagation positions in 4 mm steps from 5 mm to 17 mm. For better visualization, the beam profiles shown in Figure 4 are normalized. As can be seen, the two generated Bessel beams distribute symmetrically, moving away from the center of the beam profile with continuous propagation. While the distance between the beams in the 5 mm position is 53 µm and increases to 182 µm at a propagation distance of 17 mm, showing that the two beams are arranged at an angle of 0.62° in relation to the propagation axis. It can also be seen that the Bessel beams are surrounded by diffraction patterns, which also separate with increasing propagation. However, it is obvious that

the two Bessel beams have the same intensity in the simulation, which predicts the id
position of the components.

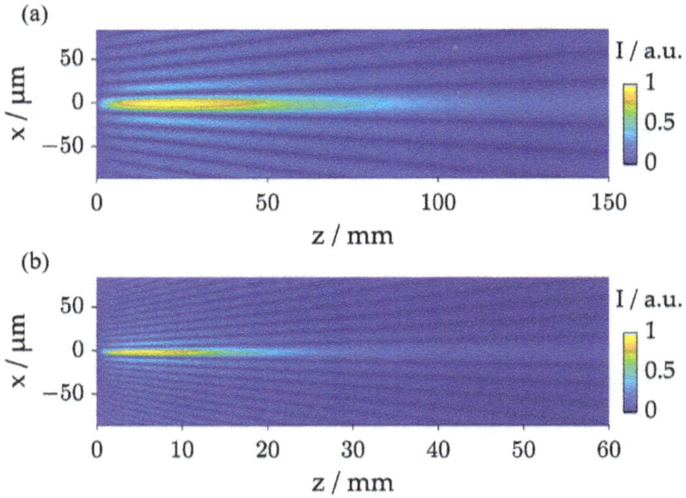


**Figure 3.** Simulated intensity distribution of the Bessel beam in propagation direction behind
axicon (**a**) and lens 3 (**b**).

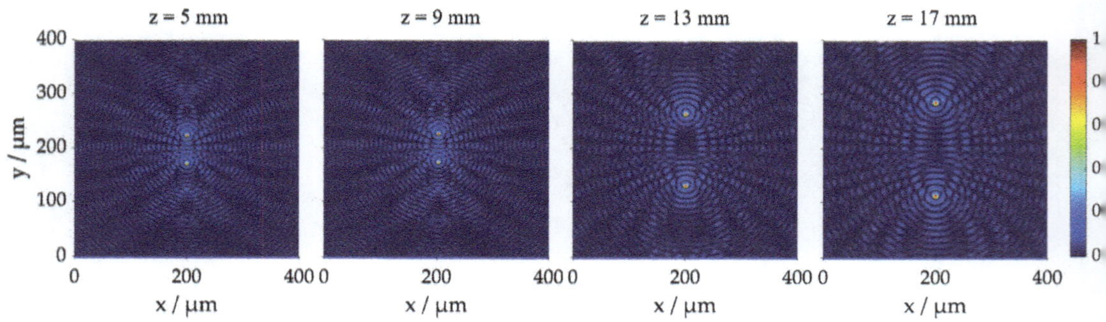


**Figure 4.** Simulated lateral intensity distributions of the multi-Bessel beam in the focal plane a
lens 3 is taken in different z-axis positions from 5 mm to 17 mm in 4 mm steps.

## 4. Experimental Results

As illustrated in Figure 1, based on the simulation results, the optical setup is tra
ferred to an optical breadboard to realize module integration into an existing ultrash
pulsed laser machine. To evaluate the optical simulation, the imaged workpiece was
placed by a camera, a neutral density filter with a transmission of 1% and a re-imaging l
setup for the detection of the lateral intensity profiles, which can be moved precisely alo
the propagation of the laser beam due to the installed axis with a length of 60 mm. F
precise imaging of the experimentally generated Bessel beam, the camera is moved in ste
of 250 µm by a motorized axis. For the single Bessel beam, the sequence of intensity dis
butions recorded in this way is shown in Figure 5. Due to the experimental measureme
by a camera and the approach of different z positions in 250 µm steps, a lower resoluti
results in the diagram shown compared to the simulation. Due to the sequencing of t
recorded beam profiles, pixel-wise positional errors occur, causing the intensity distributi
in the propagation to exhibit slight fluctuations. Nevertheless, Figure 5 highlights th

the intensity distribution of the experimentally generated Bessel beam corresponds to the simulation and has a length of about 16 mm with a diameter of 6.4 µm at the peak of the highest intensity. The intensity maximum is approximately beyond a propagation length of 6 mm. In general, it can be stated that the beam-shaping results correspond well to the simulated results, apart from minor error influences.

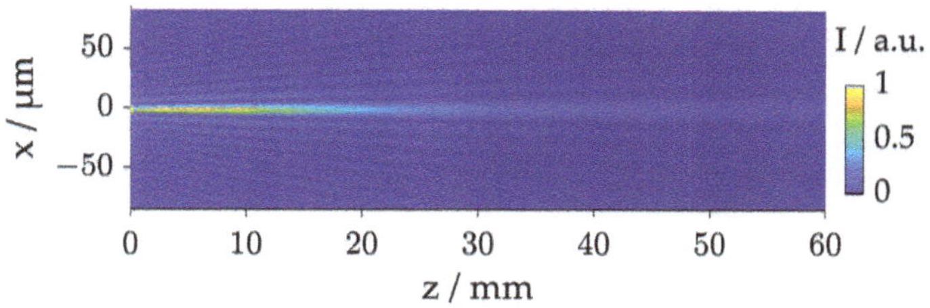


**Figure 5.** Measured intensity distribution in propagation direction of the Bessel beam behind lens 3.

In addition, we measured the multi-Bessel beam profiles at different positions behind lens 3 with an IDS camera in accordance with Figure 4. As shown in Figure 6 the Gaussian input beam with a diameter of 3.4 mm results in a 2 × 1 multi-beam with 88 µm separation, 5 mm behind lens 3. As already shown in the simulation, the Bessel beams deviate from a parallel course to the propagation axis so that they reach a separation of 258 µm after 17 mm propagation. The resulting angle is 0.81°, which is 0.2° larger than in the simulation. It can also be seen that the two Bessel beams are not identical, but they have slightly different intensities, which is particularly evident in the 5 mm and 13 mm positions. This may result from a slight misalignment and non-ideal optical components.

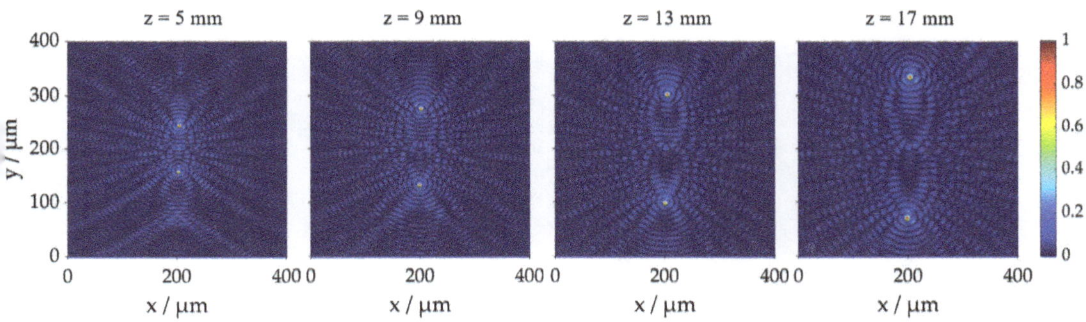


**Figure 6.** Measured lateral intensity distributions of the multi-Bessel beam in the focal plane after lens 3 taken in different z-axis positions from 5 mm to 17 mm in 4 mm steps.

In both the simulation and the experimental setup, diffraction effects occur due to the superposition of the SLM phase pattern and the axicon. Additionally, Orlov et al. [22] reported these interferences for beams with smaller separations. For example, a 2 × 1 Bessel beam profile with a separation of about 800 µm shows strong interference patterns, whereas with a double-sized separation, no interferences can be detected. Another effect is the visibility of the zeroth order in the center of the beam profile, which does not occur in the simulation results because an ideal diffraction element is used.

## 5. Drilling Study

To exemplify and initially demonstrate the experimental applicability of the multi-Bessel beam, a drilling study is performed on 5 µm-thin aluminum foils. The maximum average power of 10 W at a repetition rate of 50 kHz and a pulse duration of 220 fs was used to generate the drill holes. To achieve better drilling results in terms of the heat effects, a pulse divider of five was used. Due to power losses in the numerous opti elements as well as the SLM, the measured power in the machining plane is 7 W. Duri the study, the number of pulses varied between 200 and 200,000. For the analysis of the boreholes drilled with the 2 × 1 Bessel beam per parameter, the diameter of the boreho was measured by two perpendicular lines and the averages were calculated from th The output diameter is shown in Figure 7a for the different pulse numbers.

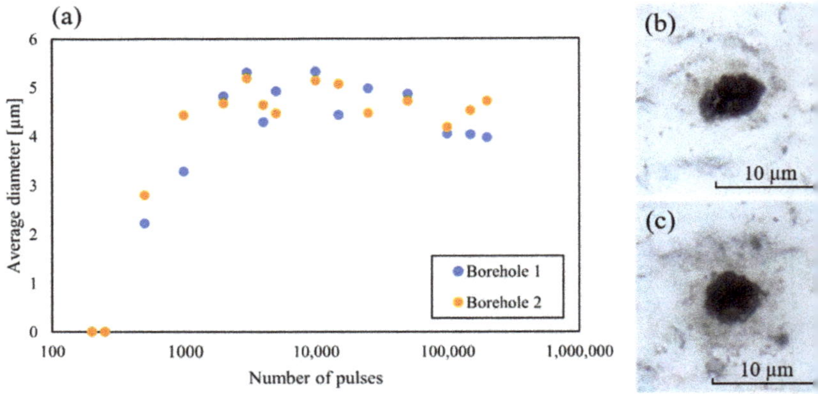


**Figure 7.** (**a**) Average diameter of the two holes drilled by a multi-Bessel beam as a function of used number of pulses. The diameter is averaged through six drill holes, which were measure x and y directions at an angle of 90 degrees. Please note, the standard deviation, of the measu quantities is in the order of 0.7 μm and is thus not considered in the diagram. (**b**) Outlet of a w 1000-pulse-drilled holes. (**c**) Outlet of with 10,000-pulse-drilled holes.

The first holes where the backside of the material is breached can be seen after ab 500 pulses, although the holes are still very deformed and thus have a low roundness, a visible in Figure 7b. This circumstance is commonly observed for drilling non-transpar materials using Bessel beams and is already known from other publications [18,19,31]. T initial diameter increases as a result of a higher pulse number of 2000 pulses to a maxim of approx. 5 μm, whereby the roundness of the holes increases significantly, as can be se in Figure 7c. A further increase in the number of pulses leads to a stagnation of the ini diameter. The diameter of the drilled holes is about 5 μm, starting from a pulse numb of 2000 with a measured diameter of the beam profile of 6.4 μm (FWHM) at the intens maximum, according to Figure 5.

It is also evident that the diameter of borehole 2 in the boundary region of the bre through is slightly larger than that of borehole 1, indicating a slightly higher fluence this Bessel beam. This may be due to a small misalignment in the beam path. Howe the diameters converge significantly ay about 2000 pulses and vary for increasing pu numbers in the same range.

## 6. Conclusions

We have developed and demonstrated the full functionality of an optical setup generating multi-Bessel beams using a spatial light modulator and an axicon. To investig the propagation properties of the generated Bessel beams, the system was simulated a experimentally characterized, where the calculated beam profiles with a length of app 17 mm show a high agreement in diameter and intensity compared to the experimenta measured beam profiles with a length of approx. 16 mm. Additionally, the analy of a multi-Bessel beam generated by a binary-blazed grating on the SLM shows a hi agreement between the simulation and the experimental setup. However, a slight deviat in the separation of the Bessel beams results from a minor difference in angle between propagation axis and the Bessel beams. A study on drilling thin aluminum foils with holes of a 5 µm outlet diameter shows the general applicability of the multi-Bessel beam module, which combines the possibility of flexible spot separation using customized phase patterns with the possibility of large axicon angles to generate small Bessel beams. These advantages enable high-throughput micromachining using multi-Bessel beams for drilling applications.

**Author Contributions:** Conceptualization, C.L. and S.S.; methodology, C.L., S.S. and R.H.; software, C.L. and S.S.; validation, C.L.; formal analysis, C.L.; investigation, C.L.; data curation, C.L.; writing—original draft preparation, C.L. and R.H; writing—review and editing, C.L., S.S., J.M., C.E. and R.H.; visualization, C.L. and S.S.; supervision, C.E. and R.H.; project administration, C.E. and R.H.; funding acquisition, R.H. All authors have read and agreed to the published version of the manuscript.

**Funding:** We gratefully acknowledge funding by the German Federal Ministry for Economic Affairs and Climate Action (grant 16KN053049) and of the Bavarian Research Foundation (grant AZ-1554-22).

**Data Availability Statement:** Data available on request due to restrictions, e.g., privacy or ethical. The data presented in this study are available on request from the corresponding author.

**Conflicts of Interest:** The authors declare no conflict of interest.

## ferences

Durnin, J. Exact solutions for nondiffracting beams I The scalar theory. *J. Opt. Soc. Am. A* **1987**, *4*, 651–654. [CrossRef]

Durnin, J.; Miceli, J.; Eberly, J. Diffraction-free beams. *Phys. Rev. Lett.* **1987**, *58*, 1499–1501. [CrossRef]

Brzobohatý, O.; Cizmár, T.; Zemánek, P. High quality quasi-Bessel beam generated by round-tip axicon. *Opt. Express* **2008**, *16*, 12688–12700. [CrossRef]

Xie, Q.; Li, X.; Jiang, L.; Xia, B.; Yan, X.; Zhao, W.; Lu, Y. High-aspect-ratio, high-quality microdrilling by electron density control using a femtosecond laser Bessel beam. *Appl. Phys. A* **2016**, *122*, 136. [CrossRef]

Tiwari, S.K.; Mishra, S.R.; Ram, S.P.; Rawat, H.S. Generation of a Bessel beam of variable spot size. *Appl. Opt.* **2012**, *51*, 3718–3725. [CrossRef]

Schwarz, S.; Rung, S.; Esen, C.; Hellmann, R. Fabrication of a high-quality axicon by femtosecond laser ablation and $CO_2$ laser polishing for quasi-Bessel beam generation. *Opt. Express* **2018**, *26*, 23287–23294. [CrossRef]

McGloin, D.; Dholakia, K. Bessel beams: Diffraction in a new light. *Contemp. Phys.* **2005**, *46*, 15–28. [CrossRef]

Zhang, H. Lens axicons illuminated by Gaussian beams for generation of uniform-axial intensity Bessel fields. *Opt. Commun.* **2000**, *39*, 803. [CrossRef]

McLeod, E.; Hopkins, A.B.; Arnold, C.B. Multiscale Bessel beams generated by a tunable acoustic gradient index of refraction lens. *Opt. Lett.* **2006**, *31*, 3155–3157. [CrossRef]

Vijayakumar, A.; Bhattacharya, S. Phase-shifted Fresnel axicon. *Opt. Lett.* **2012**, *37*, 1980–1982. [CrossRef]

Chattrapiban, N.; Rogers, E.A.; Cofield, D.; Hill, W.T.; Roy, R. Generation of nondiffracting Bessel beams by use of a spatial light modulator. *Opt. Lett.* **2003**, *28*, 2183–2185. [CrossRef]

Bhuyan, M.K.; Courvoisier, F.; Lacourt, P.-A.; Jacquot, M.; Furfaro, L.; Withford, M.J.; Dudley, J.M. High aspect ratio taper-free microchannel fabrication using femtosecond Bessel beams. *Opt. Express* **2010**, *18*, 566–574. [CrossRef]

Bowman, R.; Muller, N.; Zambrana-Puyalto, X.; Jedrkiewicz, O.; Di Trapani, P.; Padgett, M.J. Efficient generation of Bessel beam arrays by means of an SLM. *Eur. Phys. J. Spec. Top.* **2011**, *199*, 159–166. [CrossRef]

Zhai, Z.; He, X.; Yu, X.; Liu, D.; Lv, Q.; Xiong, Z.; Wang, X.; Xu, Z. Parallel Bessel beam arrays generated by envelope phase holograms. *Opt. Lasers Eng.* **2023**, *161*, 107348. [CrossRef]

Stoyanov, L.; Zhekova, M.; Stefanov, A.; Stefanov, I.; Paulus, G.G.; Dreischuh, A. Zeroth- and first-order long range non-diffracting Gauss-Bessel beams generated by annihilating multiple-charged optical vortices. *Sci. Rep.* **2020**, *10*, 21981. [CrossRef]

Dudutis, J.; Gečys, P.; Račiukaitis, G. Non-ideal axicon-generated Bessel beam application for intra-volume glass modification. *Opt. Express* **2016**, *24*, 28433–28443. [CrossRef]

Mitra, S.; Chanal, M.; Clady, R.; Mouskeftaras, A.; Grojo, D. Millijoule femtosecond micro-Bessel beams for ultra-high aspect ratio machining. *Appl. Opt.* **2015**, *54*, 7358–7365. [CrossRef]

Alexeev, I.; Leitz, K.-H.; Otto, A.; Schmidt, M. Application of Bessel beams for ultrafast laser volume structuring of non transparent media. *Phys. Procedia* **2010**, *5*, 533–540. [CrossRef]

He, F.; Yu, J.; Tan, Y.; Chu, W.; Zhou, C.; Cheng, Y.; Sugioka, K. Tailoring femtosecond 1.5-µm Bessel beams for manufacturing high-aspect-ratio through-silicon vias. *Sci. Rep.* **2017**, *7*, 40785. [CrossRef]

Boucher, P.; Hoyo, J.D.; Billet, C.; Pinel, O.; Labroille, G.; Courvoisier, F. Generation of high conical angle Bessel-Gauss beams with reflective axicons. *Appl. Opt.* **2018**, *57*, 6725–6728. [CrossRef]

Dudutis, J.; Stonys, R.; Račiukaitis, G.; Gečys, P. Aberration-controlled Bessel beam processing of glass. *Opt. Express* **2018**, *26*, 3627–3637. [CrossRef]

22. Orlov, S.; Juršėnas, A.; Baltrukonis, J.; Jukna, V. Controllable Spatial Array of Bessel-like Beams with Independent Axial Intens Distributions for Laser Microprocessing. *J. Laser Micro Nanoeng.* **2018**, *3*, 324–329.
23. Schwarz, S.; Rung, S.; Esen, C.; Hellmann, R. Rapid fabrication of precise glass axicon arrays by an all laser-based manufactur technology. *J. Laser Appl.* **2020**, *32*, 12001. [CrossRef]
24. Roth, G.-L.; Rung, S.; Esen, C.; Hellmann, R. Microchannels inside bulk PMMA generated by femtosecond laser using adapt beam shaping. *Opt. Express* **2020**, *28*, 5801–5811. [CrossRef]
25. Lutz, C.; Roth, G.; Rung, S.; Esen, C.; Hellmann, R. Efficient Ultrashort Pulsed Laser Processing by Dynamic Spatial Li Modulator Beam Shaping for Industrial Use. *J. Laser Micro Nanoeng.* **2021**, *16*, 62–67.
26. García-Martínez, P.; Sánchez-López, M.M.; Davis, J.A.; Cottrell, D.M.; Sand, D.; Moreno, I. Generation of Bessel beam arr through Dammann gratings. *Appl. Opt.* **2012**, *51*, 1375–1381. [CrossRef]
27. Cheng, H.; Xia, C.; Kuebler, S.M.; Golvari, P.; Sun, M.; Zhang, M.; Yu, X. Generation of Bessel-beam arrays for parallel fabricat in two-photon polymerization. *J. Laser Appl.* **2021**, *33*, 12040. [CrossRef]
28. Tao, S.H.; Yuan, X.-C.; Ahluwalia, B.S. The generation of an array of nondiffracting beams by a single composite compu generated hologram. *J. Opt. A Pure Appl. Opt.* **2005**, *7*, 40–46. [CrossRef]
29. Litvin, I.A.; Mhlanga, T.; Forbes, A. Digital generation of shape-invariant Bessel-like beams. *Opt. Express* **2015**, *23*, 7312–7319. [CrossR
30. Jezek, J.; Cizmár, T.; Nedela, V.; Zemánek, P. Formation of long and thin polymer fiber using nondiffracting beam. *Opt. Expr* **2006**, *14*, 8506–8515. [CrossRef]
31. Humbelani Edzani, R. Laser Drilling of Metals and Glass Using Zero-Order Bessel Beams. Master's Thesis, University of Western Cape, Cape Town, South Africa, December 2013.

photonics

MDPI

nmunication

# tatistical Mixture of Kaleidoscope States Interacting with a wo-Level Atom: Entropy and Purification

ge A. Anaya-Contreras [1,†], Arturo Zúñiga-Segundo [1,†] and Héctor M. Moya-Cessa [2,*,†]

1 ESFM Departamento de Física, Edificio 9 Unidad Profesional Adolfo López Mateos, Instituto Politécnico Nacional, México 07738, Mexico
2 Instituto Nacional de Astrofísica, Óptica y Electrónica, Calle Luis Enrique Erro No. 1, Sta. María Tonantzintla, Puebla 72840, Mexico
* Correspondence: hmmc@inaoep.mx; Tel.: +52-222-266-3100
† These authors contributed equally to this work.

**Abstract:** We investigate some of the fundamental features of the interaction of a mixture of coherent states, namely, a Kaleidoscope states mixture, with two-level atoms in the Jaynes–Cummings model framework. We begin our analysis by calculating the von Neumann entropy of the field, which is determined with the help of the virtual atom method. The oscillations appearing in the entropy indicate a state of purity greater than the initial state, i.e., a purification of the initial state due to a transfer of coherence from the atom to the field. In this oscillatory region, we obtain a negative Wigner function that hints at a (noisy) multiple Schrödinger cat.

**Keywords:** Kaleidoscope-states; mixed-states; virtual-atom

check for updates

tion: Anaya-Contreras, J.A.; iga-Segundo, A.; Moya-Cessa, I. Statistical Mixture of idoscope States Interacting with o-Level Atom: Entropy and ification. *Photonics* **2023**, *10*, 150. s://doi.org/10.3390/ tonics10020150

eived: 19 October 2022
ised: 4 January 2023
epted: 30 January 2023
ished: 31 January 2023



## 1. Introduction

The relation between classical and quantum phenomena has attracted attention over the years, becoming a very discussed problem in quantum mechanics. The main problem is that what causes, in the macroscopic world, the quantum interference of superposition states and the $n$-qubit entanglement states may not be observed [1–3].

Schrödinger cat [4–6] states, or superpositions of coherent states, have attracted the attention of researchers due to their fundamental features. Of particular interest is the case of the superposition of two (or more) coherent states [7,8], where, because of quantum interference, their properties are very different from the properties of the constituent coherent states, as well as from the incoherent superposition or statistical mixtures of such states. For example, the superposition exhibits sub-Poissonian photon statistics, higher-order squeezing, and oscillations in the photon number distribution [8]. These properties clearly differentiate the state of the superposition and statistical mixture of two coherent states [7]. Because superpositions of macroscopically distinguishable states (or Schrödinger cat-like states) may be produced by using coherent states, the problem is important for the quantum theory of measurement [9]. Several schemes have already been proposed to produce a superposition of coherent states, for instance the non-linear interaction of the field in a coherent state with a Kerr-like medium can produce their superposition [4]. Another possible way would be through the interaction between quantized fields, initially prepared in coherent states, with two-level atoms [7,8] or ion laser interactions [9].

On the other hand, entropy [9–11] is one of the main tools for measuring entanglement [2,12]. Precisely, one of the main tasks in the present manuscript is to calculate the entropy of the field for a mixture of Kaleidoscope states or a statistical mixture of coherent states, in the Jaynes–Cummings framework, which we will do with the aid of the Araki–Lieb inequality [13]. In principle, it seems impossible to use the Araki–Lieb inequality to calculate the field entropy because the field is in a statistical mixture of states [14]. However, via purification of the mixed density matrix of the quantized field [15], we will be able to use that inequality in order to calculate the field von Neumann entropy, either for the at or the field.

In the next section, we define the Kaleidoscope states that we will consider as initial state in their interaction with a two-level atom. Section 3 deals precisely with t interaction and, there, we calculate the field entropies and their Wigner functions [ where we show that the statistical mixture of coherent states, i.e., the Kaleidoscope sta mixture, goes to a minimum at half the revival time, independently of its number components. Finally, in Section 4, we summarize our conclusions.

## 2. Kaleidoscope States

Kaleidoscope states are a particular superposition of $n$ coherent states, and are defir in reference [17], as:

$$\begin{pmatrix} \sqrt{\lambda_{1\alpha}}|\psi_{1\alpha}\rangle \\ \sqrt{\lambda_{2\alpha}}|\psi_{2\alpha}\rangle \\ \sqrt{\lambda_{3\alpha}}|\psi_{3\alpha}\rangle \\ \vdots \\ \sqrt{\lambda_{n\alpha}}|\psi_{n\alpha}\rangle \end{pmatrix} = \frac{1}{n}\begin{pmatrix} 1 & 1 & 1 & \cdots & 1 \\ 1 & \omega^{*} & \omega^{*2} & \cdots & \omega^{*(n-1)} \\ 1 & \omega^{*2} & \omega^{*4} & \cdots & \omega^{*2(n-1)} \\ \vdots & \vdots & \vdots & \ddots & \vdots \\ 1 & \omega^{*(n-1)} & \omega^{*2(n-1)} & \cdots & \omega^{*(n-1)(n-1)} \end{pmatrix}\begin{pmatrix} |\alpha\rangle \\ |\omega\alpha\rangle \\ |\omega^2\alpha\rangle \\ \vdots \\ |\omega^{(n-1)}\alpha\rangle \end{pmatrix},$$

where $\sqrt{\lambda_{k\alpha}}$ are normalization constants and $\omega = \exp\left(i\frac{2\pi}{n}\right)$ with $n, k$ integers which sati $1 \leq k \leq n$.

We can observe that the Vandermonde matrix that transforms the vectors Equation (1) is proportional to the so-called discrete Fourier transform, also known the quantum Fourier transform [18].

On the other hand, an initial statistical mixture of $n$ coherent states may be written

$$\hat{\rho}_F(0) = \frac{1}{n}\sum_{k=1}^{n} |\omega^{(k-1)}\alpha\rangle\langle\omega^{(k-1)}\alpha| \, .$$

This density matrix may be diagonalized by the virtual atom method [15], where important issue is to establish a connection between this density and the virtual ($V$) dens operator $\hat{\rho}_V = |\psi_V\rangle\langle\psi_V|$, which is calculated from the pure state

$$|\psi_V\rangle = \sqrt{\lambda_{1\alpha}}|\psi_{1\alpha}\rangle|A_1\rangle + \sqrt{\lambda_{2\alpha}}|\psi_{2\alpha}\rangle|A_2\rangle + \sqrt{\lambda_{3\alpha}}|\psi_{3\alpha}\rangle|A_3\rangle + \cdots + \sqrt{\lambda_{n\alpha}}|\psi_{n\alpha}\rangle|A_n\rangle,$$

where $\{|A_k\rangle\}$ is the virtual atom basis and $|\psi_{k\alpha}\rangle$ are the Kaleidoscope states, whene $k = 1, .., n$. After tracing the density operator $\hat{\rho}_V = |\psi_V\rangle\langle\psi_V|$, over the virtual atom sta $\{|A_k\rangle\}$, we obtain the block diagonal density matrix

$$\hat{\rho}_F(0) = \sum_{k=1}^{n} \lambda_{k\alpha}|\psi_{k\alpha}\rangle\langle\psi_{k\alpha}| \, ,$$

where

$$\lambda_{k\alpha} = \frac{e^{-|\alpha|^2}}{n}\sum_{m=1}^{n} \omega^{*m(k-1)} \exp\left(\omega^m|\alpha|^2\right).$$

Similarly, by tracing the density $\hat{\rho}_V = |\psi_V\rangle\langle\psi_V|$ over the field basis, we obtain the virt atom (VA) density matrix

$$\hat{\rho}_{VA} = \frac{1}{n}\begin{pmatrix} 1 & \langle\alpha|\omega\alpha\rangle^* & \langle\alpha|\omega^2\alpha\rangle^* & \cdots & \langle\alpha|\omega^{(n-1)}\alpha\rangle^* \\ \langle\alpha|\omega\alpha\rangle & 1 & \langle\alpha|\omega\alpha\rangle^* & \cdots & \langle\alpha|\omega^{(n-2)}\alpha\rangle^* \\ \langle\alpha|\omega^2\alpha\rangle & \langle\alpha|\omega\alpha\rangle & 1 & \cdots & \langle\alpha|\omega^{(n-3)}\alpha\rangle^* \\ \vdots & \vdots & \vdots & \ddots & \vdots \\ \langle\alpha|\omega^{(n-1)}\alpha\rangle & \langle\alpha|\omega^{(n-2)}\alpha\rangle & \langle\alpha|\omega^{(n-3)}\alpha\rangle & \cdots & 1 \end{pmatrix},$$

and, taking into account that $\langle\alpha|\omega^k\alpha\rangle = \langle\alpha|\omega^{(n-k)}\alpha\rangle^*$ and $\langle\alpha|\omega^k\alpha\rangle = \exp(-|\alpha|^2)\exp(\omega^k|\alpha|^2)$, the Equation (6) may be rewritten as:

$$\hat{\rho}_{VA} = \frac{\exp(-|\alpha|^2)}{n}\begin{pmatrix} \exp(|\alpha|^2) & \exp(\omega^*|\alpha|^2) & \exp(\omega^{*2}|\alpha|^2) & \cdots & \exp(\omega^{*(n-1)}|\alpha|^2) \\ \exp(\omega^{*(n-1)}|\alpha|^2) & \exp(|\alpha|^2) & \exp(\omega^*|\alpha|^2) & \cdots & \exp(\omega^{*(n-2)}|\alpha|^2) \\ \exp(\omega^{*(n-2)}|\alpha|^2) & \exp(\omega^{*(n-1)}|\alpha|^2) & \exp(|\alpha|^2) & \cdots & \exp(\omega^{*(n-3)}|\alpha|^2) \\ \vdots & \vdots & \vdots & \ddots & \vdots \\ \exp(\omega^*|\alpha|^2) & \exp(\omega^{*2}|\alpha|^2) & \exp(\omega^{*3}|\alpha|^2) & \cdots & \exp(|\alpha|^2) \end{pmatrix}, \tag{7}$$

where

$$|A_k\rangle \to \frac{1}{\sqrt{n}}\begin{pmatrix} 1 \\ \omega^{(k-1)} \\ \omega^{2(k-1)} \\ \omega^{3(k-1)} \\ \vdots \\ \omega^{(n-1)(k-1)} \end{pmatrix}, \tag{8}$$

are eigenvectors of $\hat{\rho}_{VA}$, whose eigenvalues are given by (5). For details of the above results, we refer to the Appendixes A and B.

## 3. Interaction of Kaleidoscope States with a Two-Level Atom

The interaction between a quantized field and a two-level atom (under rotating wave approximation) is given by the Jaynes–Cummings interaction Hamiltonian [19] (for simplicity we have set $\hbar = 1$),

$$\hat{H}_I = \lambda\left(\hat{a}^\dagger\sigma_- + \hat{a}\sigma_+\right), \tag{9}$$

where we have considered on-resonance conditions (equal field and atomic transition frequencies). In the above equation, $\lambda$ is the coupling constant, $\hat{a}$ and $\hat{a}^\dagger$ and the annihilation and creation operators, respectively, and $\sigma_+$ and $\sigma_-$ are the raising and lowering Pauli operators, respectively. The evolution operator, $\hat{U} = \exp(-i\hat{H}_I t)$, in the $2\times 2$ basis is given by

$$\hat{U} = \begin{pmatrix} \cos\left(\lambda t\sqrt{\hat{a}\hat{a}^\dagger}\right) & -i\,\hat{V}\sin\left(\lambda t\sqrt{\hat{a}^\dagger\hat{a}}\right) \\ -i\,\hat{V}^\dagger\sin\left(\lambda t\sqrt{\hat{a}\hat{a}^\dagger}\right) & \cos\left(\lambda t\sqrt{\hat{a}^\dagger\hat{a}}\right) \end{pmatrix}, \tag{10}$$

with $\hat{V}$ and $\hat{V}^\dagger$ the London phase operators [20]. For details of the above and next results, we refer to the appendices.

If we consider the atom initially prepared in the state $|e\rangle$ and the field in a statistical mixture of $n$ coherent states, i.e., the system is initially prepared in $\hat{\rho}(0) = \hat{\rho}_F(0)|e\rangle\langle e|$, with $\hat{\rho}_F(0)$ defined by (2), then the time evolved density matrix is given by

$$\hat{\rho} = \begin{pmatrix} \sum_{k=1}^{n}|\mathrm{C}_k\rangle\langle\mathrm{C}_k| & |\mathrm{C}_n\rangle\langle\mathrm{S}_1| + \sum_{k=2}^{n}|\mathrm{C}_{k-1}\rangle\langle\mathrm{S}_k| \\ |\mathrm{S}_1\rangle\langle\mathrm{C}_n| + \sum_{k=2}^{n}|\mathrm{S}_k\rangle\langle\mathrm{C}_{k-1}| & \sum_{k=1}^{n}|\mathrm{S}_k\rangle\langle\mathrm{S}_k| \end{pmatrix}, \tag{11}$$

with

$$|\mathrm{C}_k\rangle = \mathrm{e}^{-\frac{|\alpha|^2}{2}}\sum_{m=0}^{+\infty}\frac{\alpha^{nm+k}}{\sqrt{(nm+k)!}}\cos\left(\lambda t\sqrt{nm+k+1}\right)|nm+k\rangle, \quad (k = 1, 2, \cdots n-1), \tag{12}$$

$$|\mathrm{C}_n\rangle = \mathrm{e}^{-\frac{|\alpha|^2}{2}}\sum_{m=0}^{+\infty}\frac{\alpha^{nm}}{\sqrt{(nm)!}}\cos\left(\lambda t\sqrt{nm+1}\right)|nm\rangle, \tag{13}$$

$$|S_k\rangle = -ie^{-\frac{|\alpha|^2}{2}} \sum_{m=0}^{+\infty} \frac{\alpha^{nm+k-1}}{\sqrt{(nm+k-1)!}} \sin\left(\lambda t\sqrt{nm+k}\right)|nm+k\rangle\,, \quad (k=1,2,\cdots n)\,.$$

In the study of the interaction between two subsystems, namely atom (A) and field
the subsystem entropies $S_A$ and $S_F$, and the total entropy $S_{AF}$ play an essential role a
they obey the Araki–Lieb inequality [13],

$$|S_A - S_F| \leq S_{AF} \leq S_A + S_F\,.$$

Particularly, when the states of the subsystems are prepared in pure states, we have t
$S_{AF} = 0$, and the entropies of the subsystems will be equal $S_A = S_F$. This situatior
precisely our case because the states of the field (3) and atom are pure states, and we c
apply the virtual atom method [15] in order to find the entropies.

Therefore, tracing over the field states, we obtain the reduced block diagonal virt
atom density matrix

$$\hat{\rho}_{VA} = \begin{pmatrix} \langle C_1|C_1\rangle & \langle C_1|S_1\rangle^* & \cdots & 0 & 0 \\ \langle C_1|S_1\rangle & \langle S_1|S_1\rangle & \cdots & 0 & 0 \\ \vdots & \vdots & \ddots & \vdots & \vdots \\ 0 & 0 & \cdots & \langle C_n|C_n\rangle & \langle C_n|S_n\rangle^* \\ 0 & 0 & \cdots & \langle C_n|S_n\rangle & \langle S_n|S_n\rangle \end{pmatrix},$$

whose eigenvalues are:

$$\lambda_k^{\pm} = \frac{1}{2}(\langle C_k|C_k\rangle + \langle S_k|S_k\rangle) \pm \frac{1}{2}\sqrt{(\langle C_k|C_k\rangle - \langle S_k|S_k\rangle)^2 + 4|\langle C_k|S_k\rangle|^2}\,,$$

for $k = 1,2,3,4,\ldots,n$, such that the field entropy can be written in the form

$$S_F = -\sum_{k=1}^{n} \lambda_k^+ \ln \lambda_k^+ - \sum_{k=1}^{n} \lambda_k^- \ln \lambda_k^-\,.$$

Finally, tracing the density (11) over the atomic states, we obtain the reduced density mat
for the field as

$$\hat{\rho}_F = \sum_{k=1}^{n} |C_k\rangle\langle C_k| + \sum_{k=1}^{n} |S_k\rangle\langle S_k|\,.$$

As two coherent states are sufficiently apart when $\alpha \approx 2$, they may be conside
orthogonal. They may be considered orthogonal as $\langle\alpha| - \alpha\rangle = \exp(-2|\alpha|^2)$ and, a
becomes larger, the exponential approaches zero. On the other hand, the Kaleidosco
states are orthogonal for any $\alpha > 0$, and $\hat{\rho}_{VA}$ will be a block diagonalizable density as w
shown in Equation (16).

The field entropy is plotted as a function of the scaled time $\lambda t$ in Figure 1,
Kaleidoscope-states with different values of statistical mixture of coherent states, nam
$n = 1,2,4,8$ and 16, and $\alpha = 6.0$. Their initial values are calculated as $\ln(n)$ and
indicated in the figure by dotted lines. It may be seen that the entropies have sim
behavior: each one possesses a global minimum of about $\lambda t \approx 19$; for all values of $n \geq$
the field entropy is below the dotted line corresponding to their initial values, makin
clear that a purification process takes place. This may be explained as follows: the phot
distribution for a mixture of coherent states is a Poissonian distribution, just as the one
a coherent state. This means that the collapses and revivals occur exactly at the same ti
for each of the components, namely $t_R/2 = \pi|\alpha|/\lambda$, with $t_R$ the so-called revival time. T
phenomenon happens because each of the components of the mixture does not interf
with any other components as the state considered is not a superposition of states b

rather a statistical mixture. Once the interaction begins, each component is divided into two counter-rotating components (in phase space, see Figure 2) that produce the purest state precisely at half the revival time [21]. In this region and in each case, the field becomes purer than its initial state and oscillations appear for $n \geq 2$. Such purification occurs because of a transfer of coherence from the atom to the field [22].

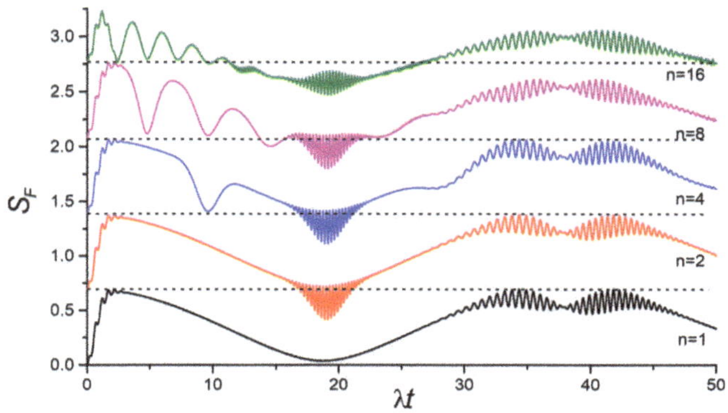


**Figure 1.** The evolution of the field entropy of Kaleidoscope-States as a function of the scaled time $\lambda t$ and different values of statistical mixture of coherent states $n = 1, 2, 4, 8$ and 16, with $\alpha = 6.0$. Their initial values are calculated as $\ln(n)$ and are indicated in the figure by dotted lines.

Here, some questions arise: what will the form of the above-mentioned purer states for different values of a statistical mixture of coherent states $n$ about $\lambda t \approx 19$ be? What will the form of the state where the field entropy reaches its maximum value for different values of a statistical mixture of coherent states $n$ be? In order to try to answer the above questions, we calculated the field Wigner function for the reduced density matrix field defined in Equation (19). The Wigner function may be written as [16]

$$\begin{aligned} W_F(\beta) &= \frac{2}{\pi} \mathrm{Tr}_F \left\{ \hat{\rho}_F \hat{D}(\beta)(-1)^{\hat{n}} \hat{D}^{\dagger}(\beta) \right\}, \\ &= \frac{2}{\pi} \sum_{k=1}^{n} \langle \mathrm{C}_k | \hat{\rho}_F \hat{D}(\beta)(-1)^{\hat{n}} \hat{D}^{\dagger}(\beta) | \mathrm{C}_k \rangle + \frac{2}{\pi} \sum_{k=1}^{n} \langle \mathrm{S}_k | \hat{\rho}_F \hat{D}(\beta)(-1)^{\hat{n}} \hat{D}^{\dagger}(\beta) | \mathrm{S}_k \rangle, \end{aligned} \tag{20}$$

where $\hat{D}(\beta) = \exp(\beta \hat{a}^{\dagger} - \beta^* \hat{a})$ is the displacement operator.

In Figure 2a–c, we show the field Wigner function at $\lambda t = 0$, and we clearly see the two, four, and eight peaks corresponding to each coherent state for $n = 2, 4$ and 8, respectively. Additionally, in Figure 2d–f, we show the field Wigner function corresponding to the Kaleidoscope-State with $n = 2, 4$ and 8 respectively when the field would become a purer state at time $\lambda t \approx 19.15$. These Wigner functions resemble a Schrödinger cat state of 2, 4 and 8 components, where we note the characteristic interference structure. We clearly see the formation of quantum interferences halfway between the $n$ humps. As we mentioned above, once the interaction starts, each of the components is divided into two counter-rotating terms and, at half the revival time, the humps recombine with the traveling contributions from other component states; that, however, is not a coherent process, as each one of the components interferes only with itself. At that time, the Wigner function becomes negative, which is a clear signature of a non-classical state. This fact, plus the purification of the field shown in Figure 1, hints that multiple Schrödinger cats are being generated. The frequency of the interference structure increases as the separation distance $\alpha$ increases [23]. For example, setting $\alpha = 4$ and $n = 2$, the entropy has a similar behavior as in Figure 1 but, as we can see in Figure 3a, now its minimum is around $\lambda t \approx 12.5$, and its corresponding field Wigner function has an interference structure with a lower frequency, as is shown in Figure 4c. Finally, when $\lambda t \approx 19$, the initial mixture of Kaleidoscope States (as we showed in Figure 2a–c) gains purity as was suggested by the entropy behavior, and the negativ of the field Wigner function are an indicator of the non-classical properties of the state [ at $\lambda t \approx 19.15$.

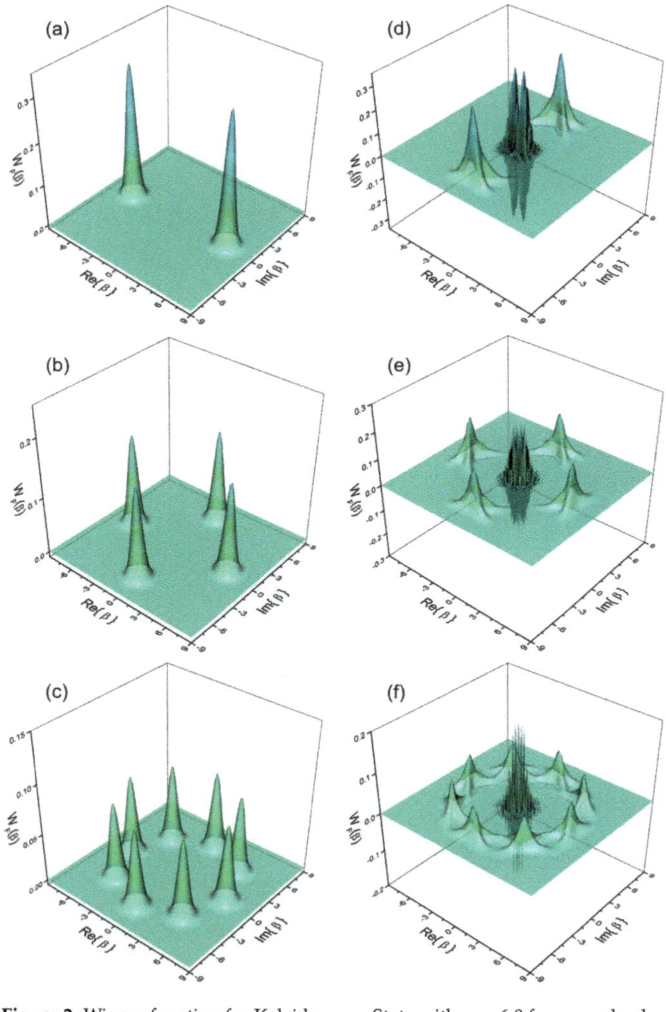


**Figure 2.** Wigner function for Kaleidoscope-State with $\alpha = 6.0$ for several values of the time $\lambda t$ an For $\lambda t = 0$: (**a**) $n = 2$, (**b**) $n = 4$ and (**c**) $n = 8$; for $\lambda t \approx 19.15$: (**d**) $n = 2$, (**e**) $n = 4$ and (**f**) $n = 8$.

On the other hand, for the case of the maximum value of the field entropy $\lambda t \approx 2.0$ it is shown in Figure 3a, we see that each peak of the initial coherent state in Figure 4a, s into two counter-rotating terms, where each one of the components interferes only w itself and spread on the circle of radius $\alpha = 4$, and the Wigner function, may have negat values, as is shown in Figure 4b, which is an indicator of the non-classical properties of state [23].

For the sake of completeness, the atom entropy is plotted in Figure 3b as a function the scaled time $\lambda t$ with $n = 2$ and $\alpha = 4.0$. This entropy quickly saturates to the value $\ln(2)$, and it may be seen that the entropy has similar behavior for other values of $n$ an

The above results can be compared with previous ones in Figure 1 for $n = 16$ and $\alpha = 6$. In that case, it was shown that the field entropy has its minimum around $\lambda t \approx 19.15$. In Figure 4d, we show the field Wigner function at $\lambda t = 0$, and we clearly see the sixteen peaks corresponding to each of $n = 16$ coherent states. For the case of the maximum value of the field entropy ($\lambda t \approx 1.15$), we see that each peak of the initial coherent state splits into two counter-rotating terms, where each component interferes only with itself, and it the formation of the quantum interference structure on the circle of radius $\alpha = 6$ may also be seen. In this case, the Wigner function has no negative values. Finally, when the time goes to $\lambda t \approx 19.15$, the peaks interfere and the formation of the quantum interference structures may be seen halfway between the 16 humps, which is a clear signature of a non-classical state.

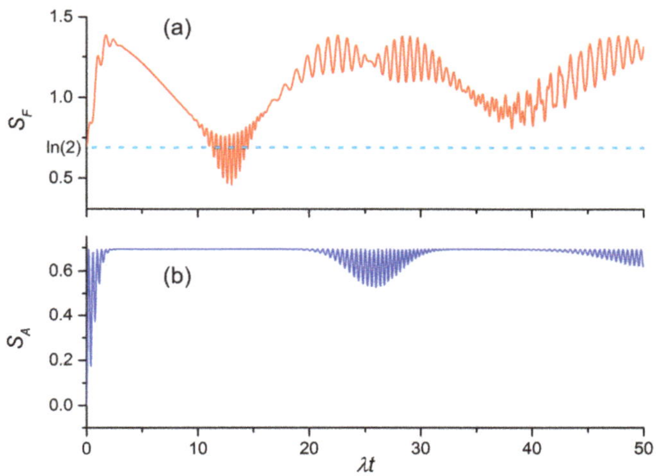


**Figure 3.** The evolution of the: (**a**) field entropy and (**b**) and atom entropy; for a Kaleidoscope-State as a function of the scaled time $\lambda t$ with $n = 2$ and $\alpha = 4.0$. The initial value of field entropy is calculated as $\ln(2)$ and is indicated in the figure by a dotted line.

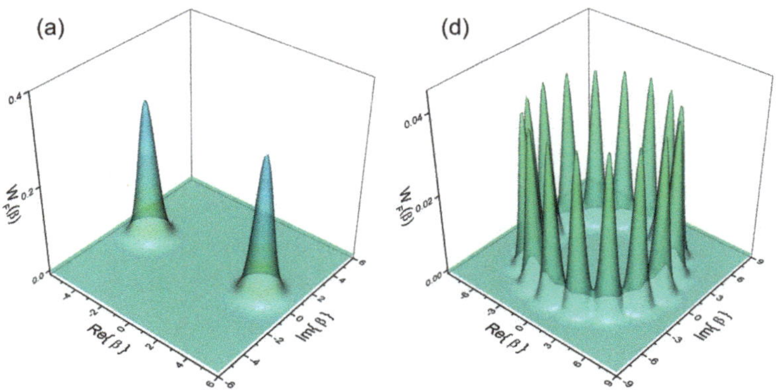


**Figure 4.** *Cont.*

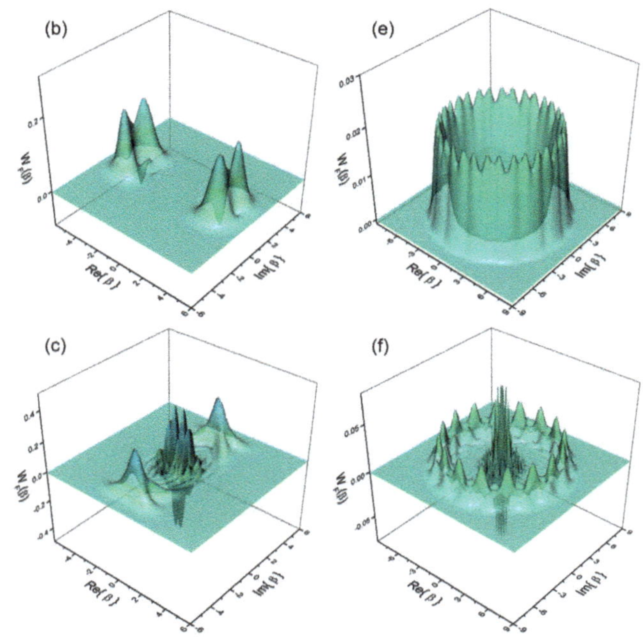


**Figure 4.** Field Wigner function for Kaleidoscope-State for several values of the time $\lambda t$ anc
For $n = 2$ (**a**) $\lambda t = 0$, (**b**) $\lambda t \approx 2.0$ and (**c**) $\lambda t \approx 12.5$. And for $n = 16$ (**d**) $\lambda t = 0$, (**e**) $\lambda t \approx 1.15$ a
(**f**) $\lambda t \approx 19.15$.

## 4. Conclusions

We have shown that the Jaynes–Cummings interaction with an initial Kaleidoscc mixture of coherent states may be modeled by the virtual Hamiltonian method by extendi the atomic Hilbert space such that a virtual pure state may be associated as an ini wavefunction. In particular, we have seen that the purification procedure takes us fror mixed field density matrix to a pure wave function that involves a virtual $2n$-level atc as can be seen in the $2n$ terms in Equation (19). Finally, we should mention that the effe presented in the field entropy for the initial field state given by a statistical mixture of constituent states are reflected in the appearance of oscillations that give rise to negati Wigner functions resembling multiple Schrödinger's cats.

**Author Contributions:** Conceptualization, J.A.A.-C.; investigation, all authors; methodology, authors; software, all authors; validation, all authors; formal analysis, all authors; writing—origi draft preparation, all authors; writing—review and editing, all authors. All authors have read a agreed to the published version of the manuscript.

**Funding:** This research received no external funding.

**Institutional Review Board Statement:** Not applicable.

**Informed Consent Statement:** Not applicable.

**Data Availability Statement:** The data presented in this study are available upon request from corresponding author.

**Conflicts of Interest:** The authors declare no conflict of interest.

## Appendix A. Initial Field Density

Following the virtual atom method to the density matrix (2) we propose:

$$|\psi_V\rangle = \frac{|\alpha\rangle}{\sqrt{n}}|a_1\rangle + \frac{|\omega\alpha\rangle}{\sqrt{n}}|a_2\rangle + \frac{|\omega^2\alpha\rangle}{\sqrt{n}}|a_3\rangle + \cdots + \frac{|\omega^{(n-1)}\alpha\rangle}{\sqrt{n}}|a_n\rangle\,, \tag{A1}$$

where for simplicity we represented the ket in the Hilbert space by the standard basis of $n$ dimensional Euclidean space as:

$$|a_1\rangle = \begin{pmatrix}1\\0\\0\\\vdots\\0\end{pmatrix}, |a_2\rangle = \begin{pmatrix}0\\1\\0\\\vdots\\0\end{pmatrix}, |a_3\rangle = \begin{pmatrix}0\\0\\1\\\vdots\\0\end{pmatrix}, \cdots, |a_n\rangle = \begin{pmatrix}0\\0\\0\\\vdots\\1\end{pmatrix}, \tag{A2}$$

(we do here as it is done when dealing with Pauli-spin operators: one may go to the $2 \times 2$ matrix form or work with the well known $\sigma$ operators). From Equation (1) and considering orthogonal properties of Discrete Fourier Transform (DFT), one has

$$\begin{pmatrix}|\alpha\rangle/\sqrt{n}\\|\omega\alpha\rangle/\sqrt{n}\\|\omega^2\alpha\rangle/\sqrt{n}\\\vdots\\|\omega^{(n-1)}\alpha\rangle/\sqrt{n}\end{pmatrix} = \frac{1}{\sqrt{n}}\begin{pmatrix}1 & 1 & 1 & \cdots & 1\\1 & \omega & \omega^2 & \cdots & \omega^{(n-1)}\\1 & \omega^2 & \omega^4 & \cdots & \omega^{2(n-1)}\\\vdots & \vdots & \vdots & \ddots & \vdots\\1 & \omega^{(n-1)} & \omega^{2(n-1)} & \cdots & \omega^{(n-1)(n-1)}\end{pmatrix}\begin{pmatrix}\sqrt{\lambda_{1\alpha}}|\psi_{1\alpha}\rangle\\\sqrt{\lambda_{2\alpha}}|\psi_{2\alpha}\rangle\\\sqrt{\lambda_{3\alpha}}|\psi_{3\alpha}\rangle\\\vdots\\\sqrt{\lambda_{n\alpha}}|\psi_{n\alpha}\rangle\end{pmatrix}. \tag{A3}$$

Thus, according to the above formula (A3), we can rewrite Equation (A1) as:

$$\begin{aligned}
|\psi_V\rangle &= \frac{1}{\sqrt{n}}\Big(\sqrt{\lambda_{1\alpha}}|\psi_{1\alpha}\rangle + \sqrt{\lambda_{2\alpha}}|\psi_{2\alpha}\rangle + \cdots + \sqrt{\lambda_{n\alpha}}|\psi_{n\alpha}\rangle\Big)|a_1\rangle\\
&+ \frac{1}{\sqrt{n}}\Big(\sqrt{\lambda_{1\alpha}}|\psi_{1\alpha}\rangle + \omega\sqrt{\lambda_{2\alpha}}|\psi_{2\alpha}\rangle + \cdots + \omega^{(n-1)}\sqrt{\lambda_{n\alpha}}|\psi_{n\alpha}\rangle\Big)|a_2\rangle\\
&+ \frac{1}{\sqrt{n}}\Big(\sqrt{\lambda_{1\alpha}}|\psi_{1\alpha}\rangle + \omega^2\sqrt{\lambda_{2\alpha}}|\psi_{2\alpha}\rangle + \cdots + \omega^{2(n-1)}\sqrt{\lambda_{n\alpha}}|\psi_{n\alpha}\rangle\Big)|a_3\rangle\\
&\vdots\\
&+ \frac{1}{\sqrt{n}}\Big(\sqrt{\lambda_{1\alpha}}|\psi_{1\alpha}\rangle + \omega^{(n-1)}\sqrt{\lambda_{2\alpha}}|\psi_{2\alpha}\rangle + \cdots + \omega^{(n-1)(n-1)}\sqrt{\lambda_{n\alpha}}|\psi_{n\alpha}\rangle\Big)|a_n\rangle\\
&= \sqrt{\lambda_{1\alpha}}|\psi_{1\alpha}\rangle\frac{1}{\sqrt{n}}(|a_1\rangle + |a_2\rangle + |a_3\rangle + \cdots + |a_n\rangle)\\
&+ \sqrt{\lambda_{2\alpha}}|\psi_{2\alpha}\rangle\frac{1}{\sqrt{n}}\Big(|a_1\rangle + \omega|a_2\rangle + \omega^2|a_3\rangle + \cdots + \omega^{(n-1)}|a_n\rangle\Big)\\
&\vdots\\
&+ \sqrt{\lambda_{n\alpha}}|\psi_{n\alpha}\rangle\frac{1}{\sqrt{n}}\Big(|a_1\rangle + \omega^{(n-1)}|a_2\rangle + \omega^{2(n-1)}|a_3\rangle + \cdots + \omega^{(n-1)(n-1)}|a_n\rangle\Big)\\
&= \sqrt{\lambda_{1\alpha}}|\psi_{1\alpha}\rangle|A_1\rangle + \sqrt{\lambda_{2\alpha}}|\psi_{2\alpha}\rangle|A_2\rangle + \cdots + \sqrt{\lambda_{n\alpha}}|\psi_{n\alpha}\rangle|A_n\rangle,
\end{aligned} \tag{A4}$$

where

$$\begin{pmatrix} |A_1\rangle \\ |A_2\rangle \\ |A_3\rangle \\ \vdots \\ |A_n\rangle \end{pmatrix} = \frac{1}{\sqrt{n}} \begin{pmatrix} 1 & 1 & 1 & \cdots & 1 \\ 1 & \omega & \omega^2 & \cdots & \omega^{(n-1)} \\ 1 & \omega^2 & \omega^4 & \cdots & \omega^{2(n-1)} \\ \vdots & \vdots & \vdots & \ddots & \vdots \\ 1 & \omega^{(n-1)} & \omega^{2(n-1)} & \cdots & \omega^{(n-1)(n-1)} \end{pmatrix} \begin{pmatrix} |a_1\rangle \\ |a_2\rangle \\ |a_3\rangle \\ \vdots \\ |a_n\rangle \end{pmatrix}, \quad \text{(A}$$

with

$$|A_k\rangle = \begin{pmatrix} 1 \\ \omega^{(k-1)} \\ \omega^{2(k-1)} \\ \vdots \\ \omega^{(n-1)(k-1)} \end{pmatrix} = \begin{pmatrix} \langle a_1|A_k\rangle \\ \langle a_2|A_k\rangle \\ \langle a_3|A_k\rangle \\ \vdots \\ \langle a_n|A_k\rangle \end{pmatrix}, \quad \text{(A}$$

then these new virtual atom states $\{|A_i\rangle_i\}$ are orthogonal kets because the discrete Fou transform (DFT) is an orthogonal matrix too.

Using the above results it can be shown that the density matrix associated with $|\psi$ defined by Equation (A1) will be

$$\hat{\rho}_V = \frac{1}{n}\sum_{i=1}^{n}\sum_{j=1}^{n} |\omega^{(i-1)}\alpha\rangle|a_i\rangle\langle a_j|\langle\omega^{(j-1)}\alpha| = \sum_{i=1}^{n}\sum_{j=1}^{n}\sqrt{\lambda_{i\alpha}\lambda^*_{j\alpha}}|\psi_{i\alpha}\rangle|A_i\rangle\langle A_j|\langle\psi_{j\alpha}|\,, \quad \text{(A}$$

where we have used relation (A4). Taking the trace of the density matrix (A7) over the fi states we get the virtual atom density operator

$$\hat{\rho}_{VA} = \frac{1}{n}\sum_{i=1}^{n}\sum_{j=1}^{n}\langle\omega^{(j-1)}\alpha|\omega^{(i-1)}\alpha\rangle|a_i\rangle\langle a_j| = \sum_{i=1}^{n}\sum_{j=1}^{n}\sqrt{\lambda_{i\alpha}\lambda^*_{j\alpha}}\langle\psi_{j\alpha}|\psi_{i\alpha}\rangle|A_i\rangle\langle A_j|\,. \quad \text{(A}$$

But in reference [17], it was remarked that Kaleidoscope states are orthogonal, then

$$\hat{\rho}_{VA} = \frac{1}{n}\sum_{i=1}^{n}\sum_{j=1}^{n}\langle\omega^{(j-1)}\alpha|\omega^{(i-1)}\alpha\rangle|a_i\rangle\langle a_j| = \sum_{i=1}^{n}\lambda_{i\alpha}|A_i\rangle\langle A_i|\,. \quad \text{(A}$$

We will now do an analysis of the eigenvalue problem for $\hat{\rho}_{VA}$, which we may start writ

$$\begin{aligned} \hat{\rho}_{VA}|A_k\rangle &= \frac{1}{n}\sum_{i=1}^{n}\sum_{j=1}^{n}\langle\omega^{(j-1)}\alpha|\omega^{(i-1)}\alpha\rangle|a_i\rangle\langle a_j|A_k\rangle = \lambda_{k\alpha}|A_k\rangle\,, \\ &= \frac{\mathrm{e}^{-|\alpha|^2}}{n}\sum_{i=1}^{n}\sum_{j=1}^{n}\exp\left(\omega^{*(j-i)}|\alpha|^2\right)\frac{1}{\sqrt{n}}\omega^{(j-1)(k-1)}|a_i\rangle = \lambda_{k\alpha}|A_k\rangle\,, \quad \text{(A} \\ &= \begin{pmatrix} b_1 \\ b_2 \\ b_3 \\ \vdots \\ b_n. \end{pmatrix}, \end{aligned}$$

where

$$b_l = \frac{e^{-|\alpha|^2}}{n} \sum_{j=1}^{n} \exp\left(\omega^{*(j-l)}|\alpha|^2\right) \frac{1}{\sqrt{n}} \omega^{(j-1)(k-1)} ,$$
$$= \frac{e^{-|\alpha|^2}}{n} \frac{1}{\sqrt{n}} \omega^{(l-1)(k-1)} \sum_{m=1-l}^{n-l} \omega^{m(k-1)} \exp\left(\omega^{*m}|\alpha|^2\right) , \tag{A11}$$
$$= \frac{e^{-|\alpha|^2}}{n} \langle a_l | A_k \rangle \sum_{m=l}^{n+l-1} \omega^{*m(k-1)} \exp\left(\omega^{m}|\alpha|^2\right) ,$$

with $1 \leq l \leq n$. When $l = 1$, we have

$$b_1 = \frac{e^{-|\alpha|^2}}{n} \langle a_1 | A_k \rangle \sum_{m=1}^{n} \omega^{*m(k-1)} \exp\left(\omega^{m}|\alpha|^2\right) = \lambda_{k\alpha} \langle a_1 | A_k \rangle , \tag{A12}$$

as $\langle a_1 | A_k \rangle \neq 0$, then we have eigenvalue

$$\lambda_{k\alpha} = \frac{e^{-|\alpha|^2}}{n} \sum_{m=1}^{n} \omega^{*m(k-1)} \exp\left(\omega^{m}|\alpha|^2\right) . \tag{A13}$$

In a similar fashion, for $l > 1$ we have

$$\begin{aligned}
b_l &= \frac{e^{-|\alpha|^2}}{n} \langle a_l | A_k \rangle \sum_{m=l}^{n+l-1} \omega^{*m(k-1)} \exp\left(\omega^{m}|\alpha|^2\right) , \\
&= \frac{e^{-|\alpha|^2}}{n} \langle a_l | A_k \rangle \sum_{m=1}^{n+l-1} \omega^{*m(k-1)} \exp\left(\omega^{m}|\alpha|^2\right) - \frac{e^{-|\alpha|^2}}{n} \langle a_l | A_k \rangle \sum_{m=1}^{l-1} \omega^{*m(k-1)} \exp\left(\omega^{m}|\alpha|^2\right) , \\
&= \frac{e^{-|\alpha|^2}}{n} \langle a_l | A_k \rangle \sum_{m=1}^{n} \omega^{*m(k-1)} \exp\left(\omega^{m}|\alpha|^2\right) - \frac{e^{-|\alpha|^2}}{n} \langle a_l | A_k \rangle \sum_{m=1}^{l-1} \omega^{*m(k-1)} \exp\left(\omega^{m}|\alpha|^2\right) \\
&+ \frac{e^{-|\alpha|^2}}{n} \langle a_l | A_k \rangle \sum_{m=n+1}^{n+l-1} \omega^{*m(k-1)} \exp\left(\omega^{m}|\alpha|^2\right) , \\
&= \frac{e^{-|\alpha|^2}}{n} \langle a_l | A_k \rangle \sum_{m=1}^{n} \omega^{*m(k-1)} \exp\left(\omega^{m}|\alpha|^2\right) - \frac{e^{-|\alpha|^2}}{n} \langle a_l | A_k \rangle \sum_{m=1}^{l-1} \omega^{*m(k-1)} \exp\left(\omega^{m}|\alpha|^2\right) \\
&+ \frac{e^{-|\alpha|^2}}{n} \langle a_l | A_k \rangle \sum_{m=1}^{l-1} \omega^{*(m+n)(k-1)} \exp\left(\omega^{m+n}|\alpha|^2\right) .
\end{aligned} \tag{A14}$$

Because $\omega^n = \left[\exp\left(i\frac{2\pi}{n}\right)\right]^n = 1$, we can cancel the last two sums in Equation (A14), and we have that

$$b_l = \frac{e^{-|\alpha|^2}}{n} \langle a_l | A_k \rangle \sum_{m=1}^{n} \omega^{*m(k-l)} \exp\left(\omega^{m}|\alpha|^2\right) = \lambda_{k\alpha} \langle a_l | A_k \rangle , \tag{A15}$$

as $\langle a_l | A_k \rangle \neq 0$, then we have eigenvalue

$$\lambda_{k\alpha} = \frac{e^{-|\alpha|^2}}{n} \sum_{m=1}^{n} \omega^{*m(k-l)} \exp\left(\omega^{m}|\alpha|^2\right) . \tag{A16}$$

With the help of Equations (A12) and (A15), the eigenvalue Equation (A10) can be written

$$\hat{\rho}_{VA}|A_k\rangle = \lambda_{k\alpha} \begin{pmatrix} \langle a_1|A_k\rangle \\ \langle a_1|A_k\rangle \\ \langle a_n|A_k\rangle \\ \vdots \\ \langle a_n|A_k\rangle \end{pmatrix} = \lambda_{k\alpha}|A_k\rangle \, . \qquad \text{(A}$$

Now, taking the trace of the density matrix (A7) over field states we get the field densi

$$\hat{\rho}_F = \frac{1}{n}\sum_{i=1}^{n} |\omega^{(i-1)}\alpha\rangle\langle\omega^{(i-1)}\alpha| = \sum_{i=1}^{n} \lambda_{i\alpha}|\psi_{i\alpha}\rangle\langle\psi_{i\alpha}| \, . \qquad \text{(A}$$

**Appendix B. Jaynes-Cummings Dynamics**

The Jaynes-Cummings time evolution operator is given by

$$\hat{U}(t) = \begin{pmatrix} \cos\left(\lambda t\sqrt{\hat{a}\hat{a}^{\dagger}}\right) & -i\,\hat{V}\sin\left(\lambda t\sqrt{\hat{a}^{\dagger}\hat{a}}\right) \\ -i\,\hat{V}^{\dagger}\sin\left(\lambda t\sqrt{\hat{a}\hat{a}^{\dagger}}\right) & \cos\left(\lambda t\sqrt{\hat{a}^{\dagger}\hat{a}}\right) \end{pmatrix}, \qquad \text{(A}$$

where $\hat{V}$ and $\hat{V}^{\dagger}$ are the London phase operators [20]. If we consider the initial dens matrix $\hat{\rho}(0) = \hat{\rho}_F(0)|e\rangle\langle e|$, where $\hat{\rho}_F(0)$ was defined in Equation (4), we obtain:

$$\begin{aligned} \hat{\rho}(t) &= \hat{U}(t)\hat{\rho}(0)\hat{U}^{\dagger}(t) = \hat{U}(t)\begin{pmatrix} \hat{\rho}_F(0) & 0 \\ 0 & 0 \end{pmatrix}\hat{U}^{\dagger}(t)\, , \\ &= \begin{pmatrix} \sum_{k=1}^{n} |C_k\rangle\langle C_k| & |C_n\rangle\langle S_1| + \sum_{k=2}^{n} |C_{k-1}\rangle\langle S_k| \\ |S_1\rangle\langle C_n| + \sum_{k=2}^{n} |S_k\rangle\langle C_{k-1}| & \sum_{k=1}^{n} |S_k\rangle\langle S_k| \end{pmatrix}, \end{aligned} \qquad \text{(A}$$

where,

$$\begin{aligned} |C_k\rangle &= \sqrt{\lambda_{(k+1)\alpha}}\cos\left(\lambda t\sqrt{\hat{a}\hat{a}^{\dagger}}\right)|\psi_{(k+1)\alpha}\rangle\, , \quad (k = 1,2,3,\ldots,n-1)\, , \\ |C_n\rangle &= \sqrt{\lambda_{1\alpha}}\cos\left(\lambda t\sqrt{\hat{a}\hat{a}^{\dagger}}\right)|\psi_{1\alpha}\rangle\, , \\ |S_k\rangle &= -i\sqrt{\lambda_{k\alpha}}\hat{V}^{\dagger}\sin\left(\lambda t\sqrt{\hat{a}\hat{a}^{\dagger}}\right)|\psi_{k\alpha}\rangle\, , \quad (k = 1,2,3,\ldots,n)\, . \end{aligned} \qquad \text{(A}$$

Nevertheless, Kaleidoscope states can be rewritten as:

$$\begin{aligned} |\psi_{k\alpha}\rangle &= \frac{e^{-|\alpha|^2/2}}{\sqrt{\lambda_{k\alpha}}}\sum_{m=0}^{+\infty}\frac{\alpha^{nm+k-1}}{(nm+k-1)!}\hat{a}^{\dagger(nm+k-1)}|0\rangle\, , \\ &= \frac{e^{-|\alpha|^2/2}}{\sqrt{\lambda_{k\alpha}}}\sum_{m=0}^{+\infty}\frac{\alpha^{nm+k-1}}{\sqrt{(nm+k-1)!}}|nm+k-1\rangle\, . \end{aligned} \qquad \text{(A}$$

With the above relation, we can write

$$|C_k\rangle = e^{-\frac{|\alpha|^2}{2}} \sum_{m=0}^{+\infty} \frac{\alpha^{nm+k}}{\sqrt{(nm+k)!}} \cos\left(\lambda t\sqrt{nm+k+1}\right)|nm+k\rangle\,, \quad (k=1,2,3,\ldots,n-1)\,,$$
$$|C_n\rangle = e^{-\frac{|\alpha|^2}{2}} \sum_{m=0}^{+\infty} \frac{\alpha^{nm}}{\sqrt{(nm)!}} \cos\left(\lambda t\sqrt{nm+1}\right)|nm\rangle$$
$$|S_k\rangle = -ie^{-\frac{|\alpha|^2}{2}} \sum_{m=0}^{+\infty} \frac{\alpha^{nm+k-1}}{\sqrt{(nm+k-1)!}} \sin\left(\lambda t\sqrt{nm+k}\right)|nm+k\rangle\,, \quad (k=1,2,3,\ldots,n)\,. \tag{A23}$$

From the time evolution density matrix (A20), we can obtain the reduced block diagonal atom density matrix by tracing over the field states, so that

$$\hat{\rho}_A = \begin{pmatrix} \langle C_1|C_1\rangle & \langle C_1|S_1\rangle^* & \cdots & 0 & 0 \\ \langle C_1|S_1\rangle & \langle S_1|S_1\rangle & \cdots & 0 & 0 \\ \vdots & \vdots & \ddots & \vdots & \vdots \\ 0 & 0 & \cdots & \langle C_n|C_n\rangle & \langle C_n|S_n\rangle^* \\ 0 & 0 & \cdots & \langle C_n|S_n\rangle & \langle S_n|S_n\rangle \end{pmatrix}, \tag{A24}$$

whose characteristic polynomial is:

$$\det(\hat{\rho}_A - \lambda\hat{I}) = \prod_{k=1}^{n} \det\begin{pmatrix} \langle C_k|C_k\rangle - \lambda & \langle C_k|S_k\rangle^* \\ \langle S_k|C_k\rangle & \langle S_k|S_k\rangle - \lambda \end{pmatrix} = 0\,, \tag{A25}$$

and we get the eigenvalues

$$\lambda_k^{\pm} = \frac{1}{2}(\langle C_k|C_k\rangle + \langle S_k|S_k\rangle) \pm \frac{1}{2}\sqrt{(\langle C_k|C_k\rangle - \langle S_k|S_k\rangle)^2 + 4|\langle C_k|S_k\rangle|^2}\,, \tag{A26}$$

for $k = 1,2,3,\ldots,n$. By recalling the properties of the trace of the block diagonal matrix we calculate the atomic von Neumann entropy defined as

$$\begin{aligned} S_A &= -\mathrm{Tr}\{\hat{\rho}_A \ln \hat{\rho}_A\}\,, \\ &= -\left(\sum_{k=1}^{n}\langle C_k|C_k\rangle\right)\ln\left(\sum_{k=1}^{n}\langle C_k|C_k\rangle\right) - \left(\sum_{k=1}^{n}\langle S_k|S_k\rangle\right)\ln\left(\sum_{k=1}^{n}\langle S_k|S_k\rangle\right), \\ &= -\sum_{k=1}^{n}\lambda_k^{+}\ln\lambda_k^{+} - \sum_{k=1}^{n}\lambda_k^{-}\ln\lambda_k^{-}\,. \end{aligned} \tag{A27}$$

Finally, by tracing the density (A20) over the atomic states, we obtain the reduced density matrix for the field as

$$\hat{\rho}_F = \sum_{k=1}^{n}|C_k\rangle\langle C_k| + \sum_{k=1}^{n}|S_k\rangle\langle S_k|\,. \tag{A28}$$

## ferences


Man, Z.X.; Zhang, Y.J.; Su, F.; Xia, Y.J. Entanglement dynamics of multiqubit system in Markovian and non-Markovian reservoirs. *Eur. Phys. J. D* **2010**, *58*, 147–151. [CrossRef]

Nourmandipour, A.; Tavassoly, M.K.; Rafiee, M. Dynamics and protection of entanglement in $n$-qubit systems within Markovian and non-Markovian environments. *Phys. Rev. A* **2016**, *93*, 022327. [CrossRef]

Doll, R.; Wubs, M.; Hänggi, P.; Kohler, S. Incomplete pure dephasing of $N$-qubit entangled W states. *Phys. Rev. B* **2007**, *76*, 045317. [CrossRef]

Yurke, B.; Stoler, D. Generating quantum mechanical superpositions of macroscopically distinguishable states via amplitude dispersion. *Phys. Rev. Lett.* **1986**, *57*, 13–16. [CrossRef] [PubMed]

Brune, M.; Haroche, S.; Raimond, J.M.; Davidovich, L.; Zagury, N. Manipulation of photons in a cavity by dispersive atom-field coupling: Quantum-nondemolition measurements and generation of "Schrödinger cat" states. *Phys. Rev. A* **1992**, *45*, 5193–5214. [CrossRef] [PubMed]

6. Davidovich, L. From quantum to classical: Schrödinger cats, entanglement, and decoherence. *Phys. Scr.* **2016**, *91*, 063( [CrossRef]
7. Moya-Cessa, H.; Vidiella-Barranco, A. On the Interaction of Two-level Atoms with Superpositions of Coherent States of Li *J. Mod. Opt.* **1995**, *42*, 1547–1552. [CrossRef]
8. Vidiella-Barranco, A.; Moya-Cessa, H.; Bužek, V. Interaction of Superpositions of Coherent States of Light with Two-level Ato *J. Mod. Opt.* **1992**, *39*, 1441–1459. [CrossRef]
9. Barnett, S.; Phoenix, S. Entropy as a measure of quantum optical correlation. *Phys. Rev. A* **1989**, *40*, 2404–2409. [CrossRef]
10. Phoenix, S. Wave-packet evolution in the damped oscillator. *Phys. Rev. A* **1995**, *41*, 5132–5138. [CrossRef]
11. Barnett, S.M.; Beige, A.; Ekert, A.; Garraway, B.M.; Keitel, C.H.; Kendon, V.; Lein, M.; Milburn, G.J.; Moya-Cessa, H. Murao, M.; et al. Journeys from quantum optics to quantum technology. *Prog. Quantum Electron.* **2017**, *54*, 19–45. [CrossRef]
12. Golkar, G.; Tavassoly, M. Atomic motion and dipole–dipole effects on the stability of atom–atom entanglement in Markovian/n Markovian reservoir. *Mod. Phys. Lett. A* **2019**, *34*, 1950077. [CrossRef]
13. Araki, H.; Lieb, E.H. Entropy inequalities. *Commun. Math. Phys.* **1970**, *18*, 160–170. [CrossRef]
14. Zúñiga-Segundo, A.; Juárez-Amaro, R.; Aguilar-Loreto, O.; Moya-Cessa, H.M. Field's entropy in the atom–field interacti Statistical mixture of coherent states. *Ann. Phys.* **2017**, *379*, 150–158. [CrossRef]
15. Anaya-Contreras, J.A.; Zúñiga-Segundo, A.; Moya-Cessa, H.M. Purification of the atom-field interaction Hamiltonian. *P Open* **2019**, *1*, 100007. [CrossRef]
16. Moya-Cessa, H.; Knight, P.L. Series representation of quantum-field quasiprobabilities. *Phys. Rev. A* **1993**, *48*, 2479–2 [CrossRef]
17. Koçak, A.; Pashaev, O.K. Special functions with mod *N* Symmetry Kaleidosc. Quantum Coherent States. *J. Phys. Conf. Ser.* **2** *1194*, 012059. [CrossRef]
18. Perez-Leija, A.; Andrade-Morales, L.A.; Soto-Eguibar, F.; Szameit, A.; Moya-Cessa, H.M. The Pegg–Barnett phase operator a the discrete Fourier transform. *Phys. Scr.* **2016**, *91*, 043008. [CrossRef]
19. Jaynes, E.; Cummings, F. Comparison of quantum and semiclassical radiation theories with application to the beam maser. *P IEEE* **1963**, *51*, 89–109. [CrossRef]
20. Susskind, L.; Glogower, J. Quantum mechanical phase and time operator. *Phys. Phys. Fiz.* **1964**, *1*, 49–61. [CrossRef]
21. Gea-Banacloche, J. Atom- and field-state evolution in the Jaynes-Cummings model for large initial fields. *Phys. Rev. A* **1** *44*, 5913–5931. [CrossRef]
22. Arancibia-Bulnes, C.A.; Moya-Cessa, H.; Sánchez-Mondragón, J.J. Purifying a thermal field in a lossless micromaser. *Phys. Rev* **1995**, *51*, 5032–5034. [CrossRef]
23. Kenfack, A.; yczkowski, K. Negativity of the Wigner function as an indicator of non-classicality. *J. Opt. B Quantum Semiclass Opt.* **2004**, *6*, 396–404. [CrossRef]

hv photonics

MDPI

ticle

# Tunable Optoelectronic Oscillator with Phase-to-Amplitude Modulation Transformation via an Acetylene Reference Cell

adimir Lebedev [1], Piotr Agruzov [1], Igor' Iliyechev [1], Andrei Varlamov [1], Ivan Tatsenko [2], Andrey Nikitin [2], exey B. Ustinov [2], Serguei Stepanov [3,*] and Alexander Shamray [1]

1 Department for Quantum Electronics, Ioffe Institute, 194021 St. Petersburg, Russia
2 Department of Physical Electronics and Technology, St. Petersburg Electrotechnical University, 197022 St. Petersburg, Russia
3 Department of Optics, CICESE, Ensenada 22860, BC, Mexico
* Correspondence: steps@cicese.mx

**Abstract:** Operation modes of the optoelectronic oscillator (OEO), based on a phase electrooptic modulator with an acetylene reference cell as a photonic filter, have been investigated. For the wideband phase-to-amplitude demodulation that was observed when the laser wavelength was tuned to one side of the acetylene absorption line, an additional tunable electronic yttrium iron garnet (YIG) filter was introduced to observe single-mode OEO generation. This configuration generated a stable monochromatic signal in the 4–12 GHz frequency range, with phase noise of −122 dBc/Hz at 10 kHz offset from the carrier frequency. In the narrowband demodulation mode (when one optical modulation sideband was tuned to the acetylene absorption line), the quasi-single-mode oscillation could be observed without additional electronic filtering. In this case, the generation frequency was controlled optically by tuning the laser wavelength.

**Keywords:** optoelectronic oscillator; phase modulator; self-reference demodulation; acetylene

check for updates

ation: Lebedev, V.; Agruzov, P.; echev, I.; Varlamov, A.; Tatsenko, Nikitin, A.; Ustinov, A.B.; oanov, S.; Shamray, A. A Tunable oelectronic Oscillator with se-to-Amplitude Modulation nsformation via an Acetylene erence Cell. *Photonics* **2023**, *10*, 196. s://doi.org/10.3390/ tonics10020196

eived: 22 December 2022
ised: 27 January 2023
epted: 3 February 2023
lished: 12 February 2023



## 1. Introduction

Optoelectronic oscillators (OEOs) present a new promising class of optoelectronic devices that have attracted the attention of many research groups in recent decades. In the general case, they can be presented by a closed optoelectronic loop [1]. One part of this loop is an optical path that ensures a high coherence length, and the other part is an electronic microwave path that enables high controllability. The proposed applications of the OEOs cover a wide variety of applications requiring high-purity, stable microwave or millimeter wave signals in the form of single monochromatic spectral lines with extremely low phase noise, frequency combs, or signals with more complicated arbitrary waveforms [1,2]. Devices of this type with integrated design are also under development now [3,4].

In the simplest basic configuration of OEO, electrooptic amplitude modulation is used to introduce a microwave signal into the optical path. In turn, direct optical detection by a fast photodiode ensures the transformation of an amplitude-modulated optical signal into the electronic path of the feedback loop. The narrowband electronic or photonic filters [5–7] are utilized to select one particular resonant frequency of the optical path of the loop. To ensure the stability of the OEO generation in this basic configuration, both the operation point of the amplitude modulator and the central frequency of the filter being used are to be locked by additional electronic circuits of the feedback control.

Below we propose an original configuration of the OEO with the phase electrooptic modulator, instead of the generally used amplitude one, and the acetylene reference cell as a phase-to-amplitude demodulator and the photonic filter simultaneously. Unlike the electrooptic amplitude modulator, the phase modulator does not require operation point stabilization and permanently ensures linear response to the modulating microwave signal.

The reference acetylene cell that is used here for demodulation is a highly stable p tonic filter in the telecommunication wavelength range (1510–1550 nm) [8]. Indeed, narr acetylene absorption lines are widely used for laser wavelength locking and stabilizati in this highly important spectral range. The low-pressure acetylene reference cells commercially available and, in particular, in the fiber-coupled configurations. Recently have demonstrated the utilization of the acetylene cell as a simple and robust optical ph demodulator [9,10] for room temperature operation.

In this paper, we study various demodulation modes, based on the acetylene cell, w respect to OEO operation and, in particular, the possibility of generating a single freque and tuning it in a broad frequency range. We show that for the broadband demodulati mode, the single mode operation and tunability of the OEO configuration can be ensu by the additional introduction of the electronic YIG spin-wave filter. On the opposite, w narrowband demodulation, the broadband tuning of the generation frequency is reach by optical means only, by tuning the laser wavelength.

The other important goal of this paper was to investigate an achievable spectral pu and phase noise performance of the proposed OEO configuration. We show, in particu that the demodulated laser phase noise contributes to the output OEO phase noise. T stability of the generated frequency is, obviously, another important characteristic of OEO operation. This requires further rather special investigations and is beyond the sc of the present paper.

## 2. OEO Configuration

A simplified diagram of the proposed OEO is shown in Figure 1a. The photograph of experimental setup is presented in Figure 1b. In the optical path of the OEO in question, utilized the "TeraXion" PS-TNL tunable semiconductor laser with low phase noise and wit spectral line width of 1 kHz as a source of the optical carrier. The integrated lithium niob electrooptic modulator with a frequency bandwidth of 12 GHz at the level of 3 dB and half-wave voltage $V_\pi$ = 5 V, fabricated in our group, was utilized for phase modulation the optical carrier. A single-mode optical fiber (SMF-28) of 2000 m in length was used as optical delay line. The demodulated output optical signal was detected by a fast photodi with the frequency bandwidth of about 13 GHz and with the saturation current of 30 mA

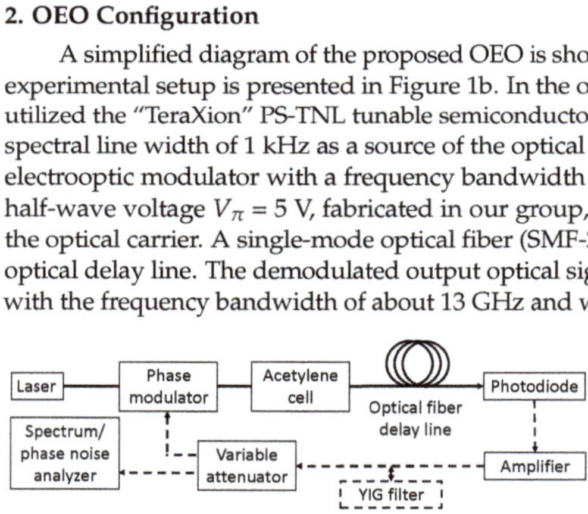


(a)

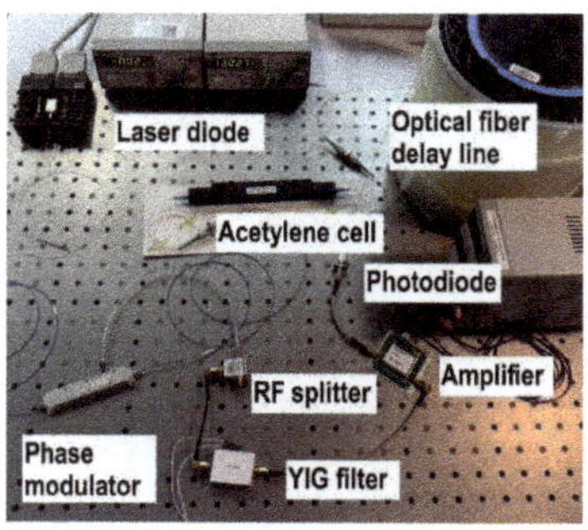


(b)

**Figure 1.** (**a**) A schematic diagram of the OEO under consideration, and (**b**) a photograph of experimental setup.

The electronic microwave path of the OEO under consideration also included a low-noise electronic amplifier, a tunable attenuator, a millimeter-wave coupler, and (as an additional frequency-selective element in some experiments) a tunable narrowband (50 MHz) yttrium iron garnet (YIG) spin-wave filter [11–14]. Spectral and noise measurements were performed using the phase noise analyzer Rohde & Schwarz FSWP26.

A commercial "WavelengthReference" bulk acetylene reference cell (with a gas pressure of about 8 Torr) with fiber-optic pigtails was used for demodulation. The most intense acetylene absorption line P9 (with absorbance $\alpha_0 L \approx 2.7$ at the central wavelength of 1530.37 nm [8]) was utilized in the reported experiments. Note that, at room temperature, the low-pressure acetylene demonstrates inhomogeneously broadened (by the Doppler effect) absorption lines of approximately Gaussian shape with a full width at half maximum (FWHM) spectral width of about 500 MHz [9]. The experimentally observed transmittance and absorption spectral curves of the utilized gas cell in the vicinity of the P9 acetylene line are shown in Figure 2.

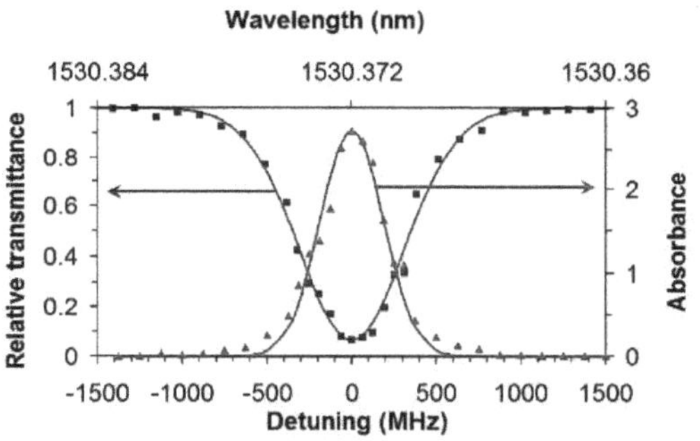


**Figure 2.** Transmittance/absorbance profiles of the utilized acetylene cell at the P9 spectral line, centered at 1530.372 nm. Solid lines present theoretical fittings with the Gaussian profile of the absorption of 500 MHz FWHM spectral width.

## 3. Demodulation Modes

The first experiments were carried out with the optical path of the OEO, in an open-loop configuration. Two different modes of phase-to-amplitude modulation conversion (demodulation) with the acetylene reference cell were probed. In the first mode, the broadband one, the demodulation was performed by introducing an additional phase shift to the phase-modulated wave carrier frequency component. To explain this process, let us consider the main difference between the amplitude and the phase sinusoidal modulations.

Equation (1) below represents the amplitude- and phase-modulated signals in the low contrast approximation ($m << 1$):

$$\begin{aligned} A_a &= A[1 + m \cdot cos(\Omega t)]; \\ A_p &= A \cdot exp[im \cdot \cos(\Omega t)] = A[1 + im \cdot \cos(\Omega t)] \end{aligned} \quad (1)$$

The main difference is, clearly, the $\pi/2$ phase shift between the carrier frequency component and the sidebands. For this reason, to transform an initial phase modulation to the desired amplitude modulation, it is necessary to introduce a similar phase shift between them. If the introduced phase shift is different from $\pm\pi/2$, this transformation (i.e., the demodulation) is, clearly, a partial one only.

In this demodulator configuration, the laser wavelength is tuned to one side of the acetylene absorption line [10], where the phase shift in the transmitted carrier frequency component is maximum; see Figure 3. If the modulation frequency $\Omega$ is assumed to be larger than the FWHM spectral width of the line, the output sideband phases are not significantly influenced by the cell.

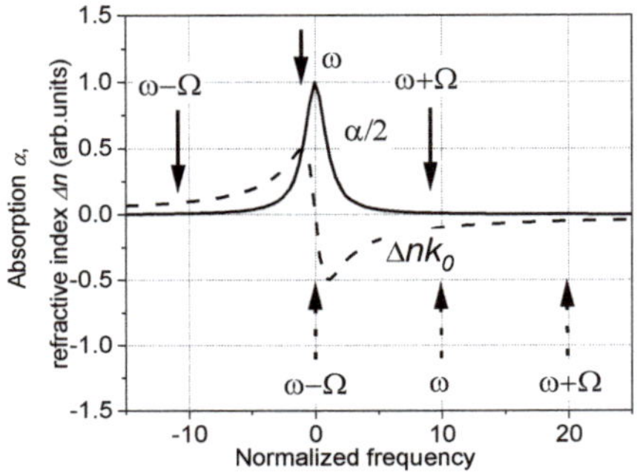


**Figure 3.** Position of the laser carrier frequency and the modulation sidebands in the acetyl absorption profile (solid line presents the gas amplitude absorption $\alpha/2$, while the dashed l presents the refractive index $\Delta nk_0$) for two utilized demodulation modes. Solid arrows correspond the wideband demodulation and dashed arrows to the narrowband demodulation.

The experimentally observed frequency response of this demodulation mode (i.e., frequency dependence of the phase-to-amplitude transfer function) is presented in Figure by curve 1. This dependence was obtained in the OEO open-loop configuration with sinusoidal modulation signal of 0 dBm, applied to the electrooptic phase modulator, a for the incident laser power of 5 mW.

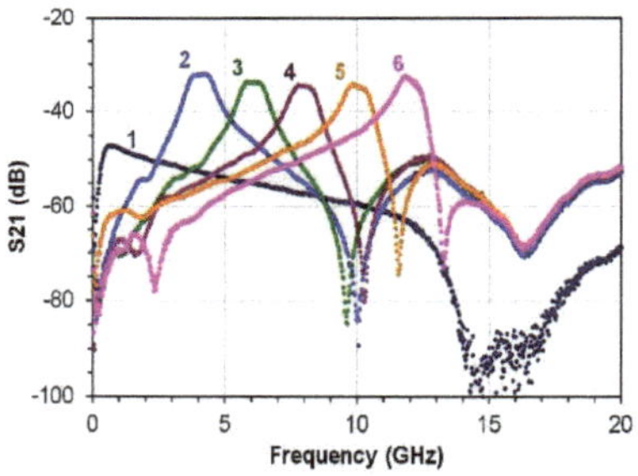


(**a**)

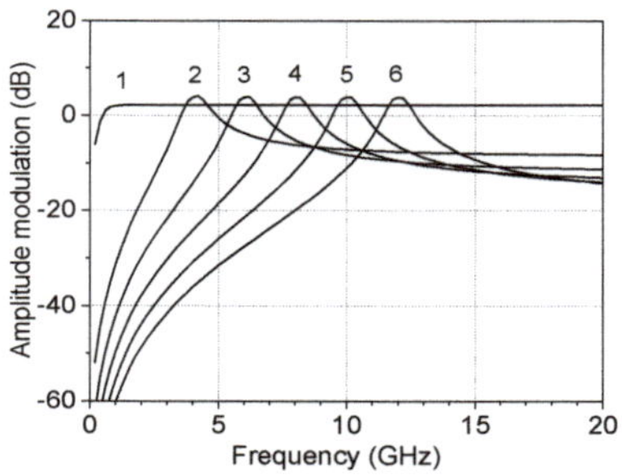


(**b**)

**Figure 4.** (**a**) Experimentally observed frequency transfer function of the demodulator for the bro band (curve 1) and the narrowband operation modes at different detuning frequencies approxima equal to 0.25 (curve 1), 4 (2), 6 (3), 8 (4), 10 (5), and 12 GHz (6). (**b**) Theoretical simulations obtair for similar detuning of the laser wavelength from the center of the absorption line with a Lorentz profile, a maximum absorbance of 3, and an FWHM spectral width of 0.5 GHz.

In the second, the narrowband or selective demodulation mode, the phase-to-amplitude conversion is performed by suppression (and, partially, by change of the phase) of one sideband of the initial phase modulated optical wave. To do this, the laser wavelength is tuned outside of the absorption line, to have one sideband of the modulated signal overlapped with the acetylene absorption/dispersion line (Figure 3).

The experimentally observed frequency dependences of the transfer functions for this demodulation mode are presented in Figure 4a (curves 2–6) for different detuning of the laser wavelength. These dependencies are, clearly, similar to the frequency response of a tunable band-pass filter with the central frequency, corresponding to the laser wavelength detuning from the maximum of the acetylene cell absorption. The FWHM spectral width of this effective filter is determined by the utilized acetylene absorption line.

Note that the frequency responses of the phase modulator, photodetector, and electronic amplifier also contribute to the experimental frequency dependence of the open-loop transfer function. In particular, the cut-off frequency (~13 GHz) in the dependence on the broadband demodulation mode relates to the frequency response of the photodiode with the electronic amplifier, and the slope to the modulator frequency response.

We have performed a theoretical simulation of the demodulation processes. In this analysis, the initial harmonic phase modulation was represented as the sum of a strong carrier frequency component and two weak sidebands shifted up and down by the modulation frequency $\Omega$. The contribution of the higher-order sideband harmonics in phase modulation was neglected in the utilized small-signal approximation [15]. The influence of the acetylene cell was taken into account by multiplying different incident spectral components of the light (the carrier frequency and the two sidebands) by the corresponding amplitude/phase transmittance coefficient of the cell. The absorption line utilized in our simplified simulation was accepted to be of the Lorentzian shape with the complex amplitude transmission coefficient:

$$T = \exp\left\{ \frac{\alpha_0 L}{2} \left[ \frac{1}{1+\left(\frac{\Delta\Omega}{\delta\Omega}\right)^2} + \frac{i\frac{\Delta\Omega}{\delta\Omega}}{1+\left(\frac{\Delta\Omega}{\delta\Omega}\right)^2} \right] \right\} \tag{2}$$

where $\alpha_0 L$ is the maximum line absorbance, $\Delta\Omega$ is the deviation of the light angular frequency from the center of the absorption line, and $\delta\Omega$ is half of the spectral FWHM width of the line. The intensity modulation amplitude with the $\Omega$ modulation frequency was evaluated from the interference of the output optical carrier and sidebands at the quadratic photodetector.

Figure 4b shows the frequency dependences of the output intensity modulation, evaluated for the maximum cell absorbance $\alpha_0 L = 3$, the spectral FWHM line width of 0.5 GHz, and the detuning of the carrier frequency by 0.25, 4, 6, 8, 10, and 12 GHz from the center of the absorption line. One can see a reasonable qualitative agreement with the obtained experimental dependences, presented in Figure 4a.

## 4. OEO Operation

Now, we go to a discussion of the operation of the closed-loop OEO configuration. Here, first of all, we point to a rather low value of demodulation (about −30 dB), observed even in the maxima of the open-loop transfer functions (Figure 4a), that was due to the low efficiency of electronic–optic–electronic transformation. For this reason, a rather high additional electronic amplification (>30 dB) was required to observe generation in a close-loop configuration of this OEO.

In the OEO with the broadband demodulation, insertion of an additional narrowband pass electronic filter is, obviously, necessary for an acceptable spectral purity generation, i.e., for a low mode-number or a single-frequency operation. Without it, the OEO in question will generate all possible resonance modes of the optical fiber delay line. They are separated by the off-set frequency $\Delta f = c/nL$, where $c$ is the speed of light in vacuum, $L$ is the fiber length, and $n$ is the effective refractive index of the fiber for the utilized laser wavelength. the reported experiments with $L \approx 2$ km, $\Delta f$ is about 0.1 MHz. These generated modes spread over the total bandwidth of the open-loop arrangement frequency range (~13 GI see Figure 4a), which is determined by characteristics of the utilized devices: the ph modulator, the photodiode, and the electronic amplifier.

To reduce the number of the generated modes in this demodulation mode, we ha additionally inserted a tunable narrowband YIG film spin-wave filter into the electro path of the loop after the amplifier; see Figure 1a [16,17]. The utilized microwave sp wave filter was fabricated in our group in the form a YIG film delay line [11,12]. consists of 5 μm-thick, 2 mm-wide, and 2 cm-long YIG film waveguide and conventio excitation/reception microstrip structures for the spin waves. An important characteris feature of such a filter is a rather weak dependence of the delay time on the operati frequency. The central frequency can be easily tuned by variation of the bias magne field. As a result, the phase noise of the OEOs, based on the YIG film filters, demonstra relatively weak dependence on the generation frequency [13], in contrast to the resonat based oscillators. Note that, recently, OEOs with the spin-wave filters are called "magno optoelectronic oscillators" [14].

Typical frequency transfer function of the open-loop configuration with a YIG fi spin-wave filter is shown in Figure 5a. The corresponding spectra of the generated sign experimentally observed for different tuning of the YIG filter, are presented in Figure 5b segment of one of these spectra is presented in Figure 5c with higher resolution. It sho that one of the delay line modes is generated mainly, with the suppression of the side mo by ≈40 dB. Note that the FWHM spectral width (Figure 5a) of the utilized microwa spin-wave filter ≈50 MHz [16] was significantly larger than the intermode spacing the optical fiber delay line (≈0.1 MHz). Such FWHM filter spectral width was sufficie however, to reach the single-mode generation. Similar single-mode spectra of the generat signals were observed for the entire tuning range (4–12 GHz) of the spin-wave filter. N that to control the generation frequency in this case, one has to tune the spin-wave filter the electronic path of the configuration only.

We have also investigated similar OEO operations with the narrowband demodulati via the acetylene cell. In this case, the YIG filter was not utilized, and the acetylene refere cell was used as the only frequency-selective element. This configuration allows us optically tune the generation frequency by changing the laser wavelength. The range of frequency tuning in this case is restricted by the frequency bandwidth of the modula photodiode, and electronic amplifier.

Typical generation spectra obtained for different tuning of the laser in a broad (0–20 GI frequency range are shown in Figure 6a. From one of such spectra, presented in Figure 6b w higher resolution, one can see the mode structure of the oscillations much better. It is cl that the additional, not central, generated modes (side modes) are more intensive than th observed with the microwave spin-wave filter (Figure 5c). The side modes suppression the narrowband demodulator operation mode was about 20 dB, while for the 50 MHz Y filter it was two orders of magnitude better (about 40 dB). It is clear that the spectral wi of the acetylene reference cell (500 MHz) is insufficient to ensure a high-purity single-mo OEO operation. One can see that the total spectral width of the generated signal is, howe significantly more narrow than the demodulator bandwidth (Figure 4).

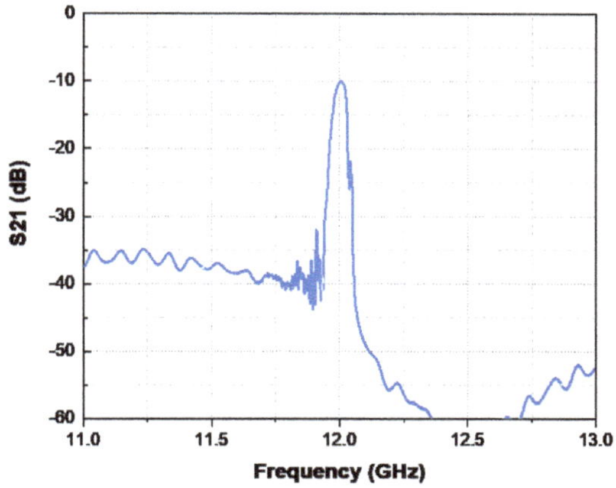


(**a**)

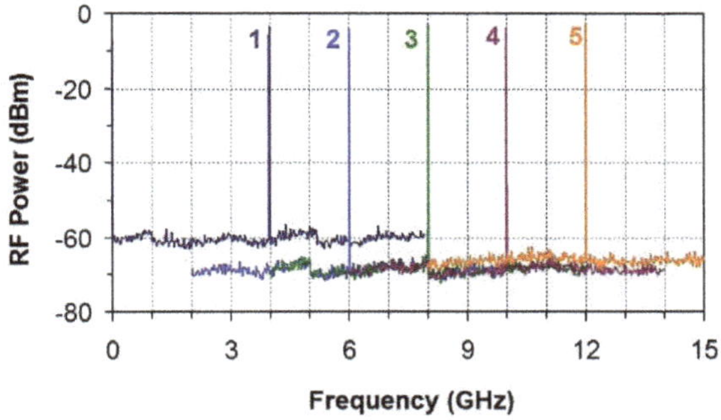


(**b**)

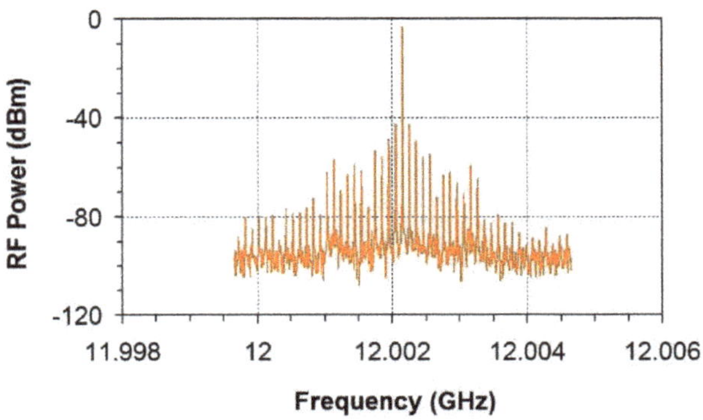


(**c**)

**Figure 5.** (**a**) Typical frequency transfer function of the open-loop configuration with the YIG film spin-wave filter, tuned at 12 GHz. (**b**) Superimposed generation spectra, obtained for OEO configuration with the broadband demodulation for different settings of the YIG filter. (**c**) Similar generation spectrum observed for tuning of the YIG film filter to ≈12 GHz and presented with higher resolution.

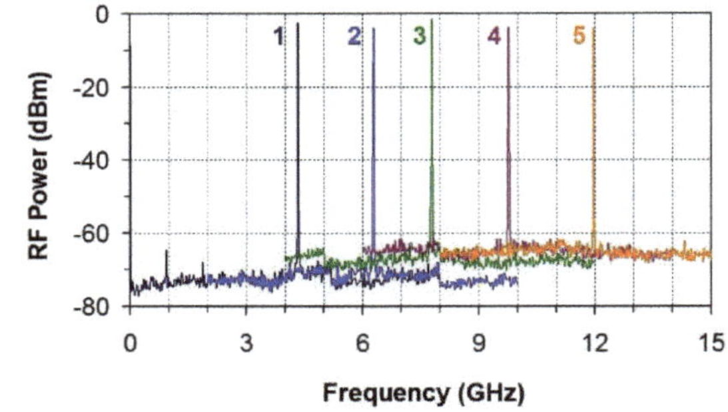


(a)

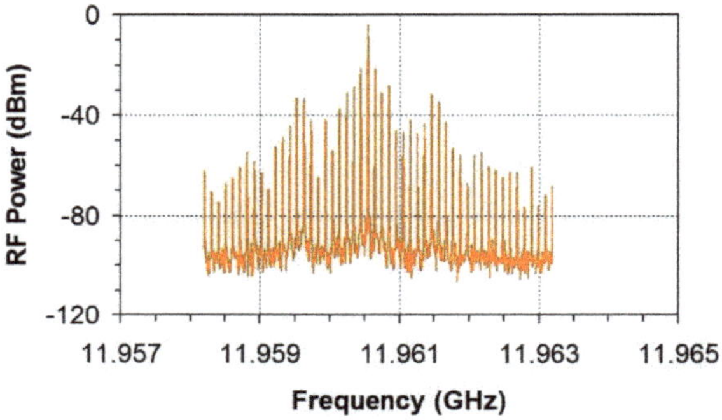


(b)

**Figure 6.** (**a**) Superimposed generation spectra obtained for the narrowband demodulation m
with different detuning of the carrier wavelength from the acetylene absorption line and (**b**)
generation spectrum, obtained with 12 GHz detuning and shown with higher resolution.

It was noted above that the spectral selectivity of the narrowband demodulator mc is governed by the inhomogeneous spectral width of the utilized acetylene absorpti line. In fact, the homogeneously broadened spectral width of low-pressure acetylene approximately an order of magnitude smaller [10]. Such narrow spectral transmission li can be formed by local saturation of the gas optical absorption via the process of a spect hole burning. Additional, more detailed analysis and investigations are, obviously, need to understand how such artificially induced narrow transmission peaks can be used improving the frequency purity of OEOs with the acetylene-based demodulation.

One can expect that simultaneous utilization of the narrowband demodulation mc and the spin-wave YIG filter can further increase the suppression of the side modes. T improvement cannot, however, be very impressive (as compared to that presented Figure 5c) since the difference in the spectral widths of the YIG filter and the narrowba acetylene-based demodulator is quite large. Additionally, tuning the generation freque in such a configuration is much more complicated; it, obviously, needs a synchronc change of the laser wavelength and the resonance frequency of the YIG filter.

As was mentioned above, similar spectra of the generated signals were observed at corresponding tuning of the laser frequency in the whole investigated range (4–12 GHz); Figure 6a. Note that for small (<10 GHz) detuning of the laser, higher harmonics of the m generation frequency were also observed; see Figure 7. We attribute them to an increas number of the modulation harmonics that appear even for linear phase modulation, as w as to the nonlinearity of the demodulation, photodiode, and electronic amplifier.

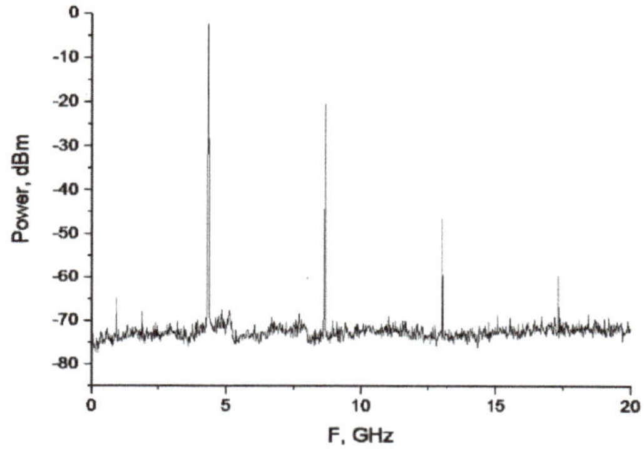


**Figure 7.** Higher harmonics in the generation spectrum that were observed with the narrowband demodulation without a YIG filter for approximately 4 GHz detuning of the laser frequency from the acetylene absorption line.

**5. OEO Noise Characteristics**

Figure 8 presents typical dependences of the generated monochromatic signal phase noise as a function of the frequency offset for the broadband demodulation mode with the tunable microwave YIG film filter in the electronic path of the OEO loop. Similar dependences, with the noise floor from –140 dBc/Hz to –130 dBc/Hz, were obtained for all generation frequencies in the range of the YIG film filter tuning (4–12 GHz). The phase noise at the 10 kHz offset was about −122 dBc/Hz. The noise peaks, observed near 100 kHz (i.e., near the intermode distance of the fiber optic delay line), are due to the contribution of the sideband modes. It is the so-called supermode noise, which is inherent to a single-loop OEO configuration.

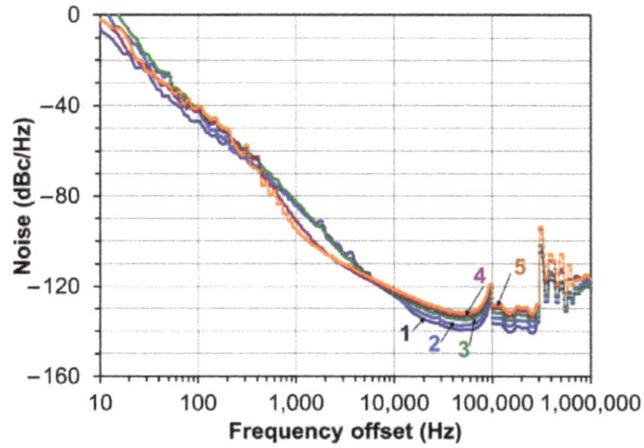


**Figure 8.** Phase noise in the OEO with the broadband demodulation and tunable YIG filter as a function of the frequency offset at different generation frequencies: curve 1–4, 2–6, 3–8, 4–10, and 5–12 GHz.

The phase noise in the OEO with the narrowband demodulation had similar dependence versus frequency offset (Figure 9). The noise floor was, however, about –132 dBc/Hz for all generation frequencies. A more significant supermode noise contribution is associated with a lower efficiency of the sideband suppression and worse generation frequency purity. Various approaches, such as multiple loops [18], can be utilized for the reduction in these noise peaks and for improving the frequency purity.

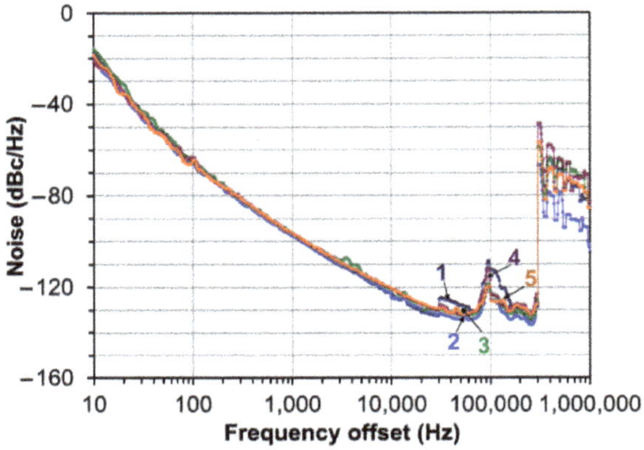


**Figure 9.** Phase noise in OEO with narrowband demodulation as a function of the frequency offse different generation frequencies: curve 1–4, 2–6, 3–8, 4–10, and 5–12 GHz.

The theoretical estimation of the oscillator phase noise power spectral density w done using the Yao–Maleki equation [19]:

$$S(f') = \frac{\delta}{\left(2 - \frac{\delta}{\tau}\right) - 2\sqrt{1 - \frac{\delta}{\tau}}\cos(2\pi f'\tau)}$$

where $\tau$ is the delay time of the optical fiber line, $f'$ is the frequency offset, and $\delta$ is input noise-to-signal ratio, i.e., the ratio of the input noise power density of the oscilla to the power of oscillations. The model, reported in [19], takes into account the followi sources of white noise: the thermal noise of the amplifier, the shot noise of the photodio and the effective relative intensity noise (RIN) of the laser. The effective RIN also inclu demodulated laser phase noise. It results in an increase in the phase noise with a slope $f^{-2}$ for the frequency offset smaller than Leeson frequency $f_L = 1/\pi\tau$.

In the OEO under consideration with the optical fiber delay line 2 km long, the va of $f_L$ equals $\approx$ 32 kHz. Indeed, the experimentally observed phase noise spectrum obta $f^{-2}$ slope at frequencies lower than this value (Figure 10). For smaller offset frequenc the experimental phase noise demonstrates the slope $f^{-3}$ from a frequency offset of ab 2.8 kHz. One can assume that this is due to the flicker noise (i.e., $1/f$ noise) of the utili microwave amplifier and the photodetector [20,21].

In the standard Yao–Maleki model, the noise-to-signal ratio of white noise can calculated in the following way:

$$\delta_{white} = \frac{\rho G_A^2}{P_{osc}}$$

where $\rho$ is the total power density of white noise, $G_A$ is the voltage gain of the microwa amplifier, and $P_{osc}$ is the oscillation power. The presence of flicker noise changes the to noise-to-signal ratio of the OEO generation, as follows:

$$\delta(f) = \delta_{white} + \frac{\delta_{1/f}}{f}$$

Typical results of the numerical simulation are shown by dashed lines in Figure From a comparison of the experimental results and simulations, we found that the noise signal ratio of the $1/f$ noise is $\delta_{1/f}$ = 5 $10^{-10}$.

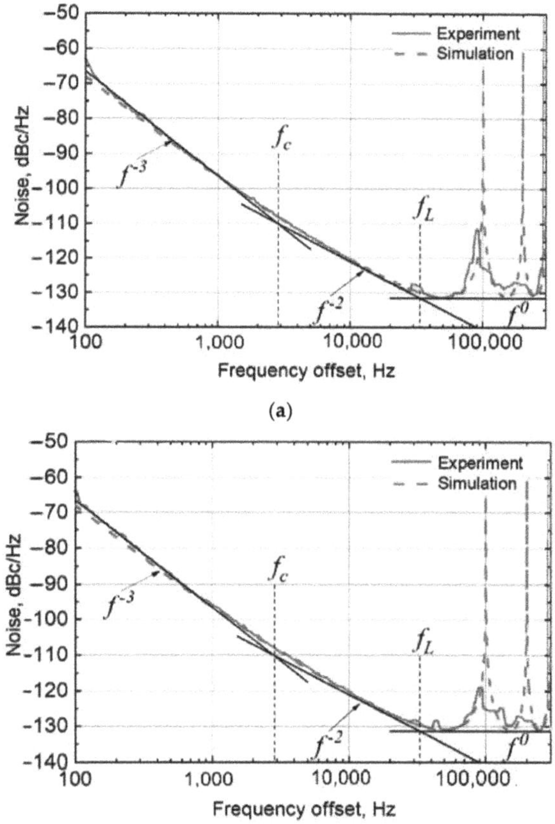


(b)

**Figure 10.** Phase noise of the OEO with narrowband demodulation as a function of the frequency offset for a generation frequency of 10 (**a**) and 12 GHz (**b**).

It is important to mention that the experimentally observed phase noise floor was approximately 10 dB below that, which was reported earlier for other configurations of OEOs with phase electrooptic modulators [7]. Unlike our experiments, the authors utilized the phase-shifted fiber Bragg grating as a narrowband optical filter. The phase noise floor, reported in [7], was determined by the residual phase noise of the active devices in the OEO loop, such as the thermal noise of the amplifier, the shot noise of the photodiode, and the RIN of the laser. The improvement in the noise performance in our experiments was reached by using a low phase noise laser source and a high saturation current photodiode, capable for the detection of high optical power.

The interested reader can find a more extended comparison of operation characteristics of different earlier investigated configurations of the OEOs in Table 1 of a recent review paper [1].

## 6. Conclusions

Summarizing, we have presented the results of an experimental study of the original configuration of tunable OEO operating in the 1.5 μm communication wavelength range. Unlike conventional OEOs with amplitude electrooptic modulators, our configuration uses the optical phase modulation and the original demodulation technique, where phase modulation is transformed into amplitude one via narrow resonance line of a referer acetylene cell.

Two regimes of the OEO operation, corresponding to different demodulation mo with a low-pressure acetylene cell, were studied. The wideband phase-to-amplitu demodulation was ensured by tuning the laser wavelength to one side of the acetyle absorption line. In the narrowband demodulation mode, one of the modulated sig sidebands was tuned to the acetylene absorption line. The microwave spin-wave Y film filter was used to reach single-frequency operation, to improve generation frequer purity, and to provide electronic tuning in case of the wideband demodulation mode. the narrowband demodulation mode, additional electronic filtering was not used. T generation frequency was controlled in this case optically by detuning the laser waveleng Various techniques, such as multiple loops, can be applied to suppress higher sideba modes and improve frequency purity in both investigated OEO configurations.

The improved phase noise characteristics (the phase noise floor of −132 dBc/Hz) ha been demonstrated. This noise floor is approximately 10 dB below that reported earlier i similar OEO with a phase electrooptic modulator and phase-shifted fiber Bragg grating a narrowband optical filter. The improvement in the noise performance of our OEO configurat was achieved by using a low phase noise laser source and a high-saturation photodiode.

From the point of view of the authors, the most promising areas of practical ap cations of the proposed OEO configurations are radars and telecommunication syste A more detailed analysis of all particular requirements (power, frequency range, etc.), obviously, needed for a more precise specification.

**Author Contributions:** Conceptualization, A.B.U., S.S. and A.S.; methodology, V.L., P.A. and A.B validation, A.B.U., S.S. and A.S.; formal analysis, I.T., A.N., A.V., A.B.U. and S.S.; investigation, V.L., A P.A., I.I., I.T., A.N., A.B.U., S.S. and A.S.; writing—original draft preparation, S.S.; writing—review a editing, A.B.U., S.S. and A.S. All authors have read and agreed to the published version of the manusc

**Funding:** The research in SPbETU was funded by the Ministry of Science and Higher Educatior the Russian Federation, grant number FSEE-2020-0005.

**Institutional Review Board Statement:** Not applicable.

**Informed Consent Statement:** Not applicable.

**Data Availability Statement:** The data are available from the first author and the correspond author upon reasonable request.

**Conflicts of Interest:** The authors declare no conflict of interest.

## References

1. Chembo, Y.K.; Brunner, D.; Jacquot, M.; Larger, L. Optoelectronic oscillators with time-delayed feedback. *Rev. Mod. Phys.* **2** *91*, 035006. [CrossRef]
2. Kondrashov, A.V.; Ustinov, A.B. Self-generation of Möbius solitons and chaotic waveforms in magnonic-optoelectronic oscillat under simultaneous action of optic and magnonic nonlinearities. *J. Appl. Phys.* **2022**, *132*, 173907. [CrossRef]
3. Tang, J.; Hao, T.; Li, W.; Domenech, D.; Baños, R.; Muñoz, P.; Zhu, N.; Capmany, J.; Li, M. Integrated optoelectronic oscillator. *Express* **2018**, *26*, 12257. [CrossRef] [PubMed]
4. Brunetti, G.; Armenise, M.N.; Ciminelli, C. Chip-Scaled Ka-Band Photonic Linearly Chirped Microwave Waveform Genera *Front. Phys.* **2022**, *10*, 785650. [CrossRef]
5. Strekalov, D.; Aveline, D.; Yu, N.; Thompson, R.; Matsko, A.; Maleki, L. Stabilizing an optoelectronic microwave oscillator w photonic filters. *J. Light. Technol.* **2003**, *21*, 3052–3061. [CrossRef]
6. Xie, X.; Zhang, C.; Sun, T.; Guo, P.; Zhu, X.; Zhu, L.; Hu, W.; Chen, Z. Wideband tunable optoelectronic oscillator based on a ph modulator and a tunable optical filter. *Opt. Lett.* **2013**, *38*, 655–657. [CrossRef] [PubMed]
7. Li, W.; Yao, J. A Wideband Frequency Tunable Optoelectronic Oscillator Incorporating a Tunable Microwave Photonic Fi Based on Phase-Modulation to Intensity-Modulation Conversion Using a Phase-Shifted Fiber Bragg Grating. *IEEE Trans. Mic Theory Tech.* **2012**, *60*, 1735–1742. [CrossRef]
8. Swann, W.C.; Gilbert, S.L. Pressure-induced shift and broadening of 1510–1540-nm acetylene wavelength calibration lines. *J. Soc. Am. B* **2000**, *17*, 1263–1270. [CrossRef]

Casillas, N.; Stepanov, S.; Ocegueda, M.; Hernández, E. Utilizing phase memory of a two-level quantum system for adaptive homodyne detection of optical phase modulation. *J. Opt.* **2019**, *21*, 045201. [CrossRef]

Casillas, N.; Stepanov, S.; Ocegueda, M.; Hernández, E. Self-referencing mW-scale detection of sub-ns optical phase modulation in acetylene at 1530 nm. *Appl. Opt.* **2019**, *58*, 6495–6503. [CrossRef] [PubMed]

Ustinov, A.B.; Drozdovskii, A.V.; A Nikitin, A.; A Kalinikos, B. Spin-wave band-pass filters based on yttrium iron garnet films for tunable microwave photonic oscillators. *J. Physics: Conf. Ser.* **2015**, *661*, 012058. [CrossRef]

Ustinov, A.B.; Nikitin, A.A.; Kalinikos, B.A. Electronically tunable spin-wave optoelectronic microwave oscillator. *Tech. Phys.* **2015**, *60*, 1392–1396. [CrossRef]

Ustinov, A.B.; Nikitin, A.A.; Kalinikos, B.A. Magnetically Tunable Microwave Spin-Wave Photonic Oscillator. *IEEE Magn. Lett.* **2015**, *6*, 3500704. [CrossRef]

Xiong, Y.; Zhang, Z.; Li, Y.; Hammami, M.; Sklenar, J.; Alahmed, L.; Li, P.; Sebastian, T.; Qu, H.; Hoffmann, A.; et al. Experimental parameters, combined dynamics, and nonlinearity of a magnonic-opto-electronic oscillator (MOEO). *Rev. Sci. Instruments* **2020**, *91*, 125105. [CrossRef] [PubMed]

Palací, J.; Villanueva, G.; Galán, J.; Martí, J.; Vidal, B. Single bandpass photonic microwave filter based on a notch ring resonator. *IEEE Photonics Technol. Lett.* **2010**, *22*, 1276–1278. [CrossRef]

Ustinov, A.B.; A Nikitin, A.; Lebedev, V.V.; A Serebrennikov, A.; Shamray, A.V.; Kondrashov, A.V.; A Kalinikos, B. A Low Phase Noise Tunable Microwave Spin Wave Optoelectronic Oscillator. *J. Physics Conf. Ser.* **2018**, *1038*, 012033. [CrossRef]

Ustinov, A.B.; Kondrashov, A.V.; A Nikitin, A.; Lebedev, V.V.; Petrov, A.N.; Shamrai, A.; A Kalinikos, B. A tunable spin wave photonic generator with improved phase noise characteristics. *J. Physics: Conf. Ser.* **2019**, *1326*, 012015. [CrossRef]

Yao, X.S.; Maleki, L. Multi-loop optoelectronic oscillator. *IEEE J. Quant. Electron.* **2000**, *36*, 79. [CrossRef]

Yao, X.S.; Maleki, L. Optoelectronic microwave oscillator. *J. Opt. Soc. Am. B* **1996**, *13*, 1725–1735. [CrossRef]

Rubiola, E.; Salik, E.; Yu, N.; Maleki, L. Flicker noise in high-speed pin photodiodes. *IEEE Trans. Microw. Theory Tech.* **2006**, *54*, 816–820. [CrossRef]

Rubiola, E. *Phase Noise and Frequency Stability in Oscillators*; Cambridge University Press: Cambridge, UK, 2008; 228p.

photonics

MDPI

*Communication*

# Simultaneous Manipulation of the Temporal and Spatial Behaviors of Nanosecond Laser Based on Hybrid Q-Switching

**Haoxi Yang [1], Yuanji Li [1,2,*], Wenrong Wang [1], Jinxia Feng [1,2] and Kuanshou Zhang [1,2,*]**

[1] State Key Laboratory of Quantum Optics and Quantum Optics Devices, Institute of Opto-Electronics, Sha University, Taiyuan 030006, China
[2] Collaborative Innovation Center of Extreme Optics, Shanxi University, Taiyuan 030006, China
* Correspondence: liyuanji@sxu.edu.cn (Y.L.); kuanshou@sxu.edu.cn (K.Z.)

**Abstract:** A hybrid Q-switching method based on a special-shaped saturated absorber was propo for simultaneous manipulation of the temporal and spatial behaviors of a solid-state pulse la The temporal–spatial rate equation model of the laser was given and used to optimize the desi parameter of the saturated absorber. Best spatial intensity homogenization performance can expected using an active-passive hybrid Q-switched laser, comprising a Pockels cell and a cylin Cr:YAG crystal with one end cut as a spherical concave surface. The optimized laser pulse wi could be narrowed to 2.39 ns and the laser radial intensity distribution became quasi-super-Gauss distribution with a radial intensity distribution ratio of 0.91, while that for the Gaussian beam v 0.84. In principle, the laser coherence can be maintained, and the laser spatial intensity distribut can be kept in a long propagation distance.

**Keywords:** hybrid Q-switching; deformed cylinder saturated absorber; laser intensity distributi manipulation; laser pulse width reduction



**Citation:** Yang, H.; Li, Y.; Wang, W.; Feng, J.; Zhang, K. Simultaneous Manipulation of the Temporal and Spatial Behaviors of Nanosecond Laser Based on Hybrid Q-Switching. *Photonics* **2023**, *10*, 227. https://doi.org/10.3390/photonics10020227

Received: 29 January 2023
Revised: 16 February 2023
Accepted: 18 February 2023
Published: 20 February 2023



## 1. Introduction

Solid-state pulse lasers with narrow pulse width (PW) and uniform intensity dis bution have found a great many applications in a wide range of fields, including la power amplification, laser material processing, imaging, high efficiency fiber injection, a lithography. In particular, for the efficient generation of ultra-high energy lasers in mas oscillator power amplifier (MOPA) configuration [1–7], which can be further applied deep ultraviolet laser generation, long distance laser ranging and laser communicati etc. [8–10], high quality seed lasers, with the properties of uniform intensity distribution a good coherence, are strongly needed to achieve efficient utilization of pump energy insi the side-pumped amplifier modules and coherent amplification. For the temporal sho ening of the laser PW, there were many mature techniques, for example the short-cav lasers employing fast electro-optical (EO) Q-switches or deflectors [11–13], the miniatu passive Q-switched lasers based on saturated absorbers [14–16], and the mode-locki lasers [17–19]. However, for the spatial homogenization of the laser energy, for instar getting the top-hat intensity distribution and super-Gaussian distribution, many practi difficulties exist [20]. The external cavity optical shaping systems based on aspheric le combination [21], microlens array [22,23], and diffractive optical elements [24–26] al the laser intensity distribution by regulating the propagation direction of each part the laser. Hence, the laser spatial coherence was usually weakened, or even destroy in principle; the top-hat intensity distribution or super-Gaussian distribution can only maintained in a centimeter level distance, and an excess power loss was inevitable. Besic that, the intracavity beam shaping method was also proposed, mainly depending o periodically swinging cavity mirror. Similarly, the laser was operating in multi-mode w poor coherence and stability.

The dual loss modulation method simultaneously employing two Q-switches, usually an active Q-switch and a saturated absorber, is another popular method for laser PW reduction [27–31]. The temporal behaviors of the laser pulse, both PW and the pulse shape symmetry, can be improved comparing with the EO Q-switched laser for the reason that the rising and falling edges of laser pulse are reduced by saturation absorption effect. In comparison with the passive Q-switched laser with a single saturated absorber, the laser energy fluctuation and time jitter are also optimized since the buildup of laser pulse is dominated by the active Q-switching.

Up to now, an effective method enabling simultaneous manipulation of the temporal and spatial behaviors of nanosecond laser, without the expense of complex structure and weak coherence, is still lacking. There is also no previous research relative to the effect of spatial saturation absorption distribution on the laser transverse mode behavior. In this paper, we propose a hybrid Q-switching method based on the deformed saturated absorber for simultaneous manipulation of the temporal and spatial behaviors of a solid-state pulse laser. The temporal–spatial theoretical model of the laser was given and the optimized design of the laser protocol was demonstrated.

## 2. Temporal–Spatial Rate Equation Model of Hybrid Q-Switched Laser

Figure 1 shows a schematic diagram of the model for a 1.06 μm hybrid Q-switched laser. Pump light with a single photon energy of $h\nu_p$ was coupled into the gain medium with an average radius of $\omega_p$. The gain medium was an *a*-cut Nd:$YVO_4$ crystal with dimensions of $2R_g \times 2R_g \times l$. The resonator consisted of five mirrors, an electro-optic (EO) Q-switch and a Cr:YAG saturated absorber. $M_1$, $M_2$ and $M_3$ were all high reflection coated at 1064 nm, while $M_1$ and $M_2$ were plane mirrors, and $M_3$ was a convex mirror with a curvature radius of Roc1. The output coupler (OC) with a curvature radius of Roc2 had an output coupling reflection of $R$ at 1064 nm. TFP was a 45° thin film polarizer (TFP) enabling *p*-polarization light high transmission and *s*-polarization light high reflection passing. A quarter-wave plate (QWP) and an $RbTiOPO_4$ (RTP) Pockels cell with a quarter-wave voltage of $V_{\lambda/4}$ played the role of EO Q-switch, which was driven by a high-voltage driver with a high-level voltage of $V_{hl}$, high-level voltage duration of $t_{qs}$, and a rise time of $t_{qr}$. SA was a deformed cylinder Cr:YAG crystal; the cylinder part had a thickness of $l_{sa,0}$ and a radius of $R_{sa}$, whilst the concave end had a curvature radius of $RS_{sa}$. $L_1$ was a match lens with the same refractive index with Cr:YAG, while the curvature radius of the convex surface was also $RS_{sa}$. The use of match lens not only simplified the resonator design, but also compensated the space-dependent diffraction introduced by the deformed SA. The whole effective cavity length was $L_{c,eff}$, leading to laser mode radii of $\omega_g$ and $\omega_{sa}$ at gain medium and SA, respectively.

To study the temporal and spatial behaviors of the laser dynamically, the photon number density evolution should be evaluated in different spatial areas. Here, we defined a mesh criterion: on the cross-section of each crystal, a radial mesh was defined consisting of $m$ rings and $q$ angular segments, since both the laser and the crystals were axisymmetric, where $m$ and $q$ were positive integers. As shown in Figure 2, in the sector with a included angle of $\theta = 2\pi/q$ and a radial variation range of 0~$R_g$ ($R_{sa}$), the gain medium was gridded along the radial direction with a step size of $\omega_g/m$; similarly, the saturated absorber was also gridded with a step size of $\omega_{sa}/m$. Based on the fact that the intracavity laser had good spatial coherence, an approximation was made that the evolution of the stimulated emission photons generated in $j$th grid on the gain medium, i.e., in the radial variation range of $(j-1)\omega_g/m$~$j\omega_g/m$, was only related to the loss provided by the $j$th grid on the saturated absorber.

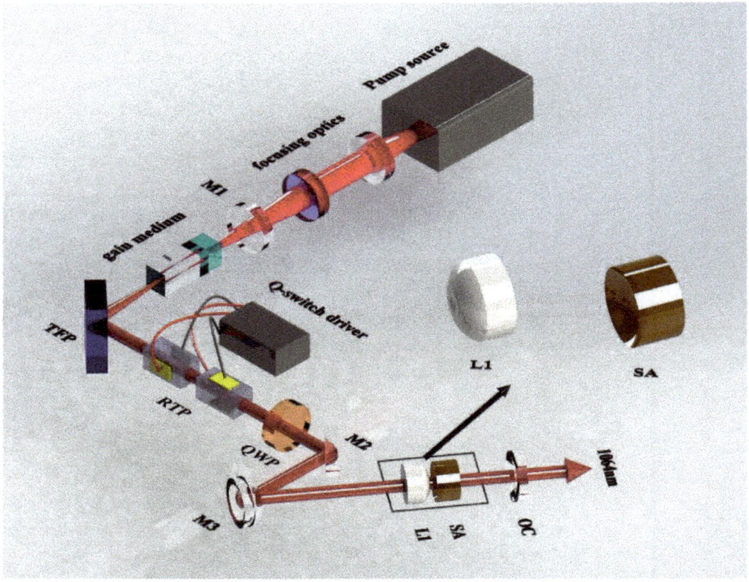


**Figure 1.** Schematic of the 1.06 μm hybrid Q-switched laser. (TFP: thin film polarizer, RTP: Rubidi titanate phosphate crystal, QWP: quarter-wave plate, SA: saturated absorber, OC: output couple

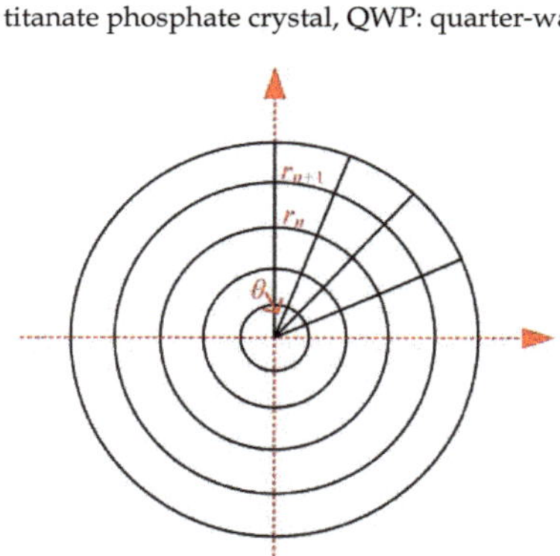

**Figure 2.** Mesh grids of the gain medium and saturated absorber.

For the saturated absorber with the shape shown in Figure 3a, the length of the sa rated absorber in the $j$th grid ($l_{sa,j}$) can be approximated to be a constant and represented

$$l_{sa,j} = l_{sa,0} + RS_{sa} - \sqrt{RS_{sa}^2 - (j\omega_{sa}/m)^2}.$$

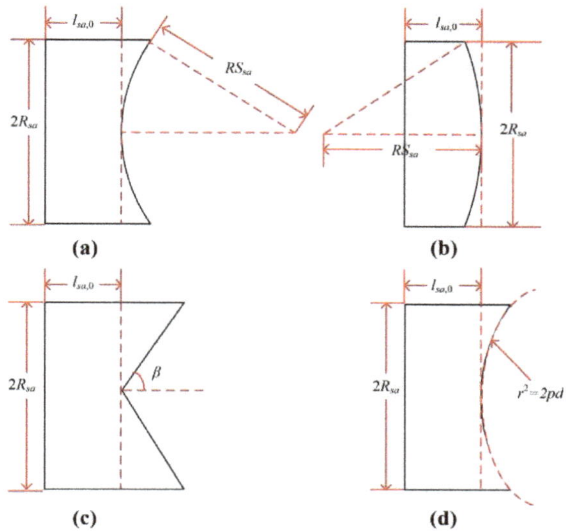


**Figure 3.** Four kinds of shapes of the saturated absorber. (**a**) the cylinder saturated absorber with one end cut as a spherical concave surface; (**b**) the cylinder saturated absorber with one end cut as a spherical convex surface; (**c**) the cylinder saturated absorber with one end cut as a conical surface; (**d**) the cylinder saturated absorber with one end cut as a parabolic surface.

When the shape of saturated absorber was changed to be a deformed cylinder with one end being a spherical convex surface, as shown in Figure 3b, the function expression came to be:

$$l_{sa,j} = l_{sa,0} - RS_{sa} + \sqrt{RS_{sa}^2 - (j\omega_{sa}/m)^2}, \tag{2}$$

For the cases that deformed end of cylinder were cut as a conical surface with an apex angle of $2\beta$, as shown in Figure 3c, and a paraboloid corresponding to a parabolic equation of $r^2 = 2pd$, as shown in Figure 3d, the function expression respectively became:

$$l_{sa,j} = l_{sa,0} + (j\omega_{sa}/m)/\tan(\beta), \tag{3}$$

$$l_{sa,j} = l_{sa,0} + (j\omega_{sa}/m)^2/(2p), \tag{4}$$

It is worth noting that in the following simulations associated with the latter three kinds of SA, the shape of the match lens $L_1$ should be changed to the complementation one.

The rate equations of the hybrid Q-switched laser in the $j$th grid can be written as:

$$d\phi_j/dt = 1/t_r\left[2N_j\sigma_{se}l\phi_{g,j} - 2N_{g,j}\sigma_g l_{sa,j}\phi_{s,j} - 2N_{e,j}\sigma_e l_{sa,j}\phi_{s,j} - \left[\ln(1/R) + \delta_0 + \delta_{QS}\right]\phi_j\right], \tag{5}$$

$$dN_j/dt = -N_j c\sigma_{se}\phi_{g,j} - N_j/\tau_f + P_p\eta_p/(\pi\omega_p^2 lh\upsilon_p), \tag{6}$$

$$dN_{g,j}/dt = -N_{g,j}\sigma_g c\phi_{s,j} + (N_{s0,j} - N_{g,j})/\tau_s, \tag{7}$$

$$N_{g,j} + N_{e,j} = N_{s0,j}, \tag{8}$$

where $N_j$ is the population inversion density in the $j$th grid of the gain medium. $N_{g,j}$, $N_{e,j}$ and $N_{s0,j}$ are the ground state ion density, excited state ion density and the total ion density in the $j$th grid of the saturated absorber, respectively. $\phi_j$, $\phi_{g,j}$ and $\phi_{s,j}$ are the average intracavity laser photon density corresponding to the $j$th grid of the gain medium, the laser photon density at the $j$th grid of the gain medium, and the laser photon density at the $j$th grid of the saturated absorber. $t_r = 2L_{c,eff}/c$ is the roundtrip time of the intracavity laser photon. $c$ is the light speed, $\sigma_{se}$ is the stimulated emission cross-section of the gain medium,

$\sigma_g$ and $\sigma_e$ are the ground state and excited state absorption cross-section of the saturat absorber. $\delta_0$ is the intrinsic loss, $\delta_{QS}$ is the time dependent loss of the EO Q-switch. $\tau_f$ is t fluorescence lifetime of Nd:YVO$_4$ crystal, $P_p$ is the incident pump power, $\eta_p = 1 - \exp(-$ is the pump laser absorption efficiency.

Assuming that the pump laser fits top-hat distribution and the intracavity laser Gaussian distribution, the initial conditions of the rate equations can be expressed as:

$$N_{0,j} = \frac{P_p T_p \eta_p \eta_f}{h\nu_p} \frac{\omega_g^2}{\omega_p^2} \frac{1}{\pi \omega_p^2 l},$$

$$\phi_{0,j} = N_{0,j} \frac{l}{c\tau_f} \frac{d\Omega}{4\pi},$$

$$N_{g0,j} = N_{s0,j},$$

where $d\Omega$ represents the solid angle of spontaneous emission that makes contribution to t stimulated emission, $T_0$ is the initial transmissivity of the cylinder saturated absorber cor sponding to a thickness of $l_{sa,0}$. $T_p$ is the pump pulse width. $\Delta t$ is the time delay betwe the rising edge of EO Q switching and the end of the falling edge of the pump pulse.

Moreover, $\delta_{QS}$ can be expressed by the equivalent transmission at the polarizer as:

$$\delta_{QS}(t) = \cos^2\left(\frac{\pi}{2}\frac{V_{hl}}{V_{\lambda/4}}\left(1 - e^{-\left(\frac{t}{t_{qr}}\right)^4}\right)^4 e^{-\left(\frac{t - t_{qs}/2}{t_{qs}}\right)^{400}}\right).$$

### 3. Simulation Results and Discussion

According to Refs [32–35], the function relations between $\sigma_{se}$ and the doped conc tration of the Nd:YVO$_4$ crystal ($D_c$), that between $\tau_f$ and $D_c$, as well as that between $\alpha$ a $D_c$ can be expressed as:

$$\sigma_{se} = \left(8.36075 + 2221.49 * D_c - 102818.3 * D_c^2 - 1490820 * D_c^3\right) * 10^{-23},$$

$$\tau_f = (-1650 * D_c + 106.67) * 10^{-6},$$

$$\alpha = (13.5 * D_c + 0.8) * 100,$$

Using Equations (1) and (5)–(15), as well as the parameter values given in Table 1, la evolution in each grid can be numerically simulated. Figure 4a shows the photon dens of the emitted pulse laser as a function of radial coordinates, i.e., the lower bound of grids, at the case of $V_{hl}/V_{\lambda/4} = 1.4$. It can be seen that the local pulse width along the cen of the cross-section of the laser is 2.76 ns. Besides that, there exists a falling edge as long 45 ns for the photons located on the wings of the cross-section. When an aperture is us to block the photons on the wings, almost 61.5% power loss will be introduced. Figure shows the results calculated using $V_{hl}/V_{\lambda/4} = 1$. One can find that better spatial intens homogenization is achieved. The local pulse width around the central part is narrowed be 2.39 ns, quasi-super-Gaussian intensity distribution can be realized with the expense 37% power loss.

**Table 1.** Parameter values used in the theoretical simulations.

| Parameter | Value | Parameter | Value |
|---|---|---|---|
| $l$ | 3 mm | $\omega_g$ | 230 μm |
| $L_{c,eff}$ | 120 mm | $\omega_{sa}$ | 2.5 mm |
| $l_{sa,0}$ | 2 mm | $\omega_p$ | 250 μm |
| $R_{sa}$ | 5.5 mm | $\delta_0$ | 0.01 |
| $RS_{sa}$ | 5.5 mm | $T_p$ | 260 μs |
| $N_{s0}$ | $3.5 \times 10^{-23}$ m$^{-3}$ | $P_p$ | 15 W |
| $h\nu_p$ | $2.4616 \times 10^{-19}$ J | $t_{qr}$ | 4 ns |
| $\sigma_g$ | $4.3 \times 10^{-22}$ m$^2$ | $t_{qs}$ | 160 ns |
| $\sigma_e$ | $8.2 \times 10^{-23}$ m$^2$ | $R$ | 0.5 |

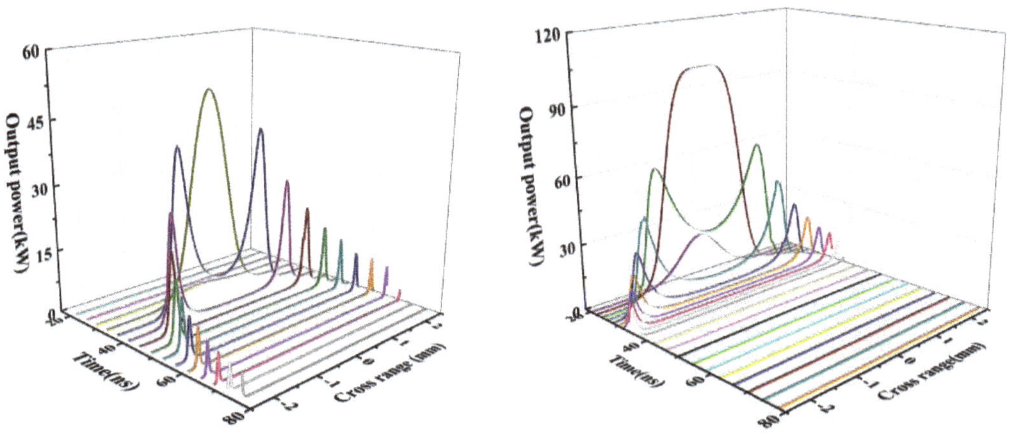


(**a**) (**b**)

**Figure 4.** Evolution of the radial intensity distribution of the laser output along with the time (**a**) $V_{hl}/V_{\lambda/4}$ = 1.4; (**b**) $V_{hl}/V_{\lambda/4}$ = 1. (Note: Different colors represent the radial intensity distribution of laser output at different times).

Considering the other three kinds of deformed cylinder saturated absorbers shown in Figure 2, namely Equation (2) with $R_{sa}$ = 3 mm, $RS_{sa}$ = 5.5 mm and $l_{sa,0}$ = 2 mm, Equation (3) with $\beta$ = 80° and $l_{sa,0}$ = 2 mm, as well as Equation (4) with $p$ = 75 and $l_{sa,0}$ = 2 mm, the simulations were repeated using these parameters and equations, respectively. Figure 5 shows the predicted radial intensity distributions at the peak of the laser output in the four cases using deformed cylinder saturated absorbers, and another case using normal cylinder saturated absorber. It is apparent that the predicted peak radial intensity distribution of the cylinder saturated absorber with one end cut as a spherical concave surface (Type I SA) is much closer to quasi-super-Gaussian intensity distribution, while the saturated absorbers with one end cut as a spherical convex surface (Type II SA) or a parabolic surface (Type III SA) exhibit weaker spatial homogenization results. Cut as a conical surface (Type IV SA) leads to an opposite effect that further spread the laser intensity to the periphery. To make a more intuitive comparison, Figure 6 shows the radial intensity distribution ratio, defined as the ratio of the central area under the radial intensity distribution curve to the whole area under the same curve, while the central area indicates the corresponding range at $1/e$ of the highest strength on the cross section. One can find that under the same thickness, Type I SA always leads to a radial intensity ratio around 0.9, which is 0.06 higher than that of Gaussian distribution and is insensitive to the variation of absorber thickness. In addition, the radial intensity distribution ratio of the laser output using Type II SA or Type III SA increases significantly with rising absorber thickness. When the absolute passive loss inside laser is enhanced enough, the radial intensity distribution ratios of these t cases approach 0.9 also. The radial intensity distribution ratio of the laser using Type SA shows opposite behavior. When the absorber thickness is 2.5 mm, a radial intens distribution ratio of 0.78 is achieved and is 0.06 lower than that of the Gaussian distributi

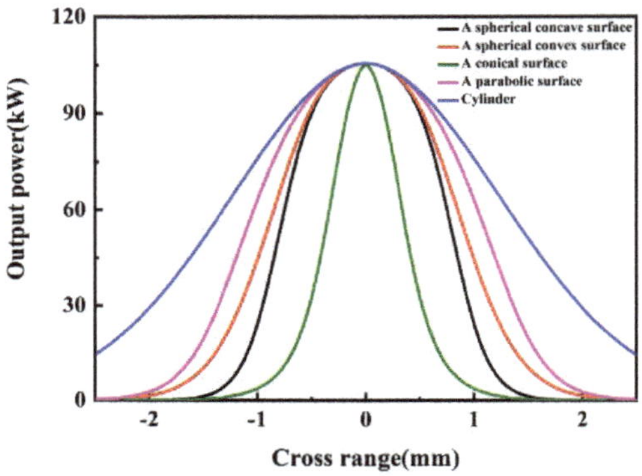


**Figure 5.** Predicted peak radial intensity distribution.

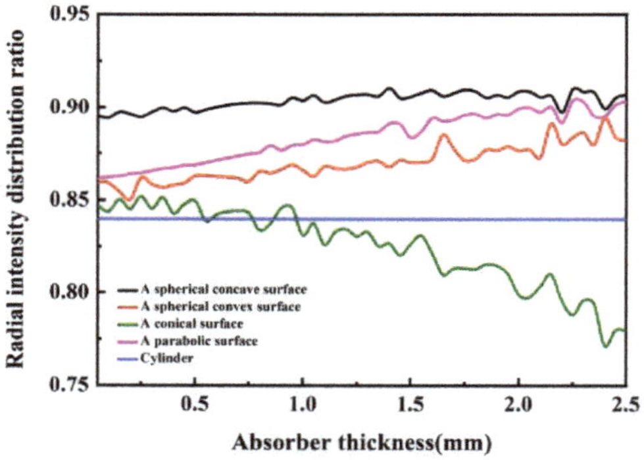


**Figure 6.** Radial intensity distribution ratio of the laser output using 5 different kinds saturated absorbers.

Pump power is the other parameter possessing global impact on the overall la behaviors. Figure 7 shows the predicted peak radial intensity distributions ratio of the la output in the five cases under different pump power when $l_{sa,0}$ is set as 2 mm. For Typ SA, the laser radial intensity distribution ratio still experiences little influence from pu power variation, and the maximum of 0.91 is achieved at a pump power of 15 W. Howe the laser radial intensity distribution ratios for Type II SA and Type III SA become wo when the pump power is raised up and quickly decayed to the same results of Gaussi distribution. This phenomenon can be understood when one notices that the satura absorbers can be saturated more easily at higher laser power. Type IV SA brings ou curious curve that contains several strong fluctuations, this may be due to the fact t the thickness of the saturated absorber changed sharply with radial position in this ca especially in the area around the center of the saturated absorber. In this case, the initial st of the laser intensity distribution can no longer be approximated as Gaussian distributi

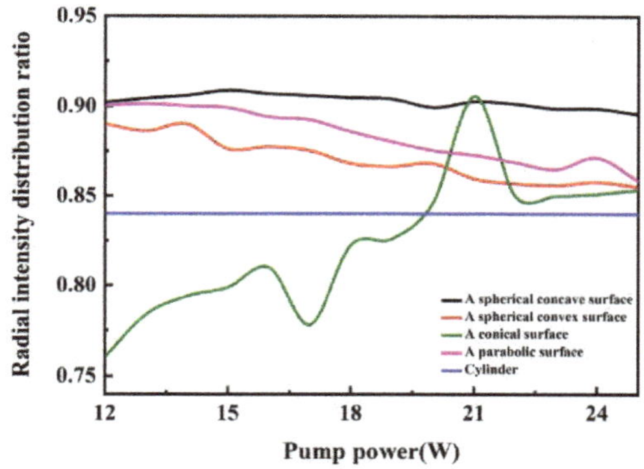


**Figure 7.** Radial intensity distribution ratio of the laser using five different kinds of saturated absorbers versus pump power.

Figure 8 shows the predicted laser peak radial intensity distribution ratio of the laser output in the five cases under different reflectance of the output coupling mirror, when $l_{sa,0}$ is set as 2 mm and $P_p$ is set as 15 W. The rough trends of the five curves are similar to that shown in Figure 7, for the reason that both higher pump power and higher output coupling reflectance will generally cause more intense intracavity laser oscillation. Note that best laser spatial homogenization performance, namely a radial intensity distribution ratio of 0.91, is obtained at a reflectance of 0.55.

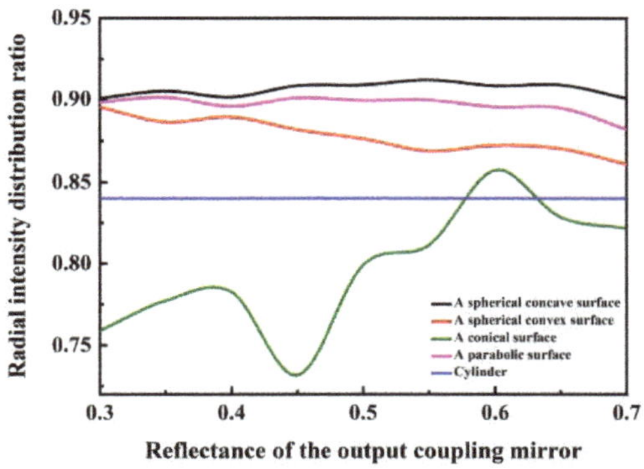


**Figure 8.** Radial intensity distribution ratio of the laser using five different kinds of saturated absorbers versus reflectance of the output coupling mirror.

## 4. Conclusions

In conclusion, a hybrid Q-switching method that relied on deformed cylinder saturated absorbers was proposed, for simultaneous manipulation of the temporal and spatial behaviors of a solid-state pulse laser. The temporal–spatial rate equation model of the laser was given and used to optimize the design parameter of the saturated absorber. The cylinder saturated absorber with one end cut as a concave surface enables the best spatial intensity homogenization. At the case of $R_{sa}$ = 5.5 mm, $RS_{sa}$ = 5.5 mm, $P_p$ = 15 W, $V_{hl}/V_{\lambda/4}$ = 1, the laser PW can be narrowed to be 2.39 ns, and the laser radial intensity distribution was altered to be quasi-super-Gaussian distribution with a radial intensity distribution ratio of 0.91. The nanosecond laser with quasi-uniform intensity distribution and good coherence paid the way for a high efficiency, high energy single mode MOPA. Better manipulation performance of the laser temporal and spatial behaviors can be expected when the parameters of the saturated absorber and resonator design are further optimized.

This kind of laser can be applied in laser amplification, laser material processing, imagi high efficiency fiber injection, and lithography.

**Author Contributions:** Conceptualization, H.Y. and Y.L.; methodology, K.Z.; software, H.Y. and Y validation, H.Y. and Y.L.; formal analysis, Y.L. and J.F.; investigation, H.Y. and W.W.; resources, K data curation, J.F. and W.W.; writing—original draft preparation, H.Y. and Y.L.; writing—review a editing, Y.L. and K.Z.; visualization, H.Y.; supervision, K.Z.; project administration, K.Z.; fundi acquisition, K.Z. All authors have read and agreed to the published version of the manuscript.

**Funding:** This research was funded by National Natural Science Foundation of China (NSF grant number 62175135, and Fundamental Research Program of Shanxi Province, grant num 202103021224025.

**Institutional Review Board Statement:** Not applicable.

**Informed Consent Statement:** Not applicable.

**Data Availability Statement:** The data underlying the results presented in this paper are not publi available at this time but may be obtained from the authors upon reasonable request.

**Conflicts of Interest:** The authors declare no conflict of interest.

## References

1. Zhao, B.-R.; Yao, B.-Q.; Qian, C.-P.; Liu, G.-Y.; Chen, Y.; Wang, R.-X.; Dai, T.-Y.; Duan, X.-M. 231 W dual-end-pumped Ho:Y MOPA system and its application to a mid-infrared ZGP OPO. *Opt. Lett.* **2018**, *43*, 5989. [CrossRef] [PubMed]
2. Liu, G.; Mi, S.; Yang, K.; Wei, D.; Li, J.; Yao, B.; Yang, C.; Dai, T.; Duan, X.; Tian, L. 161 W middle infrared ZnGeP MOPA syst pumped by 300 W-class Ho:YAG MOPA system. *Opt. Lett.* **2021**, *46*, 82–85. [CrossRef] [PubMed]
3. Yue, F.; Jambunathan, V.; David, S.P.; Mateos, X.; Šulc, J.; Smrž, M.; Mocek, T. Diode-pumped master oscillator power ampli system based on cryogenically cooled Tm:$Y_2O_3$ transparent ceramics. *Opt. Mater. Express* **2021**, *11*, 1489–1496. [CrossRef]
4. Markus, T.; Neumann, T.; Martino, A.; Abdalati, W.; Brunt, K.; Csatho, B.; Farrell, S.; Fricker, H.; Gardner, A.; Harding, D. The Cloud, and land Elevation Satellite-2 (ICESat-2): Science requirements, concept, and implementation. *Remote Sens. Environ.* **20** *190*, 260–273. [CrossRef]
5. Garcıa-López, J.H.; Aboites, V.; Kir'yanov, A.V.; Damzen, M.; Minassian, A. High repetition rate Q-switching of high pov Nd:$YVO_4$ slab laser. *Opt. Commun.* **2003**, *218*, 155–160. [CrossRef]
6. Pinto Robledo, V.J.; Lopez, G.; Espinosa, Y.M.; Pisarchik, A.N.; Jaimes Reátegui, R.; Aboites, V. Experimental study of dynamics of a diode-pumped Nd: $YVO_4$ laser under periodic modulation of losses. *Rev. Mex. Física E* **2012**, *58*, 150–155.
7. Jiao, Y.; Ma, Y.; Li, Y. All-solid-state single-longitudinal-mode pulse Nd:$YVO_4$ ring laser. *Acta Sin. Quantum Opt.* **2014**, *20*, 81– [CrossRef]
8. Wu, W.; Li, X.; Yan, R.; Zhou, Y.; Ma, Y.; Fan, R.; Dong, Z.; Chen, D. 100 kHz, 3.1 ns, 1.89 J cavity-dumped burst-mode Nd:Y MOPA laser. *Opt. Express* **2017**, *25*, 26875–26884. [CrossRef] [PubMed]
9. Wu, W.; Li, X.; Mei, F.; Chen, D.; Yan, R. 30 mJ, 1 kHz sub-nanosecond burst-mode Nd:YAG laser MOPA system. *Opt. Expr* **2019**, *27*, 36129–36136. [CrossRef]
10. Pan, L.; Geng, J.; Jiang, S. High power picosecond green and deep ultraviolet generations with an all-fiberized MOPA. *Opt. L* **2022**, *47*, 5140–5143. [CrossRef]
11. Yu, Y.-J.; Chen, X.-Y.; Wang, C.; Wu, C.-T.; Yu, M.; Jin, G.-Y. High repetition rate 880 nm diode-directly-pumped electro-op Q-switched Nd:$GdVO_4$ laser with a double-crystal RTP electro-optic modulator. *Opt. Commun.* **2013**, *304*, 39–42. [CrossRef]
12. Ma, S.; Lu, D.; Yu, H.; Zhang, H.; Han, X.; Lu, Q.; Ma, C.; Wang, J. High repetition rates optically active langasite electro-optica Q-switched laser at 1.34 µm. *Opt. Express* **2017**, *25*, 24007–24014. [CrossRef]
13. Shang, J.; Yang, J.; Hao, H.; Li, Q.; Zhang, L.; Sun, J. Compact low-voltage electro-optic Q-switch made of $LiNbO_3$. *Opt. Expr* **2020**, *28*, 22287–22296. [CrossRef]
14. Chen, Y.; Huang, J.; Lin, Y.; Gong, X.; Luo, Z.; Huang, Y. Stable passively Q-switched 1537 nm Er:Yb:$Lu_2Si_2O_7$ pulse microla with peak output power higher than 10 kW at 1–2 kHz. *Opt. Laser Technol.* **2022**, *155*, 108392. [CrossRef]
15. Cho, C.; Cheng, H.; Chang, Y.; Tang, C.; Chen, Y.-F. An energy adjustable linearly polarized passively Q-switched bulk laser w a wedged diffusion-bonded Nd:YAG/$Cr^{4+}$:YAG crystal. *Opt. Express* **2015**, *23*, 8162–8169. [CrossRef] [PubMed]
16. Sakai, H.; Kan, H.; Taira, T. >1 MW peak power single-mode high-brightness passively Q-switched $Nd^{3+}$: YAG microchip la *Opt. Express* **2008**, *16*, 19891–19899. [CrossRef]
17. Lin, Y.-N.; Fang, W.-T.; Gu, C.; Xu, L.-X. Wideband all-polarization-maintaining Yb-doped mode-locked fiber laser usin nonlinear optical loop mirror. *Chin. Phys. Lett.* **2016**, *33*, 43–45. [CrossRef]
18. Kelleher, E.J.R.; Travers, J.C.; Sun, Z.; Rozhin, A.G.; Ferrari, A.C.; Popov, S.V.; Taylor, J.R. Nanosecond-pulse fiber las mode-locked with nanotubes. *Appl. Phys. Lett.* **2009**, *95*, 111108. [CrossRef]

Qin, Z.; Xie, G.; Zhao, C.; Wen, S.; Yuan, P.; Qian, L. Mid-infrared mode-locked pulse generation with multilayer black phosphorus as saturable absorber. *Opt. Lett.* **2016**, *41*, 56–59. [CrossRef]
Wu, W.; Li, X.; Yan, R.; Chen, D.; Tang, S. Cavity-dumped burst-mode Nd:YAG laser master-oscillator power-amplifier system with a flat-top beam output realized by gain profile-controlled side pumping. *Opt. Express* **2022**, *30*, 20401–20414. [CrossRef] [PubMed]
Kasinski, J.J.; Burnham, R.L. Near-diffraction-limited laser beam shaping with diamond-turned aspheric optics. *Opt. Lett.* **1997**, *22*, 1062–1064. [CrossRef]
Xue, L.; Pang, Y.; Liu, W.; Liu, L.; Pang, H.; Cao, A.; Shi, L.; Fu, Y.; Deng, Q. Fabrication of random microlens array for laser beam homogenization with high efficiency. *Micromachines* **2020**, *11*, 338. [CrossRef] [PubMed]
Wippermann, F.; Zeitner, U.-D.; Dannberg, P.; Bräuer, A.; Sinzinger, S. Beam homogenizers based on chirped microlens arrays. *Opt. Express* **2007**, *15*, 6218–6231. [CrossRef]
Reddy, A.N.K.; Pal, V. Robust design of diffractive optical elements for forming fat-top beams with extended depth of focus. *Appl. Phy. B* **2019**, *125*, 231. [CrossRef]
Dev, V.; Reddy, A.N.K.; Pal, V. Generation of uniform-intensity light beams with controllable spatial shapes. *Opt. Commun.* **2020**, *475*, 126226. [CrossRef]
Caley, A.J.; Thomson, M.J.; Liu, J.; Waddie, A.J.; Taghizadeh, M.R. Diffractive optical elements for high gain lasers with arbitrary output beam profiles. *Opt. Express* **2007**, *15*, 10699–10704. [CrossRef]
Yang, K.; Zhao, S.; Li, G.; Zou, J.; Song, P.; Wu, W. Pulse compression in AO Q-switched diode-pumped Nd:$GdVO_4$ laser with $Cr^{4+}$:YAG saturable absorber. *Appl. Phy. B* **2005**, *80*, 687–692. [CrossRef]
Li, D.; Zhao, S.; Li, G.; Yang, K. Optimization of pulse width of double passively Q-switched lasers with GaAs and $Cr^{4+}$-doped saturable absorbers. *Opt. Laser Technol.* **2009**, *41*, 272–279. [CrossRef]
Li, G.; Zhao, S.; Yang, K.; Li, D.; Zou, J. Pulse shape symmetry and pulse width reduction in diode-pumped doubly Q-switched Nd:$YVO_4$/KTP green laser with AO and GaAs. *Opt. Express* **2005**, *13*, 1178–1187. [CrossRef]
Zhao, S.; Zhao, J.; Li, G.; Yang, K.; Sun, Y.; Li, D.; An, J.; Wang, J.; Li, M. Doubly Q-switched laser with electric-optic modulator and GaAs saturable absorber. *Laser Phys. Lett.* **2006**, *3*, 471–473. [CrossRef]
Jin, D.; Bai, Z.; Wang, Q.; Chen, Y.; Liu, Z.; Fan, R.; Qi, Y.; Ding, J.; Yang, X.; Wang, Y. Doubly Q-switched single longitudinal mode Nd: YAG laser with electro-optical modulator and $Cr^{4+}$: YAG. *Opt. Commun.* **2020**, *463*, 125500. [CrossRef]
Chen, Y.-F.; Lee, L.; Huang, T.; Wang, C. Study of high-power diode-end-pumped Nd:$YVO_4$ laser at 1.34 μm: Influence of Auger upconversion. *Opt. Commun.* **1999**, *163*, 198–202. [CrossRef]
Mukhopadhyay, P.K.; George, J.; Sharma, S.; Ranganathan, K.; Nathan, T. Experimental determination of effective stimulated emission cross-section in a diode pumped Nd:$YVO_4$ micro-laser at 1064 nm with various doping concentrations. *Opt. Laser Technol.* **2002**, *34*, 357–362. [CrossRef]
Turri, G.; Jenssen, H.P.; Cornacchia, F.; Tonelli, M.; Bass, M. Temperature-dependent stimulated emission cross section in $Nd^{3+}$:$YVO_4$ crystals. *J. Opt. Soc. Am. B* **2009**, *26*, 2084–2088. [CrossRef]
Zhao, X.; Song, Z.; Li, Y.-J.; Feng, J.-X.; Zhang, K.-S. High efficiency sub-nanosecond electro–optical Q-switched laser operating at kilohertz repetition frequency. *Chin. Phys. B* **2020**, *29*, 335–340. [CrossRef]

*photonics*

*Communication*

# Coupling of Photonic and Plasmonic Modes for Double Nanowire Cavities

**Xuanran Peng [1,2,3], Jing Liu [1,2,3], Yaru Kang [1,4], Xu Mao [1], Wei Yan [1,*], Xiaohui Wang [2,5], Kong Liu [2,5], Rui Xu [6], Fuhua Yang [1,2,3,7,8,9] and Zhaofeng Li [1,2,4,*]**

1 Engineering Research Center for Semiconductor Integrated Technology, Institute of Semiconductors, Chin Academy of Sciences, Beijing 100083, China
2 Center of Materials Science and Optoelectronics Engineering, University of Chinese Academy of Science Beijing 100049, China
3 College of Materials Science and Opto-Electronics Technology, University of Chinese Academy of Scienc Beijing 100049, China
4 School of Integrated Circuits, University of Chinese Academy of Sciences, Beijing 100049, China
5 Key Laboratory of Semiconductor Materials Science, Beijing Key Laboratory of Low Dimensional Semiconductor Materials and Devices, Institute of Semiconductors, Chinese Academy of Sciences, Beijing 100083, China
6 Fachgebiet Angewandte Nanophysik, Institut für Physik & IMN MacroNano, Technische Universität Ilme 98693 Ilmenau, Germany
7 State Key Laboratory of Superlattices and Microstructures, Institute of Semiconductors, Chinese Acader of Sciences, Beijing 100083, China
8 Beijing Academy of Quantum Information Science, Beijing 100193, China
9 Beijing Engineering Research Center of Semiconductor Micro-Nano Integrated Technology, Beijing 100083, Ch
* Correspondence: yanwei@semi.ac.cn (W.Y.); lizhaofeng@semi.ac.cn (Z.L.)

**Abstract:** We analyze the coupling between double nanowire cavities for both photonic modes a plasmonic modes. When the spacing between nanowires reduces, a redshift of the resonant freque of the symmetric mode and a blueshift of the resonant frequency of the antisymmetric mode observed. Compared to single nanowire cavity modes, the Q factors of antisymmetric supermode double nanowires can be improved by 51% for photonic modes and by 24% for plasmonic mo The mechanisms of Q factor improvement for photonic modes and plasmonic modes are stud based on the field distribution of radiations from the modes. This paper may contribute to resea and applications for double nanowire lasers and nanowire laser arrays.

**Keywords:** coupling; nanowire; photonic; plasmonic; cavity; Q factor; supermode; array

check for updates

**Citation:** Peng, X.; Liu, J.; Kang, Y.; Mao, X.; Yan, W.; Wang, X.; Liu, K.; Xu, R.; Yang, F.; Li, Z. Coupling of Photonic and Plasmonic Modes for Double Nanowire Cavities. *Photonics* **2023**, *10*, 415. https://doi.org/10.3390/photonics10040415

Received: 14 March 2023
Revised: 3 April 2023
Accepted: 4 April 2023
Published: 6 April 2023



## 1. Introduction

To achieve photonic integrated circuits (PICs), nanoscale lasers have seen great dev opment in the past two decades. Various designs were brought up in pursuit of propert including ultracompact footprints, low thresholds, and room-temperature operation [1- However, influences between multiple devices must be taken into account in more practi applications, which encouraged studies towards nano-laser arrays, including both coup and uncoupled arrays [4]. In this area, arrays consisting of two nano-lasers are essential a building block toward more complex arrays of nano-lasers.

A large number of studies focused on mode splitting, which is a universal phenomen in coupled resonators such as nanoscale rings, nanopans, and square standing-wave cav resonators [5–7]. Both coupled metallo-dielectric cylinder nanocavities [8] and coup photonic crystal nanolasers [9] could create supermodes. Some of the studies aimed mode selection and switching [10], while some other studies aimed to curb coupling [8, The coupling between nanowires has been studied in previous research, for both align and parallel nanowires [12]. However, a few studies have been conducted regardi coupled modes of double nanowire cavities (DNC), both in photonic and plasmonic mod

In fact, nanowire lasers have attracted great interest as the principal kind of nanoscale lasers [13–16]. Semiconductor nanowire acts simultaneously as an optical gain medium and an optical cavity that, in some cases, possesses the intrinsic capability to produce laser light [17]. Nanowire cavities have also manifested outstanding properties in many other aspects [18–21]. Nanowire lasers also enable wavelength multiplexing and potential electrical modulation for on-chip applications [22]. However, such lasers are restricted by the diffraction limit, both in optical mode size and physical device dimension [23].

Surface plasmon polariton (SPP) is a kind of coherent electron oscillation at metal-dielectric interfaces [24]. Electromagnetic (EM) fields of surface plasmons decrease exponentially away from the interface, which allows lightwaves to be concentrated at sub-diffraction limit scales [25]. The interaction between SPPs and exciton has been studied both in weak coupling and strong coupling [26]. Exploiting this mechanism, plasmonic lasers (or spasers) of the nanoscale have been realized based on core-shell silica nanoparticles [27], metal-coated dielectric nanowires [28], silver nanopans [29], metal-insulator-metal waveguides [30], metal-insulator-semiconductor nanowire [23], and methylene blue dye-Ag nanocavities [31,32]. Among the above plasmonic lasers, nanowire plasmonic lasers [23,24,33–37] play an important role. Therefore, in this paper, coupled DNC is studied for both photonic modes and plasmonic modes.

In the following, finite element method (FEM) simulation results for photonic modes are presented, and the coupling effect on Q factors of different modes is discussed in Section 2. Then, in Section 3, the simulation results for plasmonic modes are shown and compared to those for photonic modes. Several topics related to the research results in the earlier sections are discussed subsequently in Section 4. Finally, Section 5 concludes the mechanisms in coupled DNC for photonic and plasmonic modes and discusses possible applications.

## 2. Coupling of Double Nanowire Cavities for Photonic Mode

The simulated DNCs are shown schematically in Figure 1a, consisting of two hybrid perovskite ($CH_3NH_3PbI_3$) semiconductor nanowires placed side by side with the spacing of $S_1$. Details of the parameters can be found in the caption of Figure 1. Hybrid perovskite ($CH_3NH_3PbI_3$) is chosen as the material of the nanowires in the present study based on previous reports [38,39]. In experiments, the synthesis of high-quality single-crystal perovskite can be achieved by a surface-initiated solution growth strategy [38]. The as-grown perovskite nanowires can be subsequently transferred onto the substrate to fabricate nanowire lasers [33].

It is worth noting that the main conclusions of the following results are not limited to the specific hybrid perovskite and still hold qualitatively for nanowire cavities of other dielectric materials. Since we mainly concentrate on the coupling of nanowire cavities, the substrate is not included in the simulation here for simplicity. The FEM software COMSOL Multiphysics is used in the simulation. Tetrahedron mesh is used. The minimum mesh size is 5 nm.

Single nanowire cavity (SNC) supports only fundamental modes (two degenerate modes polarized in the y and z directions) for wavelengths larger than 750 nm. When two identical nanowire cavities are placed side by side near each other, they support symmetric supermodes and antisymmetric supermodes (or bonding and antibonding modes). Symmetric modes and antisymmetric modes have similar or opposite EM field distributions, respectively, in two nanowire cavities. Figure 1b shows the E field profile on the y–z cross section for symmetric and antisymmetric modes of z and y polarizations. Figure 1c shows the symmetric and antisymmetric modes on the x–y cross section. Those symmetric and antisymmetric modes of z polarization are similar to the case of y polarization and are not shown here. In the following parts, the modes of y and z polarizations will be referred to as $E_y$ modes and $E_z$ modes, respectively.

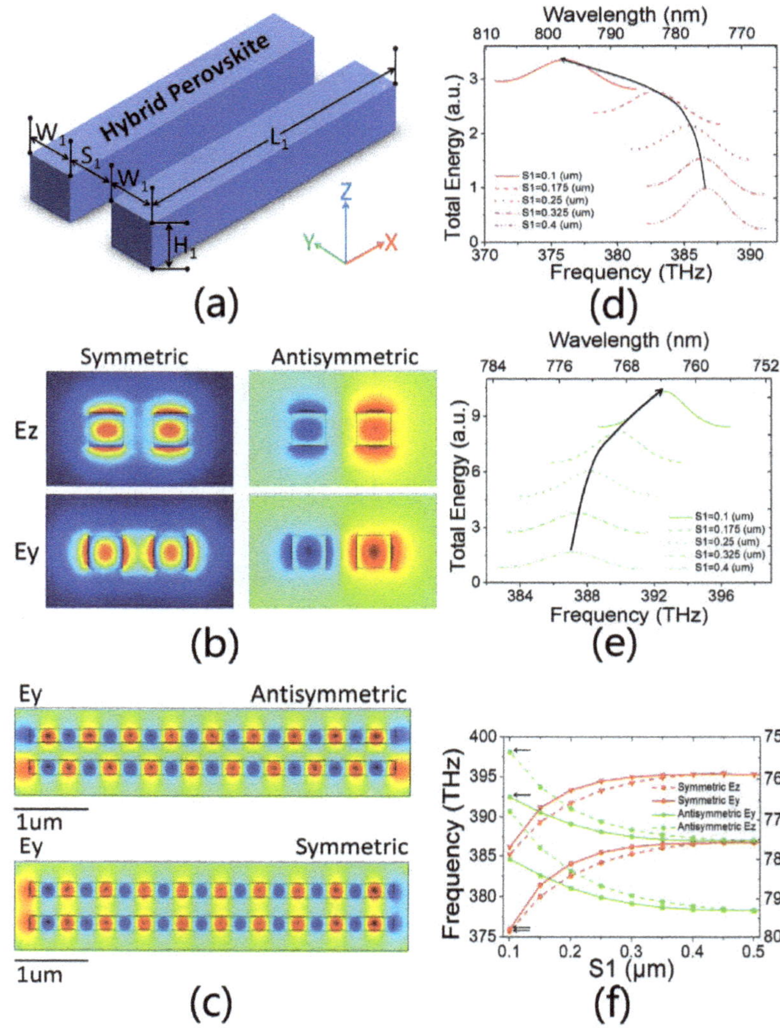


**Figure 1.** (**a**) Coupled photonic DNC. $S_1 = 0.2$ μm. $L_1 = 5$ um. $W_1 = H_1 = 0.2$ μm. The refractive ind of semiconductor nanowire $n = 2.59$. The nanowires are placed in the air. (**b**) Electric field distributi at the y–z cross section of symmetric and antisymmetric modes of z and y polarizations. Th component of the E field is shown for $E_z$ modes, while the y component of the E field is shown for modes. (**c**) E field distribution at the x–y cross section for antisymmetric and symmetric $E_y$ modes the middle of the nanowire. The y component of the electric field is shown here, at frequencies 380.3 and 385.4 THz, respectively. (**d**,**e**) Total energy spectra of symmetric $E_y$ mode and antisymme $E_y$ mode, respectively, as $S_1$ grows smaller. (**f**) Resonant frequencies versus $S_1$ for symmetric a antisymmetric modes of y and z polarizations.

To study the resonant modes, we calculate the total energy of the fields when t cavities are excited by short line sources for wavelengths between 750 nm and 800 n The peaks of the total energy spectra correspond to the resonant frequencies of mod Figure 1d,e shows the shifting of spectra for symmetric and antisymmetric $E_y$ modes wh $S_1$ decreases from 0.4 μm to 0.1 μm. A redshift of the resonant frequency of the symmet mode and a blueshift of the resonant frequency of the antisymmetric mode can be observed when $S_1$ decreases. The spectra for symmetric and antisymmetric $E_z$ modes have the same tendency when $S_1$ decreases, but are not shown here. A series of resonant frequencies are plotted versus $S_1$ in Figure 1f.

To understand Figure 1f, we start with a single nanowire cavity. As a Fabry-Perot resonator, the longitudinal modes of SNC obey the following equation [40]:

$$\frac{1}{\lambda_{eff}^{q}} - \frac{1}{\lambda_{eff}^{q+1}} = \frac{1}{2L_1} \quad (1)$$

where q is a positive integer marking the longitudinal mode index; $L_1$ is the length of the Fabry-Perot resonator; $\lambda_{eff}$ is the effective wavelength of the longitudinal mode. The effective wavelength $\lambda_{eff}$ is calculated as $\lambda_{eff} = 2\pi/\beta$, where $\beta$ is the propagation constant of the longitudinal mode. The calculated effective wavelengths of the longitudinal modes agree well with Equation (1). For SNC, which has a square cross section, longitudinal modes are two-fold degenerate in y and z polarizations. When two nanowires are placed far away from each other, they resonate just as two independent SNCs. When the two nanowires are placed near each other, the two degenerate modes of the same q index will split into four nondegenerate modes (two $E_y$ supermodes and two $E_z$ supermodes for DNC), as shown in Figure 1f. For instance, when $S_1$ = 0.5 μm, four supermodes with frequencies near 387.1 THz originate from the coupling of the two degenerate longitudinal modes of SNC. When $S_1$ decreases, the frequencies of the four modes split, i.e., the redshift of symmetric modes and the blueshift of antisymmetric modes. The same phenomenon also happens in other longitudinal modes (378.6 THz and 395.3 THz). Four supermodes that originate at 387.1 THz are depicted for clarity in Figure 1f, but only two antisymmetric supermodes that originate at 378.6 THz and two symmetric supermodes that originate at 395.3 THz are presented.

From Figure 1f, it is also seen that the difference between the resonant frequencies of symmetric and antisymmetric modes increases when $S_1$ decreases. This phenomenon was named "mode splitting" [5–7]. Due to the redshift and blueshift of resonant frequencies for symmetric and antisymmetric modes, the curves of frequency tendency from neighboring longitudinal modes can intersect, and the resonant frequencies from neighboring longitudinal modes can overlap. This overlap could cause mode competition, which may not be desired in laser operation.

To further investigate the properties of the supermodes, total energy spectra are calculated for a longitudinal mode of SNC and four corresponding supermodes of DNC (Figure 2a). These four supermodes for $S_1$ = 0.1 μm of DNC are marked by arrows in Figure 1f. The quality factor (Q factor) for these modes is calculated based on the full width at half maximum (FWHM) of the spectra. The Q factor of mode for SNC is 94.9. The Q factors of symmetric $E_y$ and $E_z$ modes for DNC are 68.7 and 83.2, respectively. The Q factors of the antisymmetric $E_y$ and $E_z$ modes for DNC are 133.2 and 143.1, respectively. Antisymmetric modes have higher Q factors than symmetric modes. This kind of Q factor splitting has been found in previous research on mode splitting [8]. Q factor splitting for $E_z$ modes is stronger than Q factor splitting for $E_y$ modes. Among the four supermodes, antisymmetric $E_z$ mode has the highest Q factor. The Q factor of antisymmetric $E_z$ mode for DNC is 51% higher than the Q factor for SNC mode.

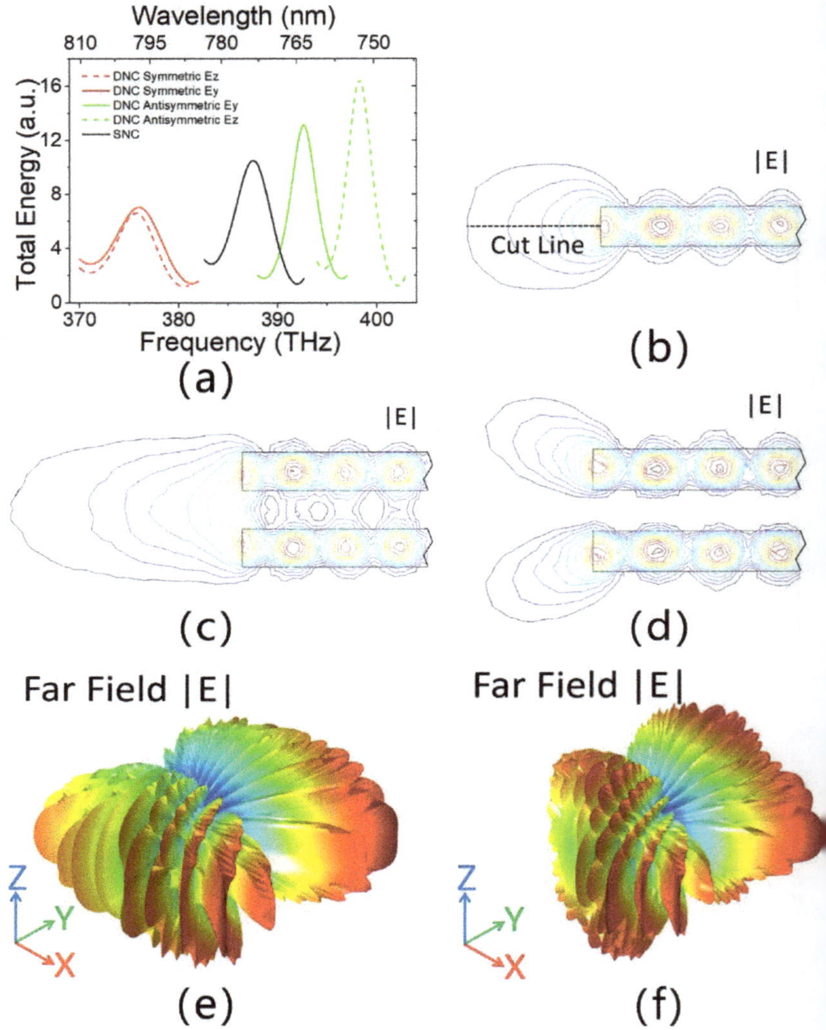


**Figure 2.** (**a**) Total energy spectra for the mode of SNC and corresponding four supermodes of DN at $S_1$ = 0.1 um. Contour plot of the electric field norm for a longitudinal mode of SNC (**b**), symme $E_z$ modes of (**c**), and antisymmetric $E_z$ modes (**d**) for DNC at $S_1$ = 0.2 um. The lobe pattern of electric field norm at the far field for antisymmetric $E_y$ mode (**e**) and antisymmetric $E_z$ mode for DNC.

Different Q factors for symmetric and antisymmetric modes and two polarizatic can be understood based on their radiation patterns. Firstly, we performed an analy based on the symmetry property of supermodes. Figure 2b shows the radiation patt of the mode of SNC, which is mainly along the direction of the longitudinal axis of nanowire. Figure 2c shows the radiation pattern of the symmetric mode of DNC. T radiation is also mainly along the direction of the longitudinal axis of the DNC, whi can be regarded as the constructive interference of two SNC modes. Figure 2d shows radiation pattern of the antisymmetric modes of DNC. The radiation deviates from direction of the longitudinal axis of the DNC. This radiation pattern can be regarded the destructive interference of two SNC modes, which is similar to the radiation pattern of a quadrupole source. For clarity, Figure 2b–d shows only the radiation patterns around the left ends of the nanowires, while the patterns around the right ends are the same due to the symmetry of the structures. In comparison to SNC modes, destructive interference increases the Q factors of antisymmetric DNC modes. This is due to the reduced radiation energy of these modes.

Secondly, we analyze from the perspective of polarizations. For antisymmetric modes, $E_z$ supermode has a higher Q factor than $E_y$ supermode. To understand this phenomenon, the lobe patterns of radiation of $E_y$ and $E_z$ supermodes are plotted in Figure 2e,f. The difference between the EM far field distribution of two polarizations shows lower radiation energy for $E_z$ supermode than radiation energy for $E_y$ supermode. This explains the higher Q factors of antisymmetric $E_z$ modes than those of antisymmetric $E_y$ modes. Consequently, the antisymmetric $E_z$ mode has the highest Q factor among the four supermodes.

## 3. Coupling of Double Nanowire Cavities for Plasmonic Mode

In the simulation for coupled plasmonic DNC, a metallic (Ag) substrate is introduced below two hybrid perovskite semiconductor nanowires placed side by side (Figure 3a). The relative permittivity of the metallic substrate is fitted with the Drude–Lorentz dispersion model (high-frequency dielectric constant $\varepsilon_\infty = 3.75$, plasma frequency $\omega_p = 1.32 \times 10^{16}$ rad/s, characteristic collision frequency $\gamma = 1.20 \times 10^{14}$ rad/s).

Plasmonic nanowire lasers work similarly to photonic nanowire lasers in some aspects and thus have similar phenomena in the coupled DNC system. A distinction between photonic modes and plasmonic modes is the EM field distribution. The EM field for plasmonic modes exists mainly at the interface between metal and dielectric and attenuates exponentially vertically (Figure 3b). Contrastively, the EM field for photonic modes lies mainly in the central area of the cavity (Figure 1b), especially for fundamental modes. Moreover, only TM mode (or $E_z$ mode) exists in plasmonic modes, contrasting with $E_z$ and $E_y$ modes in photonic modes.

Symmetric modes and antisymmetric modes of plasmonic DNC can be observed, as shown in Figure 3b. Resonant frequencies of supermodes versus $S_2$ are plotted in Figure 3c,d for $L_2 = 5$ μm and $L_2 = 3$ μm, respectively, which indicates the mode splitting of plasmonic DNC. The longitudinal modes of plasmonic SNC split into two supermodes of plasmonic DNC when two identical nanowires are placed near each other. The longitudinal spacings are smaller for modes of relatively longer DNCs, as shown in Figure 3c. And the overlap of mode frequencies happens earlier when $S_2$ decreases. This overlap still exists even when the nanowires are as short as 3 μm for plasmonic DNC, as shown in Figure 3d. Figure 3c, similar to Figure 1f, illustrates simply the antisymmetric supermode at 384.9 THz and the symmetric supermode at 397.1 THz, for the sake of clarity. For the same reason, only the antisymmetric supermode, which originates at 376.1 THz, and the symmetric supermode, which originates at 406.8 THz, are demonstrated here in Figure 3d.

A longitudinal mode of plasmonic SNC and two supermodes of plasmonic DNC are used to calculate their total energy spectra (Figure 3e). These two supermodes for $S_2 = 0.1$ μm of plasmonic DNC are marked by arrows in Figure 3d. The Q factor of mode for SNC is 59.6. The Q factors of symmetric mode and antisymmetric mode for DNC are 58.5 and 75, respectively. These results exhibit the Q factor splitting of coupled plasmonic DNC. Q factors for antisymmetric modes are higher than symmetric modes of plasmonic DNC. The mechanism of different Q factors for modes of plasmonic DNC is similar to that of photonic DNC, i.e., destructive interference and constructive interference.

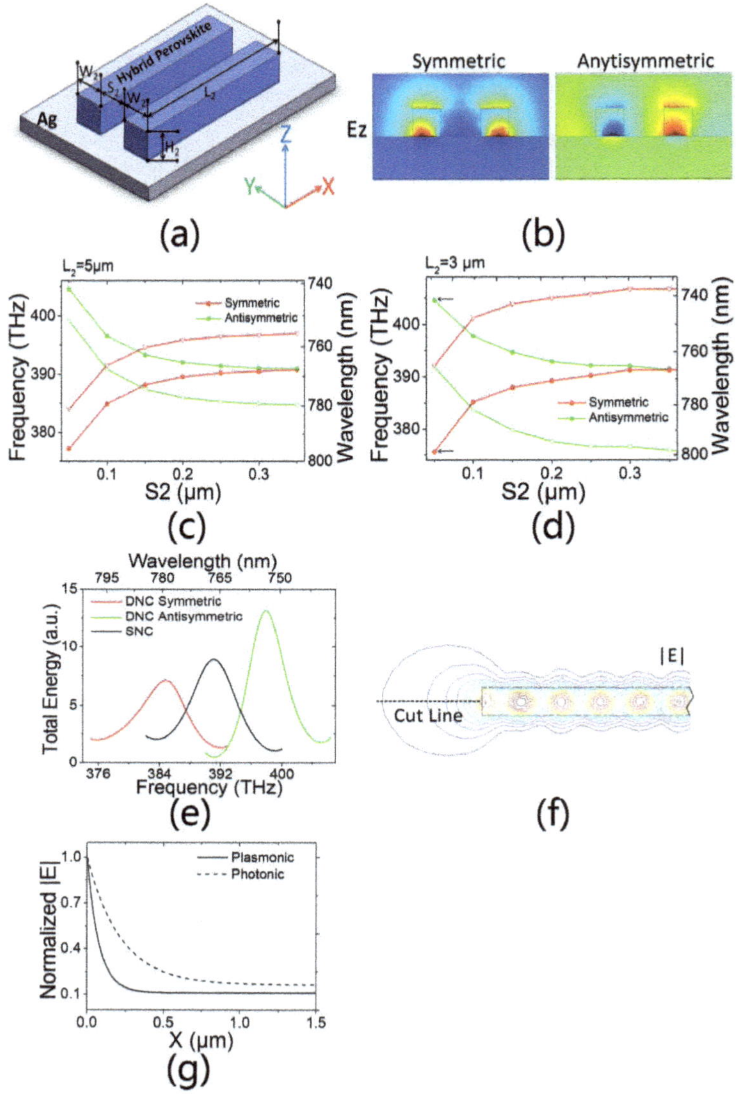


**Figure 3.** (**a**) Coupled plasmonic DNC. $S_2$ = 0.2 mm. $L_2$ = 3 μm; $W_2 = H_2$ = 0.135 mm. The refracti index of hybrid perovskite semiconductor nanowire $n$ = 2.59–0.015 i. The Ag substrate has paramet of $\varepsilon_\infty = 3.75$, $\omega_p = 1.32 \times 10^{16}$ rad/s, $\gamma = 1.20 \times 10^{14}$ rad/s. (**b**) E field distribution at y–z cro section of symmetric and antisymmetric modes. The z component of the E field is shown he (**c**) Resonant frequencies of symmetric and antisymmetric modes versus $S_2$ when $L_2$ = 5 μ (**d**) Resonant frequencies of symmetric and antisymmetric modes versus $S_2$ when $L_2$ = 3 μ (**e**) Total energy spectra for mode of plasmonic SNC and corresponding two supermodes of plasmo DNC with $S_2$ = 0.1 μm when $L_2$ = 3 μm. (**f**) Contour plot of the electric field norm for a longitudi mode of plasmonic SNC. (**g**) Normalized electric field amplitudes versus distance from the ends nanowires along the cut line (shown in (**f**) and Figure 2b) selected by main radiation directions the mode of plasmonic SNC (line) and the mode of photonic SNC (dash).

The Q factor of the antisymmetric mode of plasmonic DNC has an improvement of up to 24% over the Q factor of the mode of plasmonic SNC. This improvement in the Q factor is lower compared with the result (51%) for the case of photonic modes in Section 3. This different improvement of Q factors may be due to the different extents of destructive interference of antisymmetric modes of plasmonic DNC and photonic DNC. Additionally, the extent of interference from DNC is mainly determined by the radiation fields of SNC. The extent of the radiation fields of plasmonic SNC and photonic SNC is analyzed along the cut lines of the main radiation directions of modes as shown in Figures 3f and 2b, respectively. The normalized electric fields along the cut lines from the ends of nanowires are shown in Figure 3g, which demonstrates the attenuation of radiation fields away from the ends. The radiation electric field of plasmonic SNC attenuates faster than that of photonic SNC because of additional absorption loss due to the metallic substrate of plasmonic SNC. Consequently, the extent of interference in the antisymmetric mode of plasmonic DNC is smaller than that of photonic DNC. Moreover, the Q factor improvement in the case of plasmonic DNC is lower than that of photonic DNC.

## 4. Discussion

In Section 2, the substrate is excluded from the simulation for simplicity. In realistic applications, the presence of dielectric and metal substrates will affect the EM field distributions and Q factors of photonic modes. For instance, the Q factors of the photonic modes of DNC with $SiO_2$ substrate will decrease due to the effects of field delocalization and its displacement into the substrate, according to the simulation results. However, the main conclusions in Section 2, including longitudinal spacing, mode splitting, and Q factor splitting, are not influenced by the introduction of $SiO_2$ substrate.

However, the introduction of a metal (Ag) substrate will fundamentally change the question. For nanowires with a metal substrate, the supported modes are determined by the sizes of the cross section of nanowires with a certain refractive index for the studied wavelength band [33]. A nanowire with a height smaller than its critical value supports only the plasmonic modes [33]. When the height of the nanowire is larger than the critical value, both the plasmonic and hybrid modes will be supported. This hybrid mode is formed due to the hybridization of photonic modes and plasmonic modes [25,41,42]. According to this consideration, the dimensions of the plasmonic SNC and DNC in Section 3 are designed to support only the plasmonic modes, excluding the hybrid modes.

Additionally, anisotropy might exist for single-crystal perovskite nanowires [43]. However, the anisotropy of the material is not decisive for the main conclusions of this study. When the anisotropy is not crucial for the physical phenomena, the use of a scalar refractive index is acceptable, as has been done in other studies [38,44–46]. Therefore, the real part of the refractive index is set at $n = 2.59$ in our study based on previous research on $CH_3NH_3PbI_3$ [39]. The imaginary part of the refractive index is set at $k = -0.015$ to introduce an appropriate gain in the nanowire to compensate for the absorption loss due to the metallic substrate. This k corresponds to a gain level for laser physics $2.7 \times 10^3$ cm$^{-1}$, which is within the realizable gain level of perovskite in previous studies [47]. With this gain level, the intensity of the mode will remain approximately constant when it propagates in the nanowire.

## 5. Conclusions

The coupling of DNC is studied in this paper for both photonic modes and plasmonic modes. Our results revealed a mode splitting. A longitudinal mode of photonic SNC splits into four supermodes of photonic DNC. Moreover, a longitudinal mode of plasmonic SNC splits into two supermodes of plasmonic DNC. Due to the redshift and blueshift of resonant frequencies for symmetric and antisymmetric supermodes, the resonant frequencies of supermodes from neighboring longitudinal modes of DNC can overlap. This overlap of mode frequencies still exists even when the nanowires are as short as 3 μm for plasmonic

DNC. The possible overlap of frequencies of longitudinal modes should be taken ir account when DNC or nanowire laser arrays are designed.

The Q factors of antisymmetric modes of DNC are higher than the Q factors of moc of SNC, for both photonic and plasmonic modes. The reason is that the antisymmet modes have lower radiation loss due to the destructive interference of radiation from t SNCs. However, this improvement in Q factors for plasmonic modes (24%) is lower th that for photonic modes (51%). This difference can be explained by the faster attenuation the radiation field for plasmonic modes than for photonic modes. The improvement o factors in DNC may be applied to improve nanowire lasers.

**Author Contributions:** Conceptualization, X.P. and Z.L.; validation, X.P. and J.L.; writing—origi draft preparation, X.P. and Z.L.; writing—review and editing, J.L. and Y.K.; visualization, X supervision, Z.L.; project administration, X.M., W.Y. and R.X.; funding acquisition, X.W., K.L., and Z.L. All authors have read and agreed to the published version of the manuscript.

**Funding:** This work was mostly supported by the National Key R&D Program of China (2018YFE0204( and partially supported by the National Natural Science Foundation of China (Grant No. 619713

**Institutional Review Board Statement:** Not applicable.

**Informed Consent Statement:** Not applicable.

**Data Availability Statement:** Not applicable.

**Conflicts of Interest:** The authors declare no conflict of interest.

## References

1. Gu, Q.; Smalley, J.S.T.; Nezhad, M.P.; Simic, A.; Lee, J.H.; Katz, M.; Bondarenko, O.; Slutsky, B.; Mizrahi, A.; Lomakin, V.; e Subwavelength semiconductor lasers for dense chip-scale integration. *Adv. Opt. Photonics* **2014**, *6*, 1–56. [CrossRef]
2. Hill, M.T.; Gather, M.C. Advances in small lasers. *Nat. Photonics* **2014**, *8*, 908–918. [CrossRef]
3. Ma, R.M.; Oulton, R.F. Applications of nanolasers. *Nat. Nanotechnol.* **2019**, *14*, 12–22. [CrossRef]
4. Deka, S.S.; Jiang, S.Z.; Pan, S.H.; Fainman, Y. Nanolaser arrays: Toward application-driven dense integration. *Nanophotonics* **2(** *10*, 149–169. [CrossRef]
5. Haus, H.A.; Huang, W.P. Coupled-mode theory. *Proc. IEEE* **1991**, *79*, 1505–1518. [CrossRef]
6. Li, Q.; Wang, T.; Su, Y.K.; Yan, M.; Qiu, M. Coupled mode theory analysis of mode-splitting in coupled cavity system. *Opt. Exp* **2010**, *18*, 8367–8382. [CrossRef]
7. Popovic, M.A.; Manolatou, C.; Watts, M.R. Coupling-induced resonance frequency shifts in coupled dielectric multi-cavity filt *Opt. Express* **2006**, *14*, 1208–1222. [CrossRef]
8. Deka, S.S.; Pan, S.H.; Gu, Q.; Fainman, Y.; El Amili, A. Coupling in a dual metallo-dielectric nanolaser system. *Opt. Lett.* **2017**, 4760–4763. [CrossRef] [PubMed]
9. Hamel, P.; Haddadi, S.; Raineri, F.; Monnier, P.; Beaudoin, G.; Sagnes, I.; Levenson, A.; Yacomotti, A.M. Spontaneous mir symmetry breaking in coupled photonic-crystal nanolasers. *Nat. Photonics* **2015**, *9*, 311–315. [CrossRef]
10. Haddadi, S.; Hamel, P.; Beaudoin, G.; Sagnes, I.; Sauvan, C.; Lalanne, P.; Levenson, J.A.; Yacomotti, A.M. Photonic molecu Tailoring the coupling strength and sign. *Opt. Express* **2014**, *22*, 12359–12368. [CrossRef]
11. Atlasov, K.A.; Karlsson, K.F.; Rudra, A.; Dwir, B.; Kapon, E. Wavelength and loss splitting in directly coupled photonic-crys defect microcavities. *Opt. Express* **2008**, *16*, 16255–16264. [CrossRef] [PubMed]
12. Anikina, M.A.; Roy, P.; Kadinskaya, S.A.; Kuznetsov, A.; Kondratev, V.M.; Bolshakov, A.D. Numerical Study of GaP Nanowi Individual and Coupled Optical Waveguides and Resonant Phenomena. *Nanomaterials* **2023**, *13*, 56. [CrossRef] [PubMed]
13. Huang, M.H.; Mao, S.; Feick, H.; Yan, H.Q.; Wu, Y.Y.; Kind, H.; Weber, E.; Russo, R.; Yang, P.D. Room-temperature ultravic nanowire nanolasers. *Science* **2001**, *292*, 1897–1899. [CrossRef]
14. Johnson, J.C.; Choi, H.J.; Knutsen, K.P.; Schaller, R.D.; Yang, P.D.; Saykally, R.J. Single gallium nitride nanowire lasers. *Nat. Ma* **2002**, *1*, 106–110. [CrossRef]
15. Johnson, J.C.; Yan, H.Q.; Schaller, R.D.; Haber, L.H.; Saykally, R.J.; Yang, P.D. Single nanowire lasers. *J. Phys. Chem. B* **2001**, 11387–11390. [CrossRef]
16. Zimmler, M.A.; Capasso, F.; Muller, S.; Ronning, C. Optically pumped nanowire lasers: Invited review. *Semicond. Sci. Tech* **2010**, *25*, 12. [CrossRef]
17. Eaton, S.W.; Fu, A.; Wong, A.B.; Ning, C.Z.; Yang, P.D. Semiconductor nanowire lasers. *Nat. Rev. Mater.* **2016**, *1*, 11. [CrossRe
18. Kotal, S.; Artioli, A.; Wang, Y.J.; Osterkryger, A.D.; Finazzer, M.; Fons, R.; Genuist, Y.; Bleuse, J.; Gerard, J.M.; Gregersen, N.; e A nanowire optical nanocavity for broadband enhancement of spontaneous emission. *Appl. Phys. Lett.* **2021**, *118*, 6. [CrossRe

Kuznetsov, A.; Roy, P.; Grudinin, D.V.; Kondratev, V.M.; Kadinskaya, S.A.; Vorobyev, A.A.; Kotlyar, K.P.; Ubyivovk, E.V.; Fedorov, V.V.; Cirlin, G.E.; et al. Self-assembled photonic structure: A Ga optical antenna on GaP nanowires. *Nanoscale* **2023**, *15*, 2332–2339. [CrossRef]

Roy, P.; Bolshakov, A.D. Ga-GaP nanowire hybrid optical system for enhanced coupling, focusing and steering of light. *J. Phys. D-Appl. Phys.* **2020**, *53*, 8. [CrossRef]

Zhang, G.Q.; Takiguchi, M.; Tateno, K.; Tawara, T.; Notomi, M.; Gotoh, H. Telecom-band lasing in single InP/InAs heterostructure nanowires at room temperature. *Sci. Adv.* **2019**, *5*, 8. [CrossRef] [PubMed]

Wang, D.Q.; Wang, W.J.; Knudson, M.P.; Schatz, G.C.; Odom, T.W. Structural Engineering in Plasmon Nanolasers. *Chem. Rev.* **2018**, *118*, 2865–2881. [CrossRef]

Oulton, R.F.; Sorger, V.J.; Zentgraf, T.; Ma, R.M.; Gladden, C.; Dai, L.; Bartal, G.; Zhang, X. Plasmon lasers at deep subwavelength scale. *Nature* **2009**, *461*, 629–632. [CrossRef]

Lv, Y.C.; Xu, F.F.; Wang, K.; Li, Y.J.; Zhao, Y.S. Loss compensation of surface plasmon polaritons in organic/metal nanowire heterostructures toward photonic logic processing. *Sci. China-Mater.* **2020**, *63*, 1464–1471. [CrossRef]

Balykin, V.I. Plasmon nanolaser: Current state and prospects. *Phys. Usp.* **2018**, *61*, 846–870. [CrossRef]

He, Z.; Xu, C.; He, W.; He, J.; Zhou, Y.; Li, F. Principle and Applications of Multimode Strong Coupling Based on Surface Plasmons. *Nanomaterials* **2022**, *12*, 1242. [CrossRef]

Noginov, M.A.; Zhu, G.; Belgrave, A.M.; Bakker, R.; Shalaev, V.M.; Narimanov, E.E.; Stout, S.; Herz, E.; Suteewong, T.; Wiesner, U. Demonstration of a spaser-based nanolaser. *Nature* **2009**, *460*, 1110–1112. [CrossRef] [PubMed]

Seo, M.K.; Kwon, S.H.; Ee, H.S.; Park, H.G. Full Three-Dimensional Subwavelength High-Q Surface-Plasmon-Polariton Cavity. *Nano Lett.* **2009**, *9*, 4078–4082. [CrossRef]

Kwon, S.H.; Kang, J.H.; Seassal, C.; Kim, S.K.; Regreny, P.; Lee, Y.H.; Lieber, C.M.; Park, H.G. Subwavelength Plasmonic Lasing from a Semiconductor Nanodisk with Silver Nanopan Cavity. *Nano Lett.* **2010**, *10*, 3679–3683. [CrossRef]

Hill, M.T.; Marell, M.; Leong, E.S.P.; Smalbrugge, B.; Zhu, Y.C.; Sun, M.H.; van Veldhoven, P.J.; Geluk, E.J.; Karouta, F.; Oei, Y.S.; et al. Lasing in metal-insulator-metal sub-wavelength plasmonic waveguides. *Opt. Express* **2009**, *17*, 11107–11112. [CrossRef] [PubMed]

Han, X.; Li, F.; He, Z.; Liu, Y.; Hu, H.; Wang, K.; Lu, P. Double Rabi splitting in methylene blue dye-Ag nanocavity. *Nanophotonics* **2022**, *11*, 603–611. [CrossRef]

He, Z.C.; Li, F.; Zuo, P.; Xu, C.; He, W.H.; He, J.H.; Zhou, Y.P.; Zhang, Q.P.; Chen, K.; Huang, H.L.; et al. Strong Coupling in a Hybrid System of Silver Nanoparticles and J-Aggregates at Room Temperature. *J. Phys. Chem. C* **2022**, *126*, 17141–17151. [CrossRef]

Ren, K.K.; Wang, J.; Chen, S.Q.; Yang, Q.X.; Tian, J.; Yu, H.C.; Sun, M.F.; Zhu, X.J.; Yue, S.Z.; Sun, Y.; et al. Realization of Perovskite-Nanowire-Based Plasmonic Lasers Capable of Mode Modulation. *Laser Photon. Rev.* **2019**, *13*, 9. [CrossRef]

Sidiropoulos, T.P.H.; Roder, R.; Geburt, S.; Hess, O.; Maier, S.A.; Ronning, C.; Oulton, R.F. Ultrafast plasmonic nanowire lasers near the surface plasmon frequency. *Nat. Phys.* **2014**, *10*, 870–876. [CrossRef]

Yu, H.C.; Ren, K.K.; Wu, Q.; Wang, J.; Lin, J.; Wang, Z.J.; Xu, J.J.; Oulton, R.F.; Qu, S.C.; Jin, P. Organic-inorganic perovskite plasmonic nanowire lasers with a low threshold and a good thermal stability. *Nanoscale* **2016**, *8*, 19536–19540. [CrossRef] [PubMed]

Zhao, D.; Liu, W.; Zhu, G.B.; Zhang, Y.Y.; Wang, Y.C.; Zhou, W.Y.; Xu, C.X.; Xie, S.S.; Zou, B.S. Surface plasmons promoted single-mode polariton lasing in a subwavelength ZnO nanowire. *Nano Energy* **2020**, *78*, 10. [CrossRef]

Liu, Y.; Li, F.; Xu, C.; He, Z.; Gao, J.; Zhou, Y.; Xu, L. The Design and Research of a New Hybrid Surface Plasmonic Waveguide Nanolaser. *Materials* **2021**, *14*, 2230. [CrossRef]

Zhu, H.M.; Fu, Y.P.; Meng, F.; Wu, X.X.; Gong, Z.Z.; Ding, Q.; Gustafsson, M.V.; Trinh, M.T.; Jin, S.; Zhu, X.Y. Lead halide perovskite nanowire lasers with low lasing thresholds and high quality factors. *Nat. Mater.* **2015**, *14*, 636–642. [CrossRef]

Loper, P.; Stuckelberger, M.; Niesen, B.; Werner, J.; Filipic, M.; Moon, S.J.; Yum, J.H.; Topic, M.; De Wolf, S.; Ballif, C. Complex Refractive Index Spectra of CH3NH3PbI3 Perovskite Thin Films Determined by Spectroscopic Ellipsometry and Spectrophotometry. *J. Phys. Chem. Lett.* **2015**, *6*, 66–71. [CrossRef]

Yariv, A. *Photonics: Optical Electronics in Modern Communications*; Pub. House of Electronic: Hong Kong, China, 2007.

Goykhman, I.; Desiatov, B.; Levy, U. Experimental demonstration of locally oxidized hybrid silicon-plasmonic waveguide. *Appl. Phys. Lett.* **2010**, *97*, 3. [CrossRef]

Oulton, R.F.; Sorger, V.J.; Genov, D.A.; Pile, D.F.P.; Zhang, X. A hybrid plasmonic waveguide for subwavelength confinement and long-range propagation. *Nat. Photonics* **2008**, *2*, 496–500. [CrossRef]

Zhao, P.; Su, J.; Lin, Z.H.; Wang, J.P.; Zhang, J.C.; Hao, Y.; Ouyang, X.P.; Chang, J.J. The crystal anisotropy effect of MAPbI(3) perovskite on optoelectronic devices. *Mater. Today Energy* **2020**, *17*, 6. [CrossRef]

Chu, K.W.; Ma, F.Y.; Zhu, X.J.; Jia, X.H.; Huang, Z.T.; Dong, K.Q.; Sun, J.Q.; Liu, K.; Jin, P.; Wang, Z.J.; et al. Dimension-programmable CsPbBr3 nanowires for plasmonic lasing with PDMS templated technique. *J. Phys. D-Appl. Phys.* **2022**, *55*, 6. [CrossRef]

Jia, X.H.; Wang, J.; Huang, Z.T.; Chu, K.W.; Ren, K.K.; Sun, M.F.; Wang, Z.J.; Jin, P.; Liu, K.; Qu, S.C. Metallic cavity nanolasers at the visible wavelength based on in situ solution-grown Au-coated perovskite nanowires. *J. Mater. Chem. C* **2022**, *10*, 680–687. [CrossRef]

46. Singh, J.K.; Mandal, S.K.; Banerjee, G. Refractive index of different perovskite materials. *J. Mater. Res.* **2021**, *36*, 1773–17
[CrossRef]
47. Sutherland, B.R.; Sargent, E.H. Perovskite photonic sources. *Nat. Photonics* **2016**, *10*, 295–302. [CrossRef]

photonics

*ticle*

# eal-Time Exploration on Buildup Dynamics of Diode-Pumped assively Mode-Locked Nd:YVO$_4$ Laser with SESAM

ı-Wen Cheng [1], Yu-Hsin Hsu [1], Xiu-Wei Chang [1], Hsing-Chih Liang [2] and Yung-Fu Chen [1,*]

[1] Department of Electrophysics, National Yang Ming Chiao Tung University, Hsinchu 30010, Taiwan; windrc.sc10@nycu.edu.tw (P.-W.C.); yuhsin.sc10@nycu.edu.tw (Y.-H.H.); xiaoxiu.sc11@nycu.edu.tw (X.-W.C.)
[2] Institute of Physics, National Yang Ming Chiao Tung University, Hsinchu 30010, Taiwan; hcliang@email.ntou.edu.tw
* Correspondence: yfchen@nycu.edu.tw

**Abstract:** The buildup dynamics of diode-pumped passively mode-locked solid-state laser is thoroughly explored using the real-time measurement with temporal sampling rate of up to 40 GHz. A concise cavity is developed to ensure the transient dynamics purely arising from the gain medium and saturable absorber. Experimental results reveal that the laser output in the buildup process exhibits numerous passively Q-switched pulses followed with a damped relaxation oscillation prior to the stable mode locking. Furthermore, it is confirmed that the laser output has already displayed single clean mode-locked pulses inside the first several Q-switched envelopes before stepping into the stage of relaxation oscillation. The present real-time exploration is expected to provide important information for practical applications with temporal modulation of the pump intensity.

**Keywords:** diode-pumped; mode-locked; solid-state laser; buildup dynamics; relaxation oscillation



**ation:** Cheng, P.-W.; Hsu, Y.-H.; .ng, X.-W.; Liang, H.-C.; Chen, . Real-Time Exploration on ıdup Dynamics of Diode-Pumped sively Mode-Locked Nd:YVO$_4$ er with SESAM. *Photonics* **2023**, *10*, https://doi.org/10.3390/ tonics10070826

eived: 24 June 2023
ised: 12 July 2023
epted: 14 July 2023
lished: 15 July 2023



## 1. Introduction

Techniques used to achieve mode locking in lasers can be classified as "active" or "passive". Active techniques typically involve the use of external signals to induce modulation of light within the cavity. On the other hand, passive techniques do not use external signals, but instead rely on putting some components into the laser cavity, which induce the self-modulation of the light. Passively continuous-wave mode-locked (CWML) solid-state lasers with saturable absorbers [1–3] have been widely used in various applications of time-resolved measurements, material processing, and other precise spectroscopies [4–6]. Saturable absorbers, such as semiconductor saturable absorber mirror (SESAM) and carbon nanotube, can be deliberately introduced into the laser cavity to self-start mode locking by providing nonlinearity [7–9]. Understanding the phase locking process of a large number of longitudinal cavity modes is greatly helpful for designing passively mode-locked lasers and valuable for finding the novel applications of ultrafast laser sources [10–13]. Several theoretical models have been proposed to explain the observed threshold power for the self-starting of the mode locking and to explore the buildup dynamics using the Ginzburg–Landau equation [14,15]. In theoretical models, the initial optical wave is generally assumed to be a single-pulse fluctuation residing on a continuous background [16–19]. However, it has been recently found that the initial multiple-pulse phase and pulse-splitting phase last for a long time in the buildup dynamics in self-starting of soliton mode-locked fiber lasers [20]. Furthermore, the relaxation oscillations were observed to speed up the starting process in dissipative-soliton fiber lasers [21]. In contrast, it is worth noting that there are no previous experimental reports on the transient process of initial pulse evolutions and the role of relaxation oscillations in passively CWML solid-state lasers. Therefore, exploring the transient pulse evolution in real-time measurements is important for understanding the starting dynamics of passively CWML solid-state lasers [22–25].

So far, the cavity structures used in the passively mode-locked solid-state lasers mai comprise the V-shaped [26], Z-shaped [27,28], and W-shaped [29] configurations. In spite different configurations, the common purposes are to satisfy the designed requirements the cavity length related to the pulse repetition rate, the mode size on the gain medium pumping overlap, and the mode size of the saturable absorber for obtaining stable CW The complexity of the cavity configuration somewhat introduces additional influences the transition dynamics of the mode locking buildup. Consequently, it is greatly useful exploring the buildup dynamics to construct a passively mode-locked solid-state laser w minimal components in the cavity configuration.

In this work, we perform the real-time measurement to explore the starting dynam of diode-pumped passively CWML laser. The laser cavity is designed to contain o three components of a Nd:$YVO_4$ crystal, a SESAM device, and a focusing lens to elimin the unwanted influence on the buildup dynamics. A fast photodetector and a digi oscilloscope with 40 Giga sampling rate are used to directly measure the transient behavi in various expanded time scales. Experimental results reveal that the laser output in buildup process consists of a series of passively Q-switched pulses followed with a damp relaxation oscillation prior to the stable mode locking. The number of passively Q-switch pulses and the damping time of the relaxation oscillation decrease with increasing the pu power. Furthermore, the temporal traces inside the first several Q-switched envelo clearly manifest the transient evolution of the laser output from a random set of spikes the emergence of a single clean pulse within each round-trip period. It is indicated t the buildup time for the complete phase locking is significantly shorter than that for stable mode locking. To the best of our knowledge, this result is the first direct observati of the buildup dynamics of a passively mode-locked solid-state laser with SESAM. T present real-time measurement is believed to provide useful insights into applications w temporal modulation of the pump intensity.

## 2. Experimental Setup

Figure 1 shows the experimental setup of passively mode-locked solid-state la for exploring the buildup dynamics in the turn-on process. To investigate the transi dynamics purely arising from the gain medium and saturable absorber, the laser cavity w designed to comprise only three components of a Nd:$YVO_4$ crystal, a SESAM device, an focusing lens. The pump source was a fiber-coupled semiconductor laser with wavelen at 808 nm, where the fiber had a core diameter of 200 μm and a numerical apert of 0.16. The gain medium was a *a*-cut 0.35 at.% Nd:$YVO_4$ crystal with a dimension $3 \times 3 \times 8$ mm$^3$. One facet of the Nd:$YVO_4$ crystal was a plane surface that was coated a rear mirror with high reflection (HR) at 1064 nm (reflectance > 99.8%) and high transmissi at 808 nm (transmittance > 95%). The other facet of the laser crystal with a wedged angle 0.5° for avoiding the internal etalon effect had an anti-reflection (AR) coating at 808 nm a 1064 nm (reflectance < 0.2%). The Nd:$YVO_4$ crystal wrapped with indium foil was moun in water-cooled copper block with temperature held at 20 °C. The saturable absorber w a SESAM device that was simultaneously designed as an output coupler, abbreviated SESAMOC [4,5]. The SESAM device was monolithically grown on an undoped Ga substrate with the method of metalorganic chemical vapor deposition. The SESAM was composed of 10 pairs of distributed Bragg reflectors (DBRs) that were formed AlAs/GaAs quarter-wavelength layers with the effective reflectivity of approximat 96.5% at 1064 nm. The absorbing material of the SESAMOC was made of two 8 $In_{0.34}Ga_{0.66}As$ quantum well that provided an effective modulation depth of 0.8% with saturable fluence of 50 μJ/cm$^2$. Furthermore, the back side of the 350 μm GaAs substr was coated for AR at 1064 nm (reflectance < 0.5%). The SESAMOC device was solder on a copper holder with temperature maintained at 20 °C. A focusing lens with a fo length of $f$ = 100 mm was used to satisfy the criterion of self-starting CWML cavity. B sides of the focusing lens were coated for AR at 1064 nm (reflectance < 0.5%). Experimen optical spectra were measured using a Michelson interferometer-based Fourier spectrome

(Advantest Q8347) with a resolution of up to 0.003 nm. Temporal characteristics of mode-locked pulses were measured using a high-speed InGaAs photodetector (Electro-Optics Technology Inc., (Idar, Germany) ET-3500 with rise time of 35 ps), whose output signal was recorded using a digital oscilloscope (Teledyne LeCroy, Chestnut Ridge, NY, USA, Wave Master 820Zi-A) with a 20 GHz electrical bandwidth and the sampling rates of 25 ps. Moreover, the output signal detected by the photodetector was also connected to an RF spectrum analyzer (Agilent, 8563EC, Santa Clara, CA, USA) with a bandwidth of 26.5 GHz to measure the output power spectrum. The second-order autocorrelation traces were measured using a commercial autocorrelator (APE pulse check, Angewandte Physik and Elektrnic GmbH, Berlin, Germany) to analyze the effective pulse width.

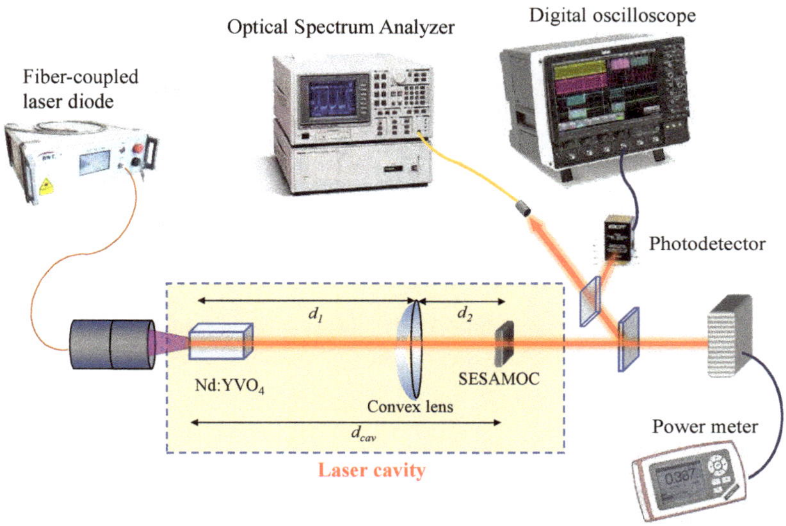


**Figure 1.** Experimental setup of passively mode-locked solid-state laser for exploring the buildup dynamics in the turn-on process.

### 3. Analysis of Cavity Mode Sizes

As shown in Figure 1, $d_1$ is the effective optical length between the HR side of the laser crystal and the focusing lens, and $d_2$ is the effective optical length between the focusing lens and the output plane of the SESAMOC. By using the ABCD matrix method, the mode radius for the $TEM_{00}$ at the HR side of Nd:YVO$_4$ crystal can be derived as

$$w_1 = \sqrt{\frac{\lambda}{\pi}\sqrt{\frac{(d_1 - f)[(d_1 + d_2)f - d_1 d_2]}{(d_2 - f)}}}, \tag{1}$$

where $\lambda$ is the laser wavelength. On the other hand, the cavity mode radius at the output plane of the SESAMOC can be found to be

$$w_2 = \sqrt{\frac{\lambda}{\pi}\sqrt{\frac{(d_2 - f)[(d_1 + d_2)f - d_1 d_2]}{(d_1 - f)}}}. \tag{2}$$

In the experiment, the distances of $d_1$ and $d_2$ are designed be 460 and 125 mm, respectively. Substituting the values of $f$, $d_1$, and $d_2$ into Equations (1) and (2), the mode radii $w_1$ and $w_2$ can be calculated to be approximately 203 and 53 μm, respectively. The values of

$w_1$ and $w_2$ will be used in the following section. The optical length of the laser cavity approximately 585 mm corresponding to the repetition rate of 256 MHz.

## 4. Criterion for the Regimes of Q-Switched and CW Mode Locking

Saturable absorbers can be used in solid-state lasers to start up the CWML; howev they also usually drive lasers into the so-called Q-switched mode-locked (QML) states. the QML operation, the laser output exhibits the characteristic of mode-locked pulses i Q-switched envelope. Due to the conventional cavity length, mode-locked pulse repetiti rates are typically in the order of 100 MHz, while typical Q-switched modulations are the order of 100 kHz. These Q-switch instabilities are undesirable for many applicatio requiring constant pulse energy and high repetition rates. However, for certain applicatio such as nonlinear wavelength conversion, laser micromachining, or ablation applicatio the QML mechanism may be attractive because the pulse energy is significantly increas but still concentrated in ultrashort mode-locked pulses. Regardless, understanding QM related solid-state laser dynamics is important to explore the transition dynamics of CWM

For a passively mode-locked solid-state laser under the CW pumping scheme, the is a critical intracavity pulse energy, $E_{P,c}$, which is the minimum intracavity pulse ener for obtaining stable CWML, i.e., for $E_P > E_{P,c}$, the CWML operation can be achieved, a for $E_P < E_{P,c}$, the laser output is displayed in the QML state, where $E_P$ is the energy o mode-locked pulse in the cavity. As shown in Figure 2, the QML state (Figure 2a) exhib that the pulse energy is modulated with a strongly peaked Q-switching envelope, wh in the CWML regime (Figure 2b), the laser output displays a train of mode-locked puls with constant pulse energy. The experimental results shown in Figure 2a,b were obtain with the present laser cavity at the pump powers of 1.0 and 2.0 W, respectively.

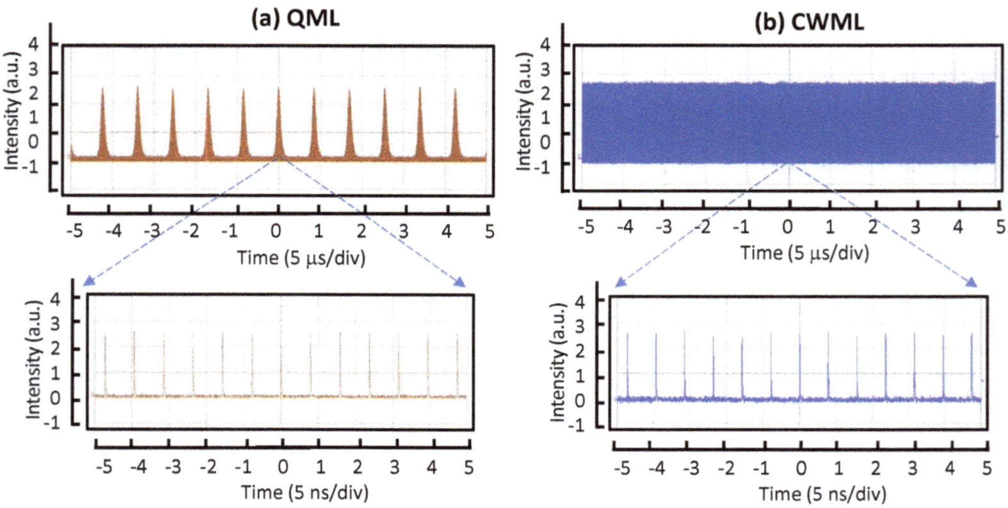


**Figure 2.** Measured oscilloscope traces at two different time spans for (**a**) the operation of QM obtained at the pump power of 1.0 W and (**b**) the operation of CWML obtained at the pump pow of 2.0 W.

Figure 3a shows the average output power versus the pump power for the las operation. At a pump power of 6.3 W, the output power for stable passive mode locki is approximately 1.3 W, corresponding to the optical-to-optical conversion efficiency 20.6%. The pump power for lasing threshold was found to be around 0.1 W, where pump power as high as 1.2 W was required for stable mode locking. Figure 3b shows t measured power spectral density at a pump power of 3.0 W for a representative case the experimental results. The power spectral density reveals a signal-to-noise ratio higher than 60 dB, indicating the excellent quality of stable mode locking. Figure 3c shows the measured oscilloscope traces at four different time spans ranging from 10 ns to 50 μs to display good stability and CW mode locking. From the autocorrelation trace, the real pulse duration of CW mode locking was estimated to be approximately 20 ps based on the assumption of the Gaussian-shaped profile.

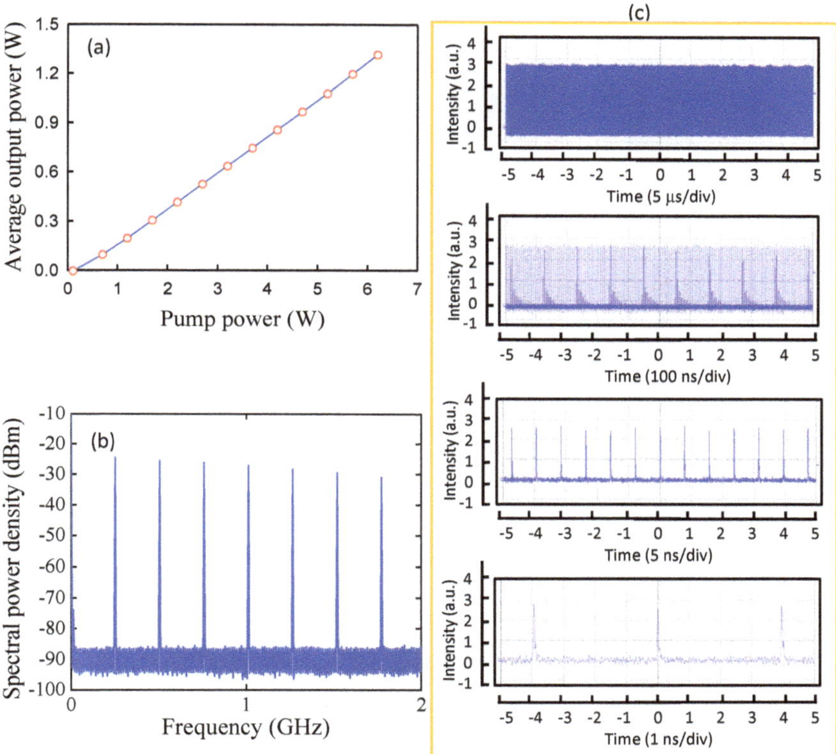


**Figure 3.** (**a**) Average output power versus the pump power for the laser operation. (**b**) Measured power spectral density at a pump power of 3.0 W for a representative case of the experimental results. (**c**) Measured oscilloscope traces at four different time spans ranging from 10 ns to 50 μs.

Hönninger et al. [30] used the rate equations for the intracavity power, gain, and saturable absorption to derive a stability criterion against QML by performing a linearized stability analysis analogous to the derivation of the Q-switching criterion [31,32]. With certain approximations, the critical intracavity pulse energy $E_{P,c}$ as the square root of the QML parameter can be derived as [30]

$$E_{P,c} = \left(F_{sat,L}A_{eff,L}F_{sat,A}A_{eff,A}\Delta R\right)^{1/2}, \tag{3}$$

where $F_{sat,L}$ is the saturation fluence of the gain medium, $A_{eff,L}$ the effective laser mode area inside the gain medium, $F_{sat,A}$ is the saturation fluence of the saturable absorber, $A_{eff,A}$ the effective laser mode area inside the saturable absorber, and $\Delta R$ is the modulation depth of the absorber. The saturation fluence of the gain medium is given by $F_{sat,L} = h\upsilon/m\sigma$, where $h\upsilon$ is the photon energy, $\sigma$ is the stimulated emission cross section, and the factor $m$

is the number of passes through the gain element per cavity round trip. For a ring cav $m = 1$, while for a simple standing wave cavity $m = 2$.

In the present experiment, the average output power for the critical transition fr QML to CWML was found to be around 0.2 W, corresponding to the intracavity pow of 5.5 W. Accordingly, the critical pulse energy $E_{P,c}$ can be experimentally determined be approximately 21.5 µJ. On the other hand, Equation (3) can be used to calculate theoretical value for the critical pulse energy $E_{P,c}$. By using the physical parameters $F_{sat,L}$= 3.66 × $10^7$ nJ/cm$^2$, $F_{sat,A}$= 5.0 × $10^4$ nJ/cm$^2$, $A_{eff,L} = \pi w_1^2/2 = 6.5 \times 10^{-4}$ c $A_{eff,A} = \pi w_2^2/2 = 4.4 \times 10^{-5}$ cm$^2$, and $\Delta R = 0.008$, the theoretical value for the criti pulse energy $E_{P,c}$ can be calculated to be 20.5 µJ. The theoretical value can be seen to ag very well with experimental result. In the following study, the transition dynamics will explored with the pulse pumping scheme.

## 5. Results and Discussion

For exploring the transient or dynamic behavior of laser oscillators, it is necessary clarify the terminologies used in the description of experimental phenomena. The terms spikes, relaxation oscillations, and gain switching refer to various kinds of comparativ slow but different amplitude oscillations that are often observed by many solid-state las either when the laser is first turned on, or when the laser is suddenly perturbed by any ki of small fluctuation in gain, cavity loss, or cavity alignment. The term "spike" is used to re to the discrete, sharp, large-amplitude pulses that typically occur during the initial turn- phase. The term "relaxation oscillation" is used to describe the small-amplitude, qu sinusoidal, exponentially damped oscillations near the steady-state amplitude, followi the initial spiking behavior. Spikes and relaxation oscillations are characteristic phenome of most solid-state lasers, semiconductor lasers, and fiber lasers, where the recovery time the population inversion of the excited state is much longer than the cavity decay time. this exploration, the occurrence of spikes includes the influence of the saturable absor Therefore, we use the term of passively Q-switched pulse instead of spikes for discussi the experimental observations.

To explore the buildup dynamics, the pump source of the laser diode was electronica operated at a pump duration of 500 µs with a pump frequency of 1 kHz. The pump- duration of 500 µs is sufficiently long to observe the pulse evolution. Furthermore, pump-off cycle of 500 µs is also long enough to clean up the residual population dens since the upper-state lifetime $t_f$ is approximately 90 µs. The laser output was direc measured using a digital oscilloscope with 40 Giga sampling rate with a fast photodetec Based on thorough experimental observations, it is confirmed that under the optimal cav alignment the buildup process of the mode locking mainly depended on the pumpi strength, rather insensitive to environmental perturbations, such as air flow, acoustic no and thermal variations. Consequently, experimental results could be utterly reproduci Figure 4a–e show the oscilloscope traces for the transient turn-on behavior measured different pump powers in the time span of 500 µs. The transient feature of the build process can be found to comprise a series of passively Q-switched pulses followed w damped relaxation oscillations. The number of passively Q-switched pulses can be clea seen to decrease with increasing the pump strength. From the results shown in Figure 4, buildup time for the stable mode locking, $t_{ML}$, can be identified as the duration from turn-on time of the pump power to the end of the damped relaxation oscillation. Referri to $t_f$ = 90 µs, the buildup time $t_{ML}$ decreases from a few times to a fraction of the upper-le lifetime for the pump power increasing from 1.5 to 5.0 W.

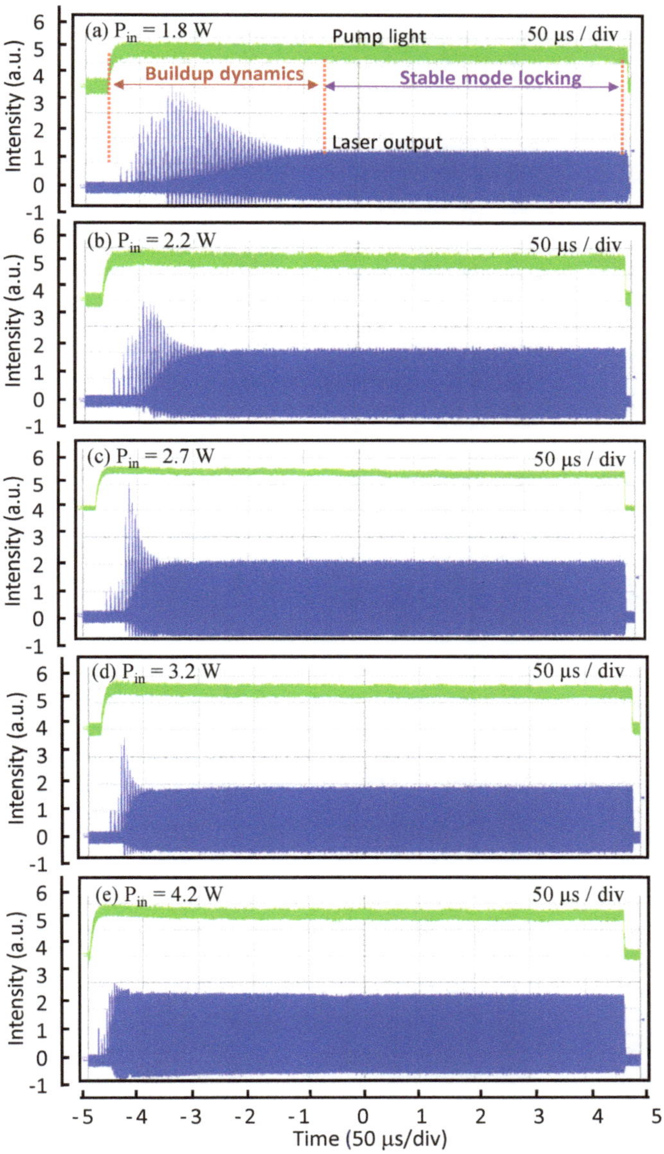


**Figure 4.** (**a**–**e**) Oscilloscope traces for the transient turn-on behavior measured at different pump powers in the time span of 500 µs; green lines: pump light; blue lines: laser output.

The formation process for the phase locking in the transient turn-on dynamics can be comprehended from the temporal characteristics inside the passively Q-switched envelopes. Figure 5a shows the oscilloscope traces for the transient dynamics measured at a pump power of 2.6 W in a time span of 55 µs. It can be found that each passively Q-switched pulse is approximately a fraction of a microsecond wide and spaced a few microseconds apart. After growing to a maximum, the Q-switched pulses are gradually transformed into a quasi-sinusoidal relaxation oscillation, then damping to the stable mode locking. The phase locking process between lasing longitudinal modes in the buildup dynamics can be further explored by looki into the temporal behaviors inside the Q-switched envelopes. The six plots with labels (1)– in Figure 5b show the temporal traces in a time span of 50 ns measured from the intensit inside the first six Q-switched envelopes shown in Figure 5a, respectively. From the left-ha column of Figure 5b for the first three Q-switched pulses, the laser output can be seen to cons not of single clean pulses but of a random set of spikes, essentially a noise-like feature, w the spikes distributed throughout the entire round-trip period. Nevertheless, it can be fou that the signal output is still strictly periodic in time, that is, it repeats exactly after one cy The signature of random spikes within each round-trip cycle indicates that the laser outp consists of a large number of longitudinal modes with random phases, regardless of wheth the amplitude distribution is random or more orderly. Precisely, the laser output in the fi three Q-switched pulses can be called a case of mode coupling that does not achieve the pha locking. The cubic–quintic complex Ginzburg–Landau equation may describe qualitatively w this observation [33]. From the right-hand column of Figure 5b for the Q-switched pulses w labels (4)–(6), the transient evolution of phase locking can be clearly observed. From the plot in Figure 5b, the laser output has already displayed a single clean pulse within each round-t period. In other words, the lasing longitudinal modes have been arrayed all in phase to reac phase-locked situation.

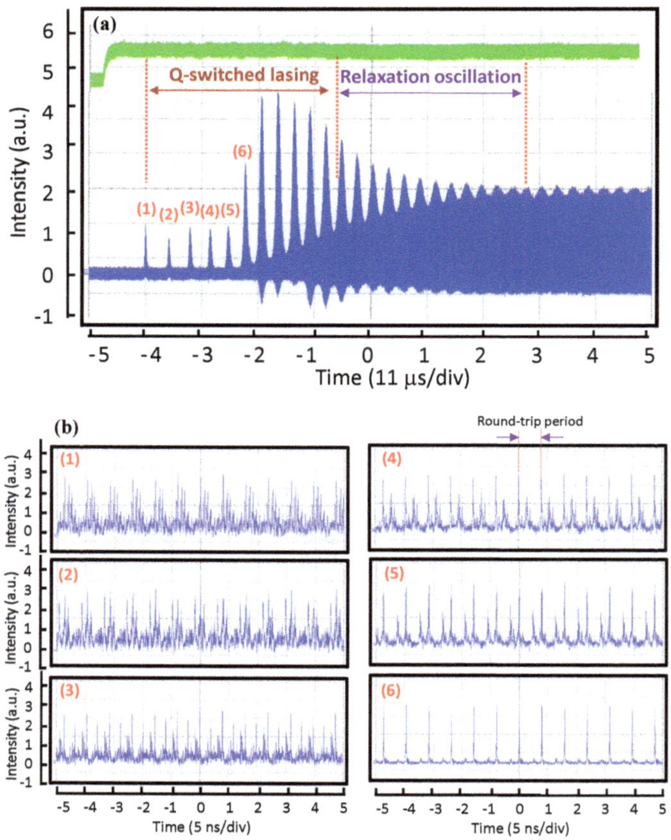


**Figure 5.** (**a**) Oscilloscope traces for the transient dynamics measured at a pump power of 2.6 W i time span of 55 μs; green lines: pump light; blue lines: laser output. (**b**) Temporal traces in a ti span of 50 ns measured from the intensities inside the first six Q-switched envelopes shown in (**a**

It is worth mentioning the distinction between mode coupling and mode locking. Any superposition of a set of longitudinal modes equally spaced in frequency, regardless of their relative magnitudes and phases, will certainly result in an output signal envelope that is periodic in time with a fundamental period equal to the reciprocal of the interval between adjacent longitudinal modes. Thus, any such state can generally be referred to as a mode-coupling situation, regardless of the signal variation over a period, since the modes are all coupled together with fixed (even if irregular) amplitude and phase. On the other hand, the term of mode locking is only used for conditions in which the longitudinal modes are all arranged in phase; under these circumstances, constructive interference between the superposed longitudinal modes can lead to the laser output to be a train of extremely short pulses. The buildup time for the complete phase locking, $t_{PL}$, can be ascertained as the duration from the turn-on time of the pump power to the emergence of a single clean pulse within each round-trip period. Figure 6 shows the experimental results for the buildup times $t_{PL}$ and $t_{ML}$ versus the pump power. Clearly, the buildup time $t_{PL}$ for the complete phase locking is significantly shorter than the time $t_{ML}$ for the stable mode locking.

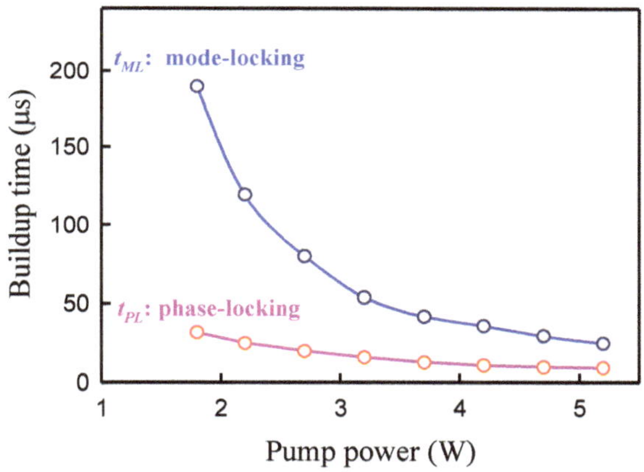


**Figure 6.** Experimental results for the buildup times for the complete phase locking and stable mode locking versus the pump power.

## 6. Conclusions

In summary, we have explored the buildup dynamics of diode-pumped passively mode-locked solid-state laser by means of the real-time measurement with temporal sampling rate up to 40 GHz. We have designed the laser cavity comprising only three components to ensure the transient dynamics purely arising from the gain medium and saturable absorber. The laser output in the buildup process was observed to exhibit a number of passively Q-switched pulses followed with a damped relaxation oscillation prior to the stable mode locking. The buildup time for the stable mode locking was found to decrease from a few times to a fraction of the upper-level lifetime for the pump power increasing from slightly to far above lasing threshold. Furthermore, we have experimentally confirmed that the laser output has already displayed single clean mode-locked pulses inside the first several Q-switched envelopes, which indicates that the buildup rate for the complete phase locking is substantially faster than that for the stable mode locking. It is believed that the present real-time exploration can offer important information for practical applications with temporal modulation of the pump intensity.

**Author Contributions:** Conceptualization, P.-W.C. and Y.-F.C.; validation, P.-W.C. and H.-C.L.; formal analysis, P.-W.C. and Y.-F.C.; resources, Y.-H.H. and X.-W.C.; writing—original draft preparation, Y.-F.C.; writing—review and editing, P.-W.C., H.-C.L. and Y.-F.C.; supervision, Y.-F.C. All authors have read and agreed to the published version of the manuscript.

**Funding:** This work is supported by National Science and Technology Council of Taiwan (contr number 109-2112-M-009-015-MY3).

**Institutional Review Board Statement:** Not applicable.

**Informed Consent Statement:** Not applicable.

**Data Availability Statement:** All data reported in the paper are presented in the main text. A other data will be provided on request.

**Conflicts of Interest:** The authors declare no conflict of interest.

## References

1. Chen, C.J.; Wai, P.K.A.; Menyuk, C.R. Self-starting of passively mode-locked lasers with fast saturable absorbers. *Opt. Lett.* **19** *20*, 350–352. [CrossRef]
2. Haus, H.A. Mode-Locking of Lasers. *IEEE J. Sel. Top. Quantum Electron.* **2000**, *6*, 1173–1185. [CrossRef]
3. Soto-Crespo, J.M.; Akhmediev, N.; Town, G. Continuous-wave versus pulse regime in a passively mode locked laser with a saturable absorber. *J. Opt. Soc. Am. B* **2002**, *19*, 234–242. [CrossRef]
4. Picqué, N.; Hänsch, T.W. Frequency comb spectroscopy. *Nat. Photonics* **2019**, *13*, 146–157. [CrossRef]
5. Phillips, K.C.; Gandhi, H.H.; Mazur, E.; Sundaram, S.K. Ultrafast laser processing of materials: A review. *Adv. Opt. Photon.* **2(** *7*, 684–712. [CrossRef]
6. Diddams, S.A.; Vahala, K.; Udem, T. Optical frequency combs: Coherently uniting the electromagnetic spectrum. *Science* **2(** *369*, eaay3676. [CrossRef] [PubMed]
7. Keller, U. Recent developments in compact ultrafast lasers. *Nature* **2003**, *424*, 831–838. [CrossRef]
8. Chernysheva, M.; Rozhin, A.; Fedotov, Y.; Mou, C.; Arif, R.; Kobtsev, S.M.; Dianov, E.M.; Turitsyn, S.K. Carbon nanotubes ultrafast fibre lasers. *Nanophotonics* **2017**, *6*, 1–30. [CrossRef]
9. Rafailov, W.S.E.U.; Cataluna, M.A. Mode-locked quantum-dot lasers. *Nat. Photonics* **2007**, *1*, 395–401. [CrossRef]
10. Herink, G.; Jalali, B.; Ropers, C.; Solli, D.R. Resolving the build-up of femtosecond mode-locking with single-shot spectroscopy 90 MHz frame rate. *Nat. Photonics* **2016**, *10*, 321–326. [CrossRef]
11. Liu, X.; Cui, Y. Revealing the behavior of soliton buildup in a mode-locked laser. *Adv. Photon.* **2019**, *1*, 016003. [CrossRef]
12. Liu, X.; Pang, M. Revealing the buildup dynamics of harmonic mode-locking states in ultrafast lasers. *Laser Photon. Rev.* **2(** *13*, 1800333. [CrossRef]
13. Liu, X.; Popa, D.; Akhmediev, N. Revealing the Transition Dynamics from Q Switching to Mode Locking in a Soliton Laser. *P* *Rev. Lett.* **2019**, *123*, 093901. [CrossRef]
14. Popov, M.; Gat, O. Pulse growth dynamics in laser mode locking. *Phys. Rev. A* **2018**, *97*, 011801. [CrossRef]
15. Uzunov, I.M.; Georgiev, Z.D.; Arabadzhiev, T.N. Transitions of stationary to pulsating solutions in the complex cubic-quir Ginzburg-Landau equation under the influence of nonlinear gain and higher-order effects. *Phys. Rev. E* **2018**, *97*, 052 [CrossRef]
16. Haus, H.A.; Ippen, E.P. Self-starting of passively mode-locked lasers. *Opt. Lett.* **1991**, *16*, 1331–1333. [CrossRef]
17. Krausz, F.; Brabec, T.; Spielmann, C. Self-starting passive mode locking. *Opt. Lett.* **1991**, *16*, 235–237. [CrossRef]
18. Krausz, F.; Brabec, T. Passive mode locking in standing-wave laser resonators. *Opt. Lett.* **1993**, *18*, 888–890. [CrossRef]
19. Hermann, J. Starting dynamic, self-starting condition and mode-locking threshold in passive, coupled-cavity or Kerr-lens mo locked solid-state lasers. *Opt. Commun.* **1993**, *98*, 111–116. [CrossRef]
20. Liu, S.; Chen, Y.; Huang, L.; Cao, T.; Qin, X.; Ning, H.; Yan, J.; Hu, K.; Guo, Z.; Peng, J. Optimal conditions for self-starting soliton mode-locked fiber lasers with a saturable absorber. *Opt. Lett.* **2021**, *46*, 2376–2379. [CrossRef]
21. Li, H.; Ouzounov, D.G.; Wise, F.W. Starting dynamics of dissipative-soliton fiber laser. *Opt. Lett.* **2010**, *35*, 2403–2405. [CrossR [PubMed]
22. Wei, W.; Liu, R.; Zhao, S.; Yang, K.; Li, D.; Guo, L.; Wang, Y. Simulation of the passively mode-locked laser with a SESAM. *O* **2012**, *123*, 2191–2194. [CrossRef]
23. Waritanant, T.; Major, A. High efficiency passively mode-locked Nd:$YVO_4$ laser with direct in-band pumping at 914 nm. *C Express* **2016**, *24*, 12851–12855. [CrossRef] [PubMed]
24. Waritanant, T.; Major, A. Discretely selectable multiwavelength operation of a semiconductor saturable absorber mirror mo locked Nd:$YVO_4$ laser. *Opt. Lett.* **2017**, *42*, 3331–3334. [CrossRef]
25. He, H.; Liu, X.; Song, Y.; Wang, C.; Cao, M.; Yan, A.; Wang, Z. LD end-pumped Nd: $YVO_4$ high energy high beam quality 1 nm picosecond laser with a semiconductor saturable absorber mirror. *Optik* **2018**, *175*, 172–176. [CrossRef]
26. Iliev, H.; Buchvarov, I.; Choi, S.Y.; Kim, K.; Rotermund, F.; Petrov, V. 1.34 μm Nd:$YVO_4$ laser mode-locked by a single-wal carbon nanotube saturable absorber. In Proceedings of the Solid State Lasers XXI: Technology and Devices, San Francisco, C USA, 22–25 January 2012; Volume 8235, p. 82350I.
27. Chen, Y.F.; Tsai, S.W.; Lan, Y.P.; Wang, S.C.; Huang, K.F. Diode-end-pumped passively mode-locked high-power Nd:$YVO_4$ la with a relaxed saturable Bragg reflector. *Opt. Lett.* **2001**, *26*, 199–201. [CrossRef] [PubMed]

Liu, Y.H.; Xie, Z.D.; Pan, S.D.; Lv, X.J.; Yuan, Y.; Hu, X.P.; Lu, J.; Zhao, L.N.; Chen, C.D.; Zhao, G.; et al. Diode-pumped passively mode-locked Nd:$YVO_4$ laser at 1342 nm with periodically poled $LiNbO_3$. *Opt. Lett.* **2011**, *36*, 698–700. [CrossRef]

Yang, Y.; Xu, J.-L.; He, J.-L.; Yang, X.-Q.; Zhang, B.-Y.; Yang, H.; Liu, S.-D.; Zhang, B.-T. Diode-pumped passively mode-locked Nd:YAG laser at 1338 nm with a semiconductor saturable absorber mirror. *Appl. Opt.* **2001**, *50*, 6713–6716. [CrossRef]

Hönninger, C.; Paschotta, R.; Morier-Genoud, F.; Moser, M.; Keller, U. Q-switching stability limits of cw passive mode locking. *J. Opt. Soc. Am. B* **1999**, *16*, 46–56. [CrossRef]

Haus, H.A. Parameter ranges for cw passive modelocking. *IEEE J. Quantum Electron.* **1976**, *12*, 169–176. [CrossRef]

Kärtner, F.X.; Brovelli, L.R.; Kopf, D.; Kamp, M.; Calasso, I.; Keller, U. Control of solid-state laser dynamics by semiconductor devices. *Opt. Eng.* **1995**, *34*, 2024–2036. [CrossRef]

Tél, T.; Lai, Y.C. Chaotic transients in spatially extended systems. *Phys. Rep.* **2008**, *460*, 245–275. [CrossRef]

photonics

Article

# $CO_2$ and Diode Lasers vs. Conventional Surgery in the Disinclusion of Palatally Impacted Canines: A Randomized Controlled Trial

**Alessandra Impellizzeri, Martina Horodynski *, Gaspare Palaia , Gerardo La Monaca , Daniele Pergolini , Antonella Polimeni , Umberto Romeo and Gabriella Galluccio**

Department of Oral and Maxillofacial Sciences, "Sapienza" University of Rome, Via Caserta 6, 00161 Rome, It

* Correspondence: martina.horodynski@uniroma1.it; Tel.: +39-3401619639

**Abstract:** Background: The aim of this RCT is to show the effectiveness of laser technology for t exposure of palatally impacted canines, using a $CO_2$ or diode laser, and to evaluate the possi bio-stimulation effect of the laser on the spontaneous eruption of the canine. Methods: This study w carried out on a sample of 27 patients, divided randomly into three groups: treated with a $CO_2$ la (Group A), treated with a diode laser (Group B), and treated with a cold blade (Group C). Monitori was performed at 1, 8, and 16 weeks after surgery, through photo and digital scans performed w a CS3500 intraoral scanner. Results: It was found that the average total eruptions are 4.55 mm Group A, 5.36 mm for Group B, and 3.01 mm for Group C. The difference in eruption between grou A and B is not significant. Comparing the laser groups with the control group, it has emerged th the difference in eruption is statistically significant. Conclusion: A significant tooth movement w observed in both Groups A and B. The response of the canine to the bio-stimulation of the laser c be considered effective, resulting in a statistically significant difference between the study grou and the control group. Both lasers have the same bio-stimulatory action on the eruption of canin

**Keywords:** impacted canines; orthodontics; intraoral scanner; digital monitoring; diode laser; C laser; palatally impacted canine



**Citation:** Impellizzeri, A.; Horodynski, M.; Palaia, G.; La Monaca, G.; Pergolini, D.; Polimeni, A.; Romeo, U.; Galluccio, G. $CO_2$ and Diode Lasers vs. Conventional Surgery in the Disinclusion of Palatally Impacted Canines: A Randomized Controlled Trial. *Photonics* **2023**, *10*, 244. https://doi.org/10.3390/photonics10030244

Received: 9 January 2023
Revised: 16 February 2023
Accepted: 22 February 2023
Published: 23 February 2023



## 1. Introduction

The permanent maxillary canine is the most impacted tooth after the third mola with an incidence of 0.9% to 2.2% [1–3]. It shows a predilection for the palatal over t labial side, it is more common in female patients than in male patients and it has a defin tendency toward being unilateral rather than bilateral [4].

Over the years, numerous researchers have focused on identifying the cause of impact upper canines. The etiology of palatally impacted canines has been suggested as polyge and multifactorial. There are two main theories: the genetic theory and the guidance thec According to the genetic theory, palatal displacement of the canine is just another associat genetic characteristic. According to the guidance theory of canine impaction, these factc create a genetically determined environment in which the developing canine is deprived of guidance, thus influencing it to adopt an abnormal eruption path [5].

The therapeutic approach involves surgical exposure of the impacted canines, follow by orthodontic traction to guide and align the tooth to the dental arch. For palata impacted canines, different surgical approaches and various orthodontic traction techniqu have been described in the literature [6]. There are two basic types of surgery for t exposure of a palatally impacted canine: the open and the closed technique.

The open technique includes the surgical exposure of the crown by complete remov of the bone and soft tissue directly overlying the impacted canine. Afterwards, a surgi pack might be used to cover the wound, while the canine can be either left to spontaneou erupt or an orthodontic attachment can be directly bonded on the canine in order to direc apply traction. The closed technique, on the other hand, involves raising a full mucoperiostal flap, exposing the canine crown, and bonding an attachment to it. Afterwards, the flap is repositioned, and orthodontic traction is applied after initial healing until the canine erupts in the oral cavity and is subsequently guided to the dental arch. Although both approaches are versatile, can be adapted to each case and have been used extensively for many years, reports about their comparative performance are mixed. Several studies have evaluated various aspects of their performance, including surgical duration and postoperative recovery time, postoperative pain, periodontal health and esthetic appearance [5–7].

Today, laser technology has achieved a more important role in dentistry, and in the last two decades, there has been an increase in research studies into laser applications [8–12].

The use of a laser offers several advantages when compared with conventional surgery, including precision, minimal intraoperative hemorrhage, sterilization of the surgical area, healing with minimal scarring and decreased postoperative pain and swelling [9].

The purpose of this study is to introduce the use of a laser to expose the crown of a palatally impacted canine, avoiding the application of an orthodontic device to traction it. At the base of this research, there is an evaluation of the literature, which reports significant studies demonstrating the biostimulation effect of the photobiomodulation therapy (PBMT) [10–12].

Biostimulation means the activation of regenerative and healing processes that depends on the ability of subcellular photoreceptors to respond to visible red and near-infrared wavelengths. Stimulation of these receptors influences the electron transport chain, the respiratory chain and oxidation, expressed as an increase in the cellular metabolic processes [12,13].

Several studies have examined the effects of low-level laser therapy, known as photobiomodulation therapy, on bone formation in vitro and in vivo [12].

Photobiomodulation is one of the most promising approaches today to stimulate dental movements. It has been found that laser light stimulates the proliferation of osteoclasts, osteoblasts and fibroblasts and thereby affects bone remodeling and accelerates tooth movement. The mechanism involved in the acceleration of tooth movement is the production of ATP and the activation of cytochrome C, and low-energy laser irradiation enhanced the velocity of tooth movement via RANK/RANKL and the macrophage-colony-stimulating factor and its receptor expression [9].

Furthermore, the literature reports that high-intensity laser therapies (HILTs), the "surgical lasers", exerting a cutting action on the soft tissues [14–18], can also be used for opercolectomy, removing the soft tissue that overlies the impacted tooth.

Soft-tissue lasers have numerous applications in orthodontics, including gingivectomy, frenectomy, operculectomy, papilla flattening, uncovering temporary anchorage devices, ablation of aphthous ulcerations, exposure of impacted teeth and even tooth whitening. As an adjunctive procedure, laser surgery has helped many orthodontists to enhance the design of a patient's smile and improve treatment efficacy [17].

However, no study has evaluated the eruptive stimulating action when a laser is applied to expose the crown of impacted canines.

The aim of this RCT is to show the effectiveness of laser technology for the exposure of palatally impacted canines using a $CO_2$ laser device (Smart US20D®, DEKA—Florence, Italy) and a diode laser device (Raffaello, DMT, Lissone, Italy, 980 nm +645 nm), and to evaluate the possible biostimulation of the laser on the spontaneous eruption movement of the canine, without orthodontic traction application. Moreover, the purpose of this study is to show the accuracy of the monitoring of the movement of the impacted tooth after exposure with a laser approach using digital technologies.

## 2. Material and Methods

### 2.1. Study Design

This study was carried out on a sample of 27 subjects referred to the Orthodontics UOC of the Department of Oral and Maxillo-Facial Sciences of "Sapienza" University of Rome. A total of 15 patients presented unilaterally impacted canines, and bilateral impactions were present in the remaining 12 patients, leaving a total of 39 impacted canines. The period of recruitment was 12 months. All the patients were informed about the content of the stu treatment methods and potential risks and benefits before providing written informed cons to take part in this study. The study received approval from the Ethical Committee of Sapien University of Rome (#4389) and was registered in the international public register.

A preliminary investigation was performed to estimate the power of the study (PS) a to establish the effect size (ES) of the population sampled for the experimental study. Statisti significance was calculated to determine the exact number of patients needed for the stud

Assuming we want to estimate the prevalence of a disease (impacted canines) in population, through the study of the sample, we want an estimate of the prevalence wit certain precision and a chosen level of confidence. The size can be calculated, with a 9 confidence level, using the following Formula (1):

$$n = \frac{1.96^2 \, P_{att} \, (1 - P_{att})}{D^2}$$

with n = sample size, $P_{att}$ = prevalence estimating and $D^2$ = absolute precision.

In our case, with a 95% confidence level and these data, when

$P_{att}$: 96.5% (0.965)
D: 10% (0.1)
then $n = (1.96^2 \times 0.965) \times (1 - 0.965)/0.1^2 = 12.97$,
leaving a value of 13.

If the recommended sample size was >5% compared to the population from which is extracted, its sample size could be reduced; with statistical analysis, it was found that significant sample size for this study was 13 canines.

The inclusion criteria considered in this study design were:

- Patients with palatally impacted canines;
- Male or female;
- Aged between 14 and 24 years;
- Patients reliable for follow-up;
- Patients that understand the protocol and can give informed consent.

The exclusion criteria were:

- Non-cooperative subjects;
- Inoperable subjects;
- Vestibular impacted canines;
- Systemic pathologies;
- Subject in drugs therapy.

Patients who did not meet all the inclusion criteria were excluded from the RCT. T age of the selected patients is between 14 and 24 years. Patient characteristics that w collected are age, sex and number and location of the impacted canines.

The 27 selected subjects were randomly divided into three groups, those treated w two different laser devices ($CO_2$ or diode laser) and a control group treated with a co blade, to compare the results obtained with the gold standard method and demonstrate the effectiveness of laser technology for the exposure of palatally impacted canines.

Experimental groups:

- GROUP A (9 patients; 13 impacted canines): treated with a $CO_2$ laser (SmartXid DEKA, Florence, Italy, 10,600 nm) with a power of 4.5 watts in superpulsed mo (frequency: 80 Hz, fluency: 44.78 J/cm$^2$, spot diameter: 400 µm).
- GROUP B (9 patients; 13 impacted canines): treated with a diode laser (Raffaello, DM Lissone, Italy, 980 nm + 645 nm) with a power of 4 watts in continuous mode (fluen from 0.1 J/cm$^2$ to 10 J/cm$^2$).

Control group:

- GROUP C (9 patients; 13 impacted canines): treated with a traditional surgical-orthodontic approach with a cold blade.

## 2.2. Analysis of Radiographs

An RX orthopantomography and a CBCT were requested from patients at the beginning of therapy (Figure 1) to accurately evaluate the cases before surgery. An evaluation of the prognosis of impacted canines was performed on OPT by two orthodontists, according to Ericson and Kurol [2,19].

The major axis of the canine and its perpendicularity to the alveolar margin, which represents the axis of the canine in its presumed ideal position, were plotted on the OPT. The examination of the CT allowed us to evaluate the three-dimensional morphology of the impacted tooth, its location and inclination in the three planes of space, the depth and the type of inclusion, and the relationships with the other elements.

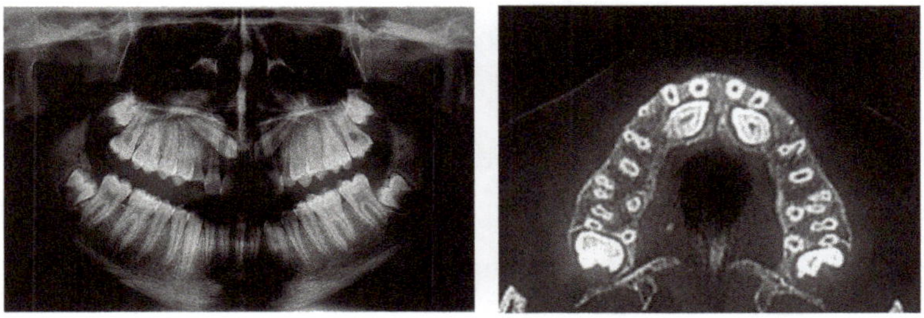

**Figure 1.** An RX orthopantomography (**left**) and a CBCT (**right**) of a patient with two palatally impacted canines.

## 2.3. Steps of the Surgery

The surgery protocol includes local anesthesia given with a Mepivacaine 2% solution with Adrenaline 1:100,000 for Groups A and C, and with Mepivacaine 3% without adrenaline, 1.8 mL solution for injection (Pierrel Spa, Milan, Italy), for Group B.

For Group A, the operculum was performed with a $CO_2$ laser, whose tip is used near to the tissue surface during the rotary movement that is necessary to expose the crown of the impacted tooth (Figure 2).

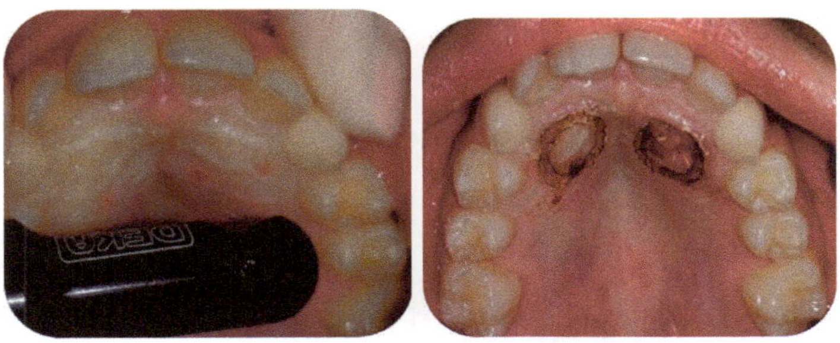

**Figure 2.** The tip of the $CO_2$ laser is used near to the tissue surface (**left**); Operculum performed with CO2 laser (**right**).

For Group B, the operculum was performed with the diode laser, whose tip is used in contact with the tissue surface during the rotary movement that is necessary to open an operculum in correspondence with the crown of the impacted tooth (Figure 3).

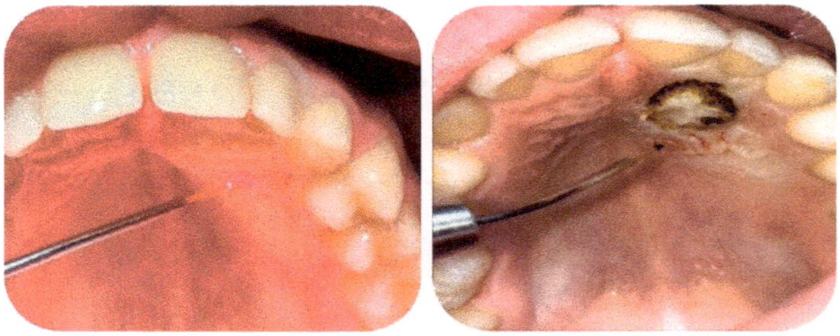

**Figure 3.** Tip of the diode laser is used in contact with the tissue surface. (**left**); Operculum perform with diode laser (**right**).

The palatal mucosa overlying the tooth is then detached with a Prichard perioste elevator and removed by the laser's cutting action (Figure 4).

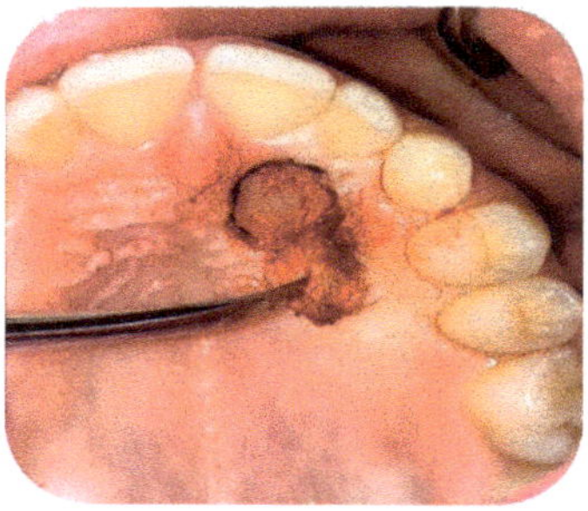

**Figure 4.** Exposure of the palatally impacted canine with a Prichard periosteal elevator.

The control group was treated with traditional cold blade surgery. An operculum performed with the scalpel to expose the canine crown, similarly to the laser groups A a B. The use of means of traction was also avoided in this group to compare, under equ conditions, the results obtained from the three groups (A, B and C).

Afterwards, since the impacted canine was covered by bone, an osteotomy was ma through a handpiece at low speed and under abundant irrigation, with a rosette bur ISO 018 diameter (Maillefer®, Ecublens, Switzerland). The drill was used with a tangent sliding movement to the bone tissue to gradually remove it until the canine's surface covered, which must not be damaged (Figure 5).

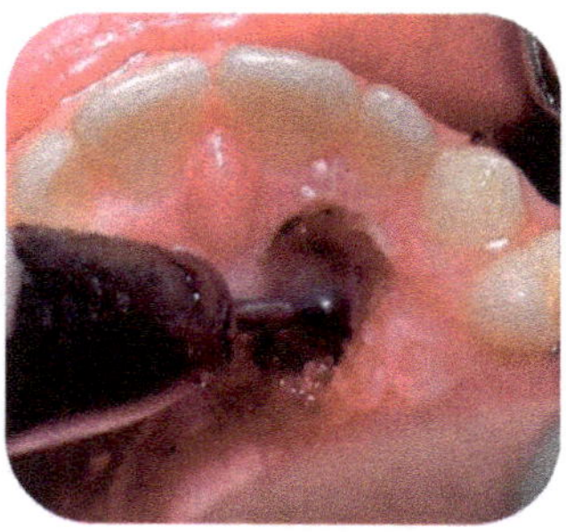

**Figure 5.** Osteotomy through a handpiece at low speed with a rosette bur by ISO 018 diame (Maillefer®, Ecublens, Switzerland).

In Groups A and B, the lasers were applied again on the tissue surrounding the canine to use the possible stimulating effect of laser light on the periodontal ligament, which activates and speeds up tooth movement.

Finally, a periodontal dressing was applied (Coe-Pak®, GC, Tokyo, Japan), to protect the treated area of the palate (Figure 6), for about 7 days, blocked by a suture point in Vicryl 3.0 (Ethicon® V311H 3/0 SH-1 70CM, J&J Medical Devices, Somerville, MA, USA).

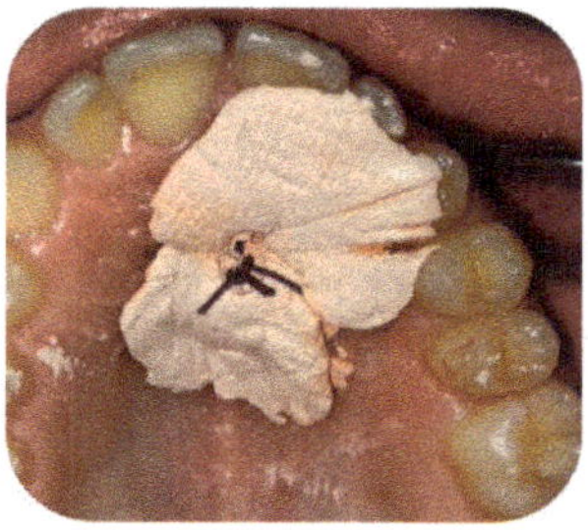

**Figure 6.** Periodontal dressing application (Coe-Pak) blocked by a suture point in Vicryl 3.0.

After 1 week, the periodontal dressing and the suture points were removed. When the tooth has erupted sufficiently into the palate, ideally in a period of four months (16 weeks), a bracket or an orthodontic button is placed on the tooth and the upper arch is bandaged with Damon System technique. In this way, the canine can be gradually guided into the dental arch.

*2.4. Monitoring of Dental Movements*

Patients underwent three check-ups at 1 week (Figure 7 left), 8 weeks (Figure 7 middle), and 16 weeks (Figure 7 right) after surgery, to evaluate and monitor, through photographic documentation and digitally colored scans through use of the intraoral scanner CS3500 (CS3500®, Carestream Health, Rochester, USA), the impacted canine's eruption or the possible early closure of the mucosa and lack of appearance of the element in the palate.

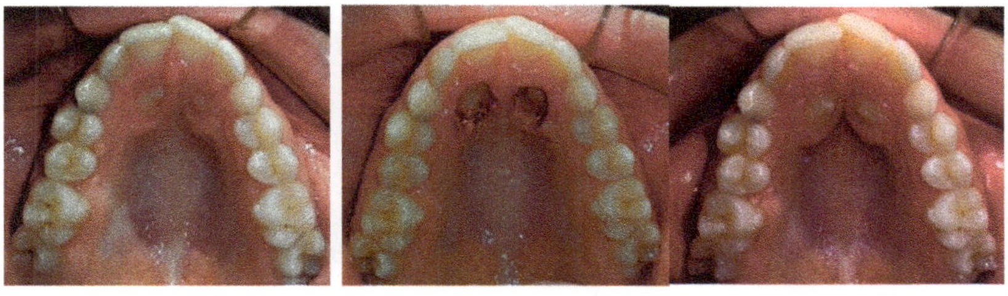

**Figure 7.** Intraoral photo at 1 week (**left**), 8 weeks (**middle**) and 16 weeks (**right**) after $CO_2$ laser surgery, that show the spontaneous eruption of the canines on the palate.

Moreover, clinical examination was carried out and questions were asked to the participants regarding the occurrence of pain, swelling or bleeding, the need to take medicine and reduced functionality of the area involved; the answers to the questions were all negative. The Visual Analog Scale (VAS) was submitted to the subjects of Groups A and B and none of them registered pain on this scale, contrary to the control group, who recorded post-operative pain in 44.4% of the cases.

The hypothesis of this work was that palatally impacted canines, after laser surgery, will undergo a "spontaneous" eruption for reactivation of the physiological eruption, without any orthodontic traction.

To obtain an accurate initial idea of the situation, Invesalius 3.1 software was used extrapolate an STL file (T0) from each CT, which could be superimposed with the intrao scans of the same patient at 8 weeks (T1) and 16 weeks (T2) after the surgery.

The 3D impressions (STL files) were imported into the open-source software MeshLa (Visual Computing Lab, Pisa, Italy). After importing and superimposing the three sca at T0, T1 and T2, through the "measuring tool", it was possible to measure the distan between the cusp of the canine at T0 and the cusp of the canine that erupted on the pal at T1 and T2 (eruption values in millimeters) (Figure 8).

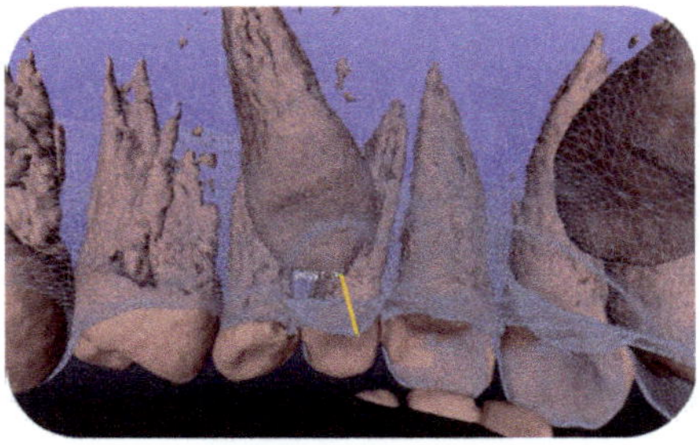

**Figure 8.** Measurement of the eruption values (distances between the cusp of the canine at T0 a the cusp of the canine erupted on the palate at T1 and T2) in mm.

The partial eruptions, obtained from T0 to T1 and from T1 to T2, were measured superimposing the scans and calculating the distances between the initial position of cusp of the canine and its new position at T1 and T2.

The maximum eruption of the canines was measured by calculating the distance mm between the cusp of the canine at T0 and the cusp at T2 on the superimposed mod

When the teeth had erupted sufficiently into the palate, in an ideal period of 16 wee an orthodontic button was placed on the vestibular surface of the canines, the upper a was bandaged with a self-ligating device [20,21] and the teeth could be gradually transla into the dental arch.

*2.5. Statistical Analysis*

The eruption values of Groups A and B were compared with the application of *t*-test (also called the Student's *t*-test), to evaluate the difference between the surg performed with the $CO_2$ and diode lasers.

The results obtained in the experimental Groups A and B were compared with eruption values of the control Group C with the One-Way ANOVA test to evaluate effectiveness of the surgery performed with the laser with respect to the conventio surgery performed with the cold blade.

## 3. Results

*3.1. Prognosis of the Canines*

The prognosis of each impacted tooth, according to the analysis of Ericson and Ku was evaluated by the OPT. The prognosis can be considered favorable when the angle " determined by the major axis of the impacted canine with respect to the axis perpendicula the occlusal plane, is equal to or less than 30° if mesio-inclinated, or 45° if it is disto-incline

We also considered the height of the crown of the canine relative to the roots of contiguous elements and the mesio-distal position of the cusp of the canine, which co be located in the four sectors.

It emerged that:

- A total of 16 of the 39 canines considered in the study had a positive prognosis, and the other 23 had a negative prognosis (22 canines with angle $\alpha > 30°$, 1 canine with $\alpha = 30°$);
- A total of 18 canines were in sectors I-II and 21 canines were in sectors III and IV;
- A total of 5 canines were located in the third apical of the incisor's root, 6 in the third coronal, and the remaining 28 in the third medium;
- A total of 26 canines were covered by bone and the other 13 were in mucosal inclusion only.

### 3.2. Eruption Values

The millimetric values of spontaneous eruption of the palatally impacted canines treated by laser or cold blade surgery were obtained through superimposition (with the MeshLab software) of the scans performed with the intraoral scanner CS3500.

Table 1 contains all the eruption data calculated for canines in T0-T1 (corresponding to eruption from CT to 8 weeks) and in T1-T2 (from 8 to 16 weeks), and the total eruption movement in T0-T2 (at 16 weeks).

**Table 1.** Eruption values for Groups A, B and C in T0-T1, T1-T2, T0-T2.

| Group A | | | Group B | | | Group C | | |
|---|---|---|---|---|---|---|---|---|
| **T0-T1** | **T1-T2** | **T0-T2** | **T0-T1** | **T1-T2** | **T0-T2** | **T0-T1** | **T1-T2** | **T0-T2** |
| 2.84 | 1.89 | 4.74 | 3.66 | 2.28 | 5.94 | 2.04 | 0.63 | 2.67 |
| 2.33 | 1.92 | 4.25 | 1.03 | 2.72 | 3.75 | 2.1 | 0.23 | 2.33 |
| 3.13 | 1.81 | 4.95 | 4.26 | 1.91 | 6.17 | 3.12 | 0 | 3.12 |
| 2.21 | 1.48 | 3.69 | 1.44 | 1.68 | 3.12 | 2.11 | 0 | 2.11 |
| 3.34 | 1.31 | 4.72 | 3.13 | 2.21 | 5.34 | 1.76 | 0.11 | 1.87 |
| 3.11 | 0.82 | 3.93 | 2.32 | 1.86 | 4.18 | 2.23 | 0.81 | 3.04 |
| 2.31 | 1.83 | 4.14 | 4.02 | 3.02 | 7.04 | 2.98 | 1.23 | 4.21 |
| 1.29 | 1.43 | 2.72 | 4.12 | 2.78 | 6.9 | 1.86 | 0.2 | 2.06 |
| 1.78 | 1.55 | 3.33 | 3.61 | 2.69 | 6.3 | 1.62 | 0.94 | 2.56 |
| 4.71 | 1.83 | 6.54 | 3.23 | 2.68 | 5.91 | 3.01 | 0.98 | 3.99 |
| 2.66 | 2.14 | 4.8 | 3.03 | 1.82 | 4.85 | 3.19 | 1.01 | 4.2 |
| 3.48 | 2.45 | 5.93 | 3.41 | 1.97 | 5.38 | 2.89 | 0.89 | 3.78 |
| 3.54 | 1.88 | 5.42 | 2.81 | 2.02 | 4.83 | 1.33 | 1.88 | 3.21 |

We also calculated the average eruption values at each time interval for each group.

$$\text{Average eruption } 0 - \text{T1} = \frac{\sum \text{eruptions T1}}{\text{N}} \quad (2)$$

$$\text{Average eruption T1} - \text{T2} = \frac{\sum (\text{eruption T2} - \text{eruption T1})}{\text{N}} \quad (3)$$

$$\text{Average Total eruption} = \frac{\sum \text{eruptions T2}}{\text{N}} \quad (4)$$

with N = 13.

The results for Group A are:

- The average eruption in T0-T1 = 2.83 mm;
- The average eruption in T1-T2 = 1.72 mm;
- The average total eruption = 4.55 mm.

The results for Group B are:

- The average eruption in 0-T1 = 3.08 mm;
- The average eruption in T1-T2 = 2.28 mm;
- The average total eruption = 5.36 mm.

The results for Group C are:

- The average eruption in 0-T1 = 2.33 mm;
- The average eruption in T1-T2 = 0.69 mm;
- The average total eruption = 3.01 mm.

*3.3. Comparison between the Effect of the $CO_2$ Laser and the Diode Laser*

Within the 26 canines that were surgically exposed by laser operculectomy, the tw experimental groups A and B, treated with the $CO_2$ laser and the diode laser, respective were compared.

For Group A (n = 13 palatally impacted canines treated by $CO_2$ laser) and Group (n = 13 palatally impacted canines treated by diode laser), the measurement of the erupti path (in millimeters) of the tooth performed over 16 weeks gave the results reported Table 1 and Figures 9 and 10.

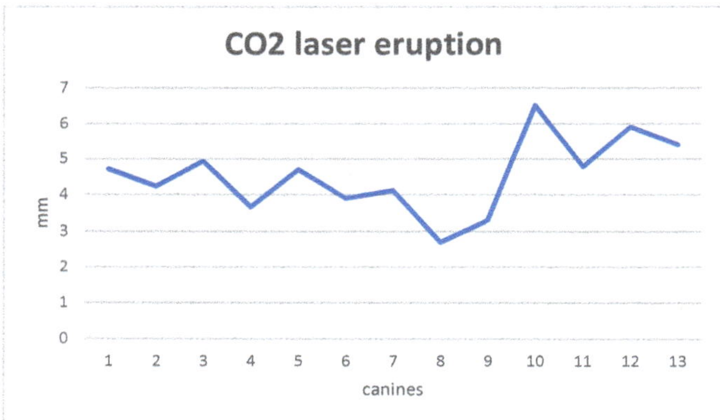


**Figure 9.** $CO_2$ laser eruption values.

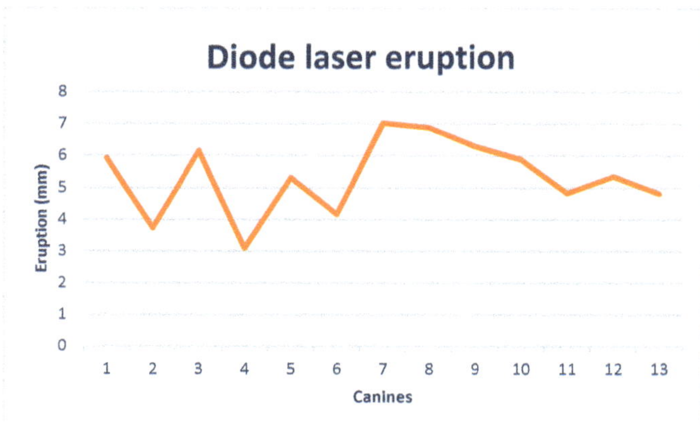


**Figure 10.** Diode laser eruption values.

To compare the two averages of the results obtained with the two different types of lasers, a *t*-test (also called the Student's *t*-Test) was applied (Tables 2 and 3).

**Table 2.** Comparison between the average eruptions obtained with the two different types of lasers.

| | $CO_2$ Laser | Diode Laser |
|---|---|---|
| Sampling number | 13 | 13 |
| Mean | 4.55 | 5.36 |
| Standard deviation (SD) | 1.0049 | 1.140 |

**Table 3.** Comparison between the average eruptions obtained with the two different types of lasers.

| T | 0.0471 |
|---|---|
| Degrees of freedom | 24 |
| P (significance level) | 0.7280 |

A significance level of $\alpha = 0.05$ was chosen.

The degrees of freedom correspond to d.f. $= N - 2 = (13 + 13) - 2 = 24$, where N corresponds to the sum of the subjects belonging to the two samples N = n (group A) + n (group B).

The difference between the observed means of the two groups is not significant for $p < 0.05$, so it is not attributable to the different action of the two different types of lasers but is considered accidental.

### 3.4. Comparison between the Study Groups (Group A + Group B) and the Control Group C

The spontaneous eruptions in mm of canines of the study groups (Groups A and B; n = 26 canines disincluded by laser surgery) were compared with the eruptions of the canines of the control group (Group C; n = 13 canines treated by the conventional surgical approach).

The total eruption values of the control groups are reported in Table 1.

To compare the results obtained with the two different types of lasers with the control group, the one-way ANOVA test was applied. The results of the ANOVA test are reported in the following tables and graphics (Tables 4–6; Figures 11–13).

**Table 4.** Comparison between Groups A, B and C in T0-T1.

| ANOVA | Sum of Squares | Degrees of Freedom | Mean Square | Contribution, % | Fisher Value |
|---|---|---|---|---|---|
| Between | 3.84 | 2 | 1.92 | 12.98 | 2.69 |
| Within | 25.76 | 36 | 0.72 | 87.02 | |
| Total | 29.60 | 38 | 2.64 | | |

**Table 5.** Comparison between Groups A, B and C in T1-T2.

| ANOVA | Sum of Squares | Degree of Freedom | Mean Square | Contribution, % | Fisher Value |
|---|---|---|---|---|---|
| Between | 17.01 | 2 | 8.50 | 67.71 | 37.74 |
| Within | 8.11 | 36 | 0.23 | 32.29 | |
| Total | 25.12 | 38 | 8.73 | | |

**Table 6.** Comparison between Groups A, B and C in T0-T2.

| ANOVA | Sum of Squares | Degree of Freedom | Mean Square | Contribution, % | Fisher Value |
|---|---|---|---|---|---|
| Between | 37.07 | 2 | 18.53 | 49.18 | 17.42 |
| Within | 38.30 | 36 | 1.06 | 50.82 | |
| Total | 75.37 | 38 | 19.60 | | |

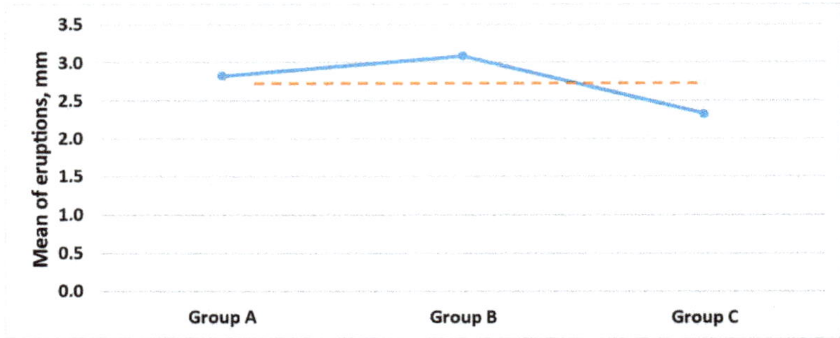


**Figure 11.** Mean of eruption in mm in the interval T0-T1.

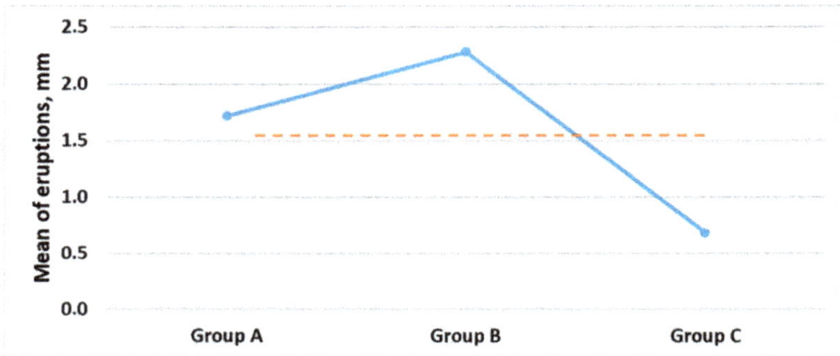


**Figure 12.** Mean of eruption in mm in the interval T1-T2.

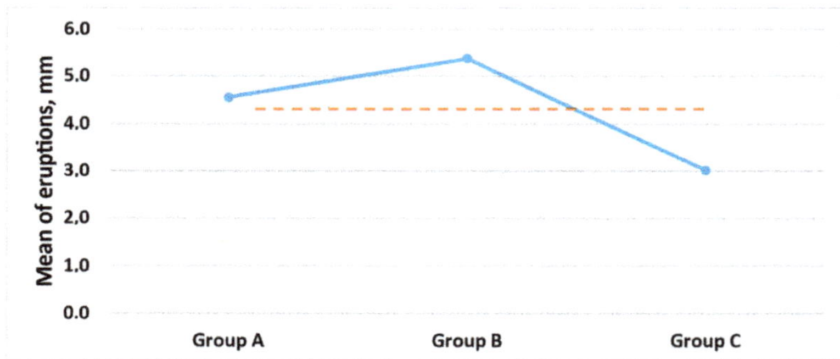


**Figure 13.** Mean of eruption in mm in the interval T0-T2.

It has emerged that the differences in the results obtained in the T1-T2 (Fisher val 37.74) and T0-T2 (Fisher value: 17.42) intervals are statistically significant, while for t T0-T1 interval, the difference is not significant.

### *3.5. Speed of Eruption*

Starting from the eruption data, the average speeds in the intervals T0-T1, T1-T2 a T0-T2 were calculated, considering T0 as the value obtained from the STL file extrapolat from the CT.

The speeds in mm/day in the time ranges considered were found with the following formula:

$$\text{Speed} = \frac{\text{eruption Tf} - \text{eruption Ti}}{\text{Tf} - \text{Ti}} \quad (5)$$

with Tf = final time of the interval and Ti = initial time of the interval.

For Group A:

The average total speed is 0.038 mm/day.

For Group B:

The average total speed is 0.045 mm/day.

For Group C:

The average total speed is 0.025 mm/day.

All the data were measured by two different operators to determine the method error of digital measurements. For each compared value, the correlation coefficient, the relative and absolute errors and the mean differences were calculated (Table 7).

**Table 7.** Range of correlation, error (%) and mean difference (%) values.

| Correlation Coefficient | Error of Method | Mean Difference |
|---|---|---|
| 0.994–0.999 | 0.20–0.99% | 0.17–0.87% |

A coefficient of correlation close to 1 indicates a "positive" and "strong" relationship between the two examined variables. In other words, when increasing one variable, the other one will increase too. In this case, in particular, a high mean coefficient of correlation (0.9965) was observed, which indicates a strong correlation, suggesting less dependence on who takes the measurement itself, i.e., it is independent of the operator. This result is also supported by the absolute error values and the mean differences, which are both positive.

## 4. Discussion

In recent years, different investigators have studied the results of low-level laser therapy (LLLT) during orthodontic tooth movement. Lasers with an output energy below 500 mW are proven to have a biostimulatory effect on tissues without increasing the temperature of the treated region above the normal body temperature. So, they have the potential to accelerate tooth movement by means of influencing the remodeling of alveolar bone without unwanted impacts on the tooth and periodontium. Histologic investigations revealed that LLLT during orthodontic tooth movement can profoundly affect cell-mediated alveolar bone remodeling [22].

Bozcurt et al. in their study evaluated the effects of diode laser biostimulation on cementoblasts and found that the biostimulation setting of a diode laser modulates the behavior of cementoblasts, inducing the mineralized-tissue-associated gene's mRNA expressions and mineralization on the molar's root [23].

The literature also reports numerous applications of lasers in oral surgery (high-intensity laser), for example for operculectomy to remove the soft tissue that overlies an impacted tooth [24].

However, the bio-stimulatory action of high-intensity lasers used to expose the crown of the impacted canines has not yet been demonstrated.

Kokich et al. [25] recommended an alternative disinclusion technique with earlier timing for uncovering palatally impacted canines. They timed the uncovering before the start of orthodontic treatment, the bone over the crown was removed down to the cementum–enamel junction and an opercolum was made on the palatal mucosa with a surgical scalpel. They hypothesized that, when the bone and tissue have been removed, these canines will erupt on their own in about 6 to 8 months. This consideration suggests that when the teeth are freed from the overlying tissue, a spontaneous movement of the impacted canine is activated.

In this RCT, at the end of the monitoring of 16 weeks, a significant tooth movement w observed in both Groups A (mean eruption T0-T2: 4.55 mm) and B (mean eruption T0- 5.36 mm). Furthermore, the exposure of part of the dental crown allowed, in all these cas the application of a bracket or a button to align the tooth in the dental arch. In Group C, t values of total eruption were lower (mean eruption T0-T2: 3.01 mm), demonstrating le movement of the canine in the surgery, with respect to the two study groups. Monitori with digital technologies has allowed us to obtain more precise measurements than t monitoring on plaster casts [26,27].

The one-way ANOVA test suggest that the response of the canine to the bio-stimula action of the laser, applied to expose the crown, can be considered effective in the period monitoring (T0-T2), in particular in the interval T1-T2, resulting in a statistically significa difference between the two study groups and the control group.

A further purpose of this study was the evaluation of the possible different acti between the $CO_2$ laser (wavelength: 10,600 nm; power: 4.5 Watts) used in superpuls emission mode and the diode laser (wavelength: 980 nm + 645 nm; power: 4 Watts) used continuous wave emission mode.

Comparing the millimeters of eruption of the canines treated with the two differe types of lasers and applying the Student's $t$-test, we found a superimposable value.

This means that both lasers have the same bio-stimulatory action on the eruption canines, regardless of their wavelength, power or emission mode, and this suggests th the action of the laser itself can stimulate the spontaneous eruption of the treated tooth.

This study has some limitations. The main criticisms of the study are the sm sample size and the lack of histological analysis of the tissues surrounding the impact canines, which could further support the hypothesis that the high-intensity laser cou have a stimulating action on tooth movement. It would be interesting to study this furth expanding the sample size, given the promising results of the current study.

## 5. Conclusions

The present study was conducted in order to evaluate the effectiveness of laser surge as an alternative approach to conventional surgical–orthodontic treatment and to evalua the possible effect of this laser on the canine's movement after the operculectomy. T results of the study are promising as it has emerged that the spontaneous eruption tim of the canines treated with laser surgery were shorter than the times described in t literature [3,5], which could be an advantage attributable to the use of the laser.

The surgical exposure of the maxillary deep impacted canines by $CO_2$ and diode las has shown undoubted advantages: an absence of bleeding during and after the procedu no stitches, relative ease and speed of execution, reduced or absent postoperative sympto and decontaminating action on the treated tissues.

Furthermore, it has emerged that post-operative wound healing was rapid, and patients treated with laser surgery reported less discomfort in the post-operative period, a only minor discomfort during speech and chewing, than patients treated with conventior surgery (control group). The young patients resumed normal school and sports activiti the day after the surgery.

One week after the surgery, no pain or discomfort was reported by patients of t experimental groups.

The results obtained indicate the effectiveness of the new laser surgery approa proposed in this study, although it was performed on a small sample of patients.

**Author Contributions:** Conceptualization, A.I. and M.H.; methodology, A.I.; software, M.H.; vali tion A.P., U.R. and G.G.; formal analysis, A.I.; investigation, M.H.; resources, G.L.M.; data curati D.P.; writing—original draft preparation, M.H.; writing—review and editing, M.H. and A.I.; visuali tion, G.L.M., A.P., U.R. amd G.G.; supervision, G.G.; project administration, G.G.; funding acquisiti U.R. and G.P. All authors have read and agreed to the published version of the manuscript.

**Funding:** This research received no external funding.

**Institutional Review Board Statement:** The study was conducted in accordance with the Declaration of Helsinki, and approved by the Ethics Committee of "Sapienza" University of Rome (#4389).

**Informed Consent Statement:** Informed consent was obtained from all subjects involved in the study.

**Data Availability Statement:** Data available on request due to ethical restrictions.

**Conflicts of Interest:** The authors declare no conflict of interest.

## ferences


Dachi, S.F.; Howell, F.V. A survey of 3874 routine full-mouth radiographs. A study of retained roots and teeth. *Oral Surg. Oral Med. Oral Pathol.* **1961**, *14*, 916–924. [CrossRef] [PubMed]

Ericson, S.; Kurol, J. Radiographic examination of ectopically *Erupting* maxillary canines. *Am. J. Orthod. Dentofacial. Orthop.* **1987**, *91*, 483–492. [CrossRef] [PubMed]

Bishara, S.E. Clinical management of impacted maxillary canines. *Semin. Orthod.* **1998**, *4*, 87–98. [CrossRef] [PubMed]

Becker, A.; Chaushu, S. Etiology of maxillary canine impaction: A review. *Am. J. Orthod. Dentofacial. Orthop.* **2015**, *148*, 557–567. [CrossRef] [PubMed]

Cassina, C.; Papageorgiou, S.N.; Eliades, T. Open versus closed surgical exposure for permanent impacted canines: A systematic review and meta-analyses. *Eur. J. Orthodontics.* **2018**, *40*, 1–10. [CrossRef] [PubMed]

Ackerman, M.B. The apically repositioned flap and unerupted teeth. *Am. J. Orthod. Dentofacial. Orthop.* **2004**, *125*, 17A–18A; discussion 18A. [CrossRef]

Becker, A.; Shpack, N.; Shteyer, A. Attachment bonding to impacted teeth at the time of surgical exposure. *Eur. J. Orthod.* **1996**, *18*, 457–463. [CrossRef]

Suter, V.G.A.; Altermatt, H.J.; Bornstein, M.M. A randomized controlled clinical and histopathological trial comparing excisional biopsies of oral fibrous hyperplasias using $CO_2$ and Er:YAG laser. *Lasers Med. Sci.* **2017**, *32*, 573–581. [CrossRef]

Nimeri, G.; Kau, C.H.; Abou-Kheir, N.S.; Corona, R. Acceleration of tooth movement during orthodontic treatment—A frontier in orthodontics. *Prog. Orthod.* **2013**, *14*, 42. [CrossRef]

Karu, T.I. Mitochondrial signaling in mammalian cells activated by red and near-IR radiation. *Photochem. Photobiol.* **2008**, *84*, 1091–1099. [CrossRef] [PubMed]

Güray, Y.; Yüksel, A.S. Effect of light-emitting photobiomodulation therapy on the rate of orthodontic tooth movement: A randomized controlled clinical trial. *J. Orofac. Orthop.* 2022, *ahead of print*. [CrossRef]

Fujita, S.; Yamaguchi, M.; Utsunomiya, T.; Yamamoto, H.; Kasai, K. Low-energy laser stimulates tooth movement velocity via expression of RANK and RANKL. *Orthod. Craniofac. Res.* **2008**, *11*, 143–155. [CrossRef] [PubMed]

Verma, S.K.; Maheshwari, S.; Singh, R.K.; Chaudhari, P.K. Laser in dentistry: An innovative tool in modern dental practice. *Natl. J. Maxillofac. Surg.* **2012**, *3*, 124–132. [CrossRef] [PubMed]

Palaia, G.; Del Vecchio, A.; Impellizzeri, A.; Tenore, G.; Visca, P.; Libotte, F.; Russo, C.; Romeo, U. Histological in vitro evaluation of periincisional thermal effect created by new-generation $CO_2$ super-pulsed laser. *Sci. Worls J.* **2014**, *2014*, 345685.

Palaia, G.; Impellizzeri, A.; Tenore, G.; Caporali, F.; Visca, P.; Del Vecchio, A.; Galluccio, G.; Polimeni, A.; Romeo, U. Ex vivo histological analysis of the thermal effects created by a 445-nm diode laser in oral soft tissue biopsy. *Clin. Oral. Investig.* **2020**, *24*, 2645–2652. [CrossRef] [PubMed]

Kravitz, N.D.; Kusnotob, B. Soft-tissue lasers in orthodontics: An overview. *Am. J. Orthod. Dentofacial. Orthop.* **2008**, *133*, S110–S114. [CrossRef]

Derikvand, N.; Chinipardaz, Z.; Ghasemi, S.; Chiniforush, N. The Versatility of 980 nm Diode Laser in Dentistry: A Case Series. *J. Lasers Med. Sci.* **2016**, *7*, 205–208. [CrossRef]

Vescovi, P.; Corcione, L.; Meleti, M.; Merigo, E.; Fornaini, C.; Manfredi, M.; Bonanini, M.; Govoni, P.; Rocca, J.-P.; Nammour, S. Nd:YAG laser versus traditional scalpel. A preliminary histological analysis of specimens from the human oral mucosa. *Lasers Med. Sci.* **2010**, *25*, 685–691. [CrossRef]

Ericson, S.; Kurol, J. Early treatment of palatally erupting maxillary canines by extraction of the primary canines. *Eur. J. Orthod.* **1988**, *10*, 283–295. [CrossRef]

Impellizzeri, A.; Putrino, A.; Barbato, E.; Galluccio, G. Treatment efficiency of Conventional vs. Self-ligating braces: Systematic Review and Meta-Analysis. *Int. J. Orthod.* **2017**, *28*, 41–47.

Impellizzeri, A.; Putrino, A.; Zangrillo, C.; Barbato, E.; Galluccio, G. Efficiency of self-ligating vs. conventional braces: Systematic review and meta-analysis. *J. Dental Cadmos* **2019**, *87*, 347–356. [CrossRef]

Impellizzeri, A.; Horodynski, M.; Fusco, R.; Palaia, G.; Polimeni, A.; Romeo, U.; Barbato, E.; Galluccio, G. Photobiomodulation therapy on orthodontic movement: Analysis of preliminary studies with a new protocol. *Int. J. Environ. Res. Public Health* **2020**, *17*, 3547. [CrossRef] [PubMed]

Bozkurt, S.B.; Hakki, E.E.; Kayis, S.A.; Dundar, N.; Hakki, S.S. Biostimulation with diode laser positively regulates cementoblast functions, in vitro. *Lasers Med. Sci.* **2017**, *32*, 911–919. [CrossRef] [PubMed]

Impellizzeri, A.; Palaia, G.; Horodynski, M.; Pergolini, D.; Vernucci, R.A.; Romeo, U.; Galluccio, G. $CO_2$ laser for surgical exposure of impacted palatally canines. *Dental. Cadmos* **2020**, *88*, 182–190. [CrossRef]

25. Kokich, V.G. Preorthodontic uncovering and autonomous eruption of palatally impacted maxillary canines. *Semin. Orthod.* **2(** *16*, 205–211. [CrossRef]
26. Impellizzeri, A.; Horodynski, M.; Serritella, E.; Romeo, U.; Barbato, E.; Galluccio, G. Three-dimensional evaluation of der movement in orthodontics. *Dent. Cadmos* **2020**, *88*, 182–190. [CrossRef]
27. Impellizzeri, A.; Horodynski, M.; Barbato, E.; Polimeni, A.; Philippe Salah Galluccio, G. Dental Monitoring Application: It i valid innovation in the Orthodontics Practice? *Clin. Ter.* **2020**, *171*, E260–E267.

photonics

MDPI

*view*

# ;volving Role of Lasers in Endourology: Past, Present and uture of Lasers

ara Cerrato [1], Victoria Jahrreiss [1,2,3], Carlotta Nedbal [1,4], Amelia Pietropaolo [1] and Bhaskar Somani [1,3,*]

1 University Hospital Southampton NHS Trust, Southampton SO16 6YD, UK; clara.cerrato@studenti.univr.it (C.C.); victoria.jahrreiss@meduniwien.ac.at (V.J.); carlotta.netbal@uhs.nhs.uk (C.N.); amelia.pietropaolo@uhs.nhs.uk (A.P.)
2 Department of Urology, Comprehensive Cancer Center, Medical University of Vienna, 1090 Vienna, Austria
3 EAU Section of Urolithiasis (EULIS), 6803 AA Arnhem, The Netherlands
4 Urology Unit, School of Urology, Faculty of Medicine, Marche Polytechnic Unviersity, 60121 Ancona, Italy
* Correspondence: bhaskar.somani@uhs.nhs.uk

**Abstract:** The use of lasers in endourology has grown exponentially, leading to technological advancement and to miniaturization of the procedures. We aim to provide an overview of the lasers used in endourology and the associated future perspectives. Using MEDLINE, a non-systematic review was performed including articles between 2006 and 2023. English language original articles, reviews and editorials were selected based on their clinical relevance. Guidelines recommend ureteroscopy in case of stones <2 cm and a percutaneous approach for renal stones ≥2 cm. High-power holmium (Ho:YAG) lasers and the new thulium fibre laser (TFL) may change the future, offering shorter procedures for complex stones, with good outcomes. Increased intrarenal temperature associated with these new technologies may be overcome with adaptive strategies and optimal settings. For upper-tract urothelial carcinoma (UTUC), the combination of laser techniques and these new lasers may reduce the risk of stenosis and allow for a more accurate tumour ablation, potentially reducing the recurrence rates. Laser enucleation procedures are gaining a major role in benign prostate enlargement (BPE), especially in patients with larger prostates or under anticoagulant therapy. However, the superiority of one laser over the other has not been established yet, and the choice of technique is mainly deferred to the surgeon's expertise. In conclusion, lasers will further expand their horizon in endourology, allowing for instrument adaptation to challenging anatomy. Prospective, randomized clinical trials are however needed to confirm available results and to provide the optimal settings for each pathology.

**Keywords:** endourology; lasers; kidney calculi; intrarenal temperature; UTUC; BPH; TFL; holmium; thulium fibre laser; ureteroscopy; PCNL

check for updates

**ation:** Cerrato, C.; Jahrreiss, V.; dbal, C.; Pietropaolo, A.; Somani, Evolving Role of Lasers in dourology: Past, Present and ure of Lasers. *Photonics* **2023**, *10*, 5. https://doi.org/10.3390/ otonics10060635

ceived: 17 April 2023
vised: 20 May 2023
cepted: 25 May 2023
blished: 31 May 2023



## 1. Introduction

The use of lasers in endourology seems has grown exponentially over the last three decades [1]. Due to equipment advancements and the improvement of surgeon skills, minimally invasive laser treatments are an alternative to traditional open surgery in the context of kidney and ureteric stones, bladder stones, upper-tract urothelial tumours and benign prostatic hyperplasia (BPH) [2–10].

While holmium lasers are still the mainstay lasers, the landscape has changed with the recent introduction of thulium fibre lasers (TFLs). Their use has also allowed minimization of percutaneous nephrolithotomy (PCNL) and enucleation of prostate procedures. Because of the elongation of life expectancy, older patients with cardiac pathologies requiring anticoagulant prevention might benefit from a safer treatment using laser technology than classic procedures [1]. Laser procedures are less invasive and safer in more fragile patients; they are sometimes more precise and accurate than traditional surgeries and allow for shorter hospitalization time [1].

In this review article we look at the role of different lasers for endourology, t usefulness of energy, frequency and pulse modulation, and the current concepts arou intrarenal temperature and the future of laser technology.

## 2. Lasers and Urolithiasis

The prevalence of kidney stone disease has increased worldwide, especially during t last two decades, with a lifetime risk in Europe of up to 14% [2]. This has led to a broadeni of numerous advancements in laser technologies, equipment, and settings. According the European Association of Urologists (EAU) Guidelines, ureteroscopy (URS), either ar or retrograde, might be the first treatment option for proximal or distal ureteral stones > <10 mm, and for kidney stones <10 mm, or lower pole stones between 10 and 20 mm patients with unfavourable factors for Shock Wave Lithotripsy (SWL) [3]. Percutaneo nephrolithotomy (PCNL) is the first-choice treatment for renal stones >20 mm and lower pole stones between 10 and 20 mm in patients with unfavourable factors for SWL However, with technological advancements and improved surgical skills, flexible URS h been increasingly used for renal stones >2 cm, with similar stone-free rates (SFR) and lower complication rate than PCNL [4].

Available lasers for stone treatment include the holmium:yttrium-aluminium-garn (Ho:YAG) and the thulium (Tm:YAG and fibre (TFL)) lasers (Table 1 and Figure 1).

**Table 1.** Summary of laser properties for urolithiasis (use, advancements, benefits and disadvantag suggested laser settings).

| Laser Type | Advancement or Technical Aspect | Benefit | Disadvantages | Techniques Used |
|---|---|---|---|---|
| **Nd:YAG (neodymium-doped yttrium aluminium garnet)** | - Wavelength 1064 nm | - Lower thermal effects on ureter than high-power lasers | - Difficulty in fragmenting "hard" stone | - Fragmentation |
| **Thulium fibre laser (TFL)** | - Highly versatile: Frequency up to 2000 Hz, wide range Energy pulse (0.005-6J);<br>- Wavelength 1940 nm (may vary between 1810 and 2100 depending on the design of TFL)<br>- High power laser | - Smaller fibres, "micro-explosion mechanisms" → instrument advantage and faster procedures<br>- Lower retropulsion than Ho:YAG laser (not when applied to MOSES) | - Higher renal and ureteral temperatures with the use of high-power lasers → to maintain the temperature ≤43 °C use irrigation, UAS, on/off laser activation intervals, power below 40 W. | - Fragmentation<br>- Dusting |

### 2.1. *Neodymium-Doped Yttrium Aluminium Garnet Laser: Nd:YAG Laser*

The Nd:YAG laser was one of the first lasers to be used for urolithiasis due to its abil to release energy in pulses [11]. In its FREDDY (frequency-doubled double pulse) mode, t Nd:YAG laser demonstrates good fragmentation potential when operated within 3 mm the target stone [12]. Due to its higher ability to fragment harder stones and to its ability release "packets" of energy, improving fragmentation speed and efficacy, the Ho:YAG las is now preferred over the Nd:YAG laser, and it is now considered the optimum standa for both URS and flexible nephroscopy [3].

### 2.2. *Holmium Laser: Ho:YAG Laser*

The first report on the Ho:YAG laser in the treatment of urolithiasis was publish about three decades ago [5]. Ho:YAG is a 2140 nm wavelength laser with a high wat absorption, that reflects in a fluid medium at a distance of about 0.3–0.5 mm. This mak the Ho:YAG laser ideal for urological use in limited spaces such as the ureter or the ren pelvis [6].

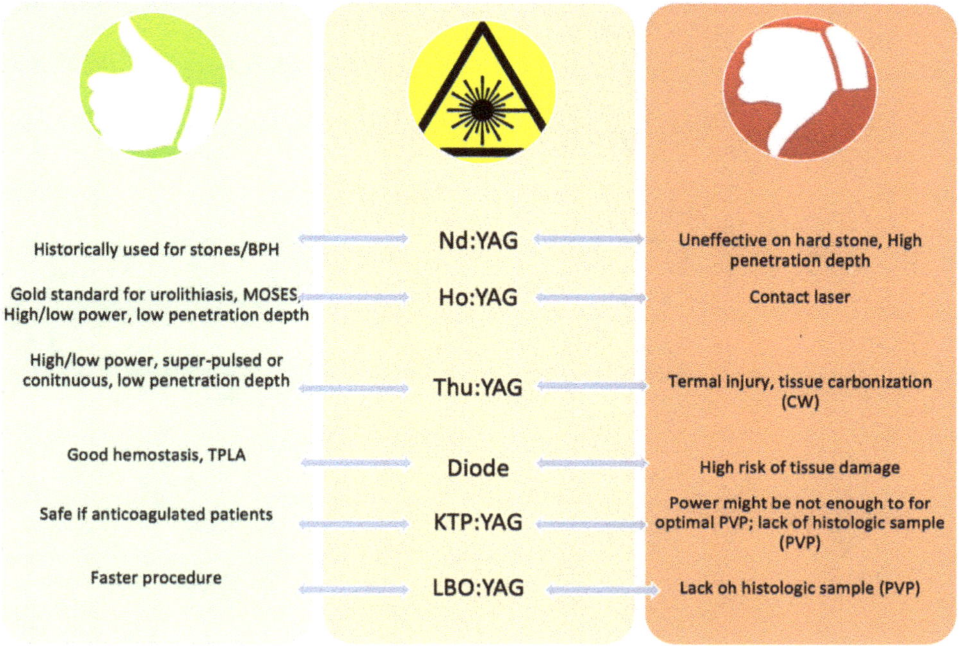


**Figure 1.** Advantages and disadvantages of each laser (BPH—benign prostatic hyperplasia, TPLA—transperineal laser ablation).

The Ho:YAG laser can be used in a high- or low-power modality. The lower power Ho:YAG devices have a maximum power up to 30–35 Watts (W) with a maximum laser pulse frequency of 25–30 Hz [7], while high-power Ho:YAG laser machines might reach higher energy output, up to 100–120 W [8]. These characteristics and variation in the settings such as pulse, energy, frequency, and pulse width can allow for other treatment aspects apart from fragmentation and the retrieval of stones.

High-energy and low-frequency settings are used for fragmentation, while low-energy and high-frequency settings are used for dusting [6]. Dusting has been demonstrated to have an advantage over fragmentation, in terms of decreased use of the ureteral access sheath (UAS), basket-associated complications, and reduced operative time [9–11].

If fragmentation is performed in the first stage of the lithotripsy procedure, a second-stage ("completion") non-contact lithotripsy may be needed, intended to produce smaller particles that might spontaneously pass [11]. Two different techniques may be used for this, the "popcorn" and the "pop-dusting". The "popcorn" technique uses high pulse energy (1.5 J), usually associated with high frequency (20–40 Hz), and long-pulse mode to produce clinically insignificant fragments [11,13]. The second one, the "pop-dusting" technique, uses instead low-power, long-pulse energy (0.5–0.8 J), resulting in finer fragments [11]. The "pop-dusting" technique has been demonstrated to enhance the speed of clearance for harder stones both in the adult and paediatric populations [11,14,15].

Additionally, the modern high-powered Ho:YAG lasers can be equipped with the MOSES "pulse modulation" technology. MOSES technology (Lumenis) divides the laser pulse into two peaks: the first pulse separates the fluid in front of the stone (Moses effect), while the second pulse is delivered directly to the stone surface by the intervening fluid, leading to better fragmentation, lower retropulsion, and less time taken for the procedure [2,6,16–18]. A recent systematic review showed that the advantage of high-power lasers was lost for larger stones [19]; however Pietropaolo et al. conducted a retrospective study demonstrating that MOSES technology with high-power lasers redu the fragmentation time and infective complications, even in patients with larger stones a those who had a stent placed, and minimized the number of patients who needed a seco procedure to achieve a stone-free status [2]. They conclude that MOSES technology ha definitive advantage, especially when used with a mid-power laser, in treating comp stones, usually without the need for secondary procedures.

### 2.3. Thulium Fibre Laser (TFL)

The super-pulsed thulium fibre lasers (TFL) use electronically modulated laser dioc allowing the delivery of a higher and more constant peak power with a wider range laser parameters through a fibre [6,20]. This makes the TFL a highly versatile laser, w pulse frequencies up to 2200 Hz, very low to very high pulse energies (0.005–6J), short- very long-pulse durations (200 μm to 12 ms), and a total power level up to 55 W [20], w optimal laser settings proposed as 0.5 J × 30 Hz for fragmentation and 0.15 J × 100 for dusting [21]. Its wavelength is 1940 nm and has an optical penetration depth in wa four-times shorter than the Ho:YAG laser [20]. This means that water absorbs TFL ene approximately four times higher than Ho:YAG, resulting in lower stone and tissue ablati thresholds. The higher water tropism of the TFL is at the base of the "micro-explosio mechanism, that is to say, an increase in the breakability of the stone due to increased wa absorption and larger stone pore size induced by laser stimulation [22]. Additionally, TFL can use smaller laser fibres (up to 50 μm core), improving irrigation, scope deflecti retropulsion reduction, accessibility, and visibility [20].

Since the approval of the TFL by the FDA and the EMA, numerous studies have be published showing that the TFL outperforms the Ho:YAG laser in terms of dusting quan and quality, lower retropulsion, and shorter operation time for any kind of stone [21,23–27]. a comparative study by Ulvik et al., the TFL showed a higher SFR (92% vs. 67%, $p$ = 0.0 lower operative time (49 vs. 57 min, $p$ = 0.008), and lower intraoperative adverse ev (5% vs. 22%, $p$ = 0.014), compared to the Ho:YAG laser [25]. In a large multicenter stu involving 4208 patients who underwent RIRS, Keat et al. showed that overall, the T was associated with lesser residual fragments (>2 mm), a finding that was confirmed a multivariate analysis for stones both more and less than 1000 HU [26]. Since resid fragments require second treatment, this is best avoided. Ryan et al. confirmed a decre in operative time for the TFL vs. the Ho:YAG without pulse modulation for any sto size (<15 mm, $p$ = 0.007; <10 mm, $p$ = 0.002), and showed that the mean 13 min reducti in operative time resulted in a saving of USD 440/case in direct operating room co resulting in a range of USD 294,000–381,900 savings per year [27].

Even if the results are promising, the TFL still needs to be tested in further randomiz prospective trials on larger sample sizes with the associated cost analysis.

## 3. Clinical Influence of Laser in Urolithiasis

### 3.1. PCNL and Laser

With the advent of laser technology and the increased ability to produce sma residual fragments or dust, PCNL underwent a process of miniaturization, that was coup to a decrease in bleeding, leading to higher chance of tubeless procedures and reduc hospital stay [28–30]. According to their size, miniaturized techniques are defined mini-PCNL (16–20 Fr), ultramini-PCNL (UMP, 11–14 Fr), and micro-PCNL (microP <10 Fr) [31].

Compared to RIRS, PCNL is associated with higher blood loss, complication ra and readmissions [32]. On the contrary, the main advantages of PCNL are that it genera ensures SFR in usually a single and faster procedure, usually with fragmentation [33,34]. T might have clinical advantages, in terms of anaesthetic length and infective complicatio One of the main problems related to the increased time of miniaturized PCNL was due stone retropulsion. However, both in vitro and in vivo studies have demonstrated that T has low or zero retropulsion [35]. With Ho:YAG lasers, retropulsion becomes evident at 0.

while with the TFL, retropulsion begins at around 1 J, being therefore clinically insignificant in many cases [35,36]. The debate is now open on whether TFL has less retropulsion with respect to Ho:YAG when combined with MOSES technology, but there is no a definitive answer as yet [35].

Interestingly, in a recent retrospective study, Patil et al. compared the high-power Ho:YAG with MOSES Technology vs. the TFL for miniPCNL [37]. They found that the TFL produced a greater proportion of fragments ≥3 mm (36% vs. 22.68%, $p = 0.002$), but a subset analysis based on stone density showed that the TFL had a shorter total operative time ($p < 0.005$) [37].

The introduction of the TFL raises a question: could the new kid on the block change the game? Taratkin et al. compared results for patients who underwent RIRS or mini-PCNL (16.5–17.5 Fr) for kidney stones > 20 mm, and found no differences in procedure duration, complications, or hospital stay [38]. In another prospective randomized comparison between RIRS and mini-PCNL for stones between 10 and 20 mm, Perri et al. demonstrated that both procedures were effective in obtaining a postoperative SFR with the TFL, but that according to stone position, either RIRS (upper calyceal stones) or mini-PCNL (lower calyceal stones) were superior [28].

It is therefore difficult to give a definitive answer, due to the lack of standardization of the studies and the heterogeneity in the reported results.

### 3.2. Role of Temperature with Laser Use

Thermal control when dealing with the upper tract and during lasering is of pivotal importance. During lithotripsy, the stone absorbs the laser energy itself, causing melting, fragmentation, chemical disintegration, and vaporization of the interstitial water [39]. Heat energy passes beyond the direct laser absorption zone, affecting the surrounding tissues. It has been demonstrated that the impact on biologic tissue ranges between 41 °C and 47 °C, but that an increase in temperature above the safe threshold of 43 °C has a detrimental cytotoxic effect [39]. This effect might cause ureteral scarring and in turn cause stricture.

Temperature rises according to power, so that high-power lasers cause higher thermal injury. According to a research by Tokas et al., when Ho:YAG and Tm:YAG are used with laser powers below 40 W, their temperature profiles are safe [40]. However, in vitro, with powers above 40 W, the temperature rises exponentially, thus making them dangerous for in vivo use [40]. For this reason, several factors might be used by the operators to keep the temperature in a safe profile [39]. With irrigation, cooler water circulates through the system, dissipating heat [40,41]. The use of a larger calibre ureteral access sheath (UAS) (12/14 F or 14/16 Fr) results in temperatures around 43 °C, and in improved irrigation and lower intrarenal pressures [39,40]. Shorter on/off laser activation intervals (5 s/5 s) are also recommended to keep temperatures low [39,42].

The TFL on the other hand has a heat generation that is four times larger than that of the Ho:YAG laser; for this reason, Tokas et al. suggest to maintain pulse rate below 500 Hz with higher saline irrigation rates [40]. Recently, several in vitro or animal studies have been conducted to investigate the thermal damage induced by the TFL [43,44]. Belle et al. have demonstrated in an in vitro silicone kidney-ureter model, that the TFL generated more heat at all settings tested (3.6, 10, 20 and 30 W) compared to the holmium laser, with higher risk of complications from thermal damage [43]. On the contrary, in their animal study, Sierra et al. compared the temperature increase and histologic changes in porcine urinary tracts treated with either TFL (left side) or Ho:YAG (right side) lasers [44]. In their experimental setting, they worked with 20 W in the kidney, 12 W in the ureter with a continuous irrigation at 40 cm $H_2O$ that provided a temperature of 36.8 °C and 34.6 °C for TFL and Ho:YAG, respectively [44]. They found that both technologies were equally safe. However, all authors agree on advocating low power settings for the urothelium, especially in the ureter, associated with continuous irrigation.

It is important to bear in mind that due to its anatomical features and lower surrounding vasculature, the ureter has a higher risk with respect to the renal pelvis to incur mucosal injury and to develop strictures. This has a special clinical value when treating UTL or stones impacted in the mucosa, where the depth of penetration of the laser acqui fundamental importance.

**4. Lasers and Urothelial Tumours**

Non-muscle-invasive bladder cancer (NMIBC) is the most prevalent of bladder cance accounting for 70–75% of cases at first presentation [45,46]. According to EAU Guidelin NMIBC can be stratified as low, intermediate, or high risk. Low-risk tumours have a lo risk of progression, but a high risk of recurrence [46]. For this reason, treatment options a aimed at decreasing the risk of recurrence, and even if the gold standard for its treatme remains the transurethral resection of the bladder tumour (TURBT), office-based las ablation or diathermy and surveillance could be offered.

Upper-tract urothelial carcinoma (UTUC) is a relatively rare cancer, representing <1 of all urothelial tumours [45]. According to EAU Guidelines, URS is considered as t first line for low-risk disease with oncological equipoise to radical nephroureterector (RNU) [46]. Usually, URS is coupled with adjuvant intracavitary therapy (with BCG MMC), but nowadays there is limited evidence [46]. Since a higher association with lo recurrence rate and risk of progression to RNU exists, for minimally invasive therapy in t context of UTUC, a careful patient selection and a strict follow-up and surveillance (bo for upper tract and bladder recurrences) according to the risk-stratified strategy of the EA guidelines, are necessary [46].

*4.1. Ho:YAG, Nd:YAG and Thulium Lasers for Urothelial Cancer*

The available lasers that might be used for urothelial ablation are the Ho:YAG, Nd:YA Thu:YAG and TFL, and the diode lasers (Table 2 and Figure 1). These lasers differ in ter of laser–tissue interaction, ablation depth, coagulation efficacy, and cutting speed [39,40,4

The Ho:YAG and the Nd:YAG lasers have been widely described and used for t management of upper-tract urothelial cancer. The Ho:YAG laser works in the pulse mo with a wavelength of 2.09 μm. Its energy is readily absorbed by water and the laser tip mu be always in contact with the tissue to achieve tumour ablation. Due to its great availabil Ho:YAG is considered the gold standard for the treatment of UTUC [48]. The Ho:YAG las provides a deeper incision and a higher tissue penetration compared to thulium lase whereas the coagulation and total laser areas are more extensive with the Tm:YAG laser [4 Compared to Ho:YAG, the Nd:YAG laser has a greater depth of penetration (4–6 mm) a provides a deeper coagulation and ablative effect on the tumour that might be useful wh treating larger tumours [50]. Direct contact with the tumour is not required and shou be avoided to prevent a decrease in the effectiveness of the fibre [51]. Additionally, t Nd:YAG laser can be used with albumin as a solder [51]. In laser tissue soldering (LTS), las energy is applied to a solder in conjunction with laser wavelength-specific chromophor to facilitate light absorption [52]. The increase in temperature denatures the solder whi forms a coagulum, at a speed of 1 min per centimetre of incision length. This property particularly useful in the context of the repair of bladder, urethra, or uretero-pelvic juncti (UPJ) [52].

**Table 2.** Summary of laser properties for urothelial cancer (use, advancements, benefits and disadvantages, techniques used).

| Laser Type | Use | Advancement or Technical Aspect | Benefit | Disadvantages | Techniques Used |
|---|---|---|---|---|---|
| **Nd:YAG (neodymium-doped yttrium aluminium garnet)** | - Urothelial lesions<br>- Repair of mucosal injuries | - Wavelength 1064 nm<br>- Coagulates tissue to depth of 4–6 up to 10 mm | - Safe in anticoagulated patients | - Should be avoided for ureteral tumors (increased risk of strictures) | - Tumor ablation (bladder)<br>- Laser tissue soldering (LTS) |
| **Holmium (Ho):YAG** | - Urothelial lesions<br>- Strictures of upper and lower urinary tract | - Pulsed laser, 2140 nm<br>- Small absorption depth (0.4 mm) → ideal for urological use in limited spaces (renal pelvis/ureter)<br>- Deep incision and high tissue penetration<br>- Simultaneous coagulation of small blood vessels<br>- Great availability | - Small penetration depth and safer for small ureteric tumours | - Direct contact with the tumour surface<br>- Can lead to some bleeding during coagulation | - Tumor ablation (bladder-UTUC), gold standard |
| **Thulium lasers: TFL QWC Tm:YAG CW TFL SuperPulsed** | - Urothelial lesions | - Tm:YAG CW: continuous ablation with low penetration<br>- TFL SuperPulased: maximum water absorption with penetration depth 0.3–0.4 mm, might be used as quasi-continuous | - Very little injury to surrounding tissue<br>- TFL QWC: very controllable, incision depth and damage increase with increasing power<br>- TFL SuperPulsed: improved cutting ability with reduced carbonization | - Tm:YAG CW: high tissue carbonization<br>- Lack of large randomized clinical trials | - Tumor ablation (UTUC) |
| **Diode** | - Bladder TCC<br>- Photodynamic therapy | - Wavelength 810 to 1064 mm<br>- Lower power<br>- Penetration depth 1–3 mm<br>- New blue diode laser, wavelength 450 nm | - Good hemostasis<br>- Smaller box size compared to Nd:YAG laser | - Higher damage to the surrounding tissue | - Tumor ablation<br>- Laser tissue soldering (LTS) |

Three machines are available nowadays using thulium lasers operating withi wavelength range of 1.94–2.0 µm: the TFL QCW mode (pulse max 120 W, 1.94 µm), Tm:YAG CW mode (pulse max 200 W, 2.0 µm), and the TFL SuperPulsed mode (pu max 500 W, 1.94 µm) [53]. The Tm:YAG CW mode ensures continuous ablation and l tissue penetration, resulting in good vaporization and coagulation properties for soft tiss However, this laser causes a higher tissue necrosis and carbonization due to the low pe power and the absence of thermal relaxation [53]. This might represent a problem in ter of both defining ablation completion and pathologic sampling. The TFL QWC mode very controllable behaviour, as both the incision depth and the damage dose increase w increasing laser power [49]. The TFL offers maximum water absorption. It can be used i quasi-continuous and in a SuperPulsed (SP) mode that allow for increased pulse pow improved cutting efficiency, and reduced carbonization [49,54]. SP TFL offers a Ho:Y incision and its coagulation depth remains nearly constant within the laser fibre–tiss distance range of 0.3–0.4 mm, without major complications [53,55].

### 4.2. Diode Laser

Semiconductor or diode lasers are made of two semiconductor material layers t can be combined to create larger more powerful versions [6]. Historically, diode las were used as an alternative to the 120 W lithium triborate (LBO) lasers, due to their hig haemostatic ability [56]. Diode lasers have been used for direct tumour ablation or a "photosensitizer" in the contest of photodynamic therapy (PDT), due to their ability match the desired wavelengths of many photosensitizer drugs (630–760 nm) and incre their penetration in the tumoral tissue [51,57]. Recently, Wu et al. conducted a n inferiority study to assess the efficacy and safety of the newer 450 nm wavelength b diode laser compared to the conventional electrocautery performed with the plasmakin loop for non-muscular invasive bladder cancer [58]. Patients in the blue-laser arm show longer operative times ($p$ = 0.001), but lower blood loss ($p$ = 0.003) and faster wou healing three months after operation [58]. However, survival outcomes were not report and authors state that longer follow-up is needed to confirm recurrence-free survi benefit [58].

### 4.3. Combination of Treatments and Oncologic Outcomes

In terms of what is the best laser for the conservative treatment of UTUC, Cornu e and Villa et al. reported recurrence rates of 60% and 76.1% of UTUC after Ho:YAG la ablation, respectively [48,59]. Other studies have demonstrated that the use of Ho:Y or Nd:YAG showed a progression to RNU in 16.6–35.9%, vs. 8.9–13.5% using Tm:Y UTUC recurrence was 24.6–90.5% with the use of Ho:YAG or Nd:YAG, vs. 19.2–4 for Tm:YAG [50]. Even if it seems that Ho:YAG lasers had a higher recurrence rate a progression than RNU, Ho-YAG laser studies have generally a longer follow-up peri (30–50 months), although the surgical view can be compromised due to their lower hae static ability compared to other lasers, thus worsening the oncological outcomes [60].

In a retrospective comparative study on 59 patients, Defidio et al. demonstra the non-inferiority of Tm:YAG vs. Ho:YAG laser ablation with better median parame performance scores in fibre-tip stability, precision in ablating tumours <1.5 cm, and redu bleeding or mucosal perforation [61]. Other reports later also corroborated these init findings [62–64].

Additionally, there is a general tendency to use a combination of dual-wavelength T Ho:YAG and/or other energies for optimal tumour ablation and coagulation. Defidio et evaluated the ablative safety and efficacy of the Tm-Ho:YAG Duo laser for UTUC, a found, at 144 months of follow-up, a recurrence-free survival rate of 69.3%, with a kidr preserving rate of 91% in the intention-to-treat population and 87.5% in imperative in cations [62]. Sanguedolce et al. used Tm:YAG for larger lesions for its higher ability ablation and coagulation, switching to Ho:YAG to cut and dislodge the carbonized tissue uncover residual neoplastic tissue to be targeted with Tm:YAG for complete ablation [6

However, these results must be carefully interpreted, since they are usually retrospective, heterogeneous in their populations, with small numbers of patients. Some lack information of tumour grading, and they generally differ in their follow-up times. As a general rule, all authors suggest that for ablative therapy in the context of UTUC, the real game-changers are with patient selection, with potentially smaller papillary lesions (<2 cm) and those able to pursue a strict endoscopic follow-up [48,66].

A further ally to decrease tumour recurrence might be confocal laser endomicroscopy. This technology, previously applied to GI tumours [67], aims to perform real-time tumour characterization. In this regard, Vanacore et al. analysed results from 186 patients who underwent URS with CLE for UTUC and ablation with Ho:YAG, Tm:YAG, or a combination of both [68]. Authors found 84.6%, 66%, and 100% concordance for low-grade, intermediate-grade, and in situ carcinoma, respectively [68]. Other authors suggest that CLE provides prompt intraoperative data with high sensitivity for high-grade tumours, but that it still has a low specificity [65]. For these reasons, more robust data and larger studies are warranted.

### 4.4. Complications: Stricture/Recurrence

Major complications (≥Clavien-Dindo IIIb) associated with flexible ureteroscopy, including major perforation, perirenal hematoma, subcapsular hematoma, ureteral avulsion, arteriovenous fistula, and acute sepsis, are generally extremely rare (<1%) [60]. The most common complication seems to be the ureteric stricture and its incidence varies between 0 and 27%, according to the follow-up period and to the energy load used for tumour ablation [50,60].

## 5. Lasers and Benign Prostatic Hyperplasia (BPH)

Lower urinary tract symptoms (LUTS) secondary to BPH are frequent in aging men and can lead to a significant loss of quality of life. While transurethral resection of the prostate (TURP) is still the standard therapy for patients with small to medium prostates in whom medical therapies were unsuccessful, the use of lasers has emerged as an alternative treatment modality in these patients. To date, a variety of lasers can be used either for the enucleation, vaporization, or resection of prostate tissue including holmium, thulium, and Greenlight lasers [69] (Table 3 and Figure 1).

### 5.1. Nd:YAG Laser

The Nd:YAG laser was used mainly in the past for non-contact "visual laser ablation of the prostate" (VLAP), contact ablation, or interstitial laser coagulation (ILC) [1]. It is characterized by more than 1 cm tissue penetration and by a 1064 wavelength. For this reason, it causes deep coagulative necrosis and thermal tissue injury, with an oedema occurring after the procedure leading to LUTS and urinary retention [1].

### 5.2. Holmium Laser

TURP remains the gold standard in the surgical management of benign prostatic enlargement (BPE), especially in small and mid-sized prostates; however, challenges in the surgical treatment of larger prostates and technological advances led to the development of new treatment strategies. Holmium laser enucleation of the prostate (HoLEP) has been found to be an efficient and safe treatment option for all prostate sizes in several randomized controlled trials [70–74]. This technique allows for the coagulation of tissue simultaneous to the enucleation of the prostate tissue from the capsule [70,71,75]. HoLEP has shown favourable functional outcomes, better haemostasis, shorter hospitalization and indwelling catheter times compared to TURP, and lower complication and morbidity rates compared to simple open prostatectomy and is therefore considered a valid alternative treatment to both procedures [70,71,75,76].

**Table 3.** Summary of laser properties for BPH (Use, advancements, benefits and disadvantages, techniques used).

| Laser Type | Use | Advancement or Technical Aspect | Benefit | Disadvantages | Techniques Used |
|---|---|---|---|---|---|
| **Nd:YAG** | - Prostate ablation<br>- Prostate coagulation | - Pulsed or continuous<br>- Wavelength 1064 nm<br>- Aborption depth: 10 mm<br>- Chromophobe: water and haemoglobin<br>- Non contact laser | - Safe in anticoagulated patients | - Deep coagulative necrosis → oedema causing irritative LUTS and urinary retention with prolonged catheterization | - Non-contact "visual laser ablation of the prostate" (VLAP)<br>- Interstitial laser coagulation |
| **Holmium (Ho):YAG** | - Prostate enucleation (HoLEP), vaporization | - Pulsed<br>- Wavelength of 2140 nm<br>- Absorption depth: 0.4 mm<br>- Simultaneous coagulation of small blood vessels<br>- Chromophobe: water<br>- Contact laser | - Favourable functional outcomes, better haemostasis, shorter hospitalization and shorter catheterization vs. TURP<br>- Power modulation: high power for enucleation, low-power for apical dissection<br>- MoLEP: shorter enucleation time, haemostasis and lower hospitalization | - Can lead to some bleeding during coagulation | - Resection (HoLRP)<br>- Low-power HoLEP<br>- High-power HoLEP<br>- MoLEP |
| **Thulium (Th): YAG** | - Optical urethrotomy<br>- Prostate vapo-resection (ThuVARP)<br>- Prostate enucleation (ThuLEP, ThuFLEP)<br>- Prostate vaporization (ThuVEP) | - Continuous<br>- Wavelength 2000 nm<br>- Absorption depth: 0.25 mm<br>- Chromophobe: water<br>- Contact and noncontact laser | - Reduced thermal damage causing irritative symptoms<br>- ThuVEP: larger prostates and safe in anticoagulated patients | - Lack of large randomized clinical trials | - ThuVARP<br>-ThuLEP/ThuFLEP<br>- ThuVEP |
| **Greenlight, Potassium titanyl phosphate (KTP), lithium borate/triborate (LBO): YAG** | - Bladder neck incision<br>- Photoselective vaporization of prostate (PVP)<br>- Photoselective sharp enucleation of the prostate (PSEP)<br>- Prostate enucleation (GreenLEP) | - Quasi-continuous<br>- Wavelength 532 nm<br>- Absorption depth: 0.8 mm<br>- Doubled frequency of Nd:YAG so shorter absorption depth<br>- Chromophobe: haemoglobin<br>- Contact and noncontact laser | - KTP: Virtually bloodless procedure → safe in anticoagulated patients<br>- Reduced catheterization rates<br>- LBO: Faster tissue ablation<br>- LBO: Can be used to treat large prostate glands<br>- PSEP: provide histologic sample | - KTP: Temperature increase of tissue not sufficient for vaporization<br>- KTP: Scattering seen through coagulated layers and reduction in intensity<br>- LBO: Reduced in haemostatic ability<br>- PVP: Lack of histologic sample with PVP | - PVP<br>- PSEP<br>- GreenLEP |
| **Diode** | - Prostate vaporization<br>- Prostate enucleation | - Pulsed or continuous<br>- Wavelength 940, 980, 1318, 1470 nm<br>- Lower power<br>- Penetration depth 0.5–5 mm<br>- Chromophobe: water and haemoglobin<br>- Contact laser | - Good haemostasis<br>- Smaller box size compared to Nd:YAG laser | - High incidence of complications and postoperative irritative symptoms | - Vaporization<br>- Enucleation<br>- TPLA |

HoLEP is usually conducted using high power settings of 80–100 W with 40–50 Hz frequency and 2 J energy and a reduction of power for coagulation and apical enucleation [77,78]. However, high-power devices require costly equipment and high-power outlets, leading to higher costs and preventing their widespread use. In comparison, low-power lasers with power settings of 20–50 W do not require special outlets and are less expensive than high-power lasers. Rassweiler et al. found the use of low power settings (24 W, 2 J, and 12 Hz; or 39.6 W, 2.2 J, and 18 Hz) to be an efficient and safe alternative to high power settings [79]. After conducting an analysis of the existing literature, Gkolezakis et al. concluded that low-power HoLEP is safe, feasible, and efficient. Furthermore, low-power HoLEP was found to reduce postoperative storage and irritative symptoms. Complications including intra- and postoperative complications were found to be independent of the laser power [77].

A recent meta-analysis included seven studies comparing HoLEP to holmium enucleation of the prostate using MOSES technology (MoLEP). The enucleation time was significantly shorter for MoLEP (mean difference [MD] −7.27 min, 95% confidence interval [CI] −11.26 to −3.28; $p = 0.0004$). Furthermore, postoperative length of hospital stay was significantly longer in the HoLEP group (MD 0.3 d, 95% CI −0.24–0.85, $p < 0.0001$). The mean maximum peak flow was higher for patients treated with HoLEP (MD 0.95 mL/s, 95% CI −1.66 to 3.57; $p = 0.47$) and the mean postvoid residual volume was lower in the MoLEP group (MD −10.08 mL, 95% CI −53.54 to 33.37; $p = 0.65$). The authors concluded that the use of MOSES technology for prostate enucleation showed advantages in enucleation, haemostasis, and procedural time and length of hospital stay, and therefore even could make the enucleation of the prostate feasible as a day surgery procedure [80].

### 5.3. Thulium Laser

Thulium laser has been reported to achieve higher ablation rates with only minimal thermal tissue injury and minimal blood loss compared to other lasers, resulting from its 2 μm wavelength which is close to the 1.92 μm water absorption peak and the possibility of using continuous wave mode [72]. Thulium laser resection was first used as a treatment alternative to TURP in 2008 and has since been established in the treatment for LUTS secondary to BPH as thulium laser enucleation of the prostate (ThuLEP) and thulium laser vapo-enucleation (ThuVEP) [72,81,82]. Endoscopic enucleation of the prostate can be conducted with several sources of energy. Enucleation of the prostate using the thulium laser is considered an efficient and safe alternative to TURP and HoLEP [69]. Like HoLEP, ThuLEP has been found to be an efficient and safe treatment modality for BPH independent of prostate size [72]. A meta-analysis comparing the enucleation of the prostate using holmium and thulium laser technology found both methods to have low morbidity and result in satisfactory micturition improvement. However, enucleation with a thulium laser (ThuLEP or ThuVEP) proved to be advantageous in efficacy, enucleation time, perioperative blood loss, flow rate and postvoid residual at one month, and international prostate symptom score at 12 months postoperatively compared to HoLEP [72]. Similarly, another recent meta-analysis found no significant differences in operating time, enucleation weight, catheterization time, functional measures and symptom scores, or hospital stay with ThuLEP and HoLEP. Furthermore, the decrease in haemoglobin was significantly lower in patients receiving ThuLEP (mean difference −0.54 g/dL, 95% CI −0.93 to −0.15; $p < 0.001$), while transient urinary incontinence was more common in patients treated with HoLEP (odds ratio 0.56, 95% CI 0.32–0.99; $p = 0.045$), only with, however, a low certainty of evidence. Therefore, the authors concluded that the choice of treatment should be made according to surgeon's expertise and local conditions [83]. A recent randomized controlled trial compared endoscopic enucleation of the prostate using a thulium: yttrium–aluminum–garnet (Tm:YAG) laser and a super-pulsed thulium fibre laser set in continuous-wave (CW) mode. In total, 110 patients were randomized and treated either with ThuLEP or CW thulium fibre laser enucleation of the prostate (CW-ThuFLEP). There was no significant difference between ThuLEP and CW-ThuFLEP in operative time (70.69

vs. 72.41 min), enucleation time (50.23 vs. 53.33 min), enucleated tissue weight (40.2 41.9 g), enucleation efficiency (0.80 vs. 0.79 g/min), catheterization time (2.45 vs. 2.57 da hospital stay (2.82 vs. 2.95 days), and haemoglobin drop (1.05 vs. 1.27 g/dL). Furtherm no significant differences were found in IPSS (5.09 vs. 5.81), peak urinary flow rates (26 vs. 27.13 mL/s), post-void residual volume (25.22 vs. 23.81 mL), and IIEF-5 (14.01 vs. 14. at the 3-month follow-up. The authors concluded that the theoretical advantages of super-pulsed thulium fibre laser, such as the shallower penetration depth and improv vaporization capacity, did not lead to significant differences in patient outcomes [84].

As with ThuLEP, the goal of ThuVEP is the enucleation of prostate tissue along capsule. Both techniques differ in energy settings and applied mechanical force [84, ThuVEP is a safe treatment option, especially in patients with large prostates or patie on anticoagulant or antiplatelet therapy [69]. Potential benefits of ThuVEP are smoot vaporization and enhanced haemostasis due to the possibility of using continuous wa and pulsed modes. Significant improvements in short-term outcomes, both subject and objective, were observed. Netsch et al. conducted a prospective, randomized tr comparing ThuVEP with HoLEP with 48 and 46 patients, respectively. In both grou peak urinary flow rates (10.7 vs. 22 mL/s), post-void residual volumes (100 vs. 20 m international prostate symptom score (20 vs. 10) and quality of life (4 vs. 3) improv significantly at 1-month follow-up. There was no significant difference between b groups [86]. Compared to the baseline parameters, the same cohort showed peak urin flow rates (10.7 vs. 25.9 mL/s), post-void residual urine (100 vs. 6.5 mL), IPSS (20 vs. quality of life (4 vs. 1), PSA (4.14 vs. 0.71 µg/L), and prostate volume (80 vs. 16 mL) h improved significantly ($p < 0.001$) at 6-month follow-up, without significant differen between both groups [87]. According to the existing evidence, ThuVEP is a safe and effici treatment option for patients with symptomatic BPE. However, large scale randomiz trials are still scarce [88].

### *5.4. Potassium-Titanyl-Phosphate (KTP) and Lithium Triborate (LBO) Lasers: Greenlight-Based P*

Photoselective vaporization of the prostate (PVP) is another possible surgical treatm alternative for patients suffering from LUTS secondary to BPE [89]. Currently, th different Greenlight lasers are in use, the 80 W KTP, the 120 W HPS LBO and the 180 W X LBO [90]. The main technical difference between the PVP and the enucleation techniq is that PVP ablates tissue by vaporization of the tissue from the prostatic urethra tow the capsule (inside-out), while enucleating techniques are used to enter the plane of prostatic capsule and dissect the prostatic lobes (outside-in) [90]. PVP is considered be a side-firing laser ablation of the prostate using a 532 nm wavelength laser which absorbed by haemoglobin, resulting in better coagulation during the procedure [91]. T main advantage of the Greenlight technology is its intraoperative haemostasis and theref a clear surgical view [72]. Therefore, PVP seems a safe treatment option for anticoagula patients, even though the RCTs were conducted on just small amounts of patients a therefore, it is recommended with a weak rating as per the Guidelines [3,91]. Disadvanta of PVP are the limitation of resectable volume and no histopathological analysis due to vaporization of the prostate tissue [72].

A meta-analysis comparing PVP using 80 W and 120 W lasers to TURP analysed 9 als with a total of 448 patients undergoing PVP (80 W in 5 trials and 120 W in 4 trials) a 441 undergoing TURP. Overall data showed shorter catheterisation time and shorter len of stay in patients treated with PVP compared to TURP, with 1.91 day (95% CI, 1.47–2. $p < 0.00001$) and 2.13 days (95% CI, 1.78–2.48; $p < 0.00001$), respectively. Operation ti was shorter in patients treated with TURP by 19.64 min (95% CI, 9.05–30.23; $p = 0.00$ The transfusion rate was significantly lower in the PVP group (risk ratio: 0.16; 95% 0.05–0.53; $p = 0.003$). No significant difference was found for other complications [ Similarly, a large PVP of the prostate was noninferior to TURP in terms of internatio prostate symptom score, peak flow rate, and freedom from complications [92]. Non feriority was sustained at the two-year follow-up [93]. A recent randomized controll trial with 154 consecutive enrolled patients investigated a new surgical procedure, an enucleation technique involving photoselective sharp enucleation of the prostate (PSEP), with a front-firing 532-nm laser for LUTS caused by BPE in order to address the limitations of PVP, such as a lack of tissue for histologic analysis and long operating times for large prostates. The authors compared this novel technique to traditional PVP. Both groups showed an improvement of lower urinary tract symptoms at one, six and twelve months postoperatively. Furthermore, both groups had equivalent international prostate symptom scores, quality-of-life scores, postvoid residual urine volume, maximum urine flow rates, as well as prostate-specific antigen at each follow-up ($p > 0.05$). The median operative time for patients treated with PSEP was significantly shorter than for patients treated with PVP, at 35 min vs. 47 min ($p < 0.001$), respectively. The median prostate volume was significantly smaller in patients treated with PSEP compared to those treated with PVP ($p < 0.05$) at six and twelve months. There was no significant difference in complication rates. While the authors found both treatments to safe and efficient, PSEP could resolve the limitations of PVP [94].

### 5.5. Diode Laser

In the context of BPH, diode lasers with a wavelength of 940, 980, 1318, and 1470 nm are marketed for vaporization and enucleation [69]. Historically, diode lasers have been associated with higher complications such as postoperative irritative symptoms, transient urinary incontinence, and epididymitis when compared to other lasers, such as the Greenlight [95]. However, more recent RCTs did not show a difference in complication rates, blood loss, IPSS, or QoL when compared to bipolar resection or enucleation of the prostate [96–98]. However, since these studies are limited and of mainly low quality with controversial data on retreatment rates, diode laser enucleation of the prostate is recommended only with a weak rating for men with moderate-to-severe LUTS, while diode laser vaporization is not recommended yet [69]. In 2022, the FDA approved the SoracteLite transperineal interstitial laser ablation (TPLA) to induce coagulative necrosis of tissue through a continuous 1064 nm wavelength for men who desire to maintain ejaculation [99]. TPLA may be performed in an office-based setting under local anaesthesia with a highly custom and precise ablation through the transperineal route [99]. Because of the novelty of this technique, the Canadian, American, and European urological guidelines have not yet endorsed it [69,100,101]. Additionally, an open-label randomized trial comparing the functional outcomes between TPLA and TURP is in the recruitment phase [99].

## 6. Limitations

The present review has some limitations. First, this is a literature review that has a narrative character. Second, we lack granular information on laser fibres, irrigation settings, activation/deactivation intervals, and laser tip/tissue distance; indeed, the included studies show heterogeneity in their endpoints. Third, most included studies are mainly retrospective with small sample sizes because of a low availability, due to the relatively recent acceptance of TFLs by the FDA and the EMA. Fourth, there is a lack of comparative clinical studies between the different technologies, especially dealing with TFLs. However, we believe that with this review we were able to give a general overview of the current state of lasers in endourology, to provide some practical information for beginners, and to give some perspectives for future studies.

## 7. Future Perspectives

Laser technology is continuously evolving. Being relatively new, a lot of attention is given to TFLs in the context of stone treatment. One advantage is that smaller laser fibre sizes that might be used with TFLs, coupled with the new miniaturized technologies and the smaller single-use flexible scopes, might extend the use of RIRS for larger and lower pole stones. Additionally, high-power lasers might widen the horizon of RIRS, since some studies have already demonstrated advantages in terms of shorter operative time, reduced use of ureteral access sheath (UAS), and postoperative stent, which in turn mea reduction in sepsis-related complication rates [102]. Large comparative studies compari TFL and Ho:YAG lasers with MOSES technology might shed new light on what the b treatment is in terms of surgical outcomes, emphasizing the outcome measures such stone-free rate and the imaging used to achieve it, quality of life, costs, and environmen impact [103–105].

In the field of UTUC, while no study has shown the superiority of one laser type o the other, over time, clinicians have reported an increasing success with thulium las especially TFLs. Proietti et al. suggested an ideal setting of 1 J and 10 Hz to provide optin haemostasis without generating excessive carbonization that might hinder the radical of tumour extirpation [53]. However, TFL laser technology is still in its infancy and lar prospective and randomized studies are needed to find the optimal settings, and lon follow-up is needed to corroborate the survival outcomes [106].

In view of the great success of lasers in the field of BPH for medium-sized prostates, the near future this technology is likely to have a role even for larger prostates. One of main biases related to these procedures is that despite the widespread use of the techniq the learning curve might be steep for laser enucleation procedures. For this reason, medi industries should improve training programs similar to what happens for robotic surg

## 8. Conclusions

Laser and laser-related techniques are very versatile and have been demonstrat to be particularly useful in the context of urolithiasis, urothelial cancer, and BPH. I urolithiasis, lasers may offer the opportunity to use ureteroscopy and lasertripsy even larger stones. When newer high-power lasers are used, the related increase in intrarenal ureteral temperature should be managed by titrating the irrigation and using a urete access sheath. However, at powers <40 W, all lasers exhibit a safe temperature profile.

In the context of UTUC, Ho:YAG has a deeper incision in comparison to thuliu while CW Tm:YAG offers fast, deep and precise cutting but has a broader coagulati area with increased carbonization. The TFL offers a Ho:YAG-like incision and at an id setting of 1 J and 10 Hz seems to provide optimal haemostasis without generating excess carbonization. TURP is still the standard therapy for patients with small to mediu prostates who require surgical treatment. However, intervention using laser technolog can be considered as a safe and efficient alternative treatment modality, especially patients with larger prostates or anticoagulated patients. For patients who want to maint ejaculation, TPLA might become an alternate option. Ultimately, the choice of treatm should be made considering the individual surgeon's expertise, local availability of vari treatment options, patient counselling, and shared decision making.

**Author Contributions:** Conceptualization, C.C. and B.S.; methodology, C.C. and B.S.; investigati C.C. and B.S.; data curation, C.C., V.J. and C.N.; writing— original draft preparation, C.C. and writing—review and editing, C.C., V.J. and B.S.; supervision, A.P. and B.S.; project administrati B.S.; funding acquisition, not applicable. All authors have read and agreed to the published vers of the manuscript.

**Funding:** This research received no external funding.

**Institutional Review Board Statement:** Not applicable.

**Informed Consent Statement:** Not applicable.

**Data Availability Statement:** All data are available in the studies included in the review and w discussed in the present manuscript.

**Conflicts of Interest:** The authors declare no conflict of interest.

## ferences


Dołowy, Ł.; Krajewski, W.; Dembowski, J.; Zdrojowy, R.; Kołodzej, A. The role of lasers in modern urology. *Cent. Eur. J. Urol.* **2015**, *68*, 175–182. [CrossRef]

Pietropaolo, A.; Hughes, T.; Mani, M.; Somani, B. Outcomes of ureteroscopy and laser stone fragmentation (URSL) for kidney stone disease (KSD): Comparative cohort study using moses technology 60 w laser system versus regular holmium 20 w laser. *J. Clin. Med.* **2021**, *10*, 2742. [CrossRef]

Skolarikos, A.; Neisius, A.; Petřík, A.; Somani, B.; Thomas, K.; Gambaro, G. EAU Guidelines on Urolithiasis. 2023. Available online: https://uroweb.org/guidelines/urolithiasis (accessed on 9 April 2023).

Barone, B.; Crocetto, F.; Vitale, R.; Di Domenico, D.; Caputo, V.; Romano, F.; De Luca, L.; Bada, M.; Imbimbo, C.; Prezioso, D. Retrograde intra renal surgery versus percutaneous nephrolithotomy for renal stones >2 cm. A systematic review and meta-analysis. *Minerva Urol. Nephrol.* **2020**, *72*, 441–450. [CrossRef]

Erhard, M.J.; Bagley, D.H. Urologic Applications of the Holmium Laser: Preliminary Experience. *J. Endourol.* **1995**, *9*, 383–386. [CrossRef]

Brewin, A.; Somani, B. What is new in lasers for endourology: Looking into the future. *Urol. News* **2021**, *25*, 2.

Schembri, M.; Sahu, J.; Aboumarzouk, O.; Pietropaolo, A.; Somani, B.K. Thulium fiber laser: The new kid on the block. *Turk. J. Urol.* **2020**, *46* (Suppl. S1), S1–S10. [CrossRef] [PubMed]

Basulto-Martínez, M.; Proietti, S.; Yeow, Y.; Rapallo, I.; Saitta, G.; Cimino, S.; Luciani, L.; Bellinzoni, P.; Gaboardi, F.; Giusti, G. Holmium laser for RIRS. Watts are we doing? *Arch. Esp. Urol.* **2020**, *73*, 735–744. [PubMed]

Matlaga, B.R.; Chew, B.; Eisner, B.; Humphreys, M.; Knudsen, B.; Krambeck, A.; Lange, D.; Lipkin, M.; Miller, N.L.; Monga, M.; et al. Ureteroscopic Laser Lithotripsy: A Review of Dusting vs. Fragmentation with Extraction. *J. Endourol.* **2018**, *32*, 1–6. [CrossRef] [PubMed]

Santiago, J.E.; Hollander, A.B.; Soni, S.D.; Link, R.E.; Mayer, W.A. To Dust or Not To Dust: A Systematic Review of Ureteroscopic Laser Lithotripsy Techniques. *Curr. Urol. Rep.* **2017**, *18*, 32. [CrossRef]

Kronenberg, P.; Somani, B. Advances in Lasers for the Treatment of Stones—A Systematic Review. *Curr. Urol. Rep.* **2018**, *19*, 45. [CrossRef]

Fuh, E.; Haleblian, G.E.; Norris, R.D.; Albala, W.D.M.; Simmons, N.; Zhong, P.; Preminger, G.M. The Effect of Frequency Doubled Double Pulse Nd: YAG Laser Fiber Proximity to the Target Stone on Transient Cavitation and Acoustic Emission. *J. Urol.* **2007**, *177*, 1542–1545. [CrossRef] [PubMed]

Emiliani, E.; Talso, M.; Cho, S.-Y.; Baghdadi, M.; Mahmoud, S.; Pinheiro, H.; Traxer, O. Optimal Settings for the Noncontact Holmium: YAG Stone Fragmentation Popcorn Technique. *J. Urol.* **2017**, *198*, 702–706. [CrossRef] [PubMed]

Pietropaolo, A.; Jones, P.; Whitehurst, L.; Somani, B.K. Role of 'dusting and pop-dusting' using a high-powered (100 W) laser machine in the treatment of large stones ($\geq$15 mm): Prospective outcomes over 16 months. *Urolithiasis* **2019**, *47*, 391–394. [CrossRef]

Reeves, T.; Griffin, S.; Pietropaolo, A.; Somani, B.K. Feasibility of dusting and pop-dusting using high-power (100W) holmium YAG (HO: YAG) laser in treatment of paediatric stones: Results of first worldwide clincial study. *Cent. Eur. J. Urol.* **2019**, *72*, 398–401.

Ibrahim, A.; Badaan, S.; Elhilali, M.M.; Andonian, S. Moses technology in a stone simulator. *Can. Urol. Assoc. J.* **2018**, *12*, 127–130. [CrossRef]

Keller, E.X.; De Coninck, V.; Audouin, M.; Doizi, S.; Bazin, D.; Daudon, M.; Traxer, O. Fragments and dust after Holmium laser lithotripsy with or without "Moses technology": How are they different? *J. Biophotonics* **2019**, *12*, e201800227. [CrossRef] [PubMed]

Winship, B.; Wollin, D.A.; Carlos, E.C.; Li, M.J.; Peters, M.C.; Simmons, W.N.; Preminger, G.M.; Lipkin, M.E. Dusting Efficiency of the Moses Holmium Laser: An Automated *In Vitro* Assessment. *J. Endourol.* **2018**, *32*, 1131–1135. [CrossRef] [PubMed]

Ventimiglia, E.; Pauchard, F.; Quadrini, F.; Sindhubodee, S.; Kamkoum, H.; Godínez, A.J.; Doizi, S.; Traxer, O. High- and Low-Power Laser Lithotripsy Achieves Similar Results: A Systematic Review and Meta-Analysis of Available Clinical Series. *J. Endourol.* **2021**, *35*, 1146–1152. [CrossRef]

Kronenberg, P.; Traxer, O. The laser of the future: Reality and expectations about the new thulium fiber laser—A systematic review. *Transl. Androl. Urol.* **2019**, *8*, S398–S417. [CrossRef]

Enikeev, D.; Grigoryan, V.; Fokin, I.; Morozov, A.; Taratkin, M.; Klimov, R.; Kozlov, V.; Gabdullina, S.; Glybochko, P. Endoscopic lithotripsy with a SuperPulsed thulium-fiber laser for ureteral stones: A single-center experience. *Int. J. Urol.* **2021**, *28*, 261–265. [CrossRef]

Hardy, L.A.; Vinnichenko, V.; Fried, N.M. High power holmium: YAG versus thulium fiber laser treatment of kidney stones in dusting mode: Ablation rate and fragment size studies. *Lasers Surg. Med.* **2019**, *51*, 522–530. [CrossRef] [PubMed]

Vaddi, C.M.; Ramakrishna, P.; Ganeshan, S.; Swamy, S.; Anandan, H.; Babu, M.; Panda, R. The clinical efficiency and safety of 60W superpulse thulium fiber laser in retrograde intrarenal surgery. *Indian J. Urol.* **2022**, *38*, 191–196. [PubMed]

Taratkin, M.; Azilgareeva, C.; Korolev, D.; Barghouthy, Y.; Tsarichenko, D.; Akopyan, G.; Chinenov, D.; Ali, S.; Kozlov, V.; Mikhailov, V.; et al. Prospective Single-Center Study of SuperPulsed Thulium Fiber Laser in Retrograde Intrarenal Surgery: Initial Clinical Data. *Urol. Int.* **2022**, *106*, 404–410. [CrossRef] [PubMed]

25. Ulvik, Ø.; Æsøy, M.S.; Juliebø-Jones, P.; Gjengstø, P.; Beisland, C. Thulium Fibre Laser versus Holmium: YAG for Ureterosco Lithotripsy: Outcomes from a Prospective Randomised Clinical Trial. *Eur. Urol.* **2022**, *82*, 73–79. [CrossRef]
26. Keat, W.O.L.; Somani, B.K.; Pietropaolo, A.; Chew, B.H.; Chai, C.A.; Inoue, T.; Ragoori, D.; Biligere, S.; Galosi, A.B.; Pavia, M et al. Do Hounsfield Units have any significance in predicting intra- and postoperative outcomes in retrograde intrarenal surg using Holmium and Thulium fiber laser? Results from the flexible ureteroscopy Outcomes Registry (FLEXOR). *World J. U* **2023**, *ahead of print*. [CrossRef]
27. Ryan, J.R.; Nguyen, M.H.; Linscott, J.A.; Nowicki, S.W.; James, E.; Jumper, B.M.; Ordoñez, M.; Ingimarsson, J.P. Ureterosc with thulium fiber laser lithotripsy results in shorter operating times and large cost savings. *World J. Urol.* **2022**, *40*, 2077–2( [CrossRef]
28. Perri, D.; Berti, L.; Pacchetti, A.; Morini, E.; Maltagliati, M.; Besana, U.; Pastore, A.L.; Romero-Otero, J.; Saredi, G.; Centrella, et al. A comparison among RIRS and MiniPerc for renal stones between 10 and 20 mm using thulium fiber laser (Fiber Dust) randomized controlled trial. *World J. Urol.* **2022**, *40*, 2555–2560. [CrossRef]
29. Rice, P.; Somani, B.K. Percutaneous laser nephrolithotripsy: Is it here to stay? Results of a systematic review. *Curr. Opin. U* **2022**, *32*, 185–191. [CrossRef]
30. Ghani, K.R.; Andonian, S.; Bultitude, M.; Desai, M.; Giusti, G.; Okhunov, Z.; Preminger, G.M.; de la Rosette, J. Percutane Nephrolithotomy: Update, Trends, and Future Directions. *Eur. Urol.* **2016**, *70*, 382–396. [CrossRef]
31. Moore, S.L.; Bres-Niewada, E.; Cook, P.; Wells, H.; Somani, B.K. Optimal management of lower pole stones: The direction future travel. *Cent. Eur. J. Urol.* **2016**, *69*, 274–279.
32. De, S.; Autorino, R.; Kim, F.J.; Zargar, H.; Laydner, H.; Balsamo, R.; Torricelli, F.C.; Di Palma, C.; Molina, W.R.; Monga, M.; e Percutaneous Nephrolithotomy versus Retrograde Intrarenal Surgery: A Systematic Review and Meta-analysis. *Eur. Urol.* **2(** *67*, 125–137. [CrossRef] [PubMed]
33. Jiang, K.; Chen, H.; Yu, X.; Chen, Z.; Ye, Z.; Yuan, H. The "all-seeing needle" micro-PCNL versus flexible ureterorenoscopy lower calyceal stones of ≤2 cm. *Urolithiasis* **2019**, *47*, 201–206. [CrossRef] [PubMed]
34. Zhang, H.; Hong, T.Y.; Li, G.; Jiang, N.; Hu, C.; Cui, X.; Chu, C.; Zhao, J. Comparison of the Efficacy of Ultra-Mini PCNL, Flexi Ureteroscopy, and Shock Wave Lithotripsy on the Treatment of 1–2 cm Lower Pole Renal Calculi. *Urol. Int.* **2019**, *102*, 153–1 [CrossRef] [PubMed]
35. Kronenberg, P.; Hameed, B.M.Z.; Somani, B. Outcomes of thulium fibre laser for treatment of urinary tract stones: Results systematic review. *Curr. Opin. Urol.* **2021**, *31*, 80–86. [CrossRef]
36. Andreeva, V.; Vinarov, A.; Yaroslavsky, I.; Kovalenko, A.; Vybornov, A.; Rapoport, L.; Enikeev, D.; Sorokin, N.; Dymov, Tsarichenko, D.; et al. Preclinical comparison of superpulse thulium fiber laser and a holmium: YAG laser for lithotripsy. *Wor Urol.* **2020**, *38*, 497–503. [CrossRef]
37. Patil, A.; Reddy, N.; Shah, D.; Singh, A.; Ganpule, A.; Sabnis, R.; Desai, M. High-Power Holmium with MOSES Technology Thulium Fiber Laser in MiniPerc with Suction: A New Curiosity. *J. Endourol.* **2022**, *36*, 1348–1354. [CrossRef]
38. Taratkin, M.; Azilgareeva, C.; Chinenov, D.; Mikhailov, V.; Inoyatov, J.; Ali, S.; Korolev, D.; Tsarichenko, D.; Corrales, M.; Enik D. Retrograde intrarenal surgery versus percutaneous nephro-lithotomy in larger kidney stones. Could superpulsed thulium-fi laser change the game? *Cent Eur. J. Urol.* **2021**, *74*, 229–234.
39. Rice, P.; Somani, B.K.; Nagele, U.; Herrmann, T.R.W.; Tokas, T. Generated temperatures and thermal laser damage dur upper tract endourological procedures using the holmium: Yttrium–aluminum-garnet (Ho: YAG) laser: A systematic review experimental studies. *World J. Urol.* **2022**, *40*, 1981–1992. [CrossRef]
40. Tokas, T.; Rice, P.; Seitz, C.; Gauhar, V.; Somani, B. Temperature change during laser upper-tract endourological procedu Current evidence and future perspective. *Curr. Opin. Urol.* **2023**, *33*, 108–115. [CrossRef]
41. Yamashita, S.; Inoue, T.; Imai, S.; Kohjimoto, Y.; Fujisawa, M.; Hara, I. Thermography-based comparison of irrigation temperatu between Moses Mode and Virtual Basket Mode: An in-vitro phantom study. *J. Endourol.* **2023**, *37*, 179–184. [CrossRef]
42. Teng, J.; Wang, Y.; Jia, Z.; Guan, Y.; Fei, W.; Ai, X. Temperature profiles of calyceal irrigation fluids during flexible ureterosco Ho: YAG laser lithotripsy. *Int. Urol. Nephrol.* **2021**, *53*, 415–419. [CrossRef] [PubMed]
43. Belle, J.D.; Chen, R.; Srikureja, N.; Amasyali, A.S.; Keheila, M.; Duane Baldwin, D. Does the Novel Thulium Fiber Laser Ha a Higher Risk of Urothelial Thermal Injury than the Conventional Holmium Laser in an in Vitro Study? *J. Endourol.* **2022**, 1249–1254. [CrossRef] [PubMed]
44. Sierra, A.; Corrales, M.; Somani, B.; Traxer, O. Laser Efficiency and Laser Safety: Holmium YAG vs. Thulium Fiber Laser. *J. C Med.* **2023**, *12*, 149. [CrossRef] [PubMed]
45. Siegel, R.L.; Miller, K.D.; Jemal, A. Cancer statistics, 2020. *CA Cancer J. Clin.* **2020**, *70*, 7–30. [CrossRef]
46. Rouprêt, M.; Babjuk, M.; Burger, M.; Compérat, E.; Cowan, N.; Gontero, P. EAU Guidelines on Upper Urinary Tract Urothe Carcinoma. 2023. Available online: https://uroweb.org/guidelines/upper-urinary-tract-urothelial-cell-carcinoma (accessed o April 2023).
47. Gkolezakis, V.; Rice, P.; Somani, B.K.; Tokas, T. Thulium Fiber Laser Behavior on Tissue During Upper- and Lower-Tr Endourology. *Curr. Urol. Rep.* **2022**, *23*, 271–278. [CrossRef]
48. Villa, L.; Haddad, M.; Capitanio, U.; Somani, B.; Cloutier, J.; Doizi, S.; Salonia, A.; Briganti, A.; Montorsi, F.; Traxer, O. Wh Patients with Upper Tract Urothelial Carcinoma Can be Safely Treated with Flexible Ureteroscopy with Holmium: YAG La Photoablation? Long-Term Results from a High Volume Institution. *J. Urol.* **2018**, *199*, 66–73. [CrossRef]

Huusmann, S.; Lafos, M.; Meyenburg, I.; Muschter, R.; Teichmann, H.O.; Herrmann, T. Tissue effects of a newly developed diode pumped pulsed Thulium: YAG laser compared to continuous wave Thulium: YAG and pulsed Holmium: YAG laser. *World J. Urol.* **2021**, *39*, 3503–3508. [CrossRef]

Ng Chieng Hin, J.; Hettiarachchilage, D.; Gravestock, P.; Rai, B.; Somani, B.K.; Veeratterapillay, R. Role of Ureteroscopy in Treatment of Upper Tract Urothelial Carcinoma. *Curr. Urol. Rep.* **2021**, *22*, 49. [CrossRef]

Zarrabi, A.; Gross, A.J. The evolution of lasers in urology. *Ther. Adv. Urol.* **2011**, *3*, 81–89. [CrossRef]

Lee, J.; Gianduzzo, T.R.J. Advances in Laser Technology in Urology. *Urol. Clin. N. Am.* **2009**, *36*, 189–198. [CrossRef]

Proietti, S.; Johnston, T.; Pupulin, M.; Di Pietro, S.; Spagna, S.; Rico, L.; Lucianò, R.; Ventimiglia, E.; Villa, L.; Gaboardi, F.; et al. Effectiveness and Safety of Thulium Fiber Laser in the Conservative Management of Patients with Upper Tract Urothelial Carcinoma. *Eur. Urol. Open Sci.* **2022**, *46*, 99–104. [CrossRef] [PubMed]

Huusmann, S.; Wolters, M.; Kramer, M.W.; Bach, T.; Teichmann, H.-O.; Eing, A.; Bardosi, S.; Herrmann, T.R.W. Tissue damage by laser radiation: An in vitro comparison between Tm: YAG and Ho: YAG laser on a porcine kidney model. *Springerplus* **2016**, *5*, 266. [CrossRef] [PubMed]

Ortner, G.; Rice, P.; Nagele, U.; Herrmann, T.R.W.; Somani, B.K.; Tokas, T. Tissue thermal effect during lithotripsy and tissue ablation in endourology: A systematic review of experimental studies comparing Holmium and Thulium lasers. *World J. Urol.* **2023**, *41*, 1–12. [CrossRef] [PubMed]

Heinrich, E.; Wendt-Nordahl, G.; Honeck, P.; Alken, P.; Knoll, T.; Michel, S.M.; Häcker, A. 120 W lithium triborate laser for photoselective vaporization of the prostate: Comparison with 80 W potassium-titanyl-phosphate laser in an ex-vivo model. *J. Endourol.* **2010**, *24*, 75–79. [CrossRef]

Rahman, K.M.M.; Giram, P.; Foster, B.A.; You, Y. Photodynamic Therapy for Bladder Cancers, A Focused Review. *Photochem. Photobiol.* **2023**, *99*, 420–436. [CrossRef]

Wu, K.; Jiang, D.; Zhang, L.; Jiang, S.; Lin, T.; Luo, Y.; Fan, J.; Yang, T.; Chen, H.; Zhang, P.; et al. Efficacy and safety of a novel 450 nm blue diode laser versus plasmakinetic electrocautery for the transurethral resection of non-muscle invasive bladder cancer: The protocol and result of a multicenter randomized con-trolled trial. *Front. Oncol.* **2023**, *12*, 1065735. [CrossRef]

Cornu, J.-N.; Rouprêt, M.; Carpentier, X.; Geavlete, B.; Medina, S.G.D.D.; Cussenot, O.; Traxer, O. Oncologic control obtained after exclusive flexible ureteroscopic management of upper urinary tract urothelial cell carcinoma. *World J. Urol.* **2010**, *28*, 151–156. [CrossRef]

Bozzini, G.; Gastaldi, C.; Besana, U.; Calori, A.; Casellato, S.; Parma, P.; Pastore, A.L.; Macchi, A.; Breda, A.; Gozen, A.; et al. Thulium-laser Retrograde Intra Renal Ablation (T-RIRA) of upper urinary tract transitional cell carcinoma: An ESUT study. *Minerva Urol. Nephrol.* **2021**, *73*, 114–121. [CrossRef]

Defidio, L.; De Dominicis, M.; Di Gianfrancesco, L.; Fuchs, G.; Patel, A. First collaborative experience with thulium laser ablation of localized upper urinary tract urothelial tumors using retrograde intra-renal surgery. *Arch. Ital. Urol. Androl.* **2011**, *83*, 147–153.

Defidio, L.; Antonucci, M.; De Dominicis, M.; Fuchs, G.; Patel, A. Thulium-Holmium: YAG Duo Laser in Conservative Upper Tract Urothelial Cancer Treatment: 13 Years Experience from a Tertiary National Referral Center. *J. Endourol.* **2019**, *33*, 902–908. [CrossRef]

Wen, J.; Ji, Z.G.; Li, H.Z. Treatment of upper tract urothelial carcinoma with ureteroscopy and thulium laser: A retrospective single center study. *BMC Cancer* **2018**, *18*, 196. [CrossRef] [PubMed]

Musi, G.; Mistretta, F.A.; Marenghi, C.; Russo, A.; Catellani, M.; Nazzani, S.; Conti, A.; Luzzago, S.; Ferro, M.; Matei, D.V.; et al. Thulium Laser Treatment of Upper Urinary Tract Carcinoma: A Multi-Institutional Analysis of Surgical and Oncological Outcomes. *J. Endourol.* **2018**, *32*, 257–263. [CrossRef] [PubMed]

Sanguedolce, F.; Fontana, M.; Turco, M.; Territo, A.; Lucena, J.B.; Cortez, J.C.; Vanacore, D.; Meneghetti, I.; Gallioli, A.; Gaya, J.M.; et al. Endoscopic Management of Upper Urinary Tract Urothelial Carcinoma: Oncologic Outcomes and Prognostic Factors in a Contemporary Cohort. *J. Endourol.* **2021**, *35*, 1593–1600. [CrossRef] [PubMed]

Scoffone, C.M.; Cracco, C.M.; Poggio, M.; Cossu, M.; Scarpa, R.M. Treatment of the pyelocalyceal tumors with laser. *Arch. Esp. Urol.* **2008**, *61*, 1080–1087. [CrossRef]

Chauhan, S.S.; Abu Dayyeh, B.K.; Bhat, Y.M.; Gottlieb, K.T.; Hwang, J.H.; Komanduri, S.; Konda, V.; Lo, S.K.; Manfredi, M.A.; Maple, J.T.; et al. Confocal laser endomicroscopy. *Gastrointest. Endosc.* **2014**, *80*, 928–938. [CrossRef]

Vanacore, D.; Sanguedolce, F.; Territo, A.; Roca, B.M.; Saitta, G.; Tallè, M.; Martinez, C.; Mosquera, L.; Meneghetti, J.; Bevilacqua, G.; et al. Evolving techniques of endoscopic UTUC management: Optimising outcomes with the appropriate use of latest technologies. *Eur. Urol. Open Sci.* **2020**, *19*, e2374–e2375. [CrossRef]

Cornu, J.N.; Gacci, M.; Hashim, H.; Herrmann, T.R.W.; Malde, S.; Netsch, C.; Rieken, M.; Sakalis, V.; Tutolo, M.; Baboudjian, M.; et al. EAU Guidelines on Non-Neurogenic Male LUTS. 2023. Available online: https://uroweb.org/guidelines/management-of-non-neurogenic-male-luts (accessed on 9 April 2023).

Herrmann, T.R.W.; Liatsikos, E.N.; Nagele, U.; Traxer, O.; Merseburger, A.S. EAU guidelines on laser technologies. *Eur. Urol.* **2012**, *61*, 783–795. [CrossRef]

Vincent, M.W.; Gilling, P.J. HoLEP has come of age. *World J. Urol.* **2015**, *33*, 487–493. [CrossRef]

Xiao, K.-W.; Zhou, L.; He, Q.; Gao, X.-S.; Chen, G.; Ma, Y.-C.; Li, H.; Wang, K.-J. Enucleation of the prostate for benign prostatic hyperplasia thulium laser versus holmium laser: A systematic review and meta-analysis. *Lasers Med. Sci.* **2019**, *34*, 815–826. [CrossRef]

73. Naspro, R.; Suardi, N.; Salonia, A.; Scattoni, V.; Guazzoni, G.; Colombo, R.; Cestari, A.; Briganti, A.; Mazzoccoli, B.; Rigatti, et al. Holmium Laser Enucleation of the Prostate versus Open Pros-tatectomy for Prostates >70 g: 24-Month Follow-up. *Eur. U* **2006**, *50*, 563–568. [CrossRef]
74. Kuntz, R.M.; Lehrich, K.; Ahyai, S.A. Holmium Laser Enucleation of the Prostate versus Open Prostatectomy for Prostates Grea than 100 Grams: 5-Year Follow-Up Results of a Randomised Clinical Trial. *Eur. Urol.* **2008**, *53*, 160–166. [CrossRef] [PubMed]
75. Cornu, J.-N.; Ahyai, S.; Bachmann, A.; de la Rosette, J.; Gilling, P.; Gratzke, C.; McVary, K.; Novara, G.; Woo, H.; Madersbach S. A Systematic Review and Meta-analysis of Functional Outcomes and Complications Following Transurethral Procedu for Lower Urinary Tract Symptoms Resulting from Benign Prostatic Obstruction: An Update. *Eur. Urol.* **2015**, *67*, 1066–1C [CrossRef] [PubMed]
76. Jones, P.; Alzweri, L.; Rai, B.P.; Somani, B.K.; Bates, C.; Aboumarzouk, O.M. Holmium laser enucleation versus simple prosta ctomy for treating large prostates: Results of a systematic review and meta-analysis. *Arab J. Urol.* **2016**, *14*, 50–58. [CrossR [PubMed]
77. Gkolezakis, V.; Somani, B.K.; Tokas, T. Low- vs. High-Power Laser for Holmium Laser Enucleation of Prostate. *J. Clin. Med.* **20** *12*, 2084. [CrossRef]
78. Gong, Y.-G.; He, D.-L.; Wang, M.-Z.; Li, X.-D.; Zhu, G.-D.; Zheng, Z.-H.; Du, Y.-F.; Chang, L.S.; Nan, X.-Y. Holmium La Enucleation of the Prostate: A Modified Enucleation Technique and Initial Results. *J. Urol.* **2012**, *187*, 1336–1340. [CrossRef]
79. Rassweiler, J.; Roder, M.; Schulze, M.; Muschter, R. Transurethral enucleation of the prostate with the holmium: YAG laser syst How much power is necessary? *Urol. A* **2008**, *47*, 441–448. [CrossRef]
80. Gauhar, V.; Gilling, P.; Pirola, G.M.; Chan, V.W.-S.; Lim, E.J.; Maggi, M.; Teoh, J.Y.-C.; Krambeck, A.; Castellani, D. Does MOS Technology Enhance the Efficiency and Outcomes of Standard Holmium Laser Enucleation of the Prostate? Results of a Systema Review and Meta-analysis of Comparative Studies. *Eur. Urol. Focus* **2022**, *8*, 1362–1369. [CrossRef]
81. Bach, T.; Netsch, C.; Haecker, A.; Michel, M.S.; Herrmann, T.R.W.; Gross, A.J. Thulium: YAG laser enucleation (VapoEnucleati of the prostate: Safety and durability during intermediate-term follow-up. *World J. Urol.* **2010**, *28*, 39–43. [CrossRef]
82. Herrmann, T.R.; Bach, T.; Imkamp, F.; Georgiou, A.; Burchardt, M.; Oelke, M.; Gross, A.J. Thulium laser enucleation of the prost (ThuLEP): Tran-surethral anatomical prostatectomy with laser support. Introduction of a novel technique for the treatment benign prostatic obstruction. *World J. Urol.* **2010**, *28*, 45–51. [CrossRef]
83. Hartung, F.O.; Kowalewski, K.-F.; von Hardenberg, J.; Worst, T.S.; Kriegmair, M.C.; Nuhn, P.; Herrmann, T.R.; Michel, M Herrmann, J. Holmium versus Thulium Laser Enucleation of the Prostate: A Systematic Review and Meta-analysis of Randomiz Controlled Trials. *Eur. Urol. Focus* **2022**, *8*, 545–554. [CrossRef]
84. Bozzini, G.; Berti, L.; Maltagliati, M.; Besana, U.; Micali, S.; Roche, J.B.; Romero-Otero, J.; Pacchetti, A.; Perri, D.; Morini, E.; et Thulium: YAG vs. continuous-wave thulium fiber laser enucleation of the prostate: Do potential advantages of thulium fil lasers translate into relevant clinical differences? *World J. Urol.* **2023**, *41*, 143–150. [CrossRef] [PubMed]
85. Herrmann, T.R.W.; Wolters, M. Transurethral anatomical enucleation of the prostate with Tm: YAG support (ThuLEP): Evoluti and variations of the technique. The inventors' perspective. *Andrologia* **2020**, *52*, e13587. [CrossRef] [PubMed]
86. Netsch, C.; Becker, B.; Tiburtius, C.; Moritz, C.; Becci, A.V.; Herrmann, T.R.W.; Gross, A.J. A prospective, randomized tr comparing thulium vapoenu-cleation with holmium laser enucleation of the prostate for the treatment of symptomatic beni prostatic obstruction: Perioperative safety and efficacy. *World J. Urol.* **2017**, *35*, 1913–1921. [CrossRef] [PubMed]
87. Becker, B.; Herrmann, T.R.W.; Gross, A.J.; Netsch, C. Thulium vapoenucleation of the prostate versus holmium laser enucleati of the prostate for the treatment of large volume prostates: Preliminary 6-month safety and efficacy results of a prospecti randomized trial. *World J. Urol.* **2018**, *36*, 1663–1671. [CrossRef]
88. Jones, P.; Rai, B.P.; Somani, B.K.; Aboumarzouk, O.M. A review of thulium laser vapo-enucleation of the prostate: A no laser-based strategy for benign prostate enlargement. *Arab J. Urol.* **2015**, *13*, 209–211. [CrossRef]
89. Broggi, E.; May, A.; Giretti, G.; Tabchouri, N.; Lorphelin, H.; Brichart, N.; Bruyère, F. Long-Term Outcomes of 80-Watt KTP a 120-Watt HPS GreenLight Photoselective Vaporization of the Prostate. *Urol. Int.* **2014**, *93*, 229–236. [CrossRef]
90. Gomez Sancha, F.; Rivera, V.C.; Georgiev, G.; Botsevski, A.; Kotsev, J.; Herrmann, T. Common trend: Move to enucleation—Is th a case for GreenLight enucleation? Development and description of the technique. *World J. Urol.* **2015**, *33*, 539–547. [CrossRe
91. Thangasamy, I.A.; Chalasani, V.; Bachmann, A.; Woo, H.H. Photoselective vaporisation of the prostate using 80-W and 120 laser versus tran-surethral resection of the prostate for benign prostatic hyperplasia: A systematic review with meta-analy from 2002 to 2012. *Eur. Urol.* **2012**, *62*, 315–323. [CrossRef]
92. Bachmann, A.; Tubaro, A.; Barber, N.; D'ancona, F.; Muir, G.; Witzsch, U.; Grimm, M.-O.; Benejam, J.; Stolzenburg, J.-U.; Riddi A.; et al. 180-W XPS greenlight laser vaporisation versus transurethral resection of the prostate for the treatment of beni prostatic obstruction: 6-month safety and efficacy results of a European multicentre ran-domised trial—The GOLIATH study. *E Urol.* **2014**, *65*, 931–942. [CrossRef]
93. Thomas, J.A.; Tubaro, A.; Barber, N.; D'Ancona, F.; Muir, G.; Witzsch, U.; Grimm, M.-O.; Benejam, J.; Stolzenburg, J.-U.; Riddi A.; et al. A Multicenter Randomized Noninferiority Trial Comparing GreenLight-XPS Laser Vaporization of the Prostate a Transurethral Resection of the Prostate for the Treatment of Benign Prostatic Obstruction: Two-yr Outcomes of the GOLIA Study. *Eur. Urol.* **2016**, *69*, 94–102. [CrossRef]

Liu, Z.; Chen, Z.; Yan, D.; Jiang, T.; Fu, J.; Zheng, J.; Zhou, Y.; Zhou, Z.; Shen, W. Photoselective sharp enucleation of the prostate with a front-firing 532-nm laser versus photoselective vaporization of the prostate in the treatment of benign prostatic hyperplasia: A randomised controlled trial with 1-year followup results. *BMC Urol.* **2022**, *22*, 173.

Chiang, P.H.; Chen, C.H.; Kang, C.H.; Chuang, Y.C. GreenLight HPS laser 120-W versus diode laser 200-W vaporization of the prostate: Comparative clinical experience. *Lasers Surg. Med.* **2010**, *42*, 624–629. [CrossRef] [PubMed]

Lusuardi, L.; Myatt, A.; Sieberer, M.; Jeschke, S.; Zimmermann, R.; Janetschek, G. Safety and efficacy of eraser laser enucleation of the prostate: Preliminary report. *J. Urol.* **2011**, *186*, 1967–1971. [CrossRef] [PubMed]

Zhang, J.; Wang, X.; Zhang, Y.; Shi, C.; Tu, M.; Shi, G. 1470 nm Diode Laser Enucleation vs. Plasmakinetic Resection of the Prostate for Benign Prostatic Hyperplasia: A Randomized Study. *J. Endourol.* **2019**, *33*, 211–217. [CrossRef]

Zou, Z.; Xu, A.; Zheng, S.; Chen, B.; Xu, Y.; Li, H.; Duan, C.; Zheng, J.; Chen, J.; Li, C.; et al. Dual-centre randomized-controlled trial comparing transurethral endoscopic enucleation of the prostate using diode laser vs. bipolar plasmakinetic for the treatment of LUTS secondary of benign prostate obstruction: 1-year follow-up results. *World J. Urol.* **2018**, *36*, 1117–1126. [CrossRef] [PubMed]

Nguyen, D.D.; Li, T.; Ferreira, R.; Berjaoui, M.B.; Nguyen, A.L.V.; Chughtai, B.; Zorn, K.Z.; Bhojani, N.; Elterman, D. Ablative minimally invasive surgical therapies for benign prostatic hyperplasia: A review of Aquablation, Rezum, and transperineal laser prostate ablation. *Prostate Cancer Prostatic Dis.* **2023**, *ahead of print*. [CrossRef]

). Lerner, L.B.; McVary, K.T.; Barry, M.J.; Bixler, B.R.; Dahm, P.; Das, A.K.; Gandhi, M.C.; Kaplan, S.A.; Kohler, T.S.; Martin, L.; et al. Management of Lower Urinary Tract Symptoms Attributed to Benign Prostatic Hyperplasia: AUA GUIDELINE PART II-Surgical Evaluation and Treatment. *J. Urol.* **2021**, *206*, 818–826. [CrossRef]

1. Elterman, D.; Aubé-Peterkin, M.; Evans, H.; Elmansy, H.; Meskawi, M.; Zorn, K.C.; Naeem, B. UPDATE—2022 Canadian Urological Association guideline on male lower urinary tract symptoms/benign prostatic hyperplasia (MLUTS/BPH). *Can. Urol. Assoc. J.* **2022**, *16*, 245–256. [CrossRef]

2. Pietropaolo, A.; Mani, M.; Hughes, T.; Somani, B.K. Role of low- *versus* high-power laser in the treatment of lower pole stones: Prospective non-randomized outcomes from a university teaching hospital. *Ther. Adv. Urol.* **2022**, *14*, 17562872221097345. [CrossRef]

3. Somani, B.K.; Robertson, A.; Kata, S.G. Decreasing the Cost of Flexible Ureterorenoscopic Procedures. *Urology* **2011**, *78*, 528–530. [CrossRef]

4. Chapman, R.A.; Somani, B.K.; Robertson, A.; Healy, S.; Kata, S.G. Decreasing the cost of flexible ureterorenoscopy: Single-use laser fiber cost analysis. *Urology* **2014**, *83*, 1003–1005. [CrossRef] [PubMed]

5. Geraghty, R.M.; Jones, P.; Herrmann, T.R.; Aboumarzouk, O.; Somani, B.K. Ureteroscopy is more cost effective than shockwave lithotripsy for stone treatment: Systematic review and meta-analysis. *World J. Urol.* **2018**, *36*, 1783–1793. [CrossRef] [PubMed]

5. Geraghty, R.M.; Davis, N.F.; Tzelves, L.; Lombardo, R.; Yuan, C.; Thomas, K.; Petrik, A.; Neisius, A.; Türk, C.; Gambaro, G.; et al. Best Practice in Interventional Management of Urolithiasis: An Update from the European Association of Urology Guidelines Panel for Urolithiasis 2022. *Eur. Urol. Focus* **2023**, *9*, 199–208. [CrossRef] [PubMed]

*photonics*

MDP

*Communication*

# Alexandrite and Nd:YAG Laser vs. IPL in the Management of Facial Hirsutism: A Retrospective Study

**Elena Zappia [1,2], Serena Federico [2], Carmen Volpe [2], Elisabetta Scali [2], Steven Paul Nisticò [2,*] and Luigi Bennard**

1 Department of Medical Sciences, University of Turin, 10126 Turin, Italy
2 Department of Health Sciences, Magna Graecia University, 88100 Catanzaro, Italy
* Correspondence: nistico@unicz.it; Tel.: +39-096-1394-7195

**Abstract:** Background and objectives: Hirsutism is characterized by the abnormal presence of hai women, which occurs in typically male areas. The causes of this symptom can be many; however, i often associated with PCOS. Among the most effective procedures that can be used for hair remo are lasers such as the Alex laser, Nd: YAG, and IPL. Materials and Methods: This study analy a total of 71 patients who were treated for hirsutism caused mainly by PCOS; the remaining ca were idiopathic. The patients were divided into two groups: the first group was treated with la sources, specifically Alex and Nd: YAG, and the second group underwent treatment with IPL. results were evaluated six months after the last treatment. Results: Out of 71 treated patients, patients experienced good or excellent hair reduction with greater efficacy in laser-treated patie The incidence of side effects was low, particularly in the IPL group. Conclusions: Both laser and are efficient, safe, and tolerable procedures in treating PCOS-related and idiopathic facial hirsuti and both have been shown to have a low rate of side effects. However, the Alex and Nd: YAG las guaranteed better hair removal results. A more extensive clinical study will be necessary to confi our study's results.

**Keywords:** hair removal; hirsutism; PCOS; Alex laser; Nd: YAG laser; IPL

check for updates

**Citation:** Zappia, E.; Federico, S.; Volpe, C.; Scali, E.; Nisticò, S.P.; Bennardo, L. Alexandrite and Nd:YAG Laser vs. IPL in the Management of Facial Hirsutism: A Retrospective Study. *Photonics* **2023**, *10*, 572. https://doi.org/10.3390/photonics10050572

Received: 27 February 2023
Revised: 10 May 2023
Accepted: 12 May 2023
Published: 14 May 2023



## 1. Introduction

Hirsutism is defined as excessive terminal hair growth in male-androgen-depend areas of women. Hirsutism affects 5–10% of women in different parts of the world. T presence of hirsutism is a distressing symptom with a negative impact on psychosoc aspects and quality of life (QOL) [1]. The underlying causes of hirsutism may be congeni adrenal hyperplasia, polycystic ovary syndrome, ovarian and adrenal tumors, or pharma logical agents, or it may be idiopathic. It is essential to determine the underlying causes Polycystic ovarian syndrome (PCOS) is defined as a combination of signs and sympto of androgen excess and ovarian dysfunction in the absence of other specific diagnos and it is the most common cause of hirsutism in women. This syndrome, first describ by Stein and Leventhal in 1935, is characterized by three major features (Rotterdam cr ria): oligo or anovulatory cycles, clinical or biochemical signs of hyperandrogenism, a ultrasonographic evidence of polycystic ovaries. A small group of women is diagnos with idiopathic hirsutism (IH). These women have no detectable hormonal abnormalit regular menses, normal ovarian appearance, and no evidence of adrenal or ovarian tum The causes of IH are genetic or ethnic polymorphism in the androgen receptor, increas peripheral activity of 5-alpha reductase in the air follicles, or mild to moderate abnorm ties in the genesis of steroids [3]. This study aims to evaluate the safety and effectiveness IPL, Alex, and Nd: YAG lasers in treating idiopathic and PCOS-related facial hirsuti dividing patients into two groups depending on the device used, and not only on the ca of the condition.

## 2. Materials and Methods

This study retrospectively analyzed and compared two groups of female patients treated for facial hirsutism: the first group of fifty women underwent laser treatment, while the second group of twenty-one patients was treated with IPL. Most subjects (fifty-six out of seventy-one) had polycystic ovaries.

Exclusion criteria included patients having performed any laser or light treatment in the last three months, reporting the use of topical medications such as retinoids, salicylic/glycolic acid, antibiotics, being pregnant or breastfeeding, and reporting hypersensitivity to light or with malignant tumors. All patients provided signed informed consent on the risks of the procedure.

- Patients of the first group with Fitzpatrick skin type I-II were treated with a 755 nm Alexandrite laser (Motus AY, DEKA M.E.L.A., Calenzano, Italy) (fluence: 11–19 J/cm$^2$, pulse duration: 10–20 ms, spot: 10–14 mm), while a 1064 nm Nd: YAG laser was used for women of the same group with Fitzpatrick skin type III-IV (Motus AY, DEKA M.E.L.A., Calenzano, Italy) (fluence: 30–40 J/cm$^2$, pulse duration: 10–20 ms, spot: 10–15 mm). Darker phototypes were treated with the YAG laser due to the longer wavelength of this laser compared to the Alexandrite laser. Longer wavelengths reduce the absorption of laser light by epidermal melanin, and as in darker skin types epidermal melanin competes as a significant chromophore and may lead to excessive heating of the surrounding tissue, and so a better margin of safety was ensured with the YAG laser.

External skin cooling was coupled with the Nd: YAG laser handpiece (Zimmer Cryo 6).

The second group of patients was treated with a traditional IPL therapy for hair removal with an optimized emission at 755 nm (Genus AX, DEKA M.E.L.A., Calenzano, Italy) (fluence: 8–12 J/cm$^2$, spot: 8.3 cm$^2$), using a transparent gel. Regardless of their group, all subjects received six treatments with an interval between sessions of 6 weeks. Immediately after each session, the appearance of side effects such as burnings, vesiculations, or folliculitis was monitored. Response to each treatment was assessed six months after the last treatment session. Statistica14.0 (TIBCO Software, Palo Alto, CA, USA) software was used for data analysis (mean, standard deviations, and rate calculations).

## 3. Results

Seventy-one patients with facial hirsutism were enrolled in the study, and six treatments were performed with a 6 week interval between sessions.

### 3.1. Laser Group

The sample treated with the Alexandrite and Nd: YAG laser consisted of 50 female patients with a mean age of 30.0 ± 6.9. Based on Fitzpatrick skin type, patients were classified as follows: 6 (12%) patients were skin type I, 22 (44%) patients were skin type II, 15 (30%) patients were skin type III, and 7 (14%) patients were skin type IV.

Therefore, the 28 (56%) patients with phototype I-II of the laser group were treated with the Alexandrite laser procedure, while the 22 (44%) patients with phototype III-IV were treated with the YAG laser.

In 10 (20%) patients facial hirsutism was idiopathic; in the remaining 40 (80%) it was determined by Polycystic Ovarian Syndrome (PCOS). The face areas treated in the laser group are shown in Table 1 as a percentage.

**Table 1.** Face areas treated in the laser group.

| Area | Number of Patients | Percentage |
|---|---|---|
| Cheek | 11 | 22% |
| Parotid | 12 | 24% |
| Mentum | 14 | 28% |
| Submental | 13 | 26% |

The effectiveness of the treatment was evaluated according to a four-point visual sc based on the percentage of hair reduction compared to the baseline: poor (<25%), f (25–50%), good (50–75%), and excellent (>75%).

A total of 16 (32%) patients achieved an excellent long-term reduction of unwant hair growth (between 51 and 75%) compared to T0, in particular, nine treated with Alexandrite laser and seven treated with the Nd:YAG laser. A total of 34 (68%) patie achieved excellent reduction (>75%) as shown in Figures 1–4; 19 were treated with Alexandrite laser and 15 were treated with the Nd:YAG laser. No patient achieved pc (<25%) or fair (25–50%) reduction as a result.

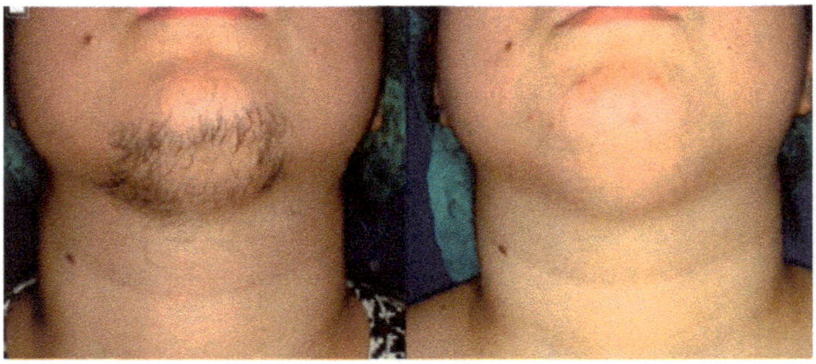

**Figure 1.** Effect of laser treatment on mental, submental, parotid, and cheek hirsutism. Comparis between baseline and six months after the last treatment session.

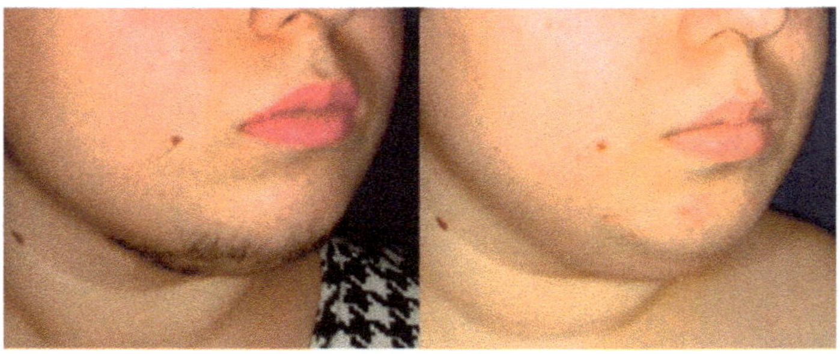

**Figure 2.** Effect of laser treatment on mental, submental, parotid, and cheek hirsutism. Comparis between baseline and six months after the last treatment session.

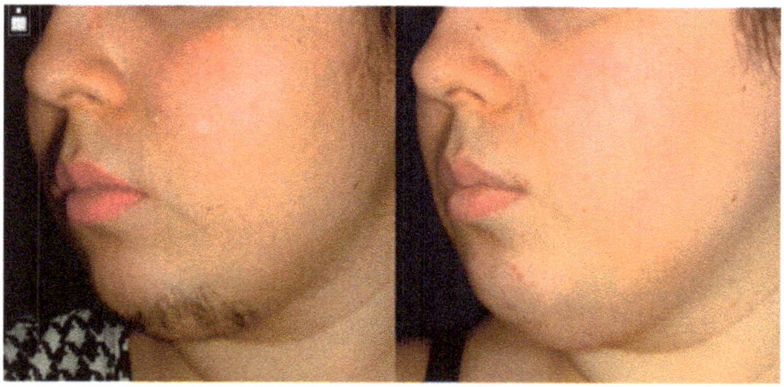

**Figure 3.** Effect of laser treatment on mental, submental, parotid, and cheek hirsutism. Comparison between baseline and six months after the last treatment session.

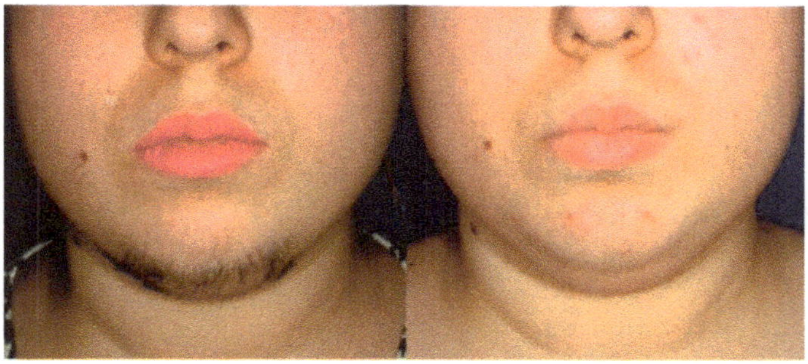

**Figure 4.** Effect of laser treatment on mental, submental, parotid, and cheek hirsutism. Comparison between baseline and six months after the last treatment session.

The side effects were low: burning occurred in 4% of cases, folliculitis in 4%, and no vesiculation occurred.

### 3.2. IPL Group

The sample of patients treated with IPL consisted of 21 women with a mean age of 29.2 ± 6.4 years. Of these, 5 (24%) were phototype I according to the Fitzpatrick scale, 12 (57%) were phototype II, and 4 (19%) were phototype III.

In 5 (24%) patients, the facial hirsutism was idiopathic; in 16 (76%) patients, it was determined by Polycystic Ovarian Syndrome (PCOS). The face areas treated in the IPL group as a percentage are shown in Table 2.

**Table 2.** Face areas treated in the IPL group.

| Area | Number of Patients | Percentage |
|---|---|---|
| Cheek | 4 | 19% |
| Parotid | 4 | 19% |
| Mentum | 7 | 33% |
| Submental | 6 | 29% |

In this case, 16 (76%) patients achieved a good reduction in long-term unwanted hair growth compared to T0 (51–75%), while 5 (24%) patients achieved an excellent reduction (>75%). Again, no patient achieved poor (<25%) or fair reduction.

Regarding the incidence of side effects, in this case it was lower than in the laser gro burning occurred in 5% of cases, while no case of vesiculation and folliculitis occurred. differences were statistically relevant among groups.

## 4. Discussion

Several types of lasers are available for hair removal along with intense pulse li (IPL) epilators, but choosing among many options according to the skin and hair type the patient took a lot of work. Patients should be evaluated for hirsutism by doing vari hormonal assays, especially for testosterone levels, as these may influence the respo to laser therapy. The hormonal dysfunction caused by Polycystic Ovarian Syndro influences hair regrowth following laser hair removal [4]. The technology used in h removal by lasers is based on the principle of selective photothermolysis. According this principle, the selective thermal destruction of a target will occur if sufficient energ delivered at a wavelength well absorbed by the target within a period less than or eq to the thermal relaxation time of the target. The thermal relaxation time is the tim takes for the target to cool (half of its baseline temperature) and transfer the heat to surrounding structures. Under these conditions, it is possible to target structures wh sparing the surrounding structures or tissues selectively. Light skin (Fitzpatrick skin ty I-IV) and dark hair are ideal for effective hair removal. The absorption is more at level of follicular melanin rather than epidermal melanin, thus reducing the chances epidermal damage. For this reason, lasers targeting melanin, such as the 755 nm laser, in phototypes selectively destroy hair follicles, sparing the surrounding skin. The long-pul Nd: YAG laser remains the recommended choice in very dark individuals and tann patients because of its longer wavelength. The safety of patients with type V-VI ski challenging for laser hair removal because of the high density of competing chromoph in the epidermis. A wavelength less absorbed by melanin may be less effective clinically, the target chromophore for hair removal laser is melanin in the hair bulb and bulge. T Alexandrite laser has been proven to be at least as effective and safe in hair reduction. T superficial penetration of a 755 nm wavelength targets the bulge of the hair follicle. It thus been found to be more effective for superficially embedded hair in areas such as eyebrows and upper lip [5]. In contrast to lasers, IPL systems are broadband flashlam that emit polychromatic incoherent light ranging from visible to infrared (500–1300 n Optical filters are used to tailor the polychromatic light to specific needs [6]. The side effe such as sensitivity, erythema, hyperpigmentation, and burns, depend on the patient's s phototype, on the device, and on the operator's expertise. It is well-known that la hair reduction causes no long-term adverse effects [7] and currently underway studies highlighting how the "in motion" use of laser sources such as long-pulsed Nd: YAG, alb with the same efficacy compared to its standard use, has fewer side effects [8].

## 5. Conclusions

Both laser and IPL are efficient, safe, and tolerable procedures in treating idiopat and PCOS-related facial hirsutism with a low rate of side effects. However, Alex a Nd: YAG laser procedures achieved better results in terms of the percentage reduction unwanted facial hair, despite the fact that the incidence of side effects was lower in IPL group than in the laser groups. Further study with long-term evaluation is needed confirm this result.

**Author Contributions:** Conceptualization, E.Z. and S.P.N.; methodology, C.V.; validation, E.S., and L.B.; formal analysis, L.B.; data curation, E.Z.; writing—original draft preparation, E.Z.; writin review and editing, L.B.; visualization, S.P.N.; supervision, L.B. All authors have read and agree the published version of the manuscript.

**Funding:** This research received no external funding.

**Institutional Review Board Statement:** The study was conducted in accordance with the Declaration of Helsinki, and approved by the Institutional Review Board of Calabria Centro (0373/2019 date of approval 17 December 2019).

**Informed Consent Statement:** Informed consent was obtained from all subjects involved in the study.

**Data Availability Statement:** Data available from the Corresponding Author upon reasonable request.

**Conflicts of Interest:** The authors declare no conflict of interest.

## eferences

Alizadeh, N.; Ayyoubi, S.; Naghipour, M.; Hassanzadeh, R.; Mohtasham-Amiri, Z.; Zaresharifi, S.; Gharaei Nejad, K. Can laser treatment improve quality of life of hirsute women? *Int. J. Womens Health* **2017**, *9*, 777–780. [CrossRef] [PubMed]
Shrimal, A.; Sardar, S.; Roychoudhury, S.; Sarkar, S. Long-pulsed Nd: YAG Laser and Intense Pulse Light-755 nm for Idiopathic Facial Hirsutism: A Comparative Study. *J. Cutan. Aesthet. Surg.* **2017**, *10*, 40–44. [CrossRef] [PubMed]
Nabi, N.; Bhat, Y.J.; Dar, U.K.; Hakeem, A.; Mir, S.A.; Shah, I.H.; Tilwani, M.R. Comparative study of the clinico-trichoscopic response to treatment of hirsutism with long pulsed (1064 nm) Nd:YAG laser in idiopathic hirsutism and polycystic ovarian syndrome patients. *Lasers Med. Sci.* **2022**, *37*, 545–553. [CrossRef] [PubMed]
Karn, D.; Kc, S.; Timalsina, M.; Gyawali, P. Hormonal profile and efficacy of long pulse Nd-YAG laser in treatment of hirsutism. *J. Nepal Health Res. Counc.* **2014**, *12*, 59–62. [PubMed]
Bhat, Y.J.; Bashir, S.; Nabi, N.; Hassan, I. Laser Treatment in Hirsutism: An Update. *Dermatol. Pract. Concept.* **2020**, *10*, e2020048. [CrossRef] [PubMed]
Schoenewolf, N.L.; Barysch, M.J.; Dummer, R. Intense pulsed light. *Curr. Probl. Dermatol.* **2011**, *42*, 166–172. [CrossRef] [PubMed]
Fayne, R.A.; Perper, M.; Eber, A.E.; Aldahan, A.S.; Nouri, K. Laser and Light Treatments for Hair Reduction in Fitzpatrick Skin Types IV-VI: A Comprehensive Review of the Literature. *Am. J. Clin. Dermatol.* **2018**, *19*, 237–252. [CrossRef] [PubMed]
Nistico, S.P.; Bennardo, L.; Bennardo, S.; Marigliano, M.; Zappia, E.; Silvestri, M.; Cannarozzo, G. Comparing Traditional and in Motion Nd:YAG Laser in Hair Removal: A Prospective Study. *Medicina* **2022**, *58*, 1205. [CrossRef] [PubMed]

*photonics*

MDP

*Systematic Review*

# Therapeutic and Adverse Effects of Lasers in Dentistry: A Systematic Review

**Giuseppina Malcangi †, Assunta Patano †, Irma Trilli, Fabio Piras, Anna Maria Ciocia, Alessio Danilo Inchingolo, Antonio Mancini, Denisa Hazballa, Daniela Di Venere, Francesco Inchingolo *, Elisabetta de Ruvo, Gianna Dipalma *,‡ and Angelo Michele Inchingolo ‡**

Department of Interdisciplinary Medicine, School of Medicine, University of Bari "Aldo Moro", 70124 Bari, It

* Correspondence: francesco.inchingolo@uniba.it (F.I.); giannadipalma@tiscali.it (G.D.);
Tel.: +39-3312-111-104 (F.I.); +39-3396-989-939 (G.D.)

† These authors contributed equally to this work.

‡ These authors contributed equally to this work.

**Abstract:** Laser therapy has become one of the gold standards of treatment in routine dentistry. the 1970s, CO2 lasers were the first lasers to be used in oral surgery on soft tissues. Subsequen other lasers (Diode, Nd YAG, Er: YAG, Argon and Erbium) have also been used in periodont implantology, orthodontics and restorative dentistry, as well as for hard tissues, such as bone, enar and dentin. The purpose of this systematic review is to analyze both the therapeutic properties a adverse effects of laser use in dentistry, related to a non-targeted choice of medical device based clinical need as well as an inadequate learning curve. A systematic review was performed followi the PRISMA guidelines using Pubmed, Scopus and Web of Science. The use of lasers require specific learning curve modifying the standard surgical approach. The use of lasers provides multi therapeutic benefits but can lead to some failures in dental treatments. In restorative dentistry has been found that lasers can cause cracks that can lead to fractures and/or affect the compos restoration. In periodontal treatments, the use of lasers can produce thermal damage to pulp tiss In surgical treatments, the use of a laser caused bleeding, post-operative pain and a burning sensati The failure of tissue cooling during laser beam emission could produce the necrosis of hard a soft tissues.

**Keywords:** dentistry; laser; low-level laser therapy; oral surgery; peri-implant mucositis thera photobiomodulation; photobiostimulation; photodynamic therapy; tissue regeneration



**Citation:** Malcangi, G.; Patano, A.; Trilli, I.; Piras, F.; Ciocia, A.M.; Inchingolo, A.D.; Mancini, A.; Hazballa, D.; Di Venere, D.; Inchingolo, F.; et al. Therapeutic and Adverse Effects of Lasers in Dentistry: A Systematic Review. *Photonics* **2023**, *10*, 650. https://doi.org/10.3390/photonics10060650

Received: 13 March 2023
Revised: 10 May 2023
Accepted: 1 June 2023
Published: 5 June 2023



## 1. Introduction

Laser (light amplification by stimulated emission of radiation) therapy has become c of the gold standards of treatment in routine dental practice [1]. Although the birth and fi applications of lasers in the medical field date back to the middle of the last century, it w not until the 1980s that the use of this therapeutic technique really became established Since the 1970s, lasers have also been used to treat oral cavity infections, a use that h become more widespread with the advent of new types of lasers and has also extend to tooth-specific therapeutic practices, such as endodontic, periodontal, orthodontic a cosmetic dentistry treatments [3]. Indeed, precisely because of its characteristics of versa ity, user-friendliness and reliability, the laser can be used as an alternative to conventio surgery, even of the oral cavity [2,3]. The biostimulating properties of laser beams at t cellular level has enabled its use in all branches of dentistry.

In periodontology, the laser removes the infected gum layer and the necrotic tissue d appears, enabling the growth of new cells [3,4]. Other dental fields have also benefited fr the use of laser therapy, including endodontic treatments, teeth whitening, implantolo orthodontic and soft tissue surgery in the oral cavity [5–8]. Recently lasers have been us also for the management of labial and perioral imperfections of a vascular or pigmentary nature and for the aesthetic treatment of wrinkles, especially in the era of modern dentistry, where aesthetics have taken on a leading role (Figure 1).

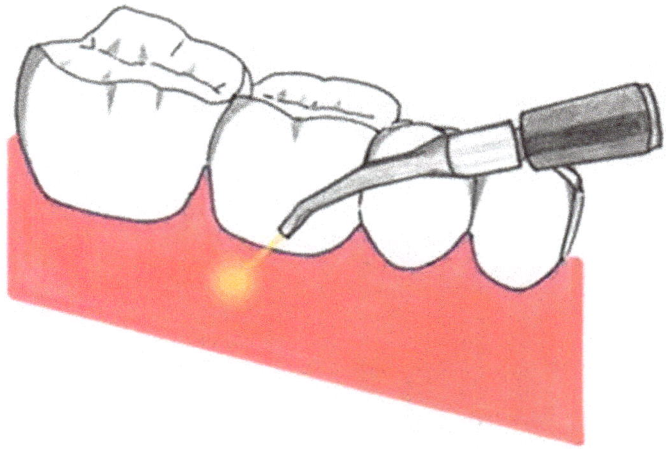

**Figure 1.** Use of a laser in dentistry as an accurate, non-invasive method with predictable results.

In endodontic treatments, the use of lasers ensures the sterilization of root canals after their proper cleaning [5]. In the tooth whitening, a technique using hydrogen peroxide gel, a laser is utilized to release free radicals, which in turns break down the dark pigmented molecules present on the teeth, facilitating their removal [6]. The discipline of implantology includes several phases in which laser use has proved useful: in the pre-surgery phase, where site preparation is of fundamental importance for the correct introduction of the implant and subsequent osseointegration; in the preprosthetic stage of operculectomy; and in peri-implantitis for pocket sterilization.

Each different type of laser can be categorized in a number of ways: the type of laser used, the specific chromophore used to activate the laser's effect, the tissue on which they can be used and the wavelength range of the laser (all lasers act within the infrared spectrum). In fact, lasers work with specific frequency and intensity and with a certain wavelength (λ) that determines a particular absorption and penetration coefficient, which differs according to the type of tissue to be treated. By modulating the intensity and power, it is possible to achieve a "customised" treatment with respect to the intrinsic characteristics of the patient and their oral health condition [9–12]. From a strictly practical point of view, the main types within the laser family can be divided into three large groups: solid-state lasers (Nd: YAG, Er: YAG, ErCr: YSGG), gaseous-state lasers (CO2) and semiconductor lasers (diode lasers). Laser applications are varied: some can be used for both soft and hard tissues (Nd: YAG, Er: YAG, ErCr: YSGG), while others act only on soft tissues (CO2 and diode lasers) [13].

Recently, a focal lens was used for the diode effects used in the low level laser therapy (LLLT) setting, which, by means of the photobiostimulation effect, allows good tissue penetration while respecting the tissues and avoiding excessive overheating and, at the same time, has a significant anti-inflammatory effect, promoting tissue healing and controlling edema and postoperative pain [14–17].

### *1.1. Laser Classifications*

#### 1.1.1. LLLT

Photobiomodulation (PBM), also known as low-level laser treatment (LLLT), is a photochemical and non-thermal biological interaction that, in most situations, uses red light with a wavelength of 600–700 nm and infrared light with a wavelength of 770–1200 r to promote healing and reduce inflammation and discomfort [18–20]. The LLLT is alternative use of the diode laser, because it is set to low power and a diffused, defocus radius. It does not cause the overheating of tissues and penetrates deeply. The biostimula effect is due to an increase in cellular metabolism through the stimulation of ATP producti by mitochondria that absorb the emitted photons, reducing inflammation, pain and heali time [18,21].

Many oral disorders have responded quite well to the use of low-intensity lasers [2 Combining these with antibiotics improves the condition of patients with bisphosphona related osteonecrosis of the jaw (BRONJ) and medication-related to the osteonecrosis of jaw (MRONJ) by reducing clinical symptoms [23,24]. In cases of hyposalivation or xeros mia, LLLT is particularly efficient at repairing cells, and it also stimulates salivary glan to produce more secretion [25]. The use of LLLT in patients with hyposalivation results an increase in mucin 5B, oligomeric mucus/gel-forming(MUC5B), unstimulated saliva flow and histatins, resulting in a protective action against fungal infections (e.g., Candi albicans). The respiratory chain of the mitochondria contains cytochrome C oxidase, wh is received from the LLLT as its chromophore [14]. This effect would appear to be due increased cellular activity, brought about via the increased ATP production of mitochond stimulated by the photons emitted by LLLT. It also reduces pain and provides comple wound-healing in cases of recurrent aphthosis and Herpes Simplex [26,27]. Trigemin neuralgia symptoms are also lessened by LLLT [28]. Periodontology has seen a signific increase in the application of LLLT [29,30]. However, the outcomes that motivate the of long-term lasers seem to offer few advantages over alternative approaches [31]. T Multinational Association of Supportive Care in Cancer/International Society of OralC cology has approved the use of LLLT for the treatment of oral mucositis in cancer patie and adults, but LLLT has been found to reduce severe mucositis in cancer patients of ages [32]. Moreover, the use of LLLT has promoted bone growth surrounding the impla and increased osseointegration [33,34].

The PBM's mode of action relies on the interaction of chromophores cells with las light, particularly cytochrome C oxidase (Cox), an enzyme of the mitochondrial respirato chain. As the final enzyme in the electron transport chain, Cox is responsible for transferri electrons from cytochrome c to molecular oxygen [35]. It appears that PBM increases number of electrons available for reducing molecular oxygen in the Cox catalytic center. T respiratory chain's electron transport velocity alteration results in boosting the amounts adenosine triphosphate (ATP), cyclic adenosine monophosphate (cAMP), reactive oxyg species (ROS) and nitric oxide (NO) as well as the mitochondrial membrane potent (MMP) [36,37].

The sodium and potassium pump in the cell membrane alters as the concentrati of intracellular H + ions increase the permeability to calcium ions in the intracellu environment. The level of cyclic nucleotides that regulate the synthesis of RNA and DN are impacted by the increased concentration of this cation [38]. In this condition, the activ of cellular receptors in the membrane increases too; as a result, endorphin producti and neural cell action potential rise while bradykinin levels and the activity of C-fib that respond to painful stimuli diminish. This entire process helps relieve symptoms pain [39].

Inhibitory nitric oxide (NO) must be photodissociated in order for the light to absorbed by cytochrome c-oxidase. This process is what causes the rapid increase vasodilation and improvement in tissue oxygenation [40,41] (Figure 2).

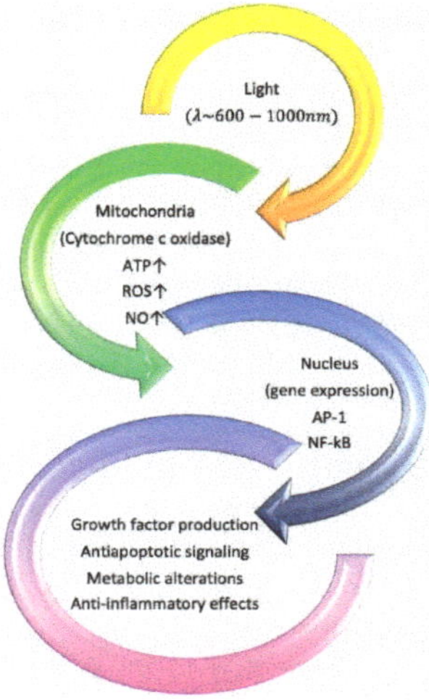


**Figure 2.** Summary of the mechanism of action and direct effects of low-level laser treatment (LLLT) on tissues.

Many transcription factors, including nuclear factor kappa-light-chain-enhancer of activated B cells (NF-kB), multifunctional protein (Ref-1), tumor protein (p53), group of transcription factors (ATF/CREB) and hypoxia-inducible factor 1 (HIF-1), are controlled by changes in cellular redox state [42]. Following that, various intracellular signaling pathways, including those for protein synthesis, enzyme activation, cell cycle progression and nucleic acid synthesis, are activated, as are transcriptional alterations [43]. Moreover, laser therapy causes some cell types to differentiate, which increases cell proliferation and migration as well as the production of growth factors, inflammatory mediators and cytokines [43,44] (Figure 3).

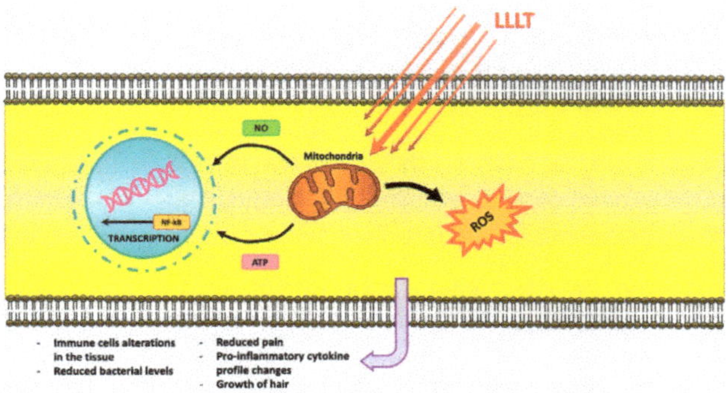


**Figure 3.** More in detail, graphical scheme of the direct effects of low-level laser (LLLT) stimulation on tissues.

The interaction of light with biological tissues is influenced by various factors, including wavelength, laser dose and the tissue's optical characteristics. The characteristics of the tissue include its structure, water content, density, heat capacity, thermal conductivity and capacity to either scatter, absorb or reflect the energy that is emitted [45].

The biphasic spectrum used by PBM has frequently produced favorable outcomes; nevertheless, the depth of penetration, which is caused by molecule absorption and scattering, is the issue. The maximum near infrared (NIR) penetration depth occurs at abo 810 nm, and as the wavelength increases, both absorption and scattering become noticeak less. At longer wavelengths, water becomes a substantial absorber, and the penetrati depth decreases once more [46].

The dose, which measures the amount of energy entering the body, is equivalent the average power (watts) applied over the course of treatment (seconds) [47]. Every tir the dosage is increased, a maximum response is obtained at some value (a higher dosa has a better response); however, there is a point up to which the dose can be increased [4 and past that maximal value, the reaction declines and disappears, and it is even possil that adverse or inhibiting effects are generated via over-dosage [20]. Overdosage c paradoxically limit cell proliferation or cause apoptosis while under-dosage has a negati impact on cellular responsiveness [49]. Thirteen out of the twenty-five published R( studies on laser tendinopathy that were the subject of a systematic review and meta-analy failed to show any benefit. Overdosing on medication was identified as the cause [5 Additionally, it seems that these cellular responses are tissue-specific. There are instanc where it seems as though a very high dose will produce a level of ROS that is cytotoxic a will destroy the tumor [51].

The definition of protocols is still considered a problem in PBM treatment; so, necessary for the World Association for Laser Therapy (WALT) to establish a methodolo for guidelines on irradiation and the form of laser prescribed for biomodulation, as t beneficial outcomes are undeniable [52].

#### 1.1.2. CO2 Laser

The CO2 laser, developed by Patel et al. in 1964, was first used in surgery in the ea 1970s [53].

The CO2 laser emits a beam of infrared light whose main wavelength is betwe 9.400 and 10.600 nm, either pulsed or as a continuous wave. It has an excellent hydrophi ity that allows it to be extremely effective in the ablation of soft, water-rich tissues (e. frenectomy or gingival remodeling for aesthetic purposes), in the treatment of oral ulc ative lesions and to ensure proper post-surgical healing. It absorbs to a lesser extent hydroxyapatite [14,54].

The absorption of laser light produces a photothermal effect, i.e., the coagulati vaporization and photoablation of soft tissue. To avoid the overheating of tissue a subsequent necrosis, the CO2 laser must be set to pulsed mode (superpulse or char-fre The CO2 laser acts at a depth of 0.1 mm and does not come in contact with the tissue. It h scalpel action when focused and, by reducing bleeding, improves the view of the operati field. When in defocused mode, the CO2 laser has photoablative action and occlud small blood vessels, thus controlling small and diffuse hemorrhages [55]. Wound-heali always occurs via second intention, without complications, without the need for sutu and with a shorter healing time. It may be necessary to administer drugs for post-operati pain, mainly related to the surgical site. Compared with standard scalpel excision thera the use of the CO2 laser has shown delayed neovascularization and re-epithelializati but better healing with reduced contraction and the modification of the newly form tissues [56].

Today, the limitation of using the CO2 laser in the operating field due to the size of t delivery arm has been overcome by the advent of the Photonic Band Gap Fiber Assemk (PBFA), which allows for flexible fiber CO2 delivery, making it easier for the surgeon visualize the head and neck areas [57,58]. The PBFA system allows for variable rate g delivery through a hollow core, allowing for cooling as well as cleaning of the site of deb and blood [57,59].

In the case of a suspected malignant lesion, excision with a scalpel is preferred cause the possible charring of tissue may alter the histopathological response of the sp imen [57,60]. The use of the CO2 PBFA laser requires a learning curve, both to optimi results and to avoid damage to the device [58].

#### 1.1.3. The Nd: YAG Laser

The wavelength of the Nd: YAG laser is 1064 nm, and energy is extinguished by using long pulses that are converted to heat in the fabric. Among the many uses of these lasers in hard and soft materials, their use in endodontics should be highlighted, where they aid in the removal of pathogenic microorganisms and obstructions in the root canal by also removing the affected materials, and thanks to their anti-inflammatory and analgesic properties, scientific studies have been conducted to further understand their role in non-surgical sulcular debridement, the management of periodontal disease and the prevention of new graft attacks [61].

The Nd: YAG laser, compared to other lasers used in dentistry, has a greater ability to penetrate tissue, a low affinity for water and a high affinity for blood vessels, resulting in good hemostatic properties. To reduce the feeling of discomfort, the light is emitted onto the tissue in super-pulsed form [61].

Studies show that the combined use of the Nd: YAG and Er: YAG lasers in non-surgical periodontal therapy results in better healing due to a greater reduction in pocket depth [62–64]. The Nd: Yag laser has proven effective in peri-implantitis due to its ability to decontaminate the titanium surfaces of implants, improved healing and pocket probing depth [65–68].

#### 1.1.4. The Er: YAG Laser

This laser type can function at two different wavelengths: Er and Cr yttrium scandium gallium garnet (YSGG) lasers at 2780 nm and Er: YAG lasers at 2940 nm [14]. Both types work with an elective affinity for hydroxyapatite, as well as hydrophilicity, which means that this laser is favored in the treatment of hard dental tissue [69]. For example, the use of the erbium laser on soft tissue allows for important advantages in routine dentistry, among which are minimized anesthesia, fewer sutures used, less post-operative sensitivity, the selective removal of diseased tissue, faster healing, less swelling and less risk of infection [70,71]. On the other hand, the use of the erbium laser on hard tissues has other advantages, including the reduced need for anesthesia for caries treatment, favoring the preservation of the healthy tooth structure, avoiding the troublesome use of a drill and facilitating the selective removal of diseased tissue [72,73]. Furthermore, treating the carious process with a Er: YAG laser in dentistry avoids the formation of micro-fractures during cavity milling, determines the decontamination and pre-treatment of the enamel-dentinal surface and avoids the generation of microcavities. This results in an improvement in ideal retention for composite resins as well as an increase in the adhesion surface, optimizing the retention effect of orthophosphoric acid to 37% [74].

The Er: YAg laser shows little tissue damage and excellent incisional capacity, making it an excellent tool in the surgical removal of lesions, such as benign oral fibroepithelial hyperplasia [75].

#### 1.1.5. The Argon Laser

This laser works at 572 nm in pulses or in continuous wave mode. It is seen particularly in traditional therapies (the polymerization of restorative resin materials; tooth whitening and remineralization) [76–78].

#### 1.1.6. The Diode Laser

The active component of this common dental laser is a solid-state semiconductor made of aluminum and gallium that emits an optical wavelength generally ranging from 810 to 980 nm. Both continuous and pulsed operation is possible, with optical fiber-based conductivity (diameter of 200–600) and programmable impulse durations, ranging from 0.1 milliseconds to infinity [79]. Diode lasers have displayed convincing qualities of accuracy and dependability since the beginning of their use [80]. This laser uses a mixture of peroxides, carbamide, urea and hydrogen, which is effective for reducing enamel discoloration to whiten both vital and non-vital teeth through the oxidation–reduction (redox) process. When utilized in pulsed mode, the diode laser carried by 200 optical fib may sterilize the apical delta and dentinal tubules to a depth of 1 mm, achieving a level clinical sterility equivalent to a reduction in germs of 0.01 CFU per $mm^2$ [81]. Indeed, dio laser therapy facilitates and optimizes the resolution of infectious processes in the too apex (granuloma) and thus assists in the execution of tooth devitalization. It also stimula osteointegration and is therefore particularly suitable for facilitating dental implants a reducing short-term postoperative complications, because its combined ablative acti along with its biostimulating effect, allows operations with a blood-free field of view a improved healing via second intention [82] (Figure 4).

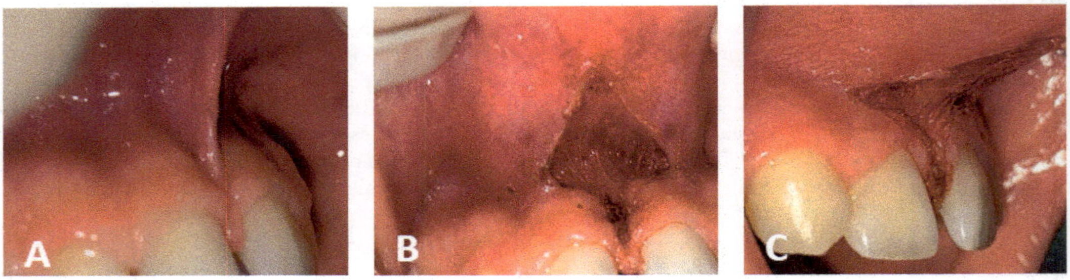


**Figure 4.** (**A**) Hypertrophic frenulum; (**B**) surgical approach; (**C**) frenectomy performed with dic laser: no bleeding, cut with defined margins, no sutures.

It also may help in the implant field by allowing a more accurate clearance of the s tissue involved in peri-implant inflammation, also allowing the better decontaminati of the implant itself. More specifically, the laser has a good effect in mucositis (the init inflammation of the peri-implant mucosa without bone loss) [31].

In addition, diode laser technology allows the dentist to treat common and widespre clinical issues, such as oral ulcers, herpes and dentinal hypersensitivity, as well as v cular lesions, such as hemangiomas (FDIP technique: forced dehydration with induc photocoagulation), and the desensitization of dentin [32].

Moreover, individuals with pacemakers, those with coagulopathies and those rece ing antiplatelet and/or anticoagulant therapy can also benefit from using the diode la because this type of laser causes a caustic effect in its 'surgery' function [83].

Localized biostimulation with a diode laser increases the release of endorphins a decreases the activity of inflammation mediators to a depth of 1 cm, and it also facilita cell replication and thus bone growth [32].

Summing up, it is obvious that lasers in dentistry have unquestionable advantag including no or the reduced use of anesthesia required for the procedure, hemostasis, post-operative swelling, greater compliance by the patient with less pain, shorter chair ti and improved overall acceptance of the procedure, reduced healing time, no or reduc use of post-operative drugs, safer operations on cardiac and coagulopathic patients [84, antimicrobial effects (mainly exploited in the periodontal and surgical-implant field) a the preservation of dental hard tissue and dental pulp itself via the careful and prudent of the laser [84].

However, the clinician must also keep in mind the possible adverse effects of ing lasers, albeit with a reduced tendency, along with potential contraindications of t technique [85–91], such as the considerable cost and sometimes the long periods of u as well as the large size of the instrument, the learning period and the safety measu necessary prior to its use. It is also important to mention that this procedure also requi the use of glasses and involves the unpleasant smell of charred flesh during laser-guid procedures [92]. It is possible to divide adverse effects into long- and short-term categor In particular, during postoperative days, tissues may appear reddened/inflamed/swol and nearby tissues may appear "tight". It is probable that exudate is present in the treat area for a brief period of time and transient tissue hyperpigmentation (or hypopigmentation) can be observed in the area, especially in dark-skinned people. Allergic reactions to anesthetics might occur, or even "fever blister" or herpes simplex dermatitis [92].

As concerns the potential long-term side effects of laser dentistry treatment, it must be said that patients could display scars (these can be reduced by carefully following the instructions for follow-up care) and the presence of soft tissue pigmentation in the treated area. There is also the possibility of infection, present in all surgical techniques, which can be reduced with appropriate postoperative care [92–94] (Figure 5).

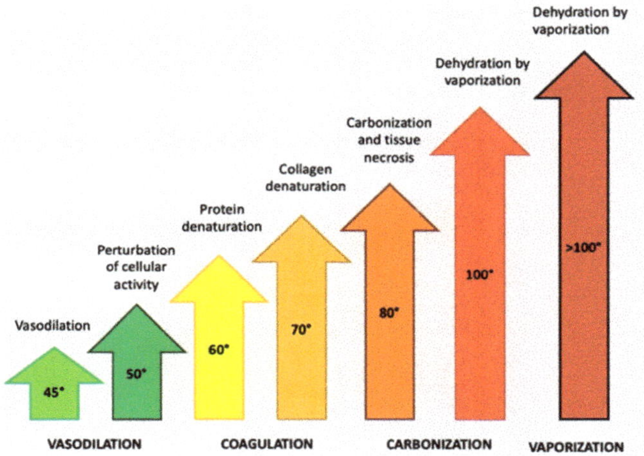


**Figure 5.** Illustration of the direct effects of the diode laser temperature on the treated tissue.

In this review, we analyze the potential side effects of laser therapy, review the literature on this topic to date and attempt to draw future perspectives on the possible mechanisms of improvement.

## 2. Materials and Methods

A systematic review was performed following the Preferred Reporting Items for Systematic Reviews and Meta-Analyses (PRISMA) guidelines using the PubMed, Scopus and Web of Sciences (WoS) databases with PROSPERO code 407324 [95]. The following keywords were used: "dentistry" AND "laser" AND "adverse effect" AND "therapeutic effect". Articles from the last 10 years (2013–2023) were searched, with particular attention paid to the relationship between the use of lasers in dentistry and adverse effects or therapeutic effects. Inclusion criteria included studies on the adverse effects of various types of lasers both in vivo and in vitro and full text articles. All papers reporting no information regarding adverse and therapeutic effects of lasers, as well as literature reviews with or without meta-analysis, were excluded. Furthermore, only articles in the English language were selected, and the last search was run on 26 February 2023. Eligible articles were assessed according to the 2011 guidelines of the Oxford Centre for Evidence-Based Medicine [96]. The references of the included literature were carefully checked to identify other papers that might be pertinent.

The review was conducted using the PICOS criteria:

- Participants: children and adults were included.
- Interventions: actions of the lasers.
- Comparisons: traditional procedures in dentistry.
- Outcomes: efficacy and disadvantages of lasers in dentistry.

- Study: randomized clinical trials, retrospective and observational studies on h man teeth.

Two researchers (I.T. and F.P.) independently extracted articles in accordance wi the inclusion criteria, and any discrepancies were eventually settled through conversati between the two review authors.

### 3. Results

A total of 1671 documents were initially identified in the literature search, of whi 275 were duplicates. After screening for eligibility and inclusion criteria, 16 publicatio were ultimately included. The study and PRISMA flowchart are summarized in Figure 6. total of 1396 documents were screened, of which 929 were excluded. Documents availab for retrieval numbered 467, and after assessment for eligibility, 451 documents were cluded. Finally, only 16 articles were suitable for inclusion. According to the Oxford Cen for Evidence-Based Medicine's 2011 standards, all of the included studies were classifi as level 4 or level 5 evidence for clinical research. The included articles are summarized detail in Table 1.

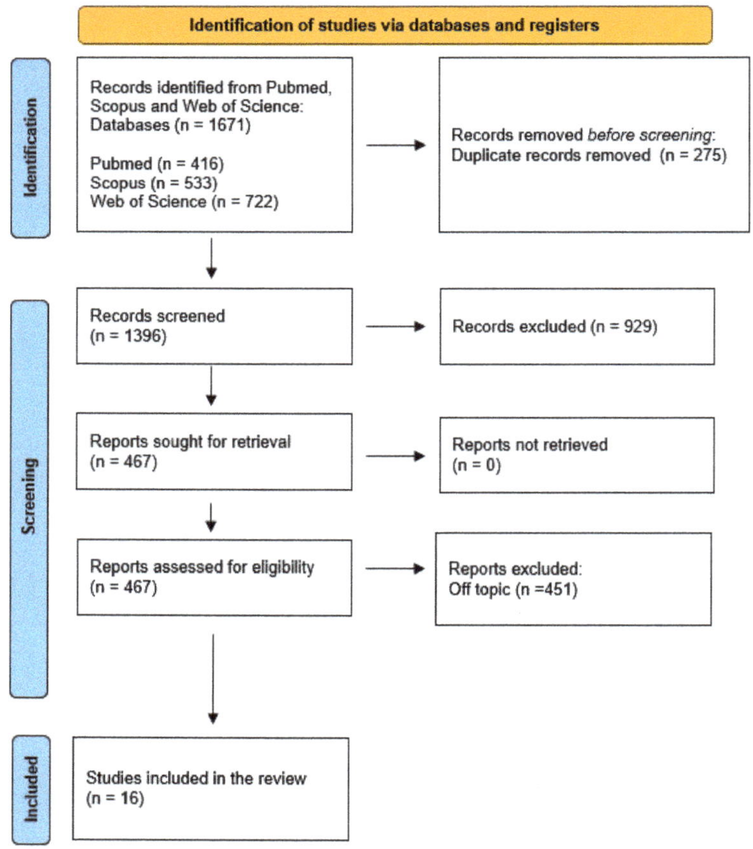


**Figure 6.** PRISMA flowchart with features of included articles.

**Table 1.** Summary of selected articles.

| Authors (Year) | Type of the Study | Aim of the Study | Materials | Results |
|---|---|---|---|---|
| Valerie, G.A., et al., 2017 [97] | Randomized controlled clinical and histopathological trial | Compare excisional biopsies using $CO_2$ and Er: YAG laser | 32 Patients with f.h. in the buccal mucosa | Intraoperative bleeding 100% excisions with Er: YAG and 56% with $CO_2$ laser |
| Saibene, A.M., et al., 2019 [98] | Study in vivo | To evaluate the outcomes of laser surgery while providing a protocol for $CO_2$ laser surgery | 78 patients treated with laser for benign and malignant lesions | 5 patients complained of marginal pain; 3 patients had post-surgery bleeding |
| Hikov, T., et al., 2017 [99] | Study in vivo | To evaluate the possibility of introducing a femtosecond laser system in restorative dentistry | Patients' molars | The FELS used in this work is promising for an efficient and controlled cavity preparation |
| Lauritano, D., et al., 2018 [100] | Case series in a group of patients | Setting up a laser-assisted protocol for the surgical excision of malignant lesions | A specially designed medical record was used. The diode laser was used | Our findings demonstrate a statistically significant reduction in the size of the lesion |
| Bahammam, M.A., et al., 2018 [101] | Case report | To raise an awareness regarding the potential complications of laser gingival depigmentation | One case report | A possible treatment approach for such complications |
| Falkenstein, F., et al., 2014 [102] | Study in vitro | To demonstrate that laser-assisted treatment can produce thermal damage to pulp tissues | Extracted upper and lower human incisors stored in 0.9% saline solution | Middle-third portion of the root seems to be the most at-risk area for pulp damage |
| Franzen, R., et al., 2015 [103] | Study in vitro | To demonstrate that laser-assisted treatment can produce thermal damage to pulp tissues | 30 single-rooted human teeth | Er, Cr:YSGG and a diode 2 W during irradiation show thermal rises on average of $1.68 \pm 0.98$ °C in the pulp chamber |
| Tonin, M.H., et al., 2022 [104] | Study in vitro | To show that the increasing dose of Er: YAG can produce damage of implant surface | Implants' surfaces | Modification of implant surface as adverse effect may not allow the correct formation of oxide layer that is important for wettability of implant and for the differentiation of osteoblasts |
| Monzavi, A., et al., 2018 [105] | Study in vitro | To investigate and compare temperature change during implant decontamination with different laser types ([$Co_2$]/Nd: YAG/Er: YAG/PDT | Sixty implants were inserted into a bone block cut from a sheep's mandible | Temperature changes over 10 °C occurred at the apical point of the implants with the $Co_2$, Nd: YAG and diode laser irradiation; however, only the Co2 laser reached statistical significance in this regard |

**Table 1.** *Cont.*

| Authors (Year) | Type of the Study | Aim of the Study | Materials | Results |
| --- | --- | --- | --- | --- |
| Hoedke, D., et al., 2018 [106] | Study in vitro | To show the removal of bacteria biofilm inside the root canals of extracted teeth through photodynamic therapy using phenothiazine chloride as photosensitizer and a diode laser with a wavelength 660 nm in combination with irrigants | 160 extracted human single-rooted teeth | Effective method for th reduction in bacterial biofilm inside the root canals of extracted teet |
| Ferreira Pires Sobral, M., et al., 2021 [107] | Longitudinal randomized and parallel clinical trial | To compare a violet light-emitting diode system and in-office dental bleaching | Sixty volunteers selected for group one bleaching with 35% hydrogen peroxide gel and for group two using bleaching using a LED | Violet light alone cause repigmentation after dental bleaching |
| Chung, S.E.V., et al., 2015 [108] | Study in vitro | To demonstrate that Er: YAG laser favors bacteria elimination in infected root canals together with sodium hypochlorite irrigation | | The adverse reactions observed in endodontic with Er: YAG laser include the risk of thermal injury to periodontal tissues, eve if it not a temperature that causes the necrosi of periodontal ligamen |
| Saluja, M., et al., 2016 [109] | Study in vitro | To show that Nd: YAG, $CO_2$ and diode lasers can be used also for treatment of dentinal hypersensitivity through the occlusion of human dental tubules | 24 extracted teeth | Risk of the expansion c dental pulp through damage |
| Mozaffari, H.R., et al., 2016 [110] | Study in vitro | To demonstrate that laser for dentinal hypersensitivity can also interfere with composite restoration | 60 extracted teeth | The application of glutaraldehyde desensitizer and the CC laser to the surface in order to complete restoration does not increase microleakage i the enamel or dentin damage tooth surface treatment, while Er: YA laser significantly increased microleakag at the dentin margins |

**Table 1.** *Cont.*

| Authors (Year) | Type of the Study | Aim of the Study | Materials | Results |
|---|---|---|---|---|
| Aljdaimi, A., et al., 2019 [111] | Study in vitro | To compare the impact of Er: YAG laser irradiation on enamel using an FIB | Coronal sections of sound enamel and dentine were machined to 50 μm thickness using an FIB | Microcracks in the peritubular dentine ran orthogonally to the tubules, whereas fractures in the intertubular area ran parallel to the tubules. The average depth of these fissures was around 10 m below the surface. As a result of preferential ablation of the less mineralized intertubular dentine on the dentine surface, a tubule-associated uneven topography was created. |
| Kuhn-Dal Magro, A., et al., 2022 [112] | Randomized double-blind clinical trial | The effect of LLLT associated with therapeutic measures from the oral mucositis | 80 subjects treated for oral mucositis with LLLT | The LLLT was effective for OM lesions |

## 4. Discussion

In 1901, Planck introduced the concept of quantum theory, which challenged classical electromagnetic theory by proposing that energy is absorbed and emitted in discrete, indivisible units known as quanta [14]. Building on this theory, in 1913, Bohr developed a model of the atom that explained how electrons occupy well-defined orbits around the nucleus, based on the atom's energy level [15]. Later, in 1916, Einstein contributed to the field with his work on stimulated emission, laying the foundation for the concept of the emission of light quanta or photons. In 1926, G. Lewis defined the term "photons" to describe particles of light. The development of the laser began with the invention of the ammonia MASER (microwave amplification by stimulated emission of radiation) by Townes, Gordon and Zeiger in 1954. The MASER works by emitting microwaves, and it is considered the precursor to the laser. Then, in 1960, Maiman created the first functioning laser device, using a ruby rod as the active medium. This achievement opened a new era in the study of light and its properties [16].

Laser therapy has increasingly been used in various fields of dentistry, showing various advantages useful for dental practice but also accompanied by certain adverse reactions [5–7]. The use of lasers in periodontics is helpful in combination with the mechanical debridement of calculus and bacteria from tooth surfaces [8]. As an adverse reaction, laser-assisted treatment can cause thermal damage in pulp tissues. The analysis of the thermal effect of irradiation with a 940 nm diode laser of 1.0 and 1.5 W in a continuous wave mode on tooth surfaces in an in vitro study shows how the middle-third portion of the root seems to be the most at risk area for pulp damage [102].

The bactericidal power of laser is beneficial also for peri-implantitis. The laser dose is fundamental: even if the removal of bacteria increases with the increase in Er: YAG dose, the increasing dose can cause damage to the implant surface [104]. In regard to the types of lasers, the Er: YAG laser and photodynamic therapy are useful in implant decontamination, while more precaution is required in the application of the Nd: YAG, diode and, especially, CO2 lasers regarding temperature elevation [105]. However, the Er: YAG laser irradiation on a sandblasted and acid-etched (SLA) titanium disk can modify the oxide layer, causing roughness on implant surface. In fact, laser treatment reduces the roughness of implant surface in order to prevent bacteria adhesion. On the other hand, as an adverse effect, the modification of the implant surface may not allow the corr formation of oxide layer, which is important for the wettability of the implant and for t differentiation of osteoblasts [104].

Regarding endodontic therapy, laser therapy has also proven effective for the purpo of root canal treatments. In fact, when it is necessary to devitalize a tooth, removing t pulp, vessels and nerves and eliminating pain, the laser can be an effective aid in t procedure. For devitalization to be effective, all cavities and tissues must, in fact, thoroughly disinfected and all root canal lumens cleansed. The laser allows for the de cleaning of these areas, favoring the successful outcome of devitalization [106]. Wh comparing the effectiveness of the violet light-emitting diode system to in-office den bleaching, it is important to consider that bleaching with violet light alone may result unwanted discoloration as an adverse effect [107].

The Nd: YAG, CO2 and diode lasers can also be used for the treatment of dentir hypersensitivity through the occlusion of dental tubules. However, in an in vitro stu about the sealing of dentinal tubules, 1.6 and 2 W lasers at a power of 1.6 or 2 on 24 extract teeth revealed the risk of the expansion of damaged dental pulp [109]. An in vitro stu highlighted that the use of laser for dentinal hypersensitivity can also interfere with co posite restoration. In particular, the Er: YAG laser significantly increased the microleaka at the dentin margins [109]. Various types of lasers have distinct clinical implications, a when it comes to composite restoration, pre-treatment with glutaraldehyde-desensiti and CO2 laser irradiation does not lead to any improvement of the enamel and den margins [110]. Another study compares the effect of Er: YAG laser irradiation on denti and enamel. The coronal sections of the healthy surfaces of enamel and dentine machin to a 50 μm thickness using a FEI-Helios Plasma (FIB) were analyzed using the Zeiss Xra 810 Ultra. Er: YAG laser irradiation causes cracks in the inter-tubular region. Cracks r orthogonally to the dentinal tubules and deepen to approximately 10 μm below the surfa Consequently, the use of lasers can determine failure in the field of restorative dentis because cracks can lead to the fracture and infiltration of the margin [111].

In surgery, a laser can be used for the excision of potentially malignant oral lesions a for the excision of the tumors of soft tissues. In oral and maxillofacial surgical procedu lasers guarantee the accurate cutting of soft tissues with good hemostasis [80]. In t case, a lack of post-operative swelling, bleeding and pain was reported. Therefore, for t excision of vascular lesions, the laser provides many advantages thanks to good coagulat properties. For the treatment of lesions occurring in conjunction with encephalotrigemi angiomatosis or Sturge–Weber syndrome, the clinical utilization of lasers is useful [7 In oral pathology, the use of lasers with different wavelengths can also be helpful the treatment of mucositis [112]. In the study carried out by Romeo, U., et al. in t evaluation of the effectiveness of using diode lasers as a protocol to excise leukoplakia light burning sensation was experienced as an adverse effect [113]. In a study in which CO2 laser was used to remove benign and malignant lesions, three patients experienc post-surgery bleeding [101,108]. Only minor side effects were described, such as therm damage or a light burning sensation [100,114]. In electronic research on the impact of las on orodental tissues, several adverse effects on the tooth surface and subcutaneous a submucosal tissues, as well as other adverse reactions, are described. The side effects lasers include pulping damage and subcutaneous and submucosal laser effects; in fa the inappropriate use of dental lasers with air cooling spray can cause subcutane and mediastinal cervicofacial emphysema. Furthermore, the use of lasers could lead cellular necrosis in the periodontal ligament, mainly due to the thermal effect. In fact, las treated teeth showed ankylosis, cement lysis and significant bone remodeling. In additi laser smoke byproducts are considered potentially toxic chemicals. These compou have harmful health effects, including the irritation of the eyes, nose and respirat tract; damage to the liver and kidneys; carcinogenic cellular abnormalities; headac dizziness; drowsiness; stomach pain; vomiting and nausea; and accelerated breathing [1 Lasers are used also in restorative dentistry to control the correct removal of enamel a dentine in cavity preparation [99]. A lower pain sensation is reported when lasers are used in orthodontics to treat soft tissues and for improving gingival margins present in finishing [17]. A possible treatment of an adverse reaction to a laser is described in a case report about the potential complications of laser gingival depigmentation [101].

Valerie, G.A., et al. discussed the clinical and histological effects of 32 excisional biopsies of the fibrous hyperplasia of the buccal mucosa conducted with CO2 and Er: YAG lasers, as analyzed in this study. The area of thermal damage in the Er: YAG laser specimens was much smaller than in histological evaluations. Clinical results revealed that the operative time did not change but bleeding and the need for electrocautery significantly increased with the use of the Er: YAG laser. No link was found between the laser technology used and the experience of postoperative pain on a VAS scale completed by patients during a 15-day period. The Er: YAG laser and electrocautery were used in 94% of cases (versus 50% for the CO2 laser), but this had no effect on the duration of surgery. The CO2 laser is known for its ability to reduce bleeding and improve visibility during surgery by closing small blood vessels. Clinical features, such as operative time, pain and analgesic use, appear to be modest and comparable with both laser procedures. Tissue integrity is better maintained after Er: YAG laser excisions than after CO2 laser excisions due to the reduced thermal impact, with the advantage of a smaller area of damage in histological examination [97].

In the research by Saibene, A.M., et al., there were only four patients who experienced postoperative bleeding, of whom only two required cautery with bipolar forceps and one needed resorbable sutures. There were no adverse drug responses or injuries caused by laser activity. In addition, the CO2 laser appears to be a painless procedure, as most patients reported a low level of pain, with a mean of 2.5 on the VAS scale after 24 h and a standard deviation of 1.78; a value of 2 after 48 h and a standard deviation of 1.33; and a value of 0 at 7 days after surgery. Two patients reported transient numbness around the surgical region, which disappeared in two and five days, respectively [98]. In contrast, the study by Hikov, T., et al. aimed to evaluate the feasibility of introducing a state-of-the-art commercial femtosecond laser system into restorative dentistry, maintaining the well-known advantages of lasers for caries removal while overcoming disadvantages such as thermal damage to the irradiated substrate [99].

By varying the irradiation parameters (pulsed laser intensity, scanning speed and pulse repetition rate), no cracks were detected in the enamel or dentin after laser ablation with the optimal settings, and the craters had precise and well-defined sides and vertical edges. The craters in both materials were evenly processed, and no side effects, such as discoloration, melting or charring, were found as a result of laser contact with the dental tissue. This indicates that there was no overheating and that the dental nerves in the pulpal cavity of a real tooth were not damaged, which is an additional benefit of the absence of collateral damage. This method seems to have the potential to promote a minimally invasive laser-assisted approach in restorative dentistry, if appropriate laser settings are used [99].

The diode laser is not recommended as the main laser for soft tissue surgery, but its adaptability led us to select it for study. The diode laser is a semiconductor laser and exists in different wavelengths: from 980 nm, which has greater cutting ability and fibers that do not need to be activated, to 810 nm, which has greater biostimulatory ability. This laser thus has a dual functionality. The protocol adopted in this investigation sought to exploit both cutting and biostimulatory capabilities; the latter, in particular, was always used after the surgical phase and for controlling orthodontic tooth movement. This helped to reduce surgical pain and edema, as well as to achieve better and faster healing without scarring or recurrent functional outcomes [56].

*Current Trends and Future Directions*

Similar to all other medical specialties, dentistry is constantly evolving in order to improve patient comfort and efficiency. Patients who require dental care often describe their time in the chair as a nightmare. The dentist has a duty to review protocols and constantly strive to stay up to date in order to guarantee, if not enhance, the quality both the services provided and, most importantly, the patient's perspective on the ent dental system. In the coming years, we will most likely discuss individualized dentis various devices, and custom-built treatments that are as minimally invasive and painful possible [116].

## 5. Conclusions

Lasers can be considered a very useful tool in all fields of dentistry. Technologi evolution may lead to a preference for the use of lasers over conventional surgical a therapeutic techniques in the near future. The adverse effects of lasers can be superimpos on a common clinical outcome with conventional therapeutic techniques. Therapeu effects differ depending on the operating protocol, the wavelengths used on differ target tissues and the individual patient response. Laser therapy has a positive effect the emotions of even the most fearful patient. The use of all types of lasers requires appropriate learning curve in order to both optimize positive therapeutic effects and av adverse ones.

**Author Contributions:** Conceptualization, A.D.I., A.M.I., G.M., F.P., A.P. and I.T.; methodolc A.M.C., D.H. and G.M., software, A.M.C., A.M., D.D.V., F.I. and G.D.; validation, G.M., A.P., D.D I.T., F.P., F.I. and G.D.; formal analysis, A.D.I., F.P., A.P., A.M.C., D.D.V., F.I. and G.D.; resour A.D.I., A.M.I., G.M., E.d.R. and D.H.; data curation, G.M., F.I., A.M. and G.D.; writing—original d preparation, A.D.I., A.M.I., D.D.V., E.d.R., F.I. and G.D.; writing—review and editing, A.P., A.M I.T., E.d.R., A.M. and D.H. visualization, E.d.R., D.H., F.I., A.P., A.M. and F.P.; supervision, A.D A.M.I., F.P., I.T., D.D.V., F.I. and G.D.; project administration, G.M., D.H., I.T., A.M.C., F.I. and G All authors have read and agreed to the published version of the manuscript.

**Funding:** This research received no external funding.

**Institutional Review Board Statement:** Not applicable.

**Informed Consent Statement:** Not applicable.

**Data Availability Statement:** Not applicable.

**Conflicts of Interest:** The authors declare no conflict of interest.

## Abbreviations

| | |
|---|---|
| ATF/CREB | Group of transcription factors |
| ATP | Adenosine triphosphate |
| BRONJ | Bisphosphonate-related osteonecrosis of the jaw |
| cAMP | Cyclic adenosine monophosphate |
| $CO_2$ | Carbone Dioxine |
| Cox | Cytochrome C oxidase |
| DNA | Deoxyribonucleic acid |
| Er: YAG | Erbium-doped yttrium aluminum garnet laser |
| FDIP | Forced Dehydration with Induced Photocoagulation |
| FIB | FEI-Helios Plasma |
| HIF-1 | Hypoxia-inducible factor 1 |
| LASER | Light Amplification by Stimulated Emission of Radiation |
| LED | Light-emitting diode |
| LLLT | Low-level laser treatment |
| MMP | Mitochondrial membrane potential |
| MRONJ | Medication related osteonecrosis of the jaw |
| Nd: YAG | Neodymium yttrium aluminum garnet |

| | |
|---|---|
| NF- kB | Nuclear factor kappa-light-chain-enhancer of activated B cells |
| NIR | Near infrared |
| NO | Nitric oxide |
| p53 | Tumor protein |
| PBM | Photobiomodulation |
| MUC5B | Oligomeric mucus/gel-forming |
| PRISMA | Preferred Reporting Items for Systematic Reviews and Meta-Analyses |
| Redox | Oxidation–reduction process |
| Ref-1 | Multifunctional protein |
| RNA | Ribonucleic acid |
| ROS | Reactive oxygen species |
| WALT | World Association for Laser Therapy |
| YSGG | Yttrium scandium gallium garnet |

## ferences


Mier y Teran Armida, M. Lasertherapy and its applications in dentistry. *Pract. Odontol.* **1989**, *10*, 9–16.

Inchingolo, A.M.; Malcangi, G.; Ferrara, I.; Viapiano, F.; Netti, A.; Buongiorno, S.; Latini, G.; Azzollini, D.; De Leonardis, N.; de Ruvo, E.; et al. Laser Surgical Approach of Upper Labial Frenulum: A Systematic Review. *Int. J. Environ. Res. Public Health* **2023**, *20*, 1302. [CrossRef]

Nammour, S. Laser Dentistry, Current Advantages, and Limits. *Photomed. Laser Surg.* **2012**, *30*, 1–4. [CrossRef]

Gholami, L.; Parsamanesh, G.; Shahabi, S.; Jazaeri, M.; Baghaei, K.; Fekrazad, R. The Effect of Laser Photobiomodulation on Periodontal Ligament Stem Cells. *Photochem. Photobiol.* **2021**, *97*, 851–859. [CrossRef]

Bordea, I.R.; Hanna, R.; Chiniforush, N.; Grădinaru, E.; Câmpian, R.S.; Sîrbu, A.; Amaroli, A.; Benedicenti, S. Evaluation of the Outcome of Various Laser Therapy Applications in Root Canal Disinfection: A Systematic Review. *Photodiagnosis Photodyn. Ther.* **2020**, *29*, 101611. [CrossRef]

Biagi, R.; Cossellu, G.; Sarcina, M.; Pizzamiglio, I.T.; Farronato, G. Laser-Assisted Treatment of Dentinal Hypersensitivity: A Literature Review. *Ann. Stomatol.* **2015**, *6*, 75–80. [CrossRef]

Farronato, D.; Mangano, F.; Briguglio, F.; Iorio-Siciliano, V.; Riccitiello, F.; Guarnieri, R. Influence of Laser-Lok Surface on Immediate Functional Loading of Implants in Single-Tooth Replacement: A 2-Year Prospective Clinical Study. *Int. J. Periodontics Restor. Dent.* **2014**, *34*, 79–89. [CrossRef]

Romanos, G.E.; Gupta, B.; Yunker, M.; Romanos, E.B.; Malmstrom, H. Lasers Use in Dental Implantology. *Implant. Dent.* **2013**, *22*, 282–288. [CrossRef] [PubMed]

Grassi, F.R.; Ciccolella, F.; D'Apolito, G.; Papa, F.; Iuso, A.; Salzo, A.E.; Trentadue, R.; Nardi, G.M.; Scivetti, M.; De Matteo, M.; et al. Effect of Low-Level Laser Irradiation on Osteoblast Proliferation and Bone Formation. *J. Biol. Regul. Homeost. Agents* **2011**, *25*, 603–614. [PubMed]

Demirsoy, K.K.; Kurt, G. Use of Laser Systems in Orthodontics. *Turk. J. Orthod.* **2020**, *33*, 133–140. [CrossRef] [PubMed]

Inchingolo, F.; Tatullo, M.; Abenavoli, F.M.; Marrelli, M.; Inchingolo, A.D.; Inchingolo, A.M.; Dipalma, G. Comparison between Traditional Surgery, $CO_2$ and Nd:Yag Laser Treatment for Generalized Gingival Hyperplasia in Sturge-Weber Syndrome: A Retrospective Study. *J. Investig. Clin. Dent.* **2010**, *1*, 85–89. [CrossRef] [PubMed]

Scarano, A.; Lorusso, F.; Inchingolo, F.; Postiglione, F.; Petrini, M. The Effects of Erbium-Doped Yttrium Aluminum Garnet Laser (Er: YAG) Irradiation on Sandblasted and Acid-Etched (SLA) Titanium, an In Vitro Study. *Materials* **2020**, *13*, 4174. [CrossRef]

Tarullo, A.; Laino, L.; Tarullo, A.; Inchingolo, F.; Flace, P.; Inchingolo, A.M.; Inchingolo, A.D.; Dipalma, G.; Podo Brunetti, S.; Cagiano, R. Use of a Diode Laser in an Excisional Biopsy of Two Spoonlike Neoformations on the Tongue Tip. *Acta Biomed.* **2011**, *82*, 63–68.

Verma, S.K.; Maheshwari, S.; Singh, R.K.; Chaudhari, P.K. Laser in Dentistry: An Innovative Tool in Modern Dental Practice. *Natl. J. Maxillofac. Surg.* **2012**, *3*, 124–132. [CrossRef]

De Santis, D.; Bertossi, D.; Zanotti, G.; Rossetto, A.; Farronato, G.; Gelpi, F.; Marconcini, S.; Covani, U. Nd-YAP Laser Assisted Frenulectomy: A Case Series on 23 Patients. *Minerva Stomatol.* **2013**, *62*, 27–36. [PubMed]

Azma, E.; Safavi, N. Diode Laser Application in Soft Tissue Oral Surgery. *J. Lasers Med. Sci.* **2013**, *4*, 206–211.

Borzabadi-Farahani, A. The Adjunctive Soft-Tissue Diode Laser in Orthodontics. *Compend. Contin. Educ. Dent.* **2017**, *38*, e18–e31. [PubMed]

Anders, J.J.; Lanzafame, R.J.; Arany, P.R. Low-Level Light/Laser Therapy Versus Photobiomodulation Therapy. *Photomed. Laser Surg.* **2015**, *33*, 183–184. [CrossRef]

Lin, F.; Josephs, S.F.; Alexandrescu, D.T.; Ramos, F.; Bogin, V.; Gammill, V.; Dasanu, C.A.; De Necochea-Campion, R.; Patel, A.N.; Carrier, E.; et al. Lasers, Stem Cells, and COPD. *J. Transl. Med.* **2010**, *8*, 16. [CrossRef] [PubMed]

Hamblin, M.R. Mechanisms and Applications of the Anti-Inflammatory Effects of Photobiomodulation. *AIMS Biophys.* **2017**, *4*, 337–361. [CrossRef] [PubMed]

21. Anders, J.J.; Arany, P.R.; Baxter, G.D.; Lanzafame, R.J. Light-Emitting Diode Therapy and Low-Level Light Therapy A Photobiomodulation Therapy. *Photobiomodulation Photomed. Laser Surg.* **2019**, *37*, 63–65. [CrossRef] [PubMed]
22. Kalhori, K.A.M.; Vahdatinia, F.; Jamalpour, M.R.; Vescovi, P.; Fornaini, C.; Merigo, E.; Fekrazad, R. Photobiomodulation in C Medicine. *Photobiomodulation Photomed. Laser Surg.* **2019**, *37*, 837–861. [CrossRef] [PubMed]
23. Tenore, G.; Zimbalatti, A.; Rocchetti, F.; Graniero, F.; Gaglioti, D.; Mohsen, A.; Caputo, M.; Lollobrigida, M.; Lamazza, De Biase, A.; et al. Management of Medication-Related Osteonecrosis of the Jaw (MRONJ) Using Leukocyte- and Platelet-R Fibrin (L-PRF) and Photobiomodulation: A Retrospective Study. *J. Clin. Med.* **2020**, *9*, 3505. [CrossRef] [PubMed]
24. Weber, J.B.B.; Camilotti, R.S.; Ponte, M.E. Efficacy of Laser Therapy in the Management of Bisphosphonate-Related Osteonecro of the Jaw (BRONJ): A Systematic Review. *Lasers Med. Sci.* **2016**, *31*, 1261–1272. [CrossRef]
25. Wibawa, A.; Sucharitakul, J.; Dansirikul, R.; Pisarnturakit, P.; Pisarnturakit, P.; Bhuridej, P.; Arirachakaran, P. Low-Le Laser Therapy to the Major Salivary Glands Increases Salivary Flow and MUC5B Protein Secretion in Diabetic Patients w Hyposalivation: A Preliminary Study. *Makara J. Health Res.* **2018**, *22*, 14–21. [CrossRef]
26. Vale, F.A.; Moreira, M.S.; de Almeida, F.C.S.; Ramalho, K.M. Low-Level Laser Therapy in the Treatment of Recurrent Aphthc Ulcers: A Systematic Review. *Sci. World J.* **2015**, *2015*, 150412. [CrossRef]
27. Honarmand, M.; Farhadmollashahi, L.; Vosoughirahbar, E. Comparing the Effect of Diode Laser against Acyclovir Cream for Treatment of Herpes Labialis. *J. Clin. Exp. Dent.* **2017**, *9*, e729–e732. [CrossRef]
28. Falaki, F.; Nejat, A.H.; Dalirsani, Z. The Effect of Low-Level Laser Therapy on Trigeminal Neuralgia: A Review of Literatu *J. Dent. Res. Dent. Clin. Dent. Prospect.* **2014**, *8*, 1–5. [CrossRef]
29. Gholami, L.; Asefi, S.; Hooshyarfard, A.; Sculean, A.; Romanos, G.E.; Aoki, A.; Fekrazad, R. Photobiomodulation in Periodon ogy and Implant Dentistry: Part 1. *Photobiomodulation Photomed. Laser Surg.* **2019**, *37*, 739–765. [CrossRef]
30. Gholami, L.; Asefi, S.; Hooshyarfard, A.; Sculean, A.; Romanos, G.E.; Aoki, A.; Fekrazad, R. Photobiomodulation in Periodon ogy and Implant Dentistry: Part 2. *Photobiomodulation Photomed. Laser Surg.* **2019**, *37*, 766–783. [CrossRef]
31. American Academy of Periodontology Statement on the Efficacy of Lasers in the Non-Surgical Treatment of Inflammato Periodontal Disease. *J. Periodontol.* **2011**, *82*, 513–514. [CrossRef]
32. He, M.; Zhang, B.; Shen, N.; Wu, N.; Sun, J. A Systematic Review and Meta-Analysis of the Effect of Low-Level Laser Thera (LLLT) on Chemotherapy-Induced Oral Mucositis in Pediatric and Young Patients. *Eur. J. Pediatr.* **2018**, *177*, 7–17. [CrossRef]
33. Mayer, L.; Gomes, F.V.; Carlsson, L.; Gerhardt-Oliveira, M. Histologic and Resonance Frequency Analysis of Peri-Implant Bo Healing After Low-Level Laser Therapy: An In Vivo Study. *Int. J. Oral. Maxillofac. Implant.* **2015**, *30*, 1028–1035. [CrossRef]
34. Rajaei Jafarabadi, M.; Rouhi, G.; Kaka, G.; Sadraie, S.H.; Arum, J. The Effects of Photobiomodulation and Low-Amplitu High-Frequency Vibration on Bone Healing Process: A Comparative Study. *Lasers Med. Sci.* **2016**, *31*, 1827–1836. [CrossRef]
35. Karu, T.I. Multiple Roles of Cytochrome c Oxidase in Mammalian Cells under Action of Red and IR-A Radiation. *IUBMB L* **2010**, *62*, 607–610. [CrossRef]
36. Wu, S.; Zhou, F.; Wei, Y.; Chen, W.R.; Chen, Q.; Xing, D. Cancer Phototherapy via Selective Photoinactivation of Respirato Chain Oxidase to Trigger a Fatal Superoxide Anion Burst. *Antioxid. Redox Signal.* **2014**, *20*, 733–746. [CrossRef] [PubMed]
37. Chung, H.; Dai, T.; Sharma, S.K.; Huang, Y.-Y.; Carroll, J.D.; Hamblin, M.R. The Nuts and Bolts of Low-Level Laser (Lig Therapy. *Ann. Biomed. Eng.* **2012**, *40*, 516–533. [CrossRef] [PubMed]
38. De Fatima Zanirato Lizarelli, R.; Bagnato, V.S. Based Orofacial Rehabilitation and Harmonization. In *Lasers in Oral and Maxillofa Surgery*; Springer: Berlin/Heidelberg, Germany, 2021.
39. Cotler, H.B.; Chow, R.T.; Hamblin, M.R.; Carroll, J. The Use of Low Level Laser Therapy (LLLT) for Musculoskeletal Pain. *M Orthop. Rheumatol.* **2015**, *2*, 00068. [CrossRef] [PubMed]
40. Lane, N. Cell Biology: Power Games. *Nature* **2006**, *443*, 901–903. [CrossRef]
41. Pannala, V.R.; Camara, A.K.S.; Dash, R.K. Modeling the Detailed Kinetics of Mitochondrial Cytochrome c Oxidase: Cataly Mechanism and Nitric Oxide Inhibition. *J. Appl. Physiol.* **2016**, *121*, 1196–1207. [CrossRef]
42. Liu, H.; Colavitti, R.; Rovira, I.I.; Finkel, T. Redox-Dependent Transcriptional Regulation. *Circ. Res.* **2005**, *97*, 967–974. [CrossR
43. Alaluf, S.; Muir-Howie, H.; Hu, H.-H.; Evans, A.; Green, M.R. Atmospheric Oxygen Accelerates the Induction of a Post-Mitc Phenotype in Human Dermal Fibroblasts: The Key Protective Role of Glutathione. *Differentiation* **2000**, *66*, 147–155. [CrossRe
44. Pastore, D.; Greco, M.; Petragallo, V.; Passarella, S. Increase in <--H+/e- Ratio of the Cytochrome c Oxidase Reaction Mitochondria Irradiated with Helium-Neon Laser. *Biochem. Mol. Biol. Int.* **1994**, *34*, 817–826. [PubMed]
45. Knappe, V.; Frank, F.; Rohde, E. Principles of Lasers and Biophotonic Effects. *Photomed. Laser Surg.* **2004**, *22*, 411–417. [CrossR [PubMed]
46. Wang, L.; Jacques, S.L.; Zheng, L. MCML—Monte Carlo Modeling of Light Transport in Multi-Layered Tissues. *Comput. Meth Programs Biomed.* **1995**, *47*, 131–146. [CrossRef]
47. Taylor, D.N.; Winfield, T.; Wynd, S. Low-Level Laser Light Therapy Dosage Variables vs Treatment Efficacy of Neuromus loskeletal Conditions: A Scoping Review. *J. Chiropr. Med.* **2020**, *19*, 119–127. [CrossRef]
48. Huang, Y.-Y.; Sharma, S.K.; Carroll, J.; Hamblin, M.R. Biphasic Dose Response in Low Level Light Therapy—An Update. *D Response* **2011**, *9*, 602–618. [CrossRef]
49. Tam, S.Y.; Tam, V.C.W.; Ramkumar, S.; Khaw, M.L.; Law, H.K.W.; Lee, S.W.Y. Review on the Cellular Mechanisms of Low-Le Laser Therapy Use in Oncology. *Front. Oncol.* **2020**, *10*, 1255. [CrossRef] [PubMed]

Tumilty, S.; Munn, J.; McDonough, S.; Hurley, D.A.; Basford, J.R.; David Baxter, G.; Longo, L. The Dose That Works: Low Level Laser Treatment of Tendinopathy. *AIP Conf. Proc.* **2010**, *1226*, 170–178.
Hamblin, M.R.; Nelson, S.T.; Strahan, J.R. Photobiomodulation and Cancer: What Is the Truth? *Photomed. Laser Surg.* **2018**, *36*, 241–245. [CrossRef]
Bjordal, J.M. Low Level Laser Therapy (LLLT) and World Association for Laser Therapy (WALT) Dosage Recommendations. *Photomed. Laser Surg.* **2012**, *30*, 61–62. [CrossRef]
Patel, C.K.N. Interpretation of $CO_2$ Optical Maser Experiments. *Phys. Rev. Lett.* **1964**, *12*, 684. [CrossRef]
Coluzzi, D.J. An Overview of Laser Wavelengths Used in Dentistry. *Dent. Clin. N. Am.* **2000**, *44*, 753–765. [CrossRef]
Lippert, B.M.; Werner, J.A.; Rudert, H. Tissue Effects of $CO_2$ Laser and Nd: YAG Laser. In *Advances in Oto-Rhino-Laryngology*; Rudert, H., Werner, J.A., Eds.; S. Karger AG: Basel, Switzerland, 1995; Volume 49, pp. 1–4. ISBN 978-3-8055-6087-0.
Chomette, G.; Auriol, M.; Labrousse, F.; Vaillant, J.M. The effect of $CO_2$ laser radiation on the morphological changes of mucocutaneous wound healing in oral surgery. A histo-enzymologic and ultrastructural study. *Rev. Stomatol. Chir. Maxillofac.* **1991**, *92*, 1–7. [PubMed]
Holsinger, F.C.; Prichard, C.N.; Shapira, G.; Weisberg, O.; Torres, D.S.; Anastassiou, C.; Harel, E.; Fink, Y.; Weber, R.S. Use of the Photonic Band Gap Fiber Assembly $CO_2$ Laser System in Head and Neck Surgical Oncology. *Laryngoscope* **2006**, *116*, 1288–1290. [CrossRef] [PubMed]
Devaiah, A.K.; Shapshay, S.M.; Desai, U.; Shapira, G.; Weisberg, O.; Torres, D.S.; Wang, Z. Surgical Utility of a New Carbon Dioxide Laser Fiber: Functional and Histological Study. *Laryngoscope* **2005**, *115*, 1463–1468. [CrossRef]
Pons, Y.; Gauthier, J.; Clément, P.; Conessa, C. Ultrasonic Partial Glossectomy. *Head Neck Oncol.* **2009**, *1*, 21. [CrossRef]
Hanby, D.F.; Gremillion, G.; Zieske, A.W.; Loehn, B.; Whitworth, R.; Wolf, T.; Kakade, A.C.; Walvekar, R.R. Harmonic Scalpel versus Flexible CO2 Laser for Tongue Resection: A Histopathological Analysis of Thermal Damage in Human Cadavers. *World J. Surg. Oncol.* **2011**, *9*, 83. [CrossRef]
Slot, D.E.; Kranendonk, A.A.; Paraskevas, S.; Van der Weijden, F. The Effect of a Pulsed Nd:YAG Laser in Non-Surgical Periodontal Therapy. *J. Periodontol.* **2009**, *80*, 1041–1056. [CrossRef]
Aoki, A.; Mizutani, K.; Schwarz, F.; Sculean, A.; Yukna, R.A.; Takasaki, A.A.; Romanos, G.E.; Taniguchi, Y.; Sasaki, K.M.; Zeredo, J.L.; et al. Periodontal and Peri-Implant Wound Healing Following Laser Therapy. *Periodontology 2000* **2015**, *68*, 217–269. [CrossRef]
Zhu, J.; Wei, R.; Lv, X.; Qu, C. Efficacy of a Combined Er:YAG Laser and Nd:YAG Laser in Non-Surgical Treatment for Severe Periodontitis. *Lasers Med. Sci.* **2022**, *37*, 1095–1100. [CrossRef] [PubMed]
Sağlam, M.; Köseoğlu, S.; Taşdemir, İ.; Erbak Yılmaz, H.; Savran, L.; Sütçü, R. Combined Application of Er:YAG and Nd:YAG Lasers in Treatment of Chronic Periodontitis. A Split-Mouth, Single-Blind, Randomized Controlled Trial. *J. Periodontal Res.* **2017**, *52*, 853–862. [CrossRef] [PubMed]
Wang, C.; Ashnagar, S.; Gianfilippo, R.D.; Arnett, M.; Kinney, J.; Wang, H. Laser-assisted Regenerative Surgical Therapy for Peri-implantitis: A Randomized Controlled Clinical Trial. *J. Periodontol.* **2021**, *92*, 378–388. [CrossRef] [PubMed]
Namour, M.; Verspecht, T.; El Mobadder, M.; Teughels, W.; Peremans, A.; Nammour, S.; Rompen, E. Q-Switch Nd:YAG Laser-Assisted Elimination of Multi-Species Biofilm on Titanium Surfaces. *Materials* **2020**, *13*, 1573. [CrossRef]
Namour; El Mobadder; Magnin; Peremans; Verspecht; Teughels; Lamard; Nammour; Rompen Q-Switch Nd:YAG Laser-Assisted Decontamination of Implant Surface. *Dent. J.* **2019**, *7*, 99. [CrossRef]
Mummolo, S.; Mancini, L.; Quinzi, V.; D'Aquino, R.; Marzo, G.; Marchetti, E. Rigenera® Autologous Micrografts in Oral Regeneration: Clinical, Histological, and Radiographical Evaluations. *Appl. Sci.* **2020**, *10*, 5084. [CrossRef]
Harashima, T.; Kinoshita, J.-I.; Kimura, Y.; Brugnera, A.; Zanin, F.; Pecora, J.D.; Matsumoto, K. Morphological Comparative Study on Ablation of Dental Hard Tissues at Cavity Preparation by Er:YAG and Er,Cr:YSGG Lasers. *Photomed. Laser Surg.* **2005**, *23*, 52–55. [CrossRef]
De Benedittis, M.; Petruzzi, M.; Pastore, L.; Inchingolo, F.; Serpico, R. Nd:YAG Laser for Gingivectomy in Sturge-Weber Syndrome. *J. Oral. Maxillofac. Surg.* **2007**, *65*, 314–316. [CrossRef]
Soft Tissue Conditions and Marginal Bone Levels of Implants with a Laser-Microtextured Collar: A 5-Year, Retrospective, Controlled Study—PubMed. Available online: https://pubmed.ncbi.nlm.nih.gov/25331762/ (accessed on 6 March 2023).
Guarnieri, R.; Serra, M.; Bava, L.; Grande, M.; Farronato, D.; Iorio-Siciliano, V. The Impact of a Laser-Microtextured Collar on Crestal Bone Level and Clinical Parameters under Various Placement and Loading Protocols. *Int. J. Oral. Maxillofac. Implants* **2014**, *29*, 354–363. [CrossRef]
Mangano, F.; Bazzoli, M.; Tettamanti, L.; Farronato, D.; Maineri, M.; Macchi, A.; Mangano, C. Custom-Made, Selective Laser Sintering (SLS) Blade Implants as a Non-Conventional Solution for the Prosthetic Rehabilitation of Extremely Atrophied Posterior Mandible. *Lasers Med. Sci.* **2013**, *28*, 1241–1247. [CrossRef]
Botsali, A.; Beksac, B.; Gahramanov, İ.; Caliskan, E. Erbium:YAG Laser Augments the Penetration of Cryotherapy. *Lasers Med. Sci.* **2022**, *37*, 353–358. [CrossRef] [PubMed]
Monteiro, L.; Delgado, M.; Garces, F.; Machado, M.; Ferreira, F.; Martins, M.; Salazar, F.; Pacheco, J. A Histological Evaluation of the Surgical Margins from Human Oral Fibrous-Epithelial Lesions Excised with $CO_2$ Laser, Diode Laser, Er:YAG Laser, Nd:YAG Laser, Electrosurgical Scalpel and Cold Scalpel. *Med. Oral.* **2019**, *24*, e271–e280. [CrossRef]

76. Nazemisalman, B.; Farsadeghi, M.; Sokhansanj, M. Types of Lasers and Their Applications in Pediatric Dentistry. *J. Lasers M Sci.* **2015**, *6*, 96–101. [CrossRef] [PubMed]
77. Yang, B.; Chen, J.; Yu, Q.; Li, H.; Lin, M.; Mustapha, A.; Hong, L.; Wang, Y. Oral Bacterial Deactivation Using a Low-Temperat Atmospheric Argon Plasma Brush. *J. Dent.* **2011**, *39*, 48–56. [CrossRef]
78. Libonati, A.; Marzo, G.; Klinger, F.G.; Farini, D.; Gallusi, G.; Tecco, S.; Mummolo, S.; De Felici, M.; Campanella, V. Embryotoxi Assays for Leached Components from Dental Restorative Materials. *Reprod. Biol. Endocrinol.* **2011**, *9*, 136. [CrossRef] [PubMe
79. Saydjari, Y.; Kuypers, T.; Gutknecht, N. Laser Application in Dentistry: Irradiation Effects of Nd:YAG 1064 Nm and Diode 810 and 980 Nm in Infected Root Canals-A Literature Overview. *Biomed Res. Int.* **2016**, *2016*, 8421656. [CrossRef] [PubMed]
80. Romanos, G.; Nentwig, G.H. Diode Laser (980 Nm) in Oral and Maxillofacial Surgical Procedures: Clinical Observations Ba on Clinical Applications. *J. Clin. Laser Med. Surg.* **1999**, *17*, 193–197. [CrossRef]
81. Saleh, H.M.; Saafan, A.M. Excision Biopsy of Tongue Lesions by Diode Laser. *Photomed. Laser Surg.* **2007**, *25*, 45–49. [Cross [PubMed]
82. Farronato, D.; Fumagalli, D.; Asa'ad, F.; Rasperini, G. Decontamination of Customized Laser-Microtextured Titanium Abutme A Comparative in Vitro Study of Different Cleaning Procedures. *Int. J. Periodontics Restor. Dent.* **2018**, *38*, e87–e95. [CrossRef
83. Takahashi, K.; Onoda, C.; Sugiyama, S.; Noro, A.; Makiishi, T.; Ishikawa, T. Clinical evaluation of Ga-Al-As semiconduc laser diode (UNI-LASER) irradiation in treatment of solitary aphtha, erosion and hypersensitive dentin. *Shikwa Gakuho* **1987**, 295–303.
84. Derikvand, N.; Chinipardaz, Z.; Ghasemi, S.; Chiniforush, N. The Versatility of 980 Nm Diode Laser in Dentistry: A Case Se *J. Lasers Med. Sci.* **2016**, *7*, 205–208. [CrossRef] [PubMed]
85. Santonocito, S.; Polizzi, A.; Cavalcanti, R.; Ronsivalle, V.; Chaurasia, A.; Spagnuolo, G.; Isola, G. Impact of Laser Therapy Periodontal and Peri-Implant Diseases. *Photobiomodulation Photomed. Laser Surg.* **2022**, *40*, 454–462. [CrossRef]
86. Merigo, E.; Rocca, J.-P.; Pinheiro, A.L.B.; Fornaini, C. Photobiomodulation Therapy in Oral Medicine: A Guide for the Practitio with Focus on New Possible Protocols. *Photobiomodulation Photomed. Laser Surg.* **2019**, *37*, 669–680. [CrossRef] [PubMed]
87. Inchingolo, F.; Hazballa, D.; Inchingolo, A.D.; Malcangi, G.; Marinelli, G.; Mancini, A.; Maggiore, M.E.; Bordea, I.R.; Scarano, Farronato, M.; et al. Innovative Concepts and Recent Breakthrough for Engineered Graft and Constructs for Bone Regenerat A Literature Systematic Review. *Materials* **2022**, *15*, 1120. [CrossRef] [PubMed]
88. Scarano, A.; Petrini, M.; Inchingolo, F.; Lorusso, F.; Amuso, D. A New Technique for the Treatment of Nasal Telangiecta Using Atmospheric Plasma (Voltaic Arc Dermabrasion): Postoperative Pain Assessment by Thermal Infrared Imaging. *J. Cos Dermatol.* **2020**, *19*, 2912–2918. [CrossRef]
89. Inchingolo, F.; Tarullo, A.; Cagiano, R.; Resta, G.; Dipalma, G.; Inchingolo, A.M.; Tarullo, A.; Scacco, S.; Marrelli, M.; Corti, L.; e Successful Use of a Topical Mixture with Ozolipoile in the Treatment of Actinic Ulcers. *Clin. Cosmet. Investig. Dermatol.* **201** 147–150. [CrossRef] [PubMed]
90. Di Ventura, A.; Lanteri, V.; Farronato, G.; Gaffuri, F.; Beretta, M.; Lanteri, C.; Cossellu, G. Three-Dimensional Evaluation of Ra Maxillary Expansion Anchored to Primary Molars: Direct Effects on Maxillary Arch and Spontaneous Mandibular Response. *J. Paediatr. Dent.* **2019**, *20*, 38–42. [CrossRef]
91. Mangano, F.G.; Caprioglio, A.; Levrini, L.; Farronato, D.; Zecca, P.A.; Mangano, C. Immediate Loading of Mandibular Overd tures Supported by One-Piece, Direct Metal Laser Sintering Mini-Implants: A Short-Term Prospective Clinical Study. *J. Periodo* **2015**, *86*, 192–200. [CrossRef]
92. Kripal, K.; Sirajuddin, S.; Rafiuddin, S.; Mp, R.; Chungkham, S. Iatrogenic Damage to the Periodontium Caused by Laser: Overview. *Open Dent. J.* **2015**, *9*, 210–213. [CrossRef]
93. Cobb, C.M. Lasers in Periodontics: A Review of the Literature. *J. Periodontol.* **2006**, *77*, 545–564. [CrossRef]
94. Coluzzi, D.J. Lasers and Soft Tissue Curettage: An Update. *Compend. Contin. Educ. Dent.* **2002**, *23*, 1104–1111. [PubMed]
95. Page, M.J.; McKenzie, J.E.; Bossuyt, P.M.; Boutron, I.; Hoffmann, T.C.; Mulrow, C.D.; Shamseer, L.; Tetzlaff, J.M.; Akl, E Brennan, S.E.; et al. The PRISMA 2020 Statement: An Updated Guideline for Reporting Systematic Reviews. *BMJ* **2021**, *372*, [CrossRef]
96. Explanation of the 2011 OCEBM Levels of Evidence—Centre for Evidence-Based Medicine (CEBM), University of Oxf Available online: https://www.cebm.ox.ac.uk/resources/levels-of-evidence/explanation-of-the-2011-ocebm-levels-of-evide (accessed on 6 March 2023).
97. Suter, V.G.A.; Altermatt, H.J.; Bornstein, M.M. A Randomized Controlled Clinical and Histopathological Trial Compar Excisional Biopsies of Oral Fibrous Hyperplasias Using $CO_2$ and Er:YAG Laser. *Lasers Med. Sci.* **2017**, *32*, 573–581. [CrossRef
98. Saibene, A.M.; Rosso, C.; Castellarin, P.; Vultaggio, F.; Pipolo, C.; Maccari, A.; Ferrari, D.; Abati, S.; Felisati, G. Managing Ben and Malignant Oral Lesions with Carbon Dioxide Laser: Indications, Techniques, and Outcomes for Outpatient Surgery. *Sur* **2019**, *5*, e69–e75. [CrossRef] [PubMed]
99. Hikov, T.; Pecheva, E.; Montgomery, P.; Antoni, F.; Leong-Hoi, A.; Petrov, T. Precise Femtosecond Laser Ablation of Dental H Tissue: Preliminary Investigation on Adequate Laser Parameters. *J. Phys. Conf. Ser.* **2017**, *794*, 012036. [CrossRef]
100. Lauritano, D.; Lucchese, A.; Gabrione, F.; Di Stasio, D.; Silvestre Rangil, J.; Carinci, F. The Effectiveness of Laser-Assisted Surg Excision of Leukoplakias and Hyperkeratosis of Oral Mucosa: A Case Series in a Group of Patients. *Int. J. Environ. Res. Pu Health* **2019**, *16*, 210. [CrossRef]

1. Bahammam, M.A. Treatment of a Gingival Injury from a Cosmetic Laser Burn: A Case Report. *Compend. Contin. Educ. Dent.* **2018**, *39*, 238–243.
2. Fabian Falkenstein; Norbert Gutknecht; René Franzen Analysis of Laser Transmission and Thermal Effects on the Inner Root Surface during Periodontal Treatment with a 940-Nm Diode Laser in an in Vitro Pocket Model. *J. Biomed. Opt.* **2014**, *19*, 128002. [CrossRef]
3. Franzen, R.; Rashidisangsary, B.; Ozturan, S.; Vanweersch, L.; Gutknecht, N. Intrapulpal Temperature Changes during Root Surface Irradiation with Dual-Wavelength Laser (2780 and 940 Nm): In Vitro Study. *J. Biomed. Opt.* **2015**, *20*, 018002. [CrossRef]
4. Tonin, M.H.; Brites, F.C.; Mariano, J.R.; Freitas, K.M.S.; Ortiz, M.A.L.; Salmeron, S. Low-Level Laser and Antimicrobial Photodynamic Therapy Reduce Peri-Implantitis-Related Microorganisms Grown In Vitro. *Eur. J. Dent.* **2022**, *16*, 161–166. [CrossRef]
5. Monzavi, A.; Fekrazad, R.; Chinipardaz, Z.; Shahabi, S.; Behruzi, R.; Chiniforush, N. Effect of Various Laser Wavelengths on Temperature Changes During Periimplantitis Treatment: An in Vitro Study. *Implant. Dent.* **2018**, *27*, 311–316. [CrossRef]
6. Hoedke, D.; Enseleit, C.; Gruner, D.; Dommisch, H.; Schlafer, S.; Dige, I.; Bitter, K. Effect of Photodynamic Therapy in Combination with Various Irrigation Protocols on an Endodontic Multispecies Biofilm Ex Vivo. *Int. Endod. J.* **2018**, *51*, e23–e34. [CrossRef]
7. Sobral, M.F.P.; Cassoni, A.; Tenis, C.A.; Steagall, W.; Brugnera Junior, A.; Bagnato, V.S.; Botta, S.B. Longitudinal, Randomized, and Parallel Clinical Trial Comparing a Violet Light-Emitting Diodes System and In-Office Dental Bleaching: 6-Month Follow-Up. *Photobiomodulation Photomed. Laser Surg.* **2021**, *39*, 403–410. [CrossRef] [PubMed]
8. Chung, S.E.V.; Tompson, B.; Gong, S.-G. The Effect of Light Emitting Diode Phototherapy on Rate of Orthodontic Tooth Movement: A Split Mouth, Controlled Clinical Trial. *J. Orthod.* **2015**, *42*, 274–283. [CrossRef] [PubMed]
9. Saluja, M. Comparative Morphologic Evaluation and Occluding Effectiveness of Nd: YAG, CO 2 and Diode Lasers on Exposed Human Dentinal Tubules: An Invitro SEM Study. *JCDR* **2016**, *10*, ZC66–ZC70. [CrossRef] [PubMed]
0. Mozaffari, H.R.; Ehteshami, A.; Zallaghi, F.; Chiniforush, N.; Moradi, Z. Microleakage in Class V Composite Restorations after Desensitizing Surface Treatment with Er:YAG and CO2 Lasers. *Laser Ther.* **2016**, *25*, 259–266. [CrossRef]
1. Aljdaimi, A.; Devlin, H.; Dickinson, M.; Burnett, T.; Slater, T.J.A. Micron-Scale Crack Propagation in Laser-Irradiated Enamel and Dentine Studied with Nano-CT. *Clin. Oral. Investig.* **2019**, *23*, 2279–2285. [CrossRef]
2. Kuhn-Dall'Magro, A.; Zamboni, E.; Fontana, T.; Dogenski, L.C.; De Carli, J.P.; Dall'Magro, E.; Fornari, F. Low-Level Laser Therapy in the Management of Oral Mucositis Induced by Radiotherapy: A Randomized Double-Blind Clinical Trial. *J. Contemp. Dent. Pract.* **2022**, *23*, 31–36.
3. Romeo, U.; Russo, C.; Palaia, G.; Lo Giudice, R.; Del Vecchio, A.; Visca, P.; Migliau, G.; De Biase, A. Biopsy of Different Oral Soft Tissues Lesions by KTP and Diode Laser: Histological Evaluation. *Sci. World J.* **2014**, *2014*, 761704. [CrossRef]
4. Launay, Y.; Mordon, S.; Cornil, A.; Brunetaud, J.M.; Moschetto, Y. Thermal Effects of Lasers on Dental Tissues. *Lasers Surg. Med.* **1987**, *7*, 473–477. [CrossRef]
5. Mortazavi, H.; Baharvand, M.; Mokhber-Dezfuli, M.; Rostami, N.; Doost-Hoseini, M.; Alavi-Chafi, O.; Nourshad, S. Lasers in Dentistry: Is It Really Safe? *Dent. Hypotheses* **2016**, *7*, 123–127. [CrossRef]
6. Fu, W.; Wo, C. The Use of Laser in Dentistry: A Narrative Review. *J. Biol. Regul. Homeost. Agents* **2021**, *35*, 11–18.

photonics

Article

# Comparison of Raman Spectra of Optically Nonlinear $LiTaO_3:Cr^{3+}$ (0.005 wt%) Crystal Laser Excited in Visible (532 nm) and Near-IR (785 nm) Areas

**Mikhail Palatnikov [1,*], Nikolay Sidorov [1], Alexander Pyatyshev [2] and Alexander Skrabatun [2]**

[1] Tananaev Institute of Chemistry—Subdivision of the Federal Research Centre "Kola Science Centre of th Russian Academy of Sciences", 184206 Apatity City, Russia; n.sidorov@ksc.ru
[2] P.N. Lebedev Physical Institute of the Russian Academy of Sciences, 119991 Moscow, Russia; jb_valensia@mail.ru (A.P.); skrabatunav@lebedev.ru (A.S.)
* Correspondence: m.palatnikov@ksc.ru

**Abstract:** Raman spectra have been excited by the radiation of lasers generatingon in the visi (532 nm) and near-IR (785 nm) ranges in a $LiTaO_3:Cr^{3+}$ (0.005 wt%) crystal. The obtained spec were compared. Recorded Raman spectra contained bands in the range of <900 $cm^{-1}$ (first ord up to 2000 $cm^{-1}$ (second order). The Raman spectra of the second order, excited by near-IR la radiation, were located against the background of a wide, luminescent halo. The frequency of a sin low-intensity band exceeded the exact frequency of an overtone corresponding to the complet symmetric fundamental $4A_1$(z)LO mode. The attribution of other second-order bands with hig frequencies remains unclear.

**Keywords:** lithium tantalate; doping; transition element; ferroelectric; Raman scattering; seco order spectra; biphonon

check for updates

**Citation:** Palatnikov, M.; Sidorov, N.; Pyatyshev, A.; Skrabatun, A. Comparison of Raman Spectra of Optically Nonlinear $LiTaO_3:Cr^{3+}$ (0.005 wt%) Crystal Laser Excited in Visible (532 nm) and Near-IR (785 nm) Areas. *Photonics* **2023**, *10*, 439. https://doi.org/10.3390/photonics10040439

Received: 10 March 2023
Revised: 5 April 2023
Accepted: 11 April 2023
Published: 12 April 2023



## 1. Introduction

The ferroelectric single crystal of lithium tantalate (LT, $LiTaO_3$) has unique elect optical, piezoelectric, pyroelectric and nonlinear optical characteristics. These propert make it a valuable and multifunctional material. The $LiTaO_3$ crystal is widely used to cre IR radiation sensors, motion detectors and thermometry, as well as to generate terahe radiation [1–7]. In addition, the development of materials based on a LT crystal for t purpose of converting and modulating optical radiation and for active nonlinear la media is promising at present [8]. Lithium tantalate has classically been used as an acti medium for parametric light generation [9–11]. For successful application, it is necessary know the features of the interaction of radiation with different wavelengths and the acti medium. In this regard, it seems relevant to study the interaction of lithium tantalate w laser radiation of various wavelengths.

LT is a nonstoichiometric oxygen octahedral phase of variable composition with a w homogeneity region on the phase diagram [12]. This circumstance helps to continuou improve the technologies for growing single crystals with different Li/Ta ratios and obtain materials with significantly better physical characteristics. These crystals can nominally pure, i.e., the charge is not doped intentionally, but crystal contains a number uncontrolled impurities in very small concentrations, or doped with a variety of eleme Features of the electrical and nonlinear optical characteristics of the $LiTaO_3$ crystal larg depend on the following factors: the polarizability of oxygen octahedral clusters $MeO_6$ (M $Ta^{5+}$, $Li^+$, dopant, impurity, vacancy V) by anharmonic interactions in the vibrations of ic and the magnitude of spontaneous polarization. Point defects are usually Me cations a V vacancies irregularly located along the polar axis of the crystal [8,12,13]. The defects p a significant role in clusters polarizability and spontaneous polarization.

When a developed functional non-linear optical material is proposed to convert IR radiation, it is necessary to minimize thermal and photo-induced (optical damage) variations in the refractive index. These changes distort the wavefront of the laser radiation propagation and significantly impair the operation of optoelectronic devices. Other distortions of the crystal structure are associated with its disordered and anharmonic structure; the structure is distorted due to compositional variability. It is widely known that Raman scattering reflects the slightest reordering of the crystal structural units [14–16]. Comparative studies are necessary for the creation of materials suitable for active nonlinear laser media, and to convert IR radiation into a visible form. Such studies can elucidate the differences between the passage of a visible laser and IR radiation through a photorefractive electro-optical crystal. The difference can be explained as follows: in micro- and macrostructures, point defects respond differently to exposure to different wavelengths. These point defects in photorefractive phases of variable composition are usually irregularly arranged main, doping (in case of doped crystals) and impurity atoms; such defects are the result of structural disorder and anharmonicity.

The optical properties of $LiTaO_3:Cr^{3+}$ crystals have been studied previously in a number of works [17–35]. According to these references, the absorption spectrum of $LiTaO_3:Cr^{3+}$ has two broad bands near 458.5 and 653 nm and a small maximum near 718.5 nm at room temperature. The two wide bands in the smaller wavelength area correspond to spin-allowed vibronic transitions $^4A_2$–$^4T_1$ and $^4A_2$–$^4T_2$. It is likely that a small maximum in the red region of the spectrum corresponds to the spin-forbidden zero-phonon transition $^2E$-$^4A_2$. The dependence of absorption on the concentration of chromium dopant was analyzed at a wavelength of 300 nm. The analysis showed that the absorption was negligible, with a chromium ion content less than 0.5 ppm. The luminescence spectrum at room temperature has a similar broad band; with a maximum at 858 nm, and shifted to 889 nm upon cooling to 5 K. This band corresponds to the $^4T_2$-$^4A_2$ vibronic transition. The dependence of polarized and unpolarized absorption and luminescence spectra on temperature (in the range of 5–390 K), pressure (up to 93 Kbar), and the lifetimes of luminescence centers were studied. A strong distortion of the beam profile of the 488 nm argon laser radiation transmitted through the $LiTaO_3:Cr^{3+}$ crystal was found. The distortion limits the application of this crystal as an active laser medium. Photoinduced and Rayleigh light scattering in a $LiTaO_3:Cr^{3+}$ crystal were analyzed in the works outlined in [34,35]. The isotropic nature of photoinduced light scattering was discovered, while the Rayleigh scattering of light had a nonstationary character. The $LiTaO_3:Cr^{3+}$ crystal was studied by nuclear and electron magnetic resonance in [36–42]. Unfortunately, data on the Raman spectra of both orders of $LiTaO_3:Cr^{3+}$ were absent from the cited works [17–42].

Comparative studies of the Raman spectra of a $LiTaO_3:Cr^{3+}$ (0.005 wt%) (LT:Cr) crystal recorded upon excitation by visible (532 nm) and near-IR (785 nm) laser radiation were performed in this work. Chromium in isomorphic $LiTaO_3$ and $LiNbO_3$ crystals has a constant valence (+3), and is not a photovoltaically active dopant. Its presence in the crystal structure does not increase the effect of photorefraction [14]. The valence (oxidation state of chromium) varies from +2 to +6 in other environments. Therefore, $LiTaO_3:Cr^{3+}$ crystals, as materials for nonlinearly active laser media, have a high resistance to the pump laser radiation. The concentration of the active $Cr^{3+}$ ion in the crystal should be low enough so as not to increase the disorder and anharmonicity of the crystal lattice and not to deteriorate the optical quality of the crystal. The fact is that the $Cr^{3+}$ ion in the cationic sublattice of $LiTaO_3$ and $LiNbO_3$ crystals can be in three positions [14]. The $Cr^{3+}$ ion is localized in the vacant $O_6$ octahedron ($Cr_V$ point defect) at low concentrations (≈0.005 wt% or less). If the concentration increases further, the $Cr^{3+}$ ion additionally occupies lithium ($Cr_{Li}$ point defect) and niobium octahedra ($Cr_{Nb}$ point defect). Doping with Cr significantly decreases the cation sublattice order along the polar axis, and optical characteristics of the crystals deteriorate. Such disordering effects can manifest themselves in the spectra of second-order Raman scattering. Thus, the study of second-order Raman spectra can be an effective tool for the control of structure perfection of the crystal. Spectra of a perfect or near-ideal LT crystal should not contain any bands in the second order range. As an LT crystal diso grows, the anharmonism of the vibrations should increase; this should widen the first-ord Raman bands and increase the intensity of the second-order bands. The changes in wic and intensity should depend on the frequency of the exciting laser radiation and manif themselves differently in the Raman spectrum when they are excited in the visible a IR regions.

## 2. Materials and Methods

This work considers the Raman spectra of a single $LiTaO_3$:Cr (0.005 wt%) crys Chromium $Cr^{3+}$ is a photorefractive dopant that increases the sensitivity of the crystal optical damage. A $LiTaO_3$:Cr (0.005 wt%) crystal was grown in an argon atmosphere usi the Czochralski method from a Pt/Rh10 ∅ 80 mm crucible, under conditions of an avera axial gradient of ~10 deg/cm in the direction of the X-axis (X-cut) at a rotation speed ~16 rpm and displacement of ~2.4 mm/h. The crystal growth rate was ~2.9–3.0 mm Crystals were grown on a growth setup with an automatic crystal diameter control syste The features of single crystal growth were described in detail in [17,20]. The congrue melt ([Li]/[Ta] ≈ 0.92) was directly doped with chromium oxide ($Cr_2O_3$). The samp used to study the Raman spectra were cut from the grown boule in the shape of rectangu parallelepipeds. Their edges coincided with the directions of the main crystallophysi axes (x and y; z is the polar axis). The faces of the parallelepipeds were carefully polish

Figure 1 shows a photograph of the test sample.

**Figure 1.** Photograph of a $LiTaO_3$:Cr (0.005 wt%) crystal sample.

Growth was completed when the weight of the $LiTaO_3$:Cr (0.005 wt%) crystal reach ~500 g. In this case, about ~28% of the total weight of the melt crystallized. The grow parameters of the $LiTaO_3$:Cr (0.005 wt%) crystal took into account the need to obtain a crystallization front. They were selected experimentally. The parameters were as follo pulling speed, rod rotation speed and temperature gradient at the crystallization front. Su an approach should have provided a sufficiently high structural perfection of the crys The grown $LiTaO_3$:Cr (0.005 wt%) crystal had a flat crystallization front, ∅~38 mm, a the length of the cylindrical part was Lc ≈ 44 mm. Dopant was introduced into the cha in the form of $Cr_2O_3$ (99.9) with a concentration of impurities at a level of $<3 \times 10^{-4}$ w After addition, dopant was mixed thoroughly with the initial charge. The dopant w intended to homogenize in the melt. In order to achieve this goal, the melt was held for at the temperature ~90 °C higher than the $LiTaO_3$ melting point ($T_{melt}$ = 1650 °C) befo crystal growth.

The crystal was in a single-domain state. High-temperature electrodiffusion annealing was performed in LANTAN (Voroshilovgradsky zavod electronnogo mashinostroeniya, Voroshilovgrad, USSR). The static piezoelectric effect controlled the degree of single-domain fraction [43]. The $LiTaO_3$:Cr (0.005 wt%) crystal was annealed at 1400 °C in a growth setup for 10 h and then cooled at a rate of ~50 K/h after growth. Prolonged post-growth annealing is required in order to homogenize the composition of the doped crystal and remove thermal and mechanical stresses.

Tantalum pentoxide $Ta_2O_5$, produced by Solikamsk magnesium works (Solikamsk, Russian Federation), and lithium carbonate $Li_2CO_3$ with an impurity concentration of $<3 \times 10^{-4}$ wt% were used to synthesize the $LiTaO_3$ charge.

A charge of congruent composition ([Li]/[Ta] $\approx$ 0.92) was synthesized from these initial components. The impurity composition of the $LiTaO_3$ charge and the grown $LiTaO_3$:Cr (0.005 wt%) crystal determined by spectral analysis is shown in Table 1. As one can see from the table, the content of trace amounts of impurities in the crystal was no more than $10^{-4}$ wt%.

**Table 1.** Content of impurities in the charge and grown $LiTaO_3$:Cr (0.005 wt%) crystal.

| Impurity | Concentration in the Charge, wt% | Concentration in $LiTaO_3$:Cr (0.005 wt%) Crystal, wt% |
|---|---|---|
| Mn, V, Mg, Sn | $<3 \times 10^{-4}$ | $<1 \times 10^{-4}$ |
| Pb, Ni | $<4 \times 10^{-4}$ | $<2 \times 10^{-4}$ |
| Co, Mo | $<4 \times 10^{-4}$ | $<3 \times 10^{-4}$ |
| Si, Fe | $<4 \times 10^{-4}$ | $<4 \times 10^{-4}$ |
| Ti | $<5 \times 10^{-4}$ | $<5 \times 10^{-4}$ |
| Al | $<6 \times 10^{-4}$ | $<6 \times 10^{-4}$ |
| Zr | $<7 \times 10^{-4}$ | $<5 \times 10^{-4}$ |
| Ca | $<3 \times 10^{-4}$ | $<5 \times 10^{-4}$ |
| Te, Sb | $<5 \times 10^{-4}$ | $<4 \times 10^{-4}$ |
| Bi | $<2 \times 10^{-4}$ | $<2 \times 10^{-4}$ |
| Rh | $<1 \times 10^{-4}$ | $<2 \times 10^{-2}$ |

Raman spectra in the visible region were recorded by a BWS465-532S i-Raman Plus spectrometer (B&W Tek, Plainsboro Township, NJ, USA). The Raman setup contained a built-in cw laser with a 532 nm wavelength. This spectrometer is able to record spectra in the range of 50–4000 $cm^{-1}$. The laser power was 30 mW when the spectra were recorded in this setup. The numerical aperture of the objective of the Raman system was $\approx$ 0.22. A BWS465-785H i-Raman Plus spectrometer (B&W Tek, Plainsboro Township, NJ, USA) was used to record Raman spectra in the near-IR region. It had a cw laser with excitation wavelength of 785 nm. This setup was able to record spectra in the range of 50–2850 $cm^{-1}$. The laser radiation power was 340 mW on the sample when the spectra were recorded in this setup. The numerical aperture of the objective of the Raman system was $\approx$0.22. The laser spot size was 85 µm at the focus in all experiments. Spectra, in both cases, were recorded in backscattering geometry at room temperature. The local influence of the exciting laser radiation was minimized, and the optimal radiation focus and accumulation time of the useful signal were adjusted for each experiment. The frequency error of the Raman bands was $\pm$3.0 $cm^{-1}$.

## 3. Results and Discussion

The rhombohedral unit cell of the ferroelectric phase of the $LiTaO_3$ crystal is characterized by a $C_{3V}^6$ space symmetry group (R3c). A LT unit cell contains two formula units made up of ten atoms. The lithium and tantalum atoms occupy the $C_3$ position. Oxygen atoms occupy the $C_1$ position. The curve of phonon dispersion has 30 vibrational branches; 27 of them are optical and 3 are acoustic. The form of optical vibrational representation in LT c be expressed by the following equation:

$$\Gamma = 5A_1(z) + 5A_2 + 10E(x,y)$$

In this representation, $4A_1(z) + 9E(x,y)$ dipole active fundamental vibrations are act in Raman and IR absorption at k = 0 (i.e., at the center of the Brillouin zone). The po nature of optical vibrations split them into longitudinal (LO) and transverse (TO) vibratic Therefore, the Raman spectra should contain 26 bands attributed to fundamental phonc given two conditions: phonons propagate along the crystallographic axes, and LO- split exists. In addition, $A_1(z) + E(x,y)$ acoustic and $5A_2$ optical fundamental vibrations inactive. These bands are expected not to appear in either the Raman or the IR absorpti spectra. The Raman tensor form [44] implies that (zz) polarization has only nondegener phonons of $A_1(z)$ symmetry, and (xy), (xz), (yx), (yz), (zx) and (zy) polarizations have o doubly degenerate E(x,y) symmetry-type phonons. Both (xx) and (yy) polarizations shou simultaneously have $A_1(z)$ and E(x, y) symmetry-type phonons.

Figures 2 and 3 show the Raman spectra of a LT:Cr crystal recorded in $x(zz, zy)$ $y(zz, zx)\,\bar{y}$ and $z(xx,\ yy,\ xy)\,\bar{z}$ scattering geometries excited by 532 and 785 nm laser ra ation, respectively. Table 2 lists the frequencies of the experimentally observed bands in Raman spectrum, corresponding to the fundamental vibrations of the crystal lattice a their attribution based on [44]. Table 2 also demonstrates alternative and ambiguous at butions (in brackets) for some of the Raman bands. According to [45], the 740 and 749 cm bands can be attributed to LT intrinsic defects. Finally, the 717–721 and 810–812 cm bands seem to correspond to two-phonon Raman processes.

**Table 2.** Real frequencies of polar vibrations (both transverse (TO) and longitudinal (LO)) of LT:Cr crystal obtained in this study, and their attribution.

| $\lambda_0$ = 532 nm | | | $\lambda_0$ = 785 nm | | | Attribution |
|---|---|---|---|---|---|---|
| | | | ν, cm$^{-1}$ | | | |
| $x(zz,zy)\bar{x}$ | $y(zz,zx)\bar{y}$ | $z(xx, yy, xy)\bar{z}$ | $x(zz,zy)\bar{x}$ | $y(zz,zx)\bar{y}$ | $z(xx, yy, xy)\bar{z}$ | |
| 140 | 140 | 143 | 142 | 142 | 142 | 1E(x,y)TO |
| | | | 190 | 190 | | 1E(x,y)LO |
| 203 | 203 | 206 | 207 | 207 | 207 | $1A_1$(z)TO |
| | | 250 | 249 | 247 | | 3E(x,y)TO |
| | | | 279 | 277 | | 3E(x,y)LO |
| 312 | 312 | 309 | 314 | 314 | 314 | 4E(x,y)TO |
| 356 | 353 | 356 | 356 | 356 | 354 | $2A_1$(z)LO |
| 377 | 380 | 377 | 380 | 380 | 381 | $3A_1$(z)TO5E(x,y)TO |
| | | 403 | | | | $3A_1$(z)LO |
| 458 | 458 | 461 | 462 | 463 | 463 | 7E(x,y)TO |
| 595 | 595 | 592 | 596 | 599 | 594 | $4A_1$(z)TO(8E(x,y)TO |
| 657 | 654 | 657 | 659 | 659 | 664 | 8E(x,y)LO(9E(x,y)TO |
| 864 | 867 | 867 | 870 | 869 | 865 | $4A_1$(z)LO(9E(x,y)LO |

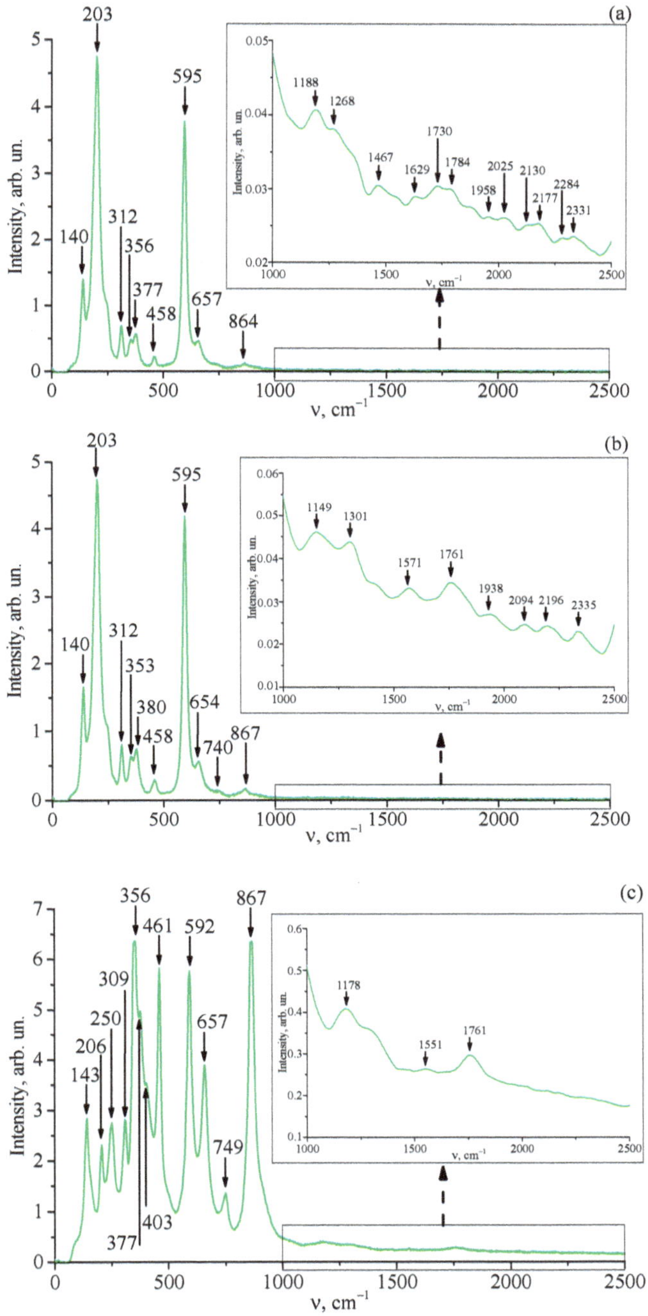


**Figure 2.** Raman spectra excited by $\lambda_0$ = 532 nm radiation in the LT:Cr crystal in $x(zz, zy)\,\bar{x}$ (**a**), $y(zz, zx)\,\bar{y}$ (**b**) and $z(xx,\ yy,\ xy)\,\bar{z}$ (**c**) scattering geometries.

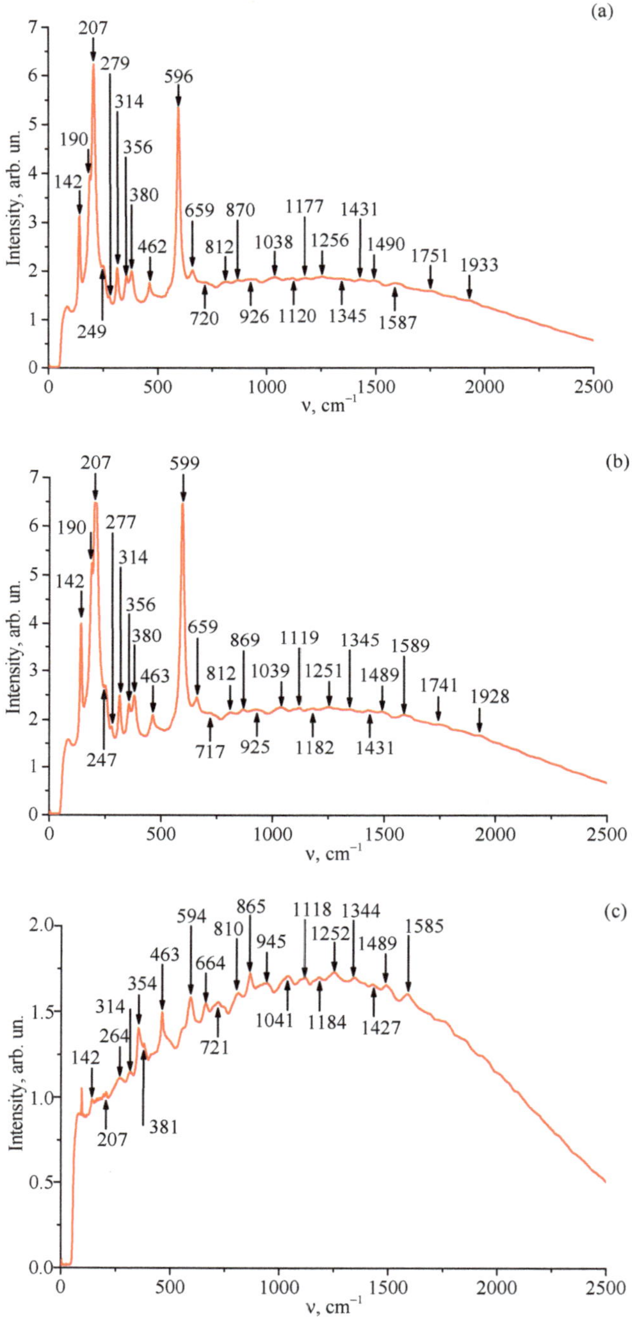


**Figure 3.** Raman spectra excited by $\lambda_0$ = 785 nm radiation in the LT:Cr crystal in scattering geometry $x(zz, zy)\bar{x}$ (**a**), $y(zz, zx)\bar{y}$ (**b**) and $z(xx, yy, xy)\bar{z}$ (**c**).

Our comparison of the data in Table 2 and Figures 2 and 3 revealed that the Raman spectra of the LT:Cr crystal significantly differed when excited by laser bands with frequencies of 532 and 785 nm. In addition, noticeable frequency differences for the TO and LO modes were observed. This indicates fundamental polar vibrations in noncentrosymmetric crystals. Figure 2 and Table 2 demonstrate that the fundamental vibrations bands were predominantly present upon excitation by 532 nm laser radiation in the Raman spectra of the crystal in the frequency range of 50–900 $cm^{-1}$. Additionally, bands with frequencies of 206–209, 250, 353–356 and 592–595 $cm^{-1}$ were attributed to the $A_1$(TO) and $A_1$(LO) symmetry-type fundamental vibrations along the polar axis of the crystal, and bands with frequencies 377–380, 458–461, 654–657, 740–749 and 864–867 $cm^{-1}$ were attributed to the fundamental doubly degenerate E(TO) and E(LO) symmetry-type vibrations perpendicular to the polar axis of the crystal. Figure 2 and Table 2 also show that unmixed (LO-TO) phonons could be observed in the spectrum of the LT:Cr crystal when it was excited by a 532 nm laser line. Mixed (LO-TO) phonons manifested in the spectrum when the phonon wave vector direction did not coincide with the direction of the main crystallographic axes. Furthermore, low-intensity extra bands were absent from the 50–900 $cm^{-1}$ region of Raman spectrum of the LT:Cr crystal. According to selection rules for the space group R3c, these bands were not meant to appear, given that the unit cell contained two formula units. Low-intensity bands appeared in the spectrum of an isomorphic $LiNbO_3$ crystal of nonstoichiometric composition in the frequency range of 50–900 $cm^{-1}$ due to the deviation of the crystal's composition from stoichiometric [14,46]. A number of broad low-intensity bands corresponding to the second-order Raman spectrum were present in the region of 900–2500 $cm^{-1}$. Some of the second-order bands were polarized. Bands appeared only in certain backscattering geometry. It is important to note that photoluminescence was absent in the entire studied frequency range of 50–2500 $cm^{-1}$ when the Raman spectrum was excited by the 532 nm line. However, it has been observed prior, according to the data of [22–25], when the spectrum of a LT:Cr crystal was excited by the 640 nm band.

Figure 3 shows that the Raman spectrum excited by a 785 nm laser line was located against the background of an intense broad photoluminescence band with a maximum of $\approx$1250 $cm^{-1}$. The luminescent halo in the near IR region may be attributable to the luminescence centers in the LT:Cr crystal. The luminescence centers in the crystal are point defects of $Cr_{Ta}$ and $Cr_{Li}$ ($Cr^{3+}$ ions located in the positions of tantalum and lithium ions). The luminescent halo can also be explained by defects in the form of small-radius polarons ($Ta_{Li}$) and more complex bipolaron bonds (in the form of $Ta_{Li}$-$Ta_{Ta}$). It was important that the Raman spectrum excited by the 785 nm laser line would contain more bands (extra bands) in the region of fundamental vibrations of the crystal lattice (50–900 $cm^{-1}$) than the selection rules allowed in the $x(zz, zy)\,\overline{x}$, $y(zz, zx)\,\overline{y}$ and $z(xx, yy, xy)\,\overline{z}$ scattering geometries. Selection rules have considered LO-TO splitting for the space group R3c ($C_{3V}^6$) in cases in which the unit cell contains two formula units. Extra bands have a low intensity; their frequencies are 264, 717–721 and 810–812 $cm^{-1}$; and they appear in various scattering geometries in the frequency range of fundamental vibrations (50–900 $cm^{-1}$) in the spectrum. Bands with frequencies of 925–926, 1038–1041, 1118–1120, 1182–1190, 1251–1256, 1344–1345, 1427, 1489, 1585–1587, 1741–1751 and 1928–1933 $cm^{-1}$ belong to the second-order Raman spectrum; the bands often appear in the >900 $cm^{-1}$ region of the LT Raman spectrum. These bands correspond to overtone processes; they can be attributed to bound states of optical phonons. In the case of LT:Cr, only one weak band, with a frequency of 1928–1933 $cm^{-1}$ in scattering geometries $y(zz, zx)\,\overline{y}$ and $z(xx, yy, xy)\,\overline{z}$, showed an anomaly: its frequency was much higher than the exact value of the overtone frequency. The initial band was located near 864–870 $cm^{-1}$, and it corresponded to the symmetrical fundamental vibration of $4A_1$(z)LO (Figures 2 and 3). The exact overtone frequency should be located near 1728–1740 $cm^{-1}$, but the real band is 200 $cm^{-1}$ higher.

The nature of the Raman bands in the >2000 $cm^{-1}$ range remains unclear and requires further research. Raman spectra of the studied LT:Cr crystal, when excited by radiation with different wavelengths, differed significantly from each other in the scattering geometry z(xx, yy, xy) $\overline{z}$. There can be two reasons for the observed difference: different distributi of trivalent chromium dopant $Cr^{3+}$ relative to the ferroelectric axis of the crystal, a different mechanisms of interaction of laser radiation with the LT:Cr crystal.

Figures 2 and 3 show that the intensity of the second-order Raman spectrum of t LT:Cr crystal was several orders of magnitude lower than the intensity of the fundamen vibrations. This might give rise to the idea that these weak bands are artifacts of t spectrometers. However, we are sure of the opposite. Previously, second-order Ram spectra were studied in a number of crystals: diamond [47], gallium phosphide [48], cesiu bromide [49], α-sulfur, etc. For example, the peak intensity of a number of lines in t diamond second-order Raman spectrum was from 250 to $10^5$ times less than the intens of the fundamental mode. The results of other works have stated that the second-ord Raman spectrum has a quasi-continuous form, with weak maxima. In addition, in o previous works, we have studied lithium niobate and tantalate crystals with other dopa and the second-order Raman spectrum had quite a low intensity [50–53]. Thus, it is n entirely correct to state that the weak bands in the 900–2000 $cm^{-1}$ region are an artifact the spectrometers used.

Thus, the LT:Cr Raman spectra excited by 532 and 785 nm laser radiation were fou to be different. This difference can be explained as follows. The mechanisms of interacti between the visible and near-IR radiation and the microstructure and structural defects the crystal are different. In addition, microstructures, together with defects, determine t bound states of phonons (i.e., phonon–phonon interaction) and, thus, the anharmonic of the crystal lattice vibrations. The formation of bound states of phonons in crystals w schematically given in theoretical works [54–56]. These works showed that one-phon Green's function $D_1(k,\omega)$ can be used to calculate the density of two-phonon states $\rho_2($ by employing five Formulas (1)–(5):

$$D_1\left(\vec{k},\omega\right) = \frac{\omega\left(\vec{k}\right)}{2}\left[\frac{1}{\omega-\omega\left(\vec{k}\right)+\frac{1}{2}i\Gamma} - \frac{1}{\omega+\omega\left(\vec{k}\right)-\frac{1}{2}i\Gamma}\right]$$

$$F(\omega) = \frac{i}{(2\pi)^4}\int d^3\vec{k}\int D_1\left(\vec{k},\omega-\omega'\right)D_1\left(\vec{k},\omega\right)d\omega'$$

The relation in (3) is correct for optical phonons with frequency $\omega_0$ near the cen of the Brillouin zone in the quasi-Newtonian approximation of the dispersion law for t density of single-particle states of phonons $\rho_1(\omega)$:

$$\rho_1(\omega) = a\sqrt{\omega_0-\omega}$$

where $a = -\frac{\omega_0 V\sqrt{2\omega_0}}{2\pi^2 s^3}$, V denotes the unit cell volume and s is the speed of sound. A result, the function $F(\omega)$ has the shape:

$$F(\omega) = \frac{1}{4}\omega_0^2 a\int_0^{\Delta}\frac{\sqrt{\omega'}}{\omega-2(\omega_0-\omega')+i\Gamma}d\omega'$$

where Δ is the range of phonon frequencies considered during the integration. The functi $F(\omega)$ depends on two factors: the effective mass of the quasiparticle and the dampi constant Γ. The constant can be associated with the inverse lifetime of the quasiparticle 1/Γ. The intensity of second-order Raman bands should then be proportional to the dens of two-phonon states $\rho_2(\omega)$:

$$\rho_2(k,\omega) \approx -\frac{2}{\pi\omega_0^2}\,\frac{\mathrm{Im}\,F(\omega)}{\left[1-\frac{1}{2}g_4\,\mathrm{Re}\,F(\omega)\right]^2+\left[\frac{1}{2}g_4\,\mathrm{Im}\,F(\omega)\right]^2}$$

Figure 4 compares theoretical calculations of the spectral distribution of the Raman intensity in the overtone region of $2\nu_0 = 1738$ cm$^{-1}$, as well as the experimentally observed intensity distribution. The dotted arrow in Figure 4 locates the exact position of the $2\nu_0$ overtone.

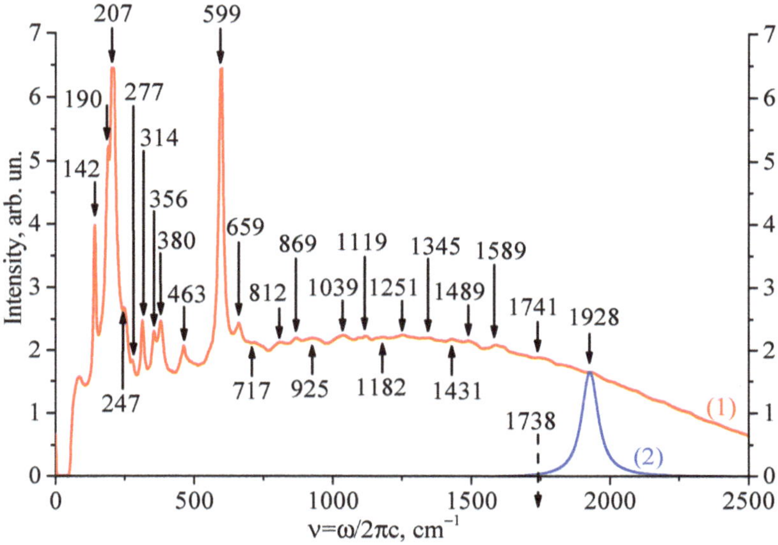


**Figure 4.** Comparison of the real Raman spectrum of the LT:Cr crystal (curve 1) with the calculated dependence of the two-phonon state density $\rho_2(\nu)$ in the overtone region (curve 2).

Figure 4 shows the bound state of the fully symmetric polar mode $4A_1$(z)LO with a frequency of 869 cm$^{-1}$, satisfactorily approximating the 1928 cm$^{-1}$ band (curve 2) at the following parameters: s = 1000, $\Delta = 0.1\omega_0$, $g_4 = 7.28 \times 10^{-18}$ and $\Gamma = 7.54 \times 10^{12}$. The $\Delta$ value was given by the dispersion curve form characteristic of optical phonons $4A_1$(z). The damping constant $\Gamma$ was equal to the width of the real Raman bands of these phonons.

## 4. Conclusions

In the present work, the full Raman spectra of a $LiTaO_3:Cr^{3+}$ (0.005 wt%) crystal with backscattering geometry in the frequency range of 50–2500 cm$^{-1}$ upon excitation by the 532 and 785 nm laser lines were recorded, and their interpretation was given. For the first time, the second-order Raman spectrum of a $LiTaO_3:Cr^{3+}$ (0.005 wt%) crystal was reported. This spectrum had weak bands in the range of 800–2000 cm$^{-1}$. Bands corresponded to overtone processes. A weak band with a maximum near 1928 cm$^{-1}$ was found among these bands in the Raman spectrum. Its frequency value exceeded the exact value of the overtone of the symmetric fundamental mode $4A_1$(z).

Comprehensive studies on nonlinear optical processes in this crystal are of interest for future studies. By nonlinear processes, we mean, for example, stimulated Raman scattering of light, generation of the second optical harmonic, generation of sum and difference optical harmonics and parametric generation of light. Since $Cr^{3+}$ chromium ions exhibit intense luminescence in the near-IR region, the $LiTaO_3:Cr^{3+}$ (0.005 wt%) crystal studied in this work may be used as an active laser medium in the future. However, for this purpose, it is necessary to grow a crystal with a much more perfect structure. It is also necessary to establish and study the optimization of the conditions in order to observe laser generation in this crystal.

**Author Contributions:** Conceptualization, N.S.; validation, N.S.; investigation, A.P. and A.S.; d curation, A.P. and A.S.; writing—original draft preparation, N.S. and A.P.; writing—review a editing, M.P.; visualization, M.P. and A.S.; supervision, M.P. All authors have read and agreed to published version of the manuscript.

**Funding:** This work was supported by the Ministry of Higher Education, Russian Federati scientific topics 0186-2022-0002 (registration FMEZ-2022-0016) and the joint Russian Foundation Basic Research and Belarus Foundation for Basic Research grant 20-52-04001 Bel_mol_a.

**Institutional Review Board Statement:** Not applicable.

**Informed Consent Statement:** Not applicable.

**Data Availability Statement:** The data obtained in this research will be available from the co sponding author, M.P., upon reasonable request.

**Conflicts of Interest:** The authors declare no conflict of interest.

## References

1. Blázquez-Castro, A.; García-Cabañes, A.; Carrascosa, M. Biological applications of ferroelectric materials. *Appl. Phys. Rev.* **2(** *5*, 041101. [CrossRef]
2. Irzaman; Siskandar, R.; Nabilah, N.; Aminullah; Yuliarto, B.; Haman, K.A.; Husin, A. Application of lithium tantalate (LiTa films as light sensor to monitor the light status in the Arduino Uno based energy-saving automatic light prototype and pass infrared sensor. *Ferroelectrics* **2018**, *524*, 44–55. [CrossRef]
3. Yang, X.; Lin, S.; Ma, D.; Long, S.; Zhu, Y.; Li, H.; Wang, B. Up-conversion luminescence of $LiTaO_3$:$Er^{3+}$ phosphors for opt thermometry. *Ceram. Intern.* **2020**, *46*, 1178–1182. [CrossRef]
4. Yao, S.; Han, H.; Jiang, S.; Xiang, B.; Chai, G.; Ruan, S. Design, simulation, and analysis of optical microring resonators in lithi tantalate on insulator. *Crystals* **2021**, *11*, 480. [CrossRef]
5. He, Y.; Wong, Y.P.; Liang, Q.; Wu, T.; Bao, J.; Hashimoto, K.Y. Transverse energy confinement and resonance suppression in S resonators using low-cut lithium tantalate. *Jap. J. Appl. Phys.* **2022**, *61*, SG1031. [CrossRef]
6. Wu, B.; Zhang, B.; Liu, W.; Lu, Q.; Wang, L.; Chen, F. Recoverable and rewritable waveguide beam splitters fabricated by tailo femtosecond laser writing of lithium tantalate crystal. *Opt. Laser Technol.* **2022**, *145*, 107500. [CrossRef]
7. Lyu, T.; Dorenbos, P.; Xiong, P.; Wei, Z. $LiTaO_3$:$Bi^{3+}$, $Tb^{3+}$, $Ga^{3+}$, $Ge^{4+}$: A smart perovskite with high charge carrier stor capacity for X-ray imaging, stress sensing, and non-real-time recording. *Adv. Funct. Mater.* **2022**, *32*, 2206024. [CrossRef]
8. Sidorov, N.V.; Palatnikov, M.N.; Biryukova, I.V.; Titov, R.A.; Makarova, O.V.; Masloboeva, S.M. *Monocrystals of Lithium Niobate Tantalate of Different Composition and Genesis*; RAN: Moscow, Russia, 2022; 288p. (In Russian)
9. Chuchumishev, D.; Gaydardzhiev, A.; Fiebig, T.; Buchvarov, I. Subnanosecond, mid-IR, 0.5 kHz periodically poled stoichiome $LiTaO_3$ optical parametric oscillator with over 1 W average power. *Opt. Lett.* **2013**, *38*, 3347–3349. [CrossRef]
10. Yin, M. Broadband quasi-phase-matching second-harmonic generation in 0.5 mol% MgO-doped periodically poled stoichiome $LiTaO_3$. *J. Korean Phys. Soc.* **2015**, *67*, 1750–1754. [CrossRef]
11. Chuchumishev, D.; Trifonov, A.; Oreshkov, B.; Xu, X.; Buchvarov, I. High-energy picosecond kHz optical parametric osci tor/amplifier tunable between 3 and 3.5 μm. *Appl. Phys. B* **2018**, *124*, 1–6. [CrossRef]
12. Prokhorov, A.M.; Kuz'minov, Y.S. *Physics and Chemistry of Crystalline Lithium Niobate*; Adam Hilger: New York, NY, USA, 1( p. 237.
13. Huband, S.; Keeble, D.S.; Zhang, N.; Glazer, A.M.; Bartasyte, A.; Thomas, P.A. Relationship between the structure and opti properties of lithium tantalate at the zero-birefringence point. *J. Appl. Phys.* **2017**, *121*, 024102. [CrossRef]
14. Sidorov, N.V.; Volk, T.R.; Mavrin, B.N.; Kalinnikov, V.T. *Lithium Niobate: Defects, Photorefraction, Vibrational Spectrum, Polarit* Nauka: Moscow, Russia, 2003; p. 255. (In Russian)
15. Fontana, M.D.; Bourson, P. Microstructure and defects probed by Raman spectroscopy in lithium niobate crystals and devi *Appl. Phys. Rev.* **2015**, *2*, 040602. [CrossRef]
16. Wu, X.-L.; Yany, F.; Zhangy, M.-S.; Jiangy, S.-S.; Fengy, D. The microstructural difference between proton-exchanged $LiNbO_3$ a $LiTaO_3$ crystals by Raman spectroscopy. *J. Phys. Condens. Matter* **1996**, *8*, 2073–2080. [CrossRef]
17. Glass, A.M. Optical spectra of $Cr^{3+}$ impurity ions in ferroelectric $LiNbO_3$ and $LiTaO_3$. *J. Chem. Phys.* **1969**, *50*, 1501–1! [CrossRef]
18. Tsuya, H. Optical damage in transition-metal-doped $LiTaO_3$. *J. Appl. Phys.* **1975**, *46*, 4323–4333. [CrossRef]
19. von der Linde, D.; Glass, A.M.; Rodgers, K.F. Optical storage using refractive index changes induced by two-step excitat *J. Appl. Phys.* **1976**, *47*, 217–220. [CrossRef]
20. Ryba-Romanowski, W.; Golab, S.; Pisarski, W.A.; Dominiak-Dzik, G.; Palatnikov, M.N.; Sidorov, N.V.; Kalinnikov, V.T. Influe of temperature on the optical properties of $LiTaO_3$:Cr. *Appl. Phys. Lett.* **1997**, *70*, 2505–2507. [CrossRef]
21. Sokolska, I.; Golab, S.; Ryba-Romanowski, W. Optical spectroscopy of doped $LiTaO_3$ crystals. *Spectrochim. Acta Part A Mol. Bio Spectrosc.* **1998**, *54*, 1685–1694. [CrossRef]

Sokolska, I.; Kuck, S. Optical characterization of $Cr^{3+}$ doped $LiTaO_3$ crystals relevant for laser application. *Spectrochim. Acta Part A Mol. Biomol. Spectrosc.* **1998**, *54*, 1695–1700. [CrossRef]
Golab, S.; Sokolska, I.; Dominiak-Dzik, G.; Palatnikov, M.N.; Sidorov, N.V.; Biryukova, I.; Kalinnikov, V.T.; Ryba-Romanowski, W. Optical absorption and luminescence of $LiTaO_3$:Cr and $LiTaO_3$:Cr,Nd crystals. *Proc. SPIE* **1999**, *3724*, 270–273. [CrossRef]
Alexandrovski, A.L.; Foulon, G.; Myers, L.E.; Route, R.K.; Fejer, M.M. UV and visible absorption in $LiTaO_3$. *Proc. SPIE* **1999**, *3610*, 44–51. [CrossRef]
Grinberg, M.; Sokolska, I.; Kuck, S.; Jaskolski, W. Temperature dependence of the luminescence decay of $Cr^{3+}$ ions in $LiTaO_3$: Confined potential model. *Phys. Rev. B* **1999**, *60*, 959–965. [CrossRef]
Shen, Y.; Bray, K.L.; Grinberg, M.; Barzowska, J.; Sokolska, I. Identification of multisite behavior in a broadly emitting transition-metal system using pressure. *Phys. Rev. B* **2000**, *61*, 14263–14266. [CrossRef]
Shen, Y.; Bray, K.L.; Grinberg, M.; Barzowska, J.; Sokolska, I. High pressure spectroscopy of chromium doped $LiTaO_3$ crystals. *High Press. Res.* **2000**, *18*, 125–130. [CrossRef]
Kaczmarek, S.; Swirkowicz, M.; Jablonski, R.; Lukasiewicz, T.; Kwasny, M. Growth and characterization of lithium tantalate single crystals doped with Ho, Tm, Nd, Yb, Pr and doped by diffusion with Cr and Cu. *J. Alloys Compd.* **2000**, *300–301*, 322–328. [CrossRef]
Grinberg, M.; Barzowska, J.; Shen, Y.; Bray, K. Crystal field model of the three principal $Cr^{3+}$ centers in $LiTaO_3$. *Radiat. Eff. Defects Solids* **2001**, *155*, 247–251. [CrossRef]
Grinberg, M.; Barzowska, J.; Bray, K.L.; Shen, Y.R. Inhomogeneous broadening of the dominant $Cr^{3+}$ sites in $LiTaO_3$ system. *J. Luminesc.* **2001**, *94–95*, 85–90. [CrossRef]
Grinberg, M.; Barzowska, J.; Shen, Y.R.; Bray, K.L. Inhomogeneous broadening of $Cr^{3+}$ luminescence in doped $LiTaO_3$. *Phys. Rev. B* **2001**, *63*, 214104. [CrossRef]
Grinberg, M. Spectroscopic characterisation of disordered materials doped with chromium. *Opt. Mater.* **2002**, *19*, 37–45. [CrossRef]
Grinberg, M.; Suchocki, A. Pressure-induced changes in the energetic structure of the $3d^3$ ions in solid matrices. *J. Luminesc.* **2007**, *125*, 97–103. [CrossRef]
Bagdasarov, K.S.; Batoev, V.B.; Uyukin, E.M. Transient isotropic photoinduced scattering of light in $LiTaO_3$:Cr. *Sov. J. Quant. Electron.* **1986**, *16*, 1295–1296. [CrossRef]
Batoev, V.B.; Uyukin, E.M. Photoinduced and Rayleigh light scattering in $LiTaO_3$:Cr. *Fiz. Tverd. Tela* **1988**, *30*, 1913–1915.
Burns, G.; O'kane, D.F. Optical and electron-spin-resonance spectra of $Yb^{3+}$, $Nd^{3+}$, and $Cr^{3+}$ in $LiNbO_3$ and $LiTaO_3$. *Phys. Rev.* **1968**, *167*, 314–319. [CrossRef]
Chen, C.Y.; Sweeney, K.L.; Halliburton, L.E. Reduction and radiation effects in lithium tantalate. *Phys. Status Solidi* **1984**, *81*, 253–257. [CrossRef]
Ahn, S.W.; Kim, J.S.; Choh, S.H.; Yeom, T.H. An induced $Cr^{3+}$ center in $\gamma$-irradiated $LiTaO_3$. *J. Kor. Phys. Soc.* **1994**, *27*, 535–537.
Ahn, S.W.; Rudowicz, C.; Choh, S.H.; Han, S.Y. EPR study of two $Cr^{3+}$ defect centers in $LiTaO_3$ single crystals. *J. Korean Phys. Soc.* **1997**, *30*, 99–102.
Yeom, T.H.; Choh, S.H. Magnetic resonance investigations of $LiNbO_3$ and $LiTaO_3$ single crystals. *J. Korean Phys. Soc.* **1998**, *32*, S672–S675.
Loyo-Menoyo, M.; Keeble, D.J.; Furukawa, Y.; Kitamura, K. Electron paramagnetic resonance of $Cr^{3+}$ in near-stoichiometric $LiTaO_3$. *J. Appl. Phys.* **2005**, *97*, 123905. [CrossRef]
Zhang, C.-X.; Kuang, X.-Y.; Mao, A.-J.; Wang, H. Optical spectrum, local lattice structure and EPR g factors for $Cr^{3+}$ impurity ions in $MgTiO_3$ and $LiTaO_3$. *Phys. B Condens. Matter* **2008**, *403*, 3114–3118. [CrossRef]
Palatnikov, M.N.; Sandler, V.A.; Sidorov, N.V.; Efremov, I.N.; Makarova, O.V. Methods for controlling the degree of unipolarity of large $LiNbO_3$ crystals. *Instrum. Exp. Tech.* **2020**, *63*, 383–387. [CrossRef]
Sanna, S.; Neufeld, S.; Rusing, M.; Berth, G.; Zrenner, A.; Schmidt, W.G. Raman scattering efficiency in $LiTaO_3$ and $LiNbO_3$ crystals. *Phys. Rev. B* **2015**, *91*, 224302. [CrossRef]
Shi, L.; Kong, Y.; Yan, W.; Liu, H.; Li, X.; Xie, X.; Zhao, D.; Sun, L.; Xu, J.; Sun, J.; et al. The composition dependence and new assignment of the Raman spectrum in lithium tantalate. *Solid State Commun.* **2005**, *135*, 251–256. [CrossRef]
Sidorov, N.V.; Palatnikov, M.N. Raman spectra of lithium niobate crystals heavily doped with zinc and magnesium. *Opt. Spectrosc.* **2016**, *121*, 842–850. [CrossRef]
Solin, S.A.; Ramdas, A.K. Raman spectrum of diamond. *Phys. Rev. B* **1970**, *1*, 1687–1698. [CrossRef]
Sushchinsky, M.M.; Gorelik, V.S.; Maximov, O.P. Higher-order Raman spectra of GaP. *J. Raman Spectrosc.* **1978**, *7*, 26–30. [CrossRef]
Narayanan, P.S. Raman spectrum of cesium bromide. *Proc. Indian Acad. Sci. A* **1955**, *42*, 303–308. [CrossRef]
Sidorov, N.; Palatnikov, M.; Pyatyshev, A.; Sverbil, P. Second-order Raman scattering in ferroelectric ceramic solid solutions $LiNb_xTa_{1-x}O_3$. *Crystals* **2022**, *12*, 456. [CrossRef]
Sidorov, N.; Palatnikov, M.; Pyatyshev, A.; Skrabatun, A. Investigation of the structural perfection of a $LiNbO_3$:$Gd^{3+}$(0.003):$Mg^{2+}$(0.65 wt.%) double-doped single crystal using the Raman spectra excited by laser lines in the visible (532 nm) and near-IR (785 nm) regions. *Appl. Sci.* **2023**, *13*, 2348. [CrossRef]
Gorelik, V.S.; Abdurakhmonov, S.D. Overtone Raman scattering in lithium niobate single crystals doped with terbium. *Crystallogr. Rep.* **2022**, *67*, 252–255. [CrossRef]

53. Sidorov, N.; Palatnikov, M.; Pyatyshev, A. Raman scattering in a double-doped single crystal $LiTaO_3$:Cr(0.2):Nd(0.45 wt *Photonics* **2022**, *9*, 712. [CrossRef]
54. Ruvalds, J.; Zawadowski, A. Two-phonon resonances and hybridization of the resonance with single-phonon states. *Phys. Rev* **1970**, *2*, 1172–1175. [CrossRef]
55. Zawadowski, A.; Ruvalds, J. Indirect coupling and antiresonance of two optic phonons. *Phys. Rev. Lett.* **1970**, *24*, 1111–11 [CrossRef]
56. Ruvalds, J.; Zawadowski, A. Resonances of two phonons from different dispersion branches. *Solid State Commun.* **1971**, *9*, 129– [CrossRef]

photonics

MDPI

ticle

# he Influence of Gadolinium Oxide Nanoparticles oncentration on the Chemical and Physical Processes Intensity uring Laser-Induced Breakdown of Aqueous Solutions

eksander V. Simakin, Ilya V. Baimler, Alexey S. Baryshev, Anastasiya O. Dikovskaya and Sergey V. Gudkov *

Prokhorov General Physics Institute of the Russian Academy of Sciences, 38 Vavilova Street, 119991 Moscow, Russia; avsimakin@gmail.com (A.V.S.); ilyabaymler@yandex.ru (I.V.B.); aleksej.baryshev@gmail.com (A.S.B.); dikovskayaao@gmail.com (A.O.D.)

* Correspondence: s_makariy@rambler.ru

**Abstract:** The paper investigates the physicochemical processes that occur during laser-induced breakdown in colloids of gadolinium oxide nanoparticles of different concentrations. A monotonic change in the number of optical breakdowns, the average distance between the nearest breakdowns in the track of a single laser pulse, the average plasma size of a single optical breakdown, the integral luminosity of an optical breakdown plasma flash, the intensity of acoustic signals, and the rate of formation of dissociation products—$O_2$, $H_2$, $OH^\bullet$, and $H_2O_2$—is demonstrated. It is shown that the rate of formation of chemical products of the decomposition of $H_2O$ molecules under the action of breakdown when using nanoparticles of rare earth metals, in particular, gadolinium oxide, is the highest compared to other materials. Based on one laser pulse, the rates of formation of chemical products formed during the dissociation of water during laser-induced breakdown of a colloid of gadolinium oxide nanoparticles are 13.13 nmol/pulse for $H_2$, 5.41 nmol/pulse for $O_2$, and 6.98 nmol/pulse for hydrogen peroxide.

**Keywords:** laser radiation; optical breakdown; nanoparticles; gadolinium oxide; rare-earth elements; breakdown plasma; acoustic signals; hydroxyl radicals; hydrogen peroxide



ation: Simakin, A.V.; Baimler, I.V.; yshev, A.S.; Dikovskaya, A.O.; lkov, S.V. The Influence of lolinium Oxide Nanoparticles centration on the Chemical and sical Processes Intensity during er-Induced Breakdown of eous Solutions. *Photonics* **2023**, 784. https://doi.org/10.3390/ tonics10070784

eived: 6 April 2023
ised: 26 June 2023
epted: 4 July 2023
lished: 5 July 2023



## 1. Introduction

The optical breakdown of various media began to be investigated almost from the time of the advent of high-power pulsed lasers [1]. It is known that the efficiency of laser-induced breakdown of condensed media, for example, organic solvents or water, increases significantly with the addition of small amounts of nanoparticles, especially metals [2]. Nanoparticles serve as seeds on which the breakdown occurs, and the nanoparticles themselves during optical breakdown also undergo chemical modification [3]. It has been shown that the dissociation of molecules in the medium in which the breakdown arise, to a greater extent occurs on NPs rather than on a metal target [4]. It is known that, at high intensities of laser radiation, plasma is formed on the surface of nanoparticles in liquids [5]. The plasma, in turn, initiates the generation of radiation in a wide spectral range from ultraviolet to infrared radiation [6]. Under the action of plasma, an intense boiling up of the liquid occurs [7], the propagation of intense acoustic oscillations [8], an intensive chemical transformation of the substance of the nanoparticles occurs [9], the size of the nanoparticles changes [10], and decomposition of the molecules of the liquid surrounding the particle is observed [11]. In the case of metallic NPs, nearly all the electromagnetic field energy of laser radiation is transferred to the free electrons in the crystal lattice of the metal; therefore, the physico-chemical effects taking place on the NPs of different metals during optical breakdown may vary considerably depending on the material of the NPs on which the optical breakdown takes place [12].

Most often, colloidal solutions of NPs are made from water. Water includes only and $O_2$ isotopes, so only oxygen–hydrogen plasma is produced by optical breakdo and further ionization. The most important short-lived products of water ionizat are •OH (hydroxyl radical), •H (hydrogen atom), and solvated electrons [13]. There only three sustainable products: $H_2$ (hydrogen molecules), $O_2$ (oxygen molecules), a $H_2O_2$ (hydrogen peroxide); occasionally additional quasi-stable ozone is released [14]. comparison, the laser optical breakdown of any organic solvent, such as ethyl alcohol, t consists of C, $H_2$ and $O_2$, already produces several tens of products in low yields. A during the optical breakdown of organic solvents, the presence of solvated electrons observed, the characteristic formation times of which are several tens of picoseconds [1

The process of laser fragmentation consists in irradiating a previously prepared loidal solution of nanoparticles with laser pulses. This technique is commonly perform to reduce the particle extent and narrow the particle size allocation [16]. It is believ that fragmentation is usually based on a photothermal mechanism or the Coulomb g model. The intensity of laser fragmentation of nanoparticles is inextricably linked to intensity of physicochemical effect taking place during optical breakdown of colloi nanoparticle solutions [17]. The fragmentation processes of nanoparticles are evalua according to their intensity. The processes observed during optical breakdown are wid known for commonly used NPs (Fe, Au, Cu, and Ni) [18–21]. Judging by the data presen in a single review, information on the processes occurring during optical breakdown nanoparticles of rare earth metals is extremely scarce [22]. Rare earth atoms have electr in the d and f orbitals, which is why the ionization threshold of rare earth metals is lov than that of previously used materials. From this it follows that the chance of devel ing an electron avalanche induced by laser radiation and the formation of plasma o nanoparticle of a rare earth metal will be higher. To test this assumption, in this wc we studied the physicochemical processes that occur during laser-induced breakdown colloidal solutions of gadolinium oxide NPs at various concentrations. In laser chemis the problem of transferring the energy of a laser pulse into a liquid is quite acute. Usua metal nanoparticles are used to solve this problem. It is known that the efficiency of la pulse energy transfer into a liquid depends, among other things, on the material of nanoparticles. With the use of gadolinium oxide nanoparticles, a record, to date, ene transfer coefficient from a laser pulse to a liquid has been registered.

## 2. Materials and Methods

### *2.1. Synthesis of Gadolinium Oxide Nanoparticles*

Gadolinium oxide NPs was synthesized using the method of laser ablation in a liq A polished solid target (Gd 99.99%) was placed at the bottom of a glass cuvette unde thin layer of working liquid 2–3 mm thick and irradiated with laser radiation (90 J/$cm^2$ wavelength of 1.064 μm, pulse duration 130 ns, frequency 10 kHz, beam diameter 100 for 15 min. MQ standard water was used as the working liquid. The structure of received particles was studied using a Libra 200 FE HR transmission electron microsc (Carl Zeiss, Jena, Germany). The size of the NPs was determined using a DC24000 analyt centrifuge (CPS Instruments, Oosterhout, Netherlands).

### *2.2. Experimental Setup*

The layout of the setup used to study the physical and chemical processes obser during optical breakdown is described in detail in [21]. Laser radiation ($\lambda$ = 1064 r $\tau$ = 10 ns, $\nu$ = 10 kHz, $\varepsilon$ = 2 mJ, beam diameter 40 μm) was delivered into a glass through a transparent bottom using a system of mirrors, and radiation was focused us an F-theta objective. The delivery of laser radiation in this way makes it possible to av defocusing of laser radiation by gas bubbles floating upwards. The cuvette was filled w an aqueous colloidal solution of gadolinium oxide nanoparticles in an amount of 10 mL a predetermined concentration. The concentration of particles was determined using d centrifuge. The change in the gadolinium oxide concentration in the irradiated soluti occurred when a various amount of a colloidal solution of nanoparticles with a known concentration (several microliters) was added to the volume of MQ water in a cuvette (approximately 10 mL). In order for the breakdown to occur each time in an unperturbed medium, the laser beam was moved along a straight line 1.5 cm long inside the cuvette using a galvano-mechanical scanner. The scanning speed was 1400 mm/s. The distance between two successive pulses at this scanning speed was about 140 μm, and 70 pulses were observed in one pass along the scan line.

Photographing of the plasma formed during optical breakdown was carried out with a Canon EOS 450D camera (macro photography, gray scale, shutter speed 100 ms, spatial resolution 8 μm/pxl). One laser pulse can have from zero to several tens of breakdowns located at different distances from each other, having different luminous intensity, cross-sectional area, amplitude, and so on. All the listed parameters were calculated by processing photographs using an automated software package, LaserImage [23].

The acoustic spectrum of the laser optical breakdown was recorded using a film piezoelectric sensor built into the cell. The sensor plane was located in parallel with the laser scan line. The signal from the piezoelectric transducer was fed to the input of a digital oscilloscope GDS-72204E (GW Instek, Xinbei, Taiwan). The oscilloscope was triggered using a signal from a pin diode that registered a laser pulse. A specially developed LaserCav program was used for data analysis. Measurements of ultrasonic signals and the acoustic processes that occur in the system have already been described in detail [24].

To register molecular hydrogen and oxygen formed during irradiation, portable hydrogen and oxygen analyzers AVP-02 and AKPM-1-02 with amperometric membrane sensors were used. The sensors of the analyzers were built into the cuvette in such a way that the concentrations of oxygen and hydrogen were detected in water online. The procedure for registering gases was described in detail earlier [25].

### 2.3. Registration of Hydrogen Peroxide and Hydroxyl Radicals

Chemical examination of the products formed during laser optical breakdown was carried out outside the experimental cuvette. Sampling was carried out using a rapid sampling system. For the quantitative determination of $H_2O_2$, a highly sensitive enhanced chemiluminescence method in the system of luminol-p-iodophenol-horseradish peroxidase was used. Luminescence detection was performed using a Biotoks-7AM chemiluminometer (Ekon, Moscow, Russia). The concentration of the formed $H_2O_2$ was determined via calibration curves, in which the chemiluminescence intensity of samples containing added hydrogen peroxide of known concentration was measured. The initial concentration of $H_2O_2$ used for calibration was determined spectrophotometrically at a wavelength of 240 nm with a molar absorption coefficient of 43.6 ($M^{-1} \times cm^{-1}$). Samples (3 mL) were placed in polypropylene vials (Beckman, Brea, CA, USA) and 0.15 mL of "counting solution" containing: 1 cM Tris-HCl buffer pH 8.5, 50 μM p-iodophenol, 50 μM luminol, and 10 nM horseradish peroxidase. These were added when determining nanomolar concentrations of $H_2O_2$. The "counting solution" was prepared immediately before the measurement. The sensitivity of the method makes it possible to determine $H_2O_2$ at a concentration of 0.1 nM [26].

The production of OH-radicals was determined using the reaction with coumarin-3-carboxylic acid (CCA), the hydroxylation product of which, 7-hydroxycoumarin-3-carboxylic acid (7-OH-CCA), is a convenient fluorescent probe for determining the formation of these radicals. A solution of CCA (0.5 mM) was prepared in phosphate-buffered saline (Sigma, St. Louis, MO, USA) pH 7.4. NPs were added to the solution just prior to laser exposure. The fluorescence of the reaction product of CCA with the hydroxyl radical, 7-OH-CCA, was measured on an FP-8300 spectrofluorimeter (JASCO, Tokyo, Japan) with $\lambda_{ex} = 400$ nm, $\lambda_{em} = 450$ nm. Calibration was performed using commercial 7-OH- CCA (Sigma, St. Louis, MO, USA) [27].

## 3. Results

### *3.1. Morphology of Gadolinium Oxide Nanoparticles*

Figure 1 shows the size distribution and TEM images of the obtained gadolinium ox
nanoparticles. The colloidal solution was analyzed using a disc centrifuge to investigate
size distribution of the nanoparticles. The size distribution of the number of nanopartic
shows that the average particle size in the colloid is about 10 nm (Figure 1A). The fig
also shows the presence of particles larger than 30 nm in the colloid. The zeta potent
of gadolinium oxide nanoparticles in the resulting colloid after ablation was −33 r
Figure 1B shows how the distribution of the resulting particles changes during long-te
exposure of the colloid to laser radiation. It is shown that the maximum distribution of
number of nanoparticles during irradiation for up to 30 min shifts towards smaller si
from 10 nm to 8 nm.

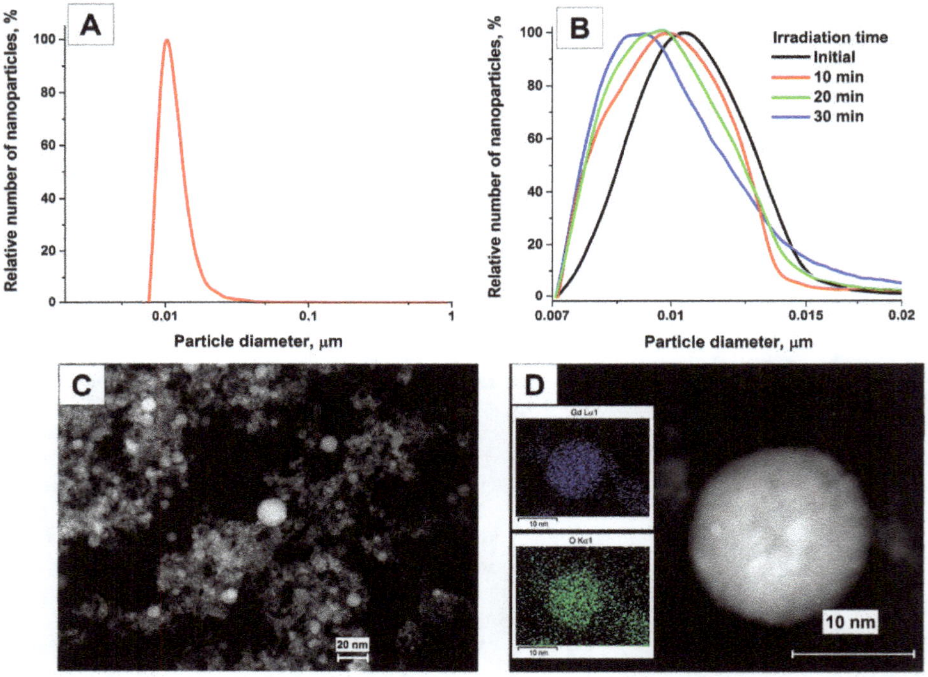


**Figure 1.** Morphology of obtained gadolinium oxide nanoparticles: (**A**) Number of gadolinium ox nanoparticles distribution depending on size; (**B**) Evolution of size distribution of gadolinium ox nanoparticles during laser fragmentation; (**C**) TEM image of gadolinium oxide nanoparticles, sc mark 20 nm; (**D**) TEM image of an individual gadolinium oxide nanoparticle, scale mark, 10 Insert: EDX data for gadolinium oxide NP with Gd and O distributions.

In Figure 1C,D TEM images of the obtained gadolinium oxide nanoparticles are sho
As can be seen from the figures, the synthesized gadolinium oxide nanoparticles hav
spherical shape; see Figure 1C. It has also been shown that a fraction of larger partic
which are not aggregated, is indeed present in the colloidal solution. However, the ove
proportion of larger particles in the colloid is insignificant, and the particle size distribut
shows that more than 99% of particles are between 9 and 22 nm in size. In addition, the in
to Figure 1D shows the results of an EDX analysis of a sample of the resulting nanopartic
It is shown that the nanoparticles mainly consist of metallic gadolinium, and there is
oxide layer around the particles.

### 3.2. Analysis of Breakdown Plasma Flashes

The dependencies between the optical parameters of the laser breakdown plasma and the concentration of gadolinium oxide nanoparticles in the colloidal solution were investigated. (Figure 2). Irradiation of a colloidal solution of gadolinium oxide nanoparticles was carried out for 2 min; therefore, it can be argued that the distribution of nanoparticles, their number, and average sizes did not change significantly upon irradiation. The concentration of nanoparticles affects the recorded plasma images. Figure 2A shows two characteristic images of plasma flashes obtained by irradiating colloidal solutions with nanoparticle concentrations of $10^9$ NPs/mL and $10^{11}$ NPs/mL.

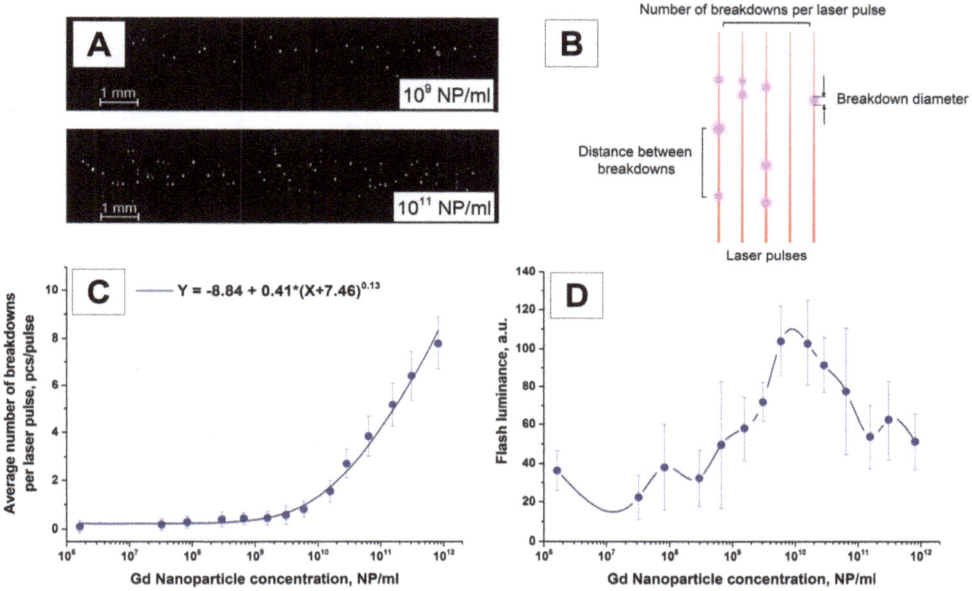


**Figure 2.** Variation of the spatial distribution of breakdowns as a function of gadolinium oxide nanoparticle concentration: (**A**) Typical images of a breakdown plasma obtained by irradiating a colloidal solution of gadolinium oxide nanoparticles with a concentration of $10^9$ NP/mL (top) and with a concentration of $10^{11}$ NP/mL (bottom) (**B**) Schematic representation of the measured parameters in the track of plasma flashes (**C**) Dependence of average count of breakdowns per laser pulse on nanoparticle concentration (**D**) Dependence of total plasma luminosity on the concentration of nanoparticles in solution (a spline was used to connect the data). The values of the measured parameters are averaged over 5 measurements; the errors correspond to the standard deviation of the mean value (SEM).

Unlike gases, in aqueous colloids of metal nanoparticles, one laser pulse, depending on the conditions, can cause from one to several thousand breakdowns. These breakthroughs are localized in space and are not connected with each other. Plasma breakdowns caused by one laser pulse and located along the line of propagation of laser radiation are usually called a track. The length between the nearest breakdowns caused by a single laser pulse, the size of the breakdowns, the number of breakdowns, and the overall brightness of the flash may significantly affect the whole process. The listed parameters measured in the analysis of images of plasma flashes are schematically shown in Figure 2B.

The change in the count of breakdowns per laser pulse as a function of the concentration of gadolinium oxide nanoparticles in a colloidal solution is shown in Figure 2C. In the range of gadolinium oxide nanoparticle concentration from $10^6$ to $10^9$ NPs/mL, the average count of breakdowns does not exceed 1 piece/pulse. At concentrations of

$10^{10}$ NPs/mL, on average, there is 1 breakdown per each laser pulse. With an incre in nanoparticle concentration above $10^{10}$ NPs/mL, the count of breakdowns begins increase monotonically and, at a NP concentration of $10^{12}$ NPs/mL, reaches approximat 7–8 pcs/pulse.

The average size per breakdown is approximately 30–37 μm over a nanoparti concentration range of $10^6$ NPs/mL to $10^9$ NPs/mL. At higher concentrations of nanop ticles, the diameter of a single breakdown starts to decrease and reaches 28–30 μm in concentration range from $10^{10}$ NPs/mL to $10^{12}$ NPs/mL.

The average distance between single breakdowns in one laser track decreases mo tonically. At concentrations from $10^6$ NPs/mL to $10^9$ NPs/mL, the distance betw breakdowns in one laser track ranges from 100 to 150 μm. At higher concentrations nanoparticles, individual breakdowns are located closer to each other. At nanoparti concentrations of $10^{11}$–$10^{12}$ NPs/mL, the distance between them decreases to 70–80 μn

The integral luminosity of the plasma over a selected range of nanoparticle conc trations varies non-monotonically (Figure 2D). With an increase in the concentration nanoparticles from $10^6$ NPs/mL to $10^{10}$ NPs/mL, the integral brightness gradually creases and reaches a maximum of 110 arbitrary units at a nanoparticle concentration $10^{10}$ NPs/mL. With a subsequent increase in the concentration of nanoparticles, the ir grated luminosity of the flash of the breakdown plasma begins to decrease monotonica

### *3.3. Analysis of Acoustic Signals during Optical Breakdown*

Changes of intensity of ultrasonic vibrations induced by optical breakdown depend on concentration of gadolinium oxide nanoparticles in colloidal solution are shown in Figu It is shown that the amplitude of ultrasonic vibrations increases, reaches a maximum and th decreases with a gradual increase in the concentration of gadolinium oxide particles in irradiated colloid, Figure 3. In the concentration range from $10^6$ NPs/mL to $10^9$ NPs/r the amplitude of acoustic vibrations does not change significantly. Once the nanoparti concentration is further increased from $10^9$ NPs/mL to about $5 \times 10^{10}$ NPs/mL, the m intense acoustic vibrations are observed in the medium. Beginning with a concentration $10^{11}$ NPs/mL, the intensity of acoustic vibrations decreases.

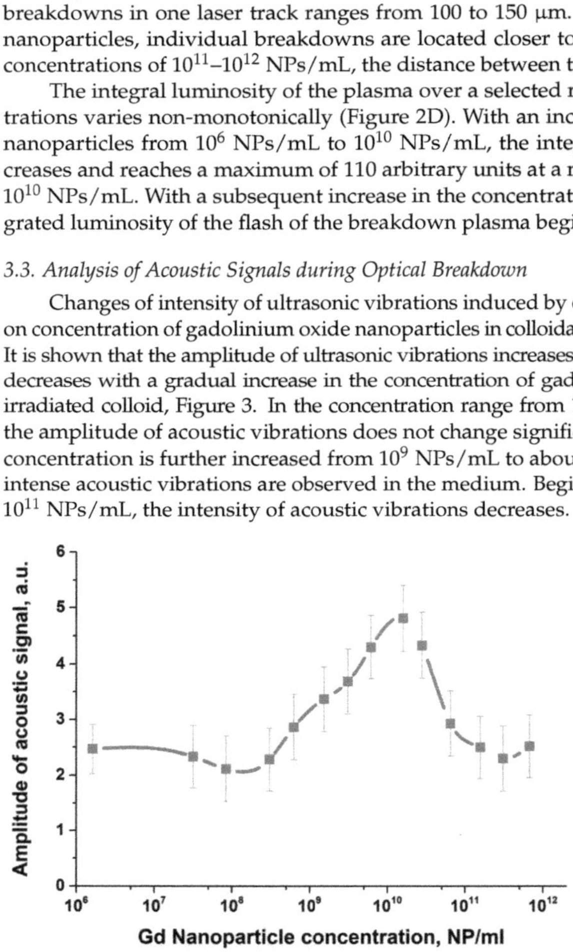


**Figure 3.** Influence of gadolinium oxide nanoparticle concentration in solution on the amplitud acoustic signals formed upon optical breakdown of colloids. The values presented in the value gr are averaged over 10 measurements. A spline was used to connect the data points. The errors den the standard deviation of the mean (SEM).

### *3.4. Formation of Molecular Oxygen and Molecular Hydrogen*

The dynamics of molecular $H_2$ and molecular $O_2$ generation by laser-induced opti breakdown as a function of gadolinium oxide nanoparticle concentration is shown

Figure 4. The insets to the graphs show the typical dependence of the concentration of molecular hydrogen and oxygen on the radiation time. The generation rate of these products was estimated from the slope of the straight line approximating the experimental points of the graphs shown in the insets in the time range from 10 to 15 min. It is shown that the maximum rate of hydrogen and oxygen formation is observed at a nanoparticle concentration of approximately $2 \times 10^{10}$ NPs/mL (Figure 4A,B). The maximum value of the hydrogen generation rate is about 13.1 nmol/pulse. The maximum molecular $O_2$ production rate is 5.1 nmol/pulse.

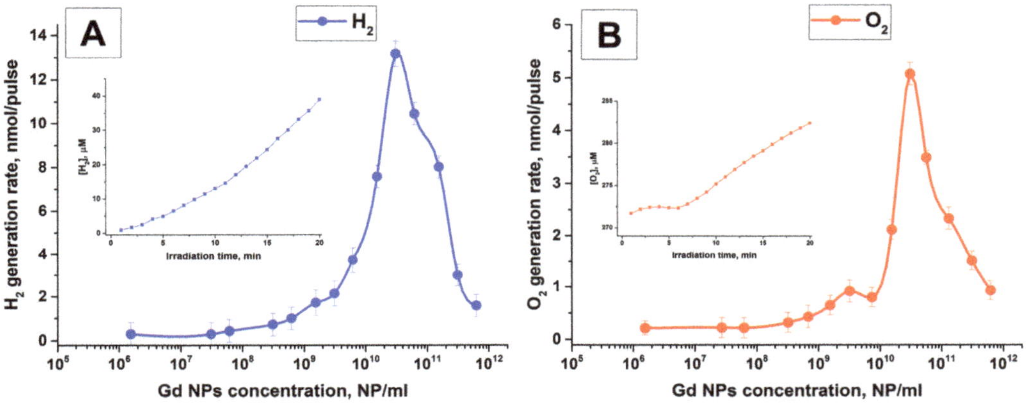


**Figure 4.** Formation of molecular $H_2$ and molecular $O_2$ by laser breakdown of aqueous solutions of gadolinium oxide nanoparticles; (**A**) Molecular hydrogen production rate versus gadolinium oxide particle concentration, inset: Dependence of $H_2$ concentration on the time of laser irradiation of a colloid (n = $10^{10}$ NP/mL) (**B**) Molecular oxygen production rate versus gadolinium oxide particle concentration, inset: Dependence of molecular $O_2$ concentration on the time of laser irradiation of a colloid (n = $10^{10}$ NP/mL). A spline was used to connect the data points. The values of the measured parameters are averaged over 5 measurements; the errors correspond to the standard deviation of the mean.

### *3.5. Formation of Hydrogen Peroxide and Hydroxyl Radicals*

The effect of gadolinium oxide nanoparticles concentration on the formation rate of •OH and $H_2O_2$ is shown in Figure 5A,B. The dependences shown in the inset to Figure 5 show the characteristic dependence of the concentration of hydroxyl radicals and hydrogen peroxide on the exposure time. The generation rate of •OH and $H_2O_2$ of these products was estimated from the slope of the straight line approximating the experimental points of the graphs shown in the insets. It has been shown that the maximum rate of $H_2O_2$ and OH formation is observed at a concentration of 3–5 $\times$ $10^{10}$ NPs/mL. The maximum hydroxyl radical generation rate is observed at a concentration of $3 \times 10^{10}$ NPs/mL and is 1.85 nmol/pulse. The highest value of the generation rate for hydrogen peroxide is observed at a concentration of $5 \times 10^{10}$ NPs/mL and is 6.98 nmol/pulse. Further increasing the concentration leads to a rapid decrease in the rate of generation of both hydrogen peroxide and hydroxyl radicals. Significant differences in the rates of production of hydroxyl radicals and hydrogen peroxide are most likely associated with the measurement technique. Since the probe molecules used to detect radicals can also decompose under the action of breakdown plasma and the process of trapping radical products occurs in a localized space around the breakdown for a short period of time, the measured rate of formation of hydroxyl radicals decreases due to these factors.

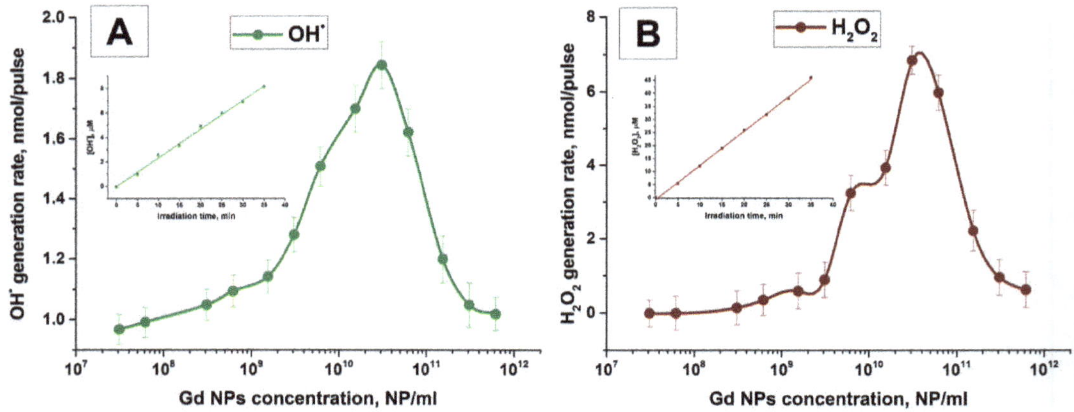


**Figure 5.** Hydroxyl radicals and hydrogen peroxide generation by laser-induced breakdowr gadolinium oxide nanoparticles in aqueous solutions; (**A**) The rate of formation of hydroxyl radi as a function of the concentration of gadolinium oxide particles in the irradiated solution, in dependence of •OH concentration on the time of laser irradiation of a colloid(n = $10^{10}$ NP/r (**B**) The rate of formation of hydrogen peroxide as a function of the concentration of gadolini oxide particles, inset: dependence of $H_2O_2$ concentration on the time of laser irradiation of a coll (n = $10^{10}$ NP/mL). A spline was used to connect the data points. The values of the measu parameters are averaged over 5 measurements, the errors correspond to the standard deviation of mean (SEM).

## 4. Discussion

The paper shows that the scenarios of the interaction of a colloidal solution with la radiation depend significantly on the concentration of gadolinium oxide nanopartic in a colloid (Figures 2–5). It has been shown that, up to a nanoparticle concentration $10^{10}$ NPs/mL, a slow increase in the breakdown probability is observed. At concentratic above $10^{10}$ NPs/mL, a rapid increase in the breakdown probability occurs, and the numl of breakdowns induced by a single laser pulse increase (Figure 2). It is important to n that the plasma luminosity intensity correlates with the behavior of acoustic signals w a change in the concentration of nanoparticles in a colloid ($R^2$ = 0.95) (Figure 3). T intensity of plasma glow gradually increases with increasing concentration and reacl maximum values at concentrations of $10^{10}$–$10^{11}$ NPs/mL. At high concentrations, the gl of plasma induced by optical breakdown decreases significantly. It can be assumed t this is due to a decrease in the laser radiation intensity in the caustic region, which becon insufficient to induce optical breakdown. It is assumed that the decrease in the intensity laser radiation is associated with intense scattering of radiation by nanoparticles in fron the focal region [28], as well as with thermal defocusing inside the focal waist [29].

It is known that the ionization of pure water produces three stable products (molecu oxygen, molecular hydrogen, and hydrogen peroxide) and about a dozen products w lifetimes of more than 1 ns (hydroxyl radical, hydroperoxide radical, superoxide ani radical, ozone, etc.) [30]. The rates of formation of all three stable products of the produ were established (Figures 4 and 5), and the generation of the hydroxyl radical was also sho (Figure 5). Using the average values of the rates of formation of stable products near maxima obtained using following parameters of laser radiation ($\lambda$ = 1064 nm, pulse durati $\tau$ = 10 ns, frequency $\nu$ = 10 kHz, average pulse energy $\varepsilon$ = 2 mJ) and nanoparticle char teristics (concentration n = $10^{10}$ NPs/mL, particle sizes d = 10 nm)—$H_2$~11.25 nmol/pu $O_2$~3 nmol/pulse, $H_2O_2$~6 nmol/pulse, it can be argued that, in the case of optical bre down of an aqueous solution on gadolinium oxide nanoparticles, the general water decc position equation is the following: $6H_2O \rightarrow 4H_2 + O_2 + 2H_2O_2$. It has been establish that with a change in the concentration of nanoparticles, the stoichiometry of the equation changes (Figures 4 and 5). It is shown that the characteristic "bell-shaped" curves have different slopes. This indicates that the breakdown of water molecules induced by optical breakdown occurs due to two or more distinct processes. It is known that one of the main processes in optical breakdown is the superexcitation of molecules. Molecules in this state have an excess energy that exceeds the ionization potential [31]. Usually, a molecule goes into a superexcited state due to the simultaneous transition to higher energy levels of several electrons. In this case, energy concentrations of all electrons in the excited state in one can be observed. It usually leads to detachment of an electron from an atom or molecule [32]. Competing with such "autoionization" are processes of internal rearrangement in molecules (dissociation) [33]. For clarity, the process of ionization of water molecules is usually represented as $H_2O \rightarrow H_2O^+ + \bar{e}$, the process of dissociation $H_2O \rightarrow H^\bullet + {}^\bullet OH$ [34]. Obviously, two charged particles are formed during ionization. The cation of a water molecule usually decomposes by the reaction $H_2O^+ \rightarrow H^+ + {}^\bullet OH$. The second particle, the resulting electron, can react with $H_2O$ (products $H^\bullet$ and ${}^-OH$), with $O_2$ (product $O_2^{\bullet -}$), and with $H^+$ (product $H^\bullet$). Also, an electron can fall into a potential energy well, which leads to the formation of a so-called hydrated electron. During dissociation, atomic hydrogen ($H^\bullet$) and hydroxyl radical (${}^\bullet OH$) are formed. These radicals can recombine with each other, resulting in the formation of molecular hydrogen $H_2$ ($H^\bullet$ recombination), hydrogen peroxide $H_2O_2$ ($HO^\bullet$ recombination), and water ($H^\bullet$ and $HO^\bullet$ recombination) [35]. With this scheme of water molecule decomposition, molecular oxygen is formed in the dismutation of hydroperoxide radicals ($HO_2^\bullet$), which are formed during the protonation of superoxide anion radicals ($O_2^{\bullet -}$) [36]. Also, molecular oxygen can be generated during the decomposition of hydrogen peroxide [37]. Thus, the primary products of the laser-induced decomposition of water are $H_2$ and $H_2O_2$, while $O_2$ is of secondary origin.

In a number of processes, it is very important to quickly evaluate the efficiency of laser fragmentation of nanoparticles [38]. Undoubtedly, to evaluate the efficiency of fragmentation, the most obvious, quick, and simple indicators from the point of view of experimental technique are the characteristics of the plasma luminosity and ultrasound. In most studies, however, the plasma luminescence intensity and the intensity of sound vibrations induced by the optical breakdown are expressed in relative units, which makes it impossible to make any quantitative comparisons of efficiency. At the same time, the concentrations of gases and hydrogen peroxide in most articles are expressed in terms of the number of molecules in the volume. We compared the efficiency of generating gases and hydrogen peroxide using gadolinium oxide nanoparticles and other nanoparticles with similar laser radiation characteristics. In this study, the maximum rate of molecular hydrogen generation of 13.1 nmol/pulse was achieved. In a study with optical breakdown of carbon nanoparticles (d = 5 μm), it was possible to achieve a molecular hydrogen generation rate of 0.7 nmol/pulse using nanosecond laser pulses at a wavelength of 532 nm and a fluence of 140 mJ/cm$^2$ [39], and it was found that the gas generation rate can change by more than a factor of two depending on the laser radiation wavelength. This is quite close to the indices registered for optical breakdown in aqueous solutions in the absence of nanoparticles using femtosecond pulses [40]. With laser radiation parameters ($\lambda$ = 1064 nm, pulse duration $\tau$ = 10 ns, frequency $\nu$ = 8–10 kHz, average pulse energy $\varepsilon$ = 1.25–2 mJ) and nanoparticles characteristics close to those of this work (concentration n = $10^{10}$ NPs/mL, particle sizes d = 10–20 nm), it was shown that, when using gold nanoparticles, it is possible to achieve a molecular hydrogen generation rate of the order of 6.75 nmol/pulse [41] and 9 nmol/pulse [42]. Under similar experimental conditions, using nanoparticles of various materials with sizes d = 20–30 nm, it was found that, upon irradiation of the most effective concentration of nanoparticles (n = $10^{10}$ NPs/mL), the rate of formation of molecular $H_2$ is 7.8 nmol/pulse for Zr NPs, Mo—8.4 nmol/pulse, Fe—10.8 nmol/pulse, Ni—12 nmol/pulse [43]. A similar picture is observed in the breakdown-induced generation of molecular $O_2$ and $H_2O_2$. In general, comparing the results of works describing the breakdown on nanoparticles of rare-earth metals ([23] a this work) and other particles [35], we can conclude that the rate at which decompositi products are generated in water is higher with the first type of nanoparticles. The maxim rates of generation of chemical products per laser pulse during irradiation of collo containing various types of nanoparticles are presented in Table 1.

**Table 1.** Maximum values of the rates of formation of chemical products during optical breakdo of colloids.

| NPs Material | $H_2$, nmol/Pulse | $O_2$, nmol/Pulse | •OH, nmol/Pulse | $H_2O_2$, nmol/Pulse |
|---|---|---|---|---|
| Au [35] | 6.75 | 2.11 | 0.38 | 4.35 |
| Tb [23] | 7.76 | 2.32 | 1.52 | 4.13 |
| Gadolinium oxide | 13.13 | 5.41 | 1.87 | 6.98 |

From the data given in Table 1, it can be seen that, under approximately equal cor tions, the rate of formation of chemical products recorded during the optical decomposit of aqueous colloids of nanoparticles is the highest for nanoparticles of rare earth met In particular, among the rare-earth materials used in the experiments, gadolinium oxi nanoparticles present in the irradiated colloid lead to the highest production of wa decomposition products. Thus, it has been shown that the decomposition of water duri laser-induced breakdown of colloids with gadolinium oxide NPs occurs more efficien than on nanoparticles made of other materials.

## 5. Conclusions

The effect of the influence of the concentration of gadolinium oxide nanoparticles the intensity of the processes occurring during the optical breakdown of aqueous colloi solutions has been studied. It has been shown that the intensity of acoustic signals, plas glow, and the rate of generation of molecular hydrogen and oxygen, hydroxyl radic and hydrogen peroxide increase with an increase in the concentration of gadolinium ox nanoparticles in the colloid and reach a maximum value at a concentration of 1010 NPs/ With a subsequent increase in the concentration of nanoparticles, a decrease in the intens of the physicochemical processes accompanying the breakdown is observed, which mainly associated with a decrease in the energy density of laser radiation due to scatteri and screening of radiation by nanoparticles and optical breakdown plasma. The dec position of water during optical breakdown on individual gadolinium oxide nanopartic occurs more efficiently than on nanoparticles from other materials. It is shown that, per laser pulse formation rates of water splitting products are following: 13.13 nmol/pulse for 5.41 nmol/pulse for $O_2$, and 1.87 nmol/pulse for •OH and 6.98 nmol/pulse for $H_2O_2$.

**Author Contributions:** A.V.S. Conceptualization, Methodology, Investigation, Writing—Review Editing; I.V.B. Investigation, Visualization, Writing—Original Draft, Writing—Review and Editi A.S.B. Investigation, Writing—Review and Editing; A.O.D. Investigation, Writing—Review Editing; S.V.G. Supervision, Conceptualization, Writing—Review and Editing. All authors have r and agreed to the published version of the manuscript.

**Funding:** This work was supported by a grant No. 22-22-00602 from the Russian Science Foundat

**Institutional Review Board Statement:** This study did not require ethical approval.

**Informed Consent Statement:** Not applicable.

**Data Availability Statement:** The data presented in this study are available on request from the corresponding author.

**Conflicts of Interest:** The authors declare no conflict of interest.

## ferences


Hecht, J. A Short History of Laser Development. *Appl. Opt.* **2010**, *49*, F99–F122. [CrossRef]

Wen, X.; Lin, Q.; Niu, G.; Shi, Q.; Duan, Y. Emission Enhancement of Laser-Induced Breakdown Spectroscopy for Aqueous Sample Analysis Based on Au Nanoparticles and Solid-Phase Substrate. *Appl. Opt.* **2016**, *55*, 6706–6712. [CrossRef]

Al-Kattan, A.; Grojo, D.; Drouet, C.; Mouskeftaras, A.; Delaporte, P.; Casanova, A.; Robin, J.D.; Magdinier, F.; Alloncle, P.; Constantinescu, C. Short-Pulse Lasers: A Versatile Tool in Creating Novel Nano-/Micro-Structures and Compositional Analysis for Healthcare and Wellbeing Challenges. *Nanomaterials* **2021**, *11*, 712. [CrossRef]

Kalus, M.-R.R.; Reimer, V.; Barcikowski, S.; Gökce, B. Discrimination of Effects Leading to Gas Formation during Pulsed Laser Ablation in Liquids. *Appl. Surf. Sci.* **2019**, *465*, 1096–1102. [CrossRef]

Smirnov, V.V.; Zhilnikova, M.I.; Barmina, E.V.; Shafeev, G.A.; Kobtsev, V.D.; Kostritsa, S.A.; Pridvorova, S.M. Laser Fragmentation of Aluminum Nanoparticles in Liquid Isopropanol. *Chem. Phys. Lett.* **2021**, *763*, 138211. [CrossRef]

Lednev, V.N.; Sdvizhenskii, P.A.; Grishin, M.Y.; Stavertiy, A.Y.; Tretyakov, R.S.; Asyutin, R.D.; Pershin, S.M. Laser Welding Spot Diagnostics by Laser-Induced Breakdown Spectrometry. *Phys. Wave Phenom.* **2021**, *29*, 221–228. [CrossRef]

Park, H.K.; Grigoropoulos, C.P.; Poon, C.C.; Tam, A.C. Optical Probing of the Temperature Transients during Pulsed-laser Induced Boiling of Liquids. *Appl. Phys. Lett.* **1996**, *68*, 596–598. [CrossRef]

Li, M.; Shen, Z.; Gusev, V.; Lomonosov, A.M.; Ni, C. Nonlinear Phenomenon in Laser-Induced Finite-Amplitude Acoustic Waves Propagating along Cracks. *J. Appl. Phys.* **2021**, *129*, 124901. [CrossRef]

Chemin, A.; Fawaz, M.W.; Amans, D. Investigation of the Blast Pressure Following Laser Ablation at a Solid–Fluid Interface Using Shock Waves Dynamics in Air and in Water. *Appl. Surf. Sci.* **2022**, *574*, 151592. [CrossRef]

Aiyyzhy, K.O.; Barmina, E.V.; Voronov, V.V.; Shafeev, G.A.; Novikov, G.G.; Uvarov, O.V. Laser Ablation and Fragmentation of Boron in Liquids. *Opt. Laser Technol.* **2022**, *155*, 108393. [CrossRef]

Frias Batista, L.M.; Nag, A.; Meader, V.K.; Tibbetts, K.M. Generation of Nanomaterials by Reactive Laser-Synthesis in Liquid. *Sci. China Phys. Mech. Astron.* **2022**, *65*, 274202. [CrossRef]

Kalus, M.-R.R.; Lanyumba, R.; Lorenzo-Parodi, N.; Jochmann, M.A.; Kerpen, K.; Hagemann, U.; Schmidt, T.C.; Barcikowski, S.; Gökce, B. Determining the Role of Redox-Active Materials during Laser-Induced Water Decomposition. *Phys. Chem. Chem. Phys.* **2019**, *21*, 18636–18651. [CrossRef]

Champion, C.; Boudrioua, O.; Cappello, C.D. Water Molecule Ionization by Charged Particles: A Short Review. In Proceedings of the Journal of Physics: Conference Series; IOP Publishing: Bristol, UK, 2008; Volume 101, p. 12010.

Crowell, R.A.; Bartels, D.M. Multiphoton Ionization of Liquid Water with 3.0–5.0 EV Photons. *J. Phys. Chem.* **1996**, *100*, 17940–17949. [CrossRef]

Toigawa, T.; Gohdo, M.; Norizawa, K.; Kondoh, T.; Kan, K.; Yang, J.; Yoshida, Y. Examination of the Formation Process of Pre-Solvated and Solvated Electron in n-Alcohol Using Femtosecond Pulse Radiolysis. *Radiat. Phys. Chem.* **2016**, *123*, 73–78. [CrossRef]

Amendola, V.; Amans, D.; Ishikawa, Y.; Koshizaki, N.; Scirè, S.; Compagnini, G.; Reichenberger, S.; Barcikowski, S. Room-temperature Laser Synthesis in Liquid of Oxide, Metal-oxide Core-shells, and Doped Oxide Nanoparticles. *Chem. Eur. J.* **2020**, *26*, 9206–9242. [CrossRef] [PubMed]

Serkov, A.A.; Kuzmin, P.G.; Rakov, I.I.; Shafeev, G.A. Influence of Laser-Induced Breakdown on the Fragmentation of Gold Nanoparticles in Water. *Quantum Electron.* **2016**, *46*, 713–718. [CrossRef]

Maximova, K.; Aristov, A.; Sentis, M.; Kabashin, A.V. Size-Controllable Synthesis of Bare Gold Nanoparticles by Femtosecond Laser Fragmentation in Water. *Nanotechnology* **2015**, *26*, 65601. [CrossRef] [PubMed]

Pandey, B.K.; Shahi, A.K.; Shah, J.; Kotnala, R.K.; Gopal, R. Optical and Magnetic Properties of $Fe_2O_3$ Nanoparticles Synthesized by Laser Ablation/Fragmentation Technique in Different Liquid Media. *Appl. Surf. Sci.* **2014**, *289*, 462–471. [CrossRef]

Ali, M.; Remalli, N.; Yehya, F.; Chaudhary, A.K.; Srikanth, V.V.S.S. Picosecond Laser Induced Fragmentation of Coarse $Cu_2O$ Particles into Nanoparticles in Liquid Media. *Appl. Surf. Sci.* **2015**, *357*, 1601–1605. [CrossRef]

Baimler, I.V.; Simakin, A.V.; Gudkov, S.V. Investigation of the Laser-Induced Breakdown Plasma, Acoustic Vibrations and Dissociation Processes of Water Molecules Caused by Laser Breakdown of Colloidal Solutions Containing Ni Nanoparticles. *Plasma Sources Sci. Technol.* **2021**, *30*, 125015. [CrossRef]

Kang, K.; Rho, Y.; Park, H.K.; Hwang, D.J. Investigation of Elemental Composition Change by Laser Ablation of a Rare-Earth Containing Material. *Phys. Status Solidi* **2018**, *215*, 1700947. [CrossRef]

Gudkov, S.V.; Astashev, M.E.; Baimler, I.V.; Uvarov, O.V.; Voronov, V.V.; Simakin, A.V. Laser-Induced Optical Breakdown of an Aqueous Colloidal Solution Containing Terbium Nanoparticles: The Effect of Oxidation of Nanoparticles. *J. Phys. Chem. B* **2022**, *126*, 5678–5688. [CrossRef]

Baimler, I.V.; Lisitsyn, A.B.; Serov, D.A.; Astashev, M.E.; Gudkov, S.V. Analysis of Acoustic Signals during the Optical Breakdown of Aqueous Solutions of Fe Nanoparticles. *Front. Phys.* **2020**, *8*, 622791. [CrossRef]

Baimler, I.V.; Lisitsyn, A.B.; Gudkov, S.V. Influence of Gases Dissolved in Water on the Process of Optical Breakdown of Aqueous Solutions of Cu Nanoparticles. *Front. Phys.* **2020**, *8*, 622775. [CrossRef]

Gudkov, S.V.; Lyakhov, G.A.; Pustovoy, V.I.; Shcherbakov, I.A. Influence of Mechanical Effects on the Hydrogen Peroxide Concentration in Aqueous Solutions. *Phys. Wave Phenom.* **2019**, *27*, 141–144. [CrossRef]

27. Gudkov, S.V.; Penkov, N.V.; Baimler, I.V.; Lyakhov, G.A.; Pustovoy, V.I.; Simakin, A.V.; Sarimov, R.M.; Scherbakov, I.A. Effec Mechanical Shaking on the Physicochemical Properties of Aqueous Solutions. *Int. J. Mol. Sci.* **2020**, *21*, 8033. [CrossRef]
28. Zvyagin, A.I.; Perepelitsa, A.S.; Lavlinskaya, M.S.; Ovchinnikov, O.V.; Smirnov, M.S.; Ganeev, R.A. Demonstration of Variatio the Nonlinear Optical Absorption of Non-Spherical Silver Nanoparticles. *Optik* **2018**, *175*, 93–98. [CrossRef]
29. Richerzhagen, B.; Delacretaz, G.P.; Salathe, R.-P. Complete Model to Simulate the Thermal Defocusing of a Laser Beam Focuse Water. *Opt. Eng.* **1996**, *35*, 2058–2066. [CrossRef]
30. Gudkov, S.V.; Grinberg, M.A.; Sukhov, V.; Vodeneev, V. Effect of Ionizing Radiation on Physiological and Molecular Processe Plants. *J. Environ. Radioact.* **2019**, *202*, 8–24. [CrossRef] [PubMed]
31. Fano, U. On the Theory of Ionization Yield of Radiations in Different Substances. *Phys. Rev.* **1946**, *70*, 44. [CrossRef]
32. Bagratashvili, V.N.; Letokhov, V.S.; Makarov, A.A.; Ryabov, E.A. Multiple Photon IR Laser Photophysics and Photochemistry *Laser Chem.* **1984**, *5*, 53–105. [CrossRef]
33. Woodin, R.L.; Bomse, D.S.; Beauchamp, J.L. Multiphoton Dissociation of Molecules with Low Power Continuous Wave Infra Laser Radiation. *J. Am. Chem. Soc.* **1978**, *100*, 3248–3250. [CrossRef]
34. Grant, E.R.; Schulz, P.A.; Sudbo, A.S.; Shen, Y.R.; Lee, Y.-T. Is Multiphoton Dissociation of Molecules a Statistical Thermal Proc *Phys. Rev. Lett.* **1978**, *40*, 115. [CrossRef]
35. Simakin, A.V.; Astashev, M.E.; Baimler, I.V.; Uvarov, O.V.; Voronov, V.V.; Vedunova, M.V.; Sevost'Yanov, M.A.; Belosludtsev, K Gudkov, S.V.; Sevost'yanov, M.A.; et al. The Effect of Gold Nanoparticle Concentration and Laser Fluence on the Laser-Indu Water Decomposition. *J. Phys. Chem. B* **2019**, *123*, 1869–1880. [CrossRef] [PubMed]
36. Gudkov, S.V.; Lyakhov, G.A.; Pustovoy, V.I.; Shcherbakov, I.A. Vibration–Vortex Mechanism of Radical-Reaction Activation in Aqueous Solution: Physical Analogies. *Phys. Wave Phenom.* **2021**, *29*, 108–113. [CrossRef]
37. Rice, F.O.; Reiff, O.M. The Thermal Decomposition of Hydrogen Peroxide. *J. Phys. Chem.* **2002**, *31*, 1352–1356. [CrossRef]
38. Kawasaki, M.; Masuda, K. Laser Fragmentation of Water-Suspended Gold Flakes via Spherical Submicroparticles to F Nanoparticles. *J. Phys. Chem. B* **2005**, *109*, 9379–9388. [CrossRef]
39. Akimoto, I.; Maeda, K.; Ozaki, N. Hydrogen Generation by Laser Irradiation of Carbon Powder in Water. *J. Phys. Chem. C* 2 *117*, 18281–18285. [CrossRef]
40. Kierzkowska-Pawlak, H.; Tyczkowski, J.; Jarota, A.; Abramczyk, H. Hydrogen Production in Liquid Water by Femtosec Laser-Induced Plasma. *Appl. Energy* **2019**, *247*, 24–31. [CrossRef]
41. Barmina, E.V.; Simakin, A.V.; Shafeev, G.A. Hydrogen Emission under Laser Exposure of Colloidal Solutions of Nanopartic *Chem. Phys. Lett.* **2016**, *655*, 35–38. [CrossRef]
42. Barmina, E.V.; Simakin, A.V.; Shafeev, G.A. Balance of $O_2$ and $H_2$ Content under Laser-Induced Breakdown of Aqueous Colloi Solutions. *Chem. Phys. Lett.* **2017**, *678*, 192–195. [CrossRef]
43. Baimler, I.V.; Lisitsyn, A.B.; Gudkov, S.V. Water Decomposition Occurring during Laser Breakdown of Aqueous Soluti Containing Individual Gold, Zirconium, Molybdenum, Iron or Nickel Nanoparticles. *Front. Phys.* **2020**, *8*, 620938. [CrossRef

photonics MDPI

ticle

# emperature Changes during Er:YAG Laser Activation with the ide-Firing Spiral Endo Tip in Root Canal Treatment

aronit Sahar-Helft [1,2], Nathanyel Sebbane [1,3,4], Adi Farber [1,3], Ronit Vogt Sionov [3], Roni Dakar [2] d Doron Steinberg [3,*]

1 Department of Endodontics, The Faculty of Dental Medicine, The Hebrew University-Hadassah, 9112102 Jerusalem, Israel; sharonit.sahar-helft@mail.huji.ac.il (S.S.-H.); nathanye.sebbane@mail.huji.ac.il (N.S.); adi.farber@mail.huji.ac.il (A.F.)

2 Endodontics Department, School of Graduate Dentistry, Rambam Health Care Campus, The Ruth and Bruce Rappaport Faculty of Medicine, Technion-Israel Institute of Technology, 3200003 Haifa, Israel; dr.ronidakar@gmail.com

3 Biofilm Research Laboratory, The Institute of Biomedical and Oral Research (IBOR), The Faculty of Dental Medicine, The Hebrew University of Jerusalem, 9112102 Jerusalem, Israel; ronit.sionov@mail.huji.ac.il

4 "Bina" Program, Faculty of Dental Medicine, The Hebrew University-Hadassah, 9112102 Jerusalem, Israel

* Correspondence: dorons@ekmd.huji.ac.il; Tel.: +972-52-468-1999

**Abstract:** Erbium-doped yttrium aluminum garnet (Er:YAG) laser-activated irrigation (LAI) is used in endodontic treatment to remove the smear layer and kill bacteria in the root canal. However, this procedure can cause photo-thermal effects that harm the surrounding tissue. Therefore, it was important to study the temperature changes that occur at the outer tooth surface during activation of the Er:YAG laser using a side-firing spiral Endo tip in the root canal. Laser treatment was performed either in the absence of fluid in the root canal or in the presence of a 17% EDTA solution. Irrigation with 17% EDTA was either performed in a continuous mode for 60 s or in a segmented mode of 4 rinses with 17% EDTA for 15 s each. The temperature was measured every second during the treatment at three tooth surface sites: the cementoenamel junction, the middle region and the apical region. Our data show that the greatest temperature changes occurred when the laser was used alone without an irrigation solution, while minor temperature changes were observed with continuous irrigation. In conclusion, we would recommend applying the laser treatment with an irrigation solution to avoid excessive heating.

**Keywords:** Er:YAG laser; smear layer; bacteria; temperature; endodontic treatment; new side-firing spiral Endo tip



tion: Sahar-Helft, S.; Sebbane, N.; ber, A.; Sionov, R.V.; Dakar, R.; nberg, D. Temperature Changes ing Er:YAG Laser Activation with Side-Firing Spiral Endo Tip in t Canal Treatment. *Photonics* **2023**, 488. https://doi.org/10.3390/ tonics10050488

eived: 22 March 2023
ised: 17 April 2023
epted: 21 April 2023
lished: 24 April 2023



## 1. Introduction

The primary goal of root canal treatment (RCT) is to remove all microorganisms from the inner surface of the root canal system, prevent reinfection and establish healthy periapical tissue [1,2]. One approach is the biomechanical instrumentation of the root canal system, in which centrifugal forces are generated by the movement of the instrument and its proximity to the dentin wall, where a thicker layer called the smear layer is formed. The smear layer contains organic and inorganic substances, including remnants of odontoblastic processes, vital and necrotic pulp tissue, microorganisms and blood cells [3,4]. In 1975, McComb and Smith [5] were the first to describe the presence of the smear layer on the surface of an instrumented root. Mader et al. [6] discovered that the smear material is present in two layers: the outer superficial smear layer and the inner layer of materials densely packed in the dentinal tubules, reaching a depth of 40 μm [6]. When surface-active reagents are used, this depth can reach 110 μm [7]. Bacteria can survive [8] and proliferate in the dentinal tubules [9]. The smear layer clogs the dentinal tubules and prevents the penetration and activation of disinfectant irrigations [10]. Moreover, it can impair the sealing and adaptation of root canal obturation materials [11].

There are several methods of removing the smear layer. The most studied are chelat agents such as ethylenediaminetetraacetic acid (EDTA). Application of EDTA for 1–5 n is considered optimal. EDTA chelates calcium ions in dentin, resulting in soluble calci chelates. Moodnik and Sulewski [12,13] showed that endodontic instruments' action a chemical irrigation cannot completely remove the smear layer from root canal walls. T conventional irrigation systems using NaOCl and EDTA solutions are unable to remo smear layers and organic debris in the apical third of the root canal [14]. Syringe irrigat is limited by the inability to extend the needle more than 1 mm from the tip [15]. T flushing is affected by the diameter of the canal, the cross-sectional shape of the can the depth of needle insertion and the diameter of the needle. The presence of air bubb and vapor locks can further affect the efficiency of irrigation. These prevent the smo flow of fluid into the narrow spaces of the fins, isthmuses and lateral canals. This is w neither conventional techniques such as hand or powered files nor irrigation routines completely clean the root canal [16,17].

An innovative new modality of endodontic treatment combines the erbium–yttri aluminum garnet (Er:YAG) laser with irrigants [18]. The physical effect of the laser root canals depends on the absorption of its wavelengths in biological components a chromophores, such as water and apatite. Laser-activated irrigation (LAI) carried by tools such as erbium lasers (Er:YAG/2980 nm and Er,Cr:YSGG/2780 nm) is becon increasingly popular due to its effective removal of dentin smear layers and ability disinfect root canals [19]. The use of laser irradiation in the root canal system has be shown to be effective in removing the smear layer and bacteria [20]. For this purpose, a si firing spiral Endo tip was designed by Prof. Adam Shtabhols (The Hebrew Univers Jerusalem, Israel) for specifically cleaning and disinfecting the root canal system dur endodontic treatments and endodontic retreatments [20]. The tip is designed to fit the sha and volume of root canals prepared with the NiTi Protaper Gold (Dentspy Sirona, Charlo NC, USA) rotary instrument. The Endo tip is designed with a flexible, hollow, conical a round cross-section with circumferential spiral slits along its entire length (Figure 1). T end of the Endo tip is sealed, which prevents radiation transmission through the api foramen. This unique design allows for efficient cleaning and disinfection. The tip is 25 n long and has three zones: 1. the cylindrical sleeve zone (4 mm), which is inserted into handpiece; 2. the flexibility zone (3 mm), which allows additional flexibility when the tip in action in the root canal and 3. the functional area (18 mm) with six spiral slits. The wi of each slit and the distance between them varies so that the width of the furthest api side slit is narrower than the width of the slit located on the coronal (handpiece) side.

**Figure 1.** An image of the side-firing spiral Endo tip. The tip is hollow, conical and has a round cr section with circumferential spiral slits along its entire length. It has three zones: A. the cylindr sleeve zone (4 mm), which is inserted into the handpiece; B. the flexibility zone (3 mm), which allo additional flexibility when the tip is in action in the root canal; and C. the functional area (18 m with six spiral slits. The width of each slit and the distance between them varies so that the width the furthest apical side slit is narrower (0.4 mm) than the width of the slit located on the coronal s (1.1 mm). The end of the Endo tip is sealed.

LAI primarily operates via cavitation shock, which causes the formation of vapor bubbles due to the absorption of laser energy by the liquid irrigation solution. These robust cavitation bubbles in the liquid enlarge during the pulse and then collapse inward, producing shock waves that lead to cleaning and disinfection of the root canal [21].

During root canal treatment, using an Er:YAG laser beam to activate the irrigation solution results in photo-thermal, photo-chemical and photo-ablative effects. This process can lead to a rise in the temperature and potentially damage the surrounding tissue, including the periodontal and bone areas. To avoid thermal damage to surrounding tissues, it is crucial to establish a safety limit for laser activation [22].

In 1983, Eriksson and Albrektsson determined a temperature of 47 °C as the critical limit for bone survival in rabbits [22]. In later studies, a temperature rise of 10 °C above body temperature was found to be the critical limit [23]. According to a 2001 dissertation by Mazaheri in BioMed Research International 7 at RWTH Aachen College, the maximum average temperature occurring during irradiation of root canals with a 3 W diode laser would fall below the critical limit if the optical fiber was continuously moved in a circular motion both coronally and apically [24]. Gutknecht et al. found that at a 3 W CW setting, bacterial reduction was observed in bovine teeth at a depth of 500 microns [25]. However, the temperature limit was exceeded at 4 W with an irradiation time of 15 s, resulting in thermal damage [25].

Gutknecht observed that using lasers below 1 W is irrelevant in endodontics because they do not completely remove the smear layer or seal the dentinal tubules [26]. However, at settings of 1.25 W to 1.5 W, the organic material is entirely removed, and the surface of the inorganic substance is fused [26]. These observations are important as they help define the conditions required for preventing reinfection in the root canal.

The present study aimed to examine the temperature changes occurring at the outer tooth surface during activation of the Er:YAG laser using the new side-firing spiral Endo tip in the root canal under the following treatment conditions: laser activation without fluid in the root canal and laser activation in the presence of a 17% EDTA solution in a continuous or segmented mode.

## 2. Materials and Methods

### 2.1. Tooth Preparation

Thirty single-rooted human teeth (central incisors) extracted for periodontal reasons were used in this study. The study was conducted with the approval of the Hadassah Hospital Ethics Committee, No. 0118-14-HMO.

The average working length of the root canal was 16.69 mm. After enlarging the canal openings with Gates-Glidden, the root canals were mechanically prepared with ProTaper files, similar to the routine clinical procedure. The root canals were enlarged to the apex with five different files for 10 cycles. The endodontic procedure was completed with file No. 40. The canal was washed with 1 mL of 2.5% sodium hypochlorite between the steps.

### 2.2. Laser Specification with the Side-Firing Spiral Tip for Endodontics

An Er:YAG laser that delivers a powerful 2940 nm wavelength (Light Instruments, Yokne'am, Israel) was used in this study. The energy used was 150 mJ, 10 Hz, for 60 s. According to the data obtained, a side-firing spiral Endo tip was used [27]. The Endo tip was inserted into the root canal to cover the entire area. The tip was moved up and down in the coronal–apical direction at 1–2 mm intervals.

### 2.3. Treatment Groups

Thirty single-rooted teeth were divided into three categories of ten teeth each:

Group A: laser irradiation for 60 s without fluid in the root canal.

Group B: laser irradiation for 60 s combined with continuous rinsing of the root canal with 10 mL of a 17% EDTA irrigation solution (DSI—Dental Solutions Israel Ltd., Ashdod, Israel).

Group C: four segments of laser irradiation for 15 s (total of 60 s), each cycle with 1 of a 17% EDTA irrigation solution in the root canal (total of 4 mL).

*2.4. Temperature Measurements*

The temperature rise at the outer tooth surface was recorded during laser activat when the Endo tip was operated inside the root canal. The temperature rise was measu using a digital readout thermometer coupled to a surface measurement probe (DAQPI Fourtec, Fourier Technology, Chatswood, NSW, Australia) with a sensitivity of 0.10 The thermocouple was placed directly on the tooth surface at the cementoenamel junct (coronal area), in the middle of the root canal and at the apical area, and the temperat was recorded at 1 s intervals during the laser activity in the root canal (Figure 2). Ea measurement was performed by the same operator for each source evaluation. The data lected through the computer and Microsoft Excel software were used to calculate the me temperature rise, the range and the standard deviation for each tooth group. A statisti analysis of variance was used to determine the significance between the three treatm groups. The temperature was measured for 60 s each time; a total of 5400 s of temperat data was collected.

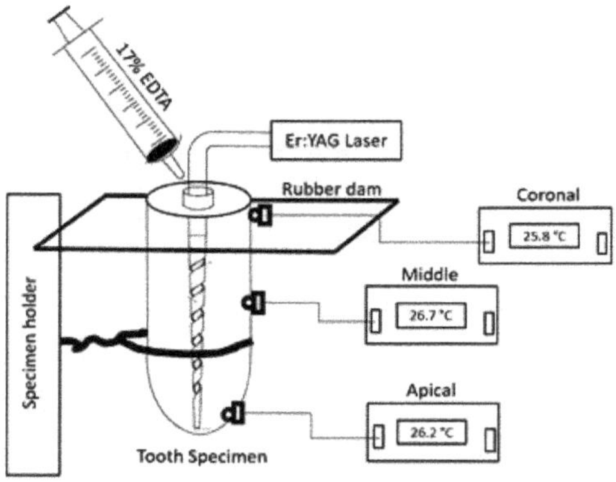


**Figure 2.** Illustration of temperature measurements. The image shows the Endo tip inserted into root canal of a tooth specimen which is fixed in a specimen holder. The blue lines present laser fir A 17% EDTA solution was injected to the canal simultaneously with the activation of the Er:Y laser. The locations of the temperature measurements (coronal, middle and apical) are shown.

*2.5. Statistical Analysis*

ANOVA and post hoc Tukey's HSD test were used to determine the statistical dif ence in the temperature changes at the outer surface of the root canal between the th treatment groups as well as the temperature changes between the three areas (cement namel junction, middle area and apical area) within the same group. The statistical analy was conducted using JUSP computer software (JASP Team (2022), Version 0.16.2). A *p*-va below 0.05 was considered statistically significant.

## 3. Results

*3.1. Comparison of the Temperature Changes during Each Treatment Procedure*

We compared the temperature changes ($\Delta$T) at the cementoenamel junction (coro region), in the middle and apical regions of the tooth surface during laser treatment in root canal under the following conditions: laser alone without irrigation solution (Contr Group A); laser in combination with continuous 17% EDTA irrigation (Group B); and la in combination with four discontinued (segmented) 17% EDTA irrigations for 15 s ea (Group C), resulting in a total treatment time of 60 s (Figure 3).

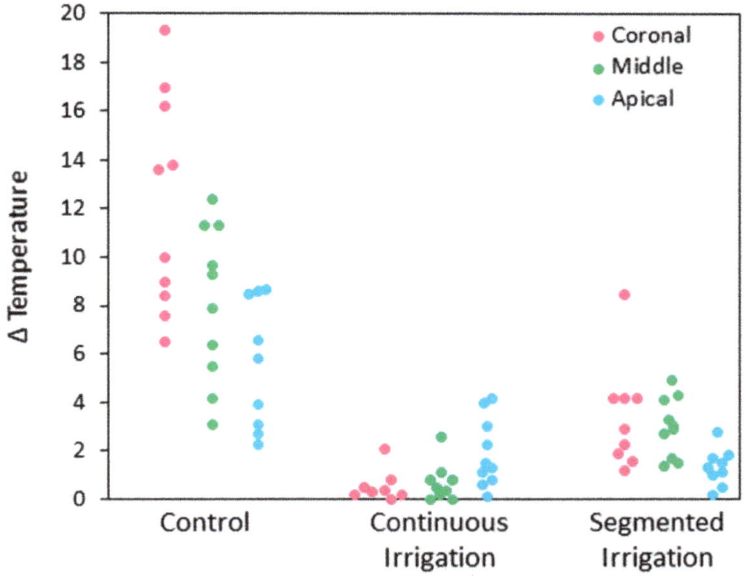


**Figure 3.** Temperature measurements of the outer tooth surface at the indicated locations (coronal, middle, apical) during the 60 s activation of the Er:YAG laser with the side-firing spiral Endo tip in the root canal for the three treatment conditions: Control was laser treatment without fluid in the root canal. Continuous irradiation was laser with 17% EDTA irrigation solution for 60 s. Segmented irritation was performed 4 times applying the laser with 17% EDTA irrigation solution for 15 s per cycle. Each circle represents the ΔT (Δ Temperature) of one sample. The pink circles represent the ΔT measurements at the coronal (cementoenamel junction) region. The green circles represent the ΔT measurements in the middle part, while the blue circles represent the ΔT measurements in the apical region.

The control group, which received the laser alone without irrigation solution (Group A), showed the greatest temperature changes, which were significantly higher ($p < 0.01$) than in the continuous (Group B) and segmented irrigation mode (Group C) (Figure 3 and Table 1). Minor temperature changes were observed in the continuous irrigation method (Group B), although there was no significant difference with the segmented method (Group C) (Figures 3–6 and Table 1).

**Table 1.** Post hoc Tukey's comparisons show statistically significant differences between the control Group A and the irrigation Groups B and C.

| Group | Compared to | Mean Difference | SE | $t$ | $P_{Tukey}$ |
|---|---|---|---|---|---|
| Continued Irrigation | Control | 7.794 | 0.652 | −11.948 | <0.001 |
| Continued Irrigation | Segmented Irrigation | 1.526 | 0.664 | −2.300 | 0.062 |
| Control | Segmented Irrigation | 6.268 | 0.645 | 9.715 | <0.001 |

Notes: $p$-value is adjusted for comparing a family of 3. Results are averaged over the levels of Measrement_Point. Mean difference: average ΔT of the group of interest—average ΔT of the compared group. Standard error (SE) = the common error factor for all three groups. $t$ = the sample statistical value under the T distribution. The $P_{Tukey}$ value stands for probability and measures the likelihood that any observed difference between groups is due to chance.

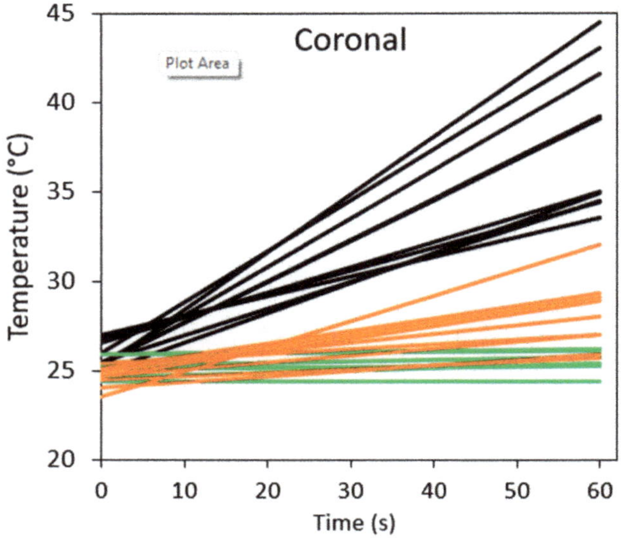


**Figure 4.** Temperature measurements of the outer tooth surface at the coronal site during the Er:YAG laser activation with the side-firing spiral Endo tip in the root canal for the three treatm conditions: Control is laser treatment without fluid in the root canal (black lines). Continu irradiation was laser with 17% EDTA irrigation solution for 60 s (green lines). Segmented irritat was performed 4 times applying the laser with 17% EDTA irrigation solution for 15 s per cycle (ora lines). Each line is a separate sample.

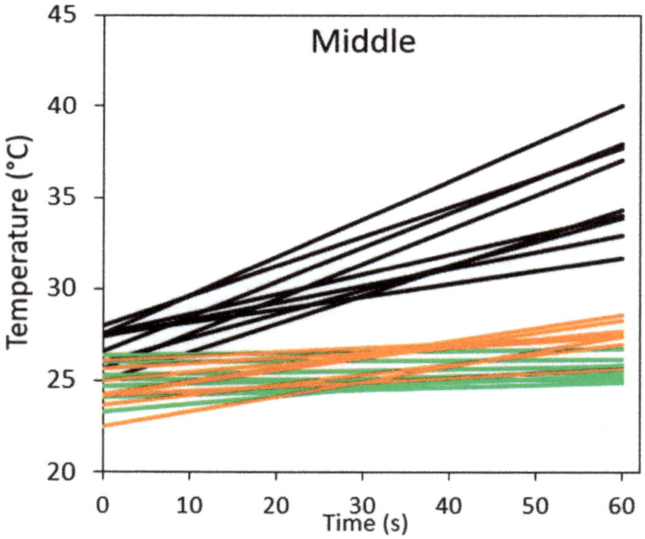


**Figure 5.** Temperature measurements of the outer tooth surface at the middle region during the Er:YAG laser activation with the side-firing spiral Endo tip in the root canal for the three treatm conditions: Control is laser treatment without fluid in the root canal (black lines). The continu irradiation is laser with 17% EDTA irrigation solution for 60 s (green lines). The segmented irritat was performed 4 times applying the laser with 17% EDTA irrigation solution for 15 s per cycle (ora lines). Each line is a separate sample.

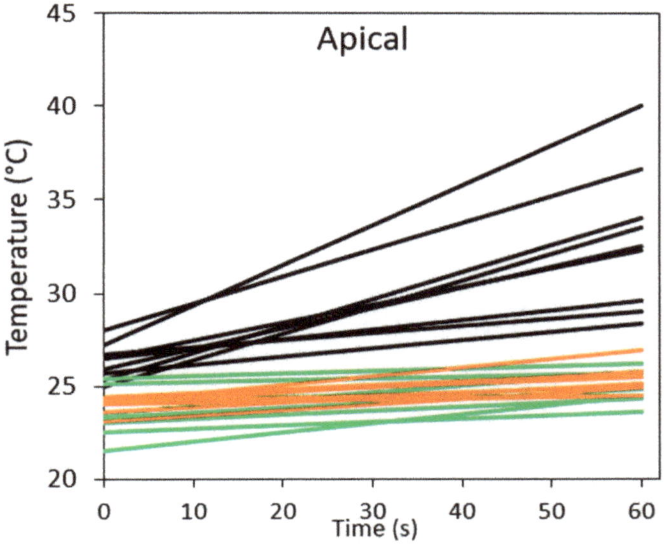


**Figure 6.** Temperature measurements of the outer tooth surface at the apical region during the 60 s Er:YAG laser activation with the side-firing spiral Endo tip in the root canal for the three treatment conditions: Control is laser treatment without fluid in the root canal (black lines). Continuous irradiation is laser with 17% EDTA irrigation solution for 60 s (green lines). Segmented irritation was performed 4 times applying the laser with 17% EDTA irrigation solution for 15 s per cycle (orange lines). Each line is a separate sample.

Results of the analyses of the statistical significance of the different temperature changes (ΔT) observed in the three treatment groups and at the three surface sites (cementoenamel junction, middle and apical area) are presented in Tables 1–3.

**Table 2.** Post hoc Tukey's comparisons showing statistically significant differences between the apical and coronal regions.

| Group | Compared to | Mean Difference | SE | $t$ | $P_{\text{Tukey}}$ |
|---|---|---|---|---|---|
| Apical | Coronal | −2.210 | 0.658 | −3.359 | <0.003 |
| Apical | Middle | −0.762 | 0.645 | −1.181 | 0.468 |
| Coronal | Middle | 1.449 | 0.658 | 2.202 | 0.077 |

Notes: $p$-value is adjusted for comparing a family of 3. Results are averaged over the levels of Measrement_Point. Mean difference: average ΔT of the group of interest—average ΔT of the compared group. Standard error (SE) = the common error factor for all three groups. $t$ = the sample statistical value under the T distribution. The $P_{\text{Tukey}}$ value stands for probability and measures the likelihood that any observed difference between groups is due to chance.

Post hoc Tukey's HSD test for multiple comparisons showed that the mean ΔT values significantly differed between the three groups. Table 1 presents the ΔT and the statistical significance between the different treatment methods: the control group of laser treatment without fluid, the continuous 17% EDTA irrigation method with the Er:YAG laser and the segmented 17% EDTA irrigation method with the Er:YAG laser. Table 2 presents the mean values of ΔT in different measurement areas (cementoenamel junction, middle and apical) and the statistical significance. Table 3 shows the treatment method, measurement points and statistical significance ($p < 0.001$ for irrigation groups versus control; Table 3). Moreover, the mean values of ΔT between the control and other groups were significantly different ($p < 0.001$ for irrigation groups versus control; Table 3).

**Table 3.** Post hoc Tukey's comparisons between the different treatment methods and measurem locations.

| Method | Surface Site | Compared to Method | Surface Site | Mean Difference | SE | *t* | $P_{Tukey}$ |
|---|---|---|---|---|---|---|---|
| Continued | Apical | Control | Apical | −4.406 | 1.097 | −4.015 | 0.004 |
| | | Segmented | Apical | 0.572 | 1.127 | 0.507 | 1.000 |
| | | Continued | Coronal | 1.331 | 1.164 | 1.144 | 0.965 |
| | | Control | Coronal | −10.246 | 1.097 | −9.337 | <0.001 |
| | | Segmented | Coronal | −1.550 | 1.127 | −1.375 | 0.904 |
| | | Continued | Middle | 1.183 | 1.127 | 1.049 | 0.979 |
| | | Control | Middle | −6.216 | 1.097 | −5.665 | <0.001 |
| | | Segmented | Middle | −1.086 | 1.097 | −0.990 | 0.985 |
| Control | Apical | Segmented | Apical | 4.978 | 1.127 | 4.415 | 0.001 |
| | | Continued | Coronal | 5.737 | 1.164 | 4.930 | <0.001 |
| | | Control | Coronal | −5.840 | 1.097 | −5.322 | <0.001 |
| | | Segmented | Coronal | 2.856 | 1.127 | 2.533 | 0.233 |
| | | Continued | Middle | 5.589 | 1.127 | 4.957 | <0.00 |
| | | Control | Middle | −1.810 | 1.097 | −1.649 | 0.774 |
| | | Segmented | Middle | 3.320 | 1.097 | 3.026 | 0.077 |
| Segmented | Apical | Continued | Coronal | 0.760 | 1.192 | 0.637 | 0.999 |
| | | Control | Coronal | −10.818 | 1.127 | −9.595 | <0.001 |
| | | Segmented | Coronal | −2.122 | 1.157 | −1.835 | 0.659 |
| | | Continued | Middle | 0.611 | 1.157 | 0.528 | 1.000 |
| | | Control | Middle | −6.788 | 1.127 | −6.021 | <0.001 |
| | | Segmented | Middle | −1.658 | 1.127 | −1.470 | 0.865 |
| Continued | Coronal | Control | Coronal | −11.577 | 1.164 | −9.947 | <0.001 |
| | | Segmented | Coronal | −2.882 | 1.192 | −2.417 | 0.290 |
| | | Continued | Middle | −0.149 | 1.192 | −0.125 | 1.000 |
| | | Control | Middle | −7.547 | 1.164 | −6.485 | <0.001 |
| | | Segmented | Middle | −2.417 | 1.164 | −2.077 | 0.496 |
| Control | Coronal | Segmented | Coronal | 8.696 | 1.127 | 7.713 | <0.001 |
| | | Continued | Middle | 11.429 | 1.127 | 10.138 | <0.001 |
| | | Control | Middle | 4.030 | 1.097 | 3.673 | 0.013 |
| | | Segmented | Middle | 9.160 | 1.097 | 8.348 | <0.001 |
| Segmented | Coronal | Continued | Middle | 2.733 | 1.157 | 2.363 | 0.319 |
| | | Control | Middle | −4.666 | 1.127 | −4.138 | 0.003 |
| | | Segmented | Middle | 0.464 | 1.127 | 0.412 | 1.000 |
| Continued | Middle | Control | Middle | −7.399 | 1.127 | −6.563 | <0.001 |
| | | Segmented | Middle | −2.269 | 1.127 | −2.013 | 0.540 |
| Control | Middle | Segmented | Middle | 5.130 | 1.097 | 4.675 | <0.001 |

Notes: *p*-value is adjusted for comparing a family of 3. Results are averaged over the levels of Measrement_P Mean difference: average ΔT of the group of interest—average ΔT of the compared group. Standard e (SE) = The common error factor for all three groups. *t* = the sample statistical value under the T distribution. $P_{Tukey}$ value stands for probability and measures the likelihood that any observed difference between group due to chance.

### 3.2. *Temperature Changes in the Coronal, Middle and Apical Tooth Sites*

In the coronal region, the lowest temperature changes (ΔT) were measured duri the continuous mode, with a median temperature increase of 0.56 ± 0.67 °C compared 3.44 ± 2.22 °C using the segmented procedure (Figures 3 and 4). The highest temperat changes (ΔT) (12.14 ± 4.43 °C) were measured in the laser method without irrigati solution in the root canal (Figures 3 and 4 and Tables 2 and 3).

In the middle region of the root canal, the following temperature changes (ΔT) w measured: 0.71 ± 0.80 °C for the continuous irrigation method; 2.98 ± 1.21°C for segmented method; and 8.11 ± 3.12 °C for the control method without irrigation soluti (Figures 3 and 5 and Tables 2 and 3).

At the apical area of the root canal, the following temperature changes (ΔT) w measured: 1.89 ± 1.42 °C for the continuous irrigation method; 1.32 ± 0.76 °C for segmented method; and 6.30 ± 3.39 °C for the control method without irrigation solution in the root canal (Figures 3 and 6 and Tables 2 and 3).

## 4. Discussion

The most important endodontic treatment is the removal of microorganisms and debris from the root canal system. Er:YAG LAI can achieve this effect through the potential energy of the light beam, which is strongly absorbed by the liquid. The cleaning mechanism involves cavitation bubbles that generate shock waves [28,29]. Due to the closed space of the root canals, there is a reduced bubble effect which results in a diminished shock waves effect [28]. To overcome the obstacle, Gregorcic et al. [29] investigated the possibility of increasing the laser effect using a liquid reservoir. The results show an acoustic effect that could be achieved by applying another laser pulse shortly after the first beam in the liquid reservoir. The collapse of the cavitation bubble increases the mechanical energy of the second oscillation [30]. However, in the limited root canal space, the laser light can only generate cavitation bubbles 2–3 mm near the fiber tip, so the liquid reservoir effect in the root canal is attenuated. Our study used the side-firing spiral Endo tip in root canal treatment. This unique design of the Endo tip allows the proximity of the laser beam to all parts of the canal walls. The laser generates efficient shock waves and oscillating fluid movement directly to the canal walls. With the Endo tip, we could efficiently remove the smear layer and bacterial biofilm and improve the cleaning mechanism inside the root canal [27].

There are two major challenges in root canal treatment. The first is the limited ability to clean the root canal using conventional methods and technologies. The second is the need to overcome obstacles such as anatomical complexities: dentinal tubules, lateral canals, isthmus and ramifications. Chemo-mechanical procedures include rotary and hand files with an irrigation solution. The traditional way of delivering the irrigating solution into the root canal is with a syringe and a 27-gauge or 30-gauge needle [31]. However, Fraser [32] found that the chelating effect was almost negligible in the apical third of the root canals. The irrigation solution can only reach 1–3 mm beyond the needle tip, depending on the needle type and irrigation flow. Several methods have been introduced in recent years to irrigate root canals better. We investigated the efficiency of different methods in our endodontic treatment [20]. Those methods included irrigation with 17% EDTA with positive pressure, passive ultrasound and laser-activated irrigations. The results show that using Er:YAG LAI laser with a 17% EDTA solution provided the best performance [20]. According to our studies, the proximity of the laser beam to the root canal wall is crucial. Sebbane et al. [27] showed that using an Er:YAG Endo tip with 17% EDTA effectively removed the biofilm from the entire root canal.

The present research focused on how the laser Endo tip affects the temperature in its vicinity. When the liquid absorbs the laser beam, it is transformed into heat, which generates internal pressure and vaporization, ultimately removing unwanted material from the root canal. Using the Er:YAG laser can cause thermal changes in the outer surface of the root canal, which may cause harm to the periodontal ligament and bone. Bahcall et al. [33] found that the use of a neodymium-doped yttrium aluminum garnet (Nd: YAG) laser in canine root canal treatment resulted in ankylosis, cementum lysis and significant bone remodeling 30 days after treatment. The potential temperature effects on surrounding structures are the primary concern when using lasers for root canal treatment. It is generally accepted that a temperature rise of about 10 °C above the average body temperature is considered critical [22]. Therefore, it is essential to analyze the temperature changes on the outer tooth surface following LAI, to ensure that the surrounding tissues are not damaged.

The temperature may vary between different segments of the tooth due to differences such as conductivity, depth and density. Therefore, we measured the temperature at various locations. When irrigation was applied with the laser, the temperature changes between the coronal, middle and apical zones were similar. The conductivity, material properties and irrigation minimized the temperature changes at the different locations of the tooth surface.

We investigated the thermal effect of the Er:YAG laser with the Endo tip on the ou
surface of the root canal at three sites using three methods. The first method was a la
treatment without irrigation solution inside the root canal. The second approach w
a continuous injection of a 17% EDTA irrigation solution into the root channel during la
activation for 60 s. In the third method, the 17% EDTA irrigation solution was applied f
times as a reservoir into the pulp chamber for 15 s each time, for a total of 60 s of treatm
of the root canal.

The results of our study show a significantly higher temperature increase (12.14 ± 4.43
can be achieved by activating the Er:YAG laser without liquid inside the root canal in
coronal part (Group A; $p < 0.01$) compared to the continuous (Group B) and segmen
irrigation methods (Group C) (Figure 3), in which the temperature increased by o
0.5–2 °C on average. Minor temperature changes were observed in the continuous irrigat
method (Group B). The ΔT in the continuous mode and the segmented method was lo
than the safety limit of 10 °C. Therefore, using the Er:YAG laser with the Endo tip in r
canal treatment is probably safe.

Using the continuous irrigation method (Group B), we observed that the temperat
increased more in the apical region (ΔT of 1.89 ± 1.42 °C) than in the coronal (ΔT
0.56 ± 0.66 °C) and middle regions (ΔT of 0.71 ± 0.80 °C) ($p = 0.02$). The explanation
this result could lie in the morphology of the root canal. The dentin structure is thin
in the apical part than in the coronal region. The results are similar to those observed
Kimura et al. [34]. Altogether, our findings demonstrate only a minimal thermal effect
the outer surface of the root canal during activation of the Er:YAG laser with the En
tip when applied at 150 mJ, 10 Hz, for 60 s, along with 17% EDTA irrigation during r
canal treatment. These parameters were chosen as they have been shown to provide go
antibacterial and antibiofilm effects [27]. Other settings might result in a higher temperat
of dentin tissue [35].

The purpose of this ex vivo study was to simulate the oral cavity of a patient. Althou
it is an ex vivo model, it is expected that the results can be translated into clinical tri
and that the in vivo results in patients would be similar to our findings. The same la
properties, parameters and irrigation solutions used in the dental clinic were applied
the extracted teeth to mimic actual clinical conditions as much as possible. Howeve
is important to note that the clinical results may differ from those achieved with the
vivo model due to variable environmental factors. For example, the teeth we used w
extracted, not fresh live teeth, and the clinic's body temperature and room temperat
may influence the in vivo results. Patients of different ages, oral hygiene and cultu
backgrounds are treated in clinical practice. These factors may influence the size of the r
canal, dentin tubules and the dentin and enamel structures. Using randomly extracted te
from different patients, we found that the results were unaffected by the tooth divers
Other variables that can differ from clinic to clinic are the temperature of the irrigati
solution and the suction conditions, both of which can impact the ΔT. Nevertheless, wh
used correctly, lasers are considered safe for ablating dental hard tissue [36].

Many different laser machines and Endo tips are available on the dental market. I
important to mention that an endodontics specialist in laser treatment conducted this stu
A special training session on the use of lasers in the dental field is required to perform t
kind of endodontic treatment. Given these efforts, it is likely that the results obtained
our study would be representative of what would also be observed in real clinical pract
Therefore, this study is a first step toward animal and subsequently human studies.

## 5. Conclusions

When the Er:YAG laser Endo tip was applied in the root canal without any fluid, th
was a significant increase in the temperature (6–12 °C), which is intolerable from a clini
point of view. However, this temperature increase was prevented when conducting sim
taneous irrigation in the root canal, with a minimal temperature change of 0.5–2 °C. Th

this study recommends that the laser treatment be performed with an irrigation solution to optimally clean the root canal and avoid thermal damage to the surrounding tissue.

**Author Contributions:** Data curation: S.S.-H., N.S. and A.F.; formal analysis: S.S.-H.; investigation: N.S. and A.F.; resources: S.S.-H. and D.S.; supervision: D.S., S.S.-H. and R.V.S.; writing—original draft: N.S., A.F. and S.S.-H.; writing—review and editing: R.D., D.S., S.S.-H., R.V.S. and N.S. All authors have read and agreed to the published version of the manuscript.

**Funding:** This study was partially funded by the Cabakoff Foundation, grant number 3175006340.

**Institutional Review Board Statement:** Not applicable.

**Informed Consent Statement:** Not applicable.

**Data Availability Statement:** Not applicable.

**Acknowledgments:** We thank Light Instruments, Yokne'am, Israel, for supplying the laser machine and the spiral Endo tip.

**Conflicts of Interest:** The authors declare no conflict of interest.

## ferences


Versiani, M.A.; Ordinola-Zapata, R. Root Canal Anatomy: Implications in Biofilm Disinfection. In *The Root Canal Biofilm;* Chavez de Paz, L., Sedgley, C., Kishen, A., Eds.; Springer: Toronto, ON, Canada, 2015; pp. 155–187.

Torabinejad, M.; Handysides, R.; Khademi, A.A.; Bakland, L.K. Clinical implications of the smear layer in endodontics: A review. *Oral Surg. Oral Med. Oral Pathol.* **2002**, *94*, 658–666. [CrossRef] [PubMed]

Hülsmann, M. Effects of mechanical instrumentation and chemical irrigation on the root canal dentin and surrounding tissues. *Endod. Top.* **2013**, *29*, 55–86. [CrossRef]

Cunningham, W.T.; Martin, H. A scanning electron microscope evaluation of root canal debridement with the endosonic ultrasonic synergistic system. *Oral Surg. Oral Med. Oral Pathol.* **1982**, *53*, 527–531. [CrossRef]

McComb, D.; Smith, D.C. A preliminary scanning electron microscopic study of root canals after endodontic procedures. *J. Endod.* **1975**, *1*, 238–242. [CrossRef]

Mader, C.L.; Baumgartner, J.C.; Peters, D.D. Scanning electron microscopic investigation of the smeared layer on root canal walls. *J. Endod.* **1984**, *10*, 477–483. [CrossRef]

Aktener, B.O.; Cengiz, T.; Piskin, B. The penetration of smear material into dentinal tubules during instrumentation with surface-active reagents: A scanning electron microscopic study. *J. Endod.* **1989**, *15*, 588–590. [CrossRef]

Bra, M.; Nyborg, H. Cavity treatment with a microbicidal fluoride solution: Growth of bacteria and effect on the pulp. *J. Prosth. Dent.* **1973**, *30*, 303–310.

Akpata, E.S.; Blechman, H. Bacterial invasion of pulpal dentin wall in vitro. *J. Dent. Res.* **1982**, *61*, 435–438. [CrossRef]

Shahravan, A.; Haghdoost, A.A.; Adl, A.; Rahimi, H.; Shadifar, F. Effect of smear layer on sealing ability of canal obturation: A systematic review and meta-analysis. *J. Endod.* **2007**, *33*, 96–105. [CrossRef]

Violich, D.R.; Chandler, N.P. The smear layer in endodontics—A review. *Int. Endod. J.* **2010**, *43*, 2–15. [CrossRef]

Moodnik, R.M.; Dorn, S.O.; Feldman, M.J.; Levey, M.; Borden, B.G. Efficacy of biomechanical instrumentation: A scanning electron microscopic study. *J. Endod.* **1976**, *2*, 261–266. [CrossRef] [PubMed]

Sulewski, J.G. Historical survey of laser dentistry. *Dent. Clin. N. Am.* **2000**, *44*, 717–752. [CrossRef]

Kiran, S.; Prakash, S.; Siddharth, P.R.; Saha, S.; Geojan, N.E.; Ramachandran, M. Comparative Evaluation of Smear Layer and Debris on the Canal Walls prepared with a Combination of Hand and Rotary ProTaper Technique using Scanning Electron Microscope. *J. Contemp. Dent. Pract.* **2016**, *17*, 574–581. [PubMed]

Ram, Z. Effectiveness of Root Canal Irrigation. *Oral Surg. Oral Med. Oral Pathol.* **1977**, *44*, 306–312. [CrossRef] [PubMed]

Raisingani, D.; Meshram, G.K. Cleanliness in the Root Canal System: A Scanning Electron Microscopic Evaluation of Manual and Automated Instrumentation using 4% Sodium Hypochlorite and EDTA (Glyde File Prep)—An in vitro Study. *Int. J. Clin. Pediatr. Dent.* **2010**, *3*, 173–182. [CrossRef] [PubMed]

Siqueira, J.F., Jr.; Rôças, I.N. Optimising Single-visit Disinfection with Supplementary Approaches: A Quest for Predictability. *Aust. Endod. J.* **2011**, *37*, 92–98. [CrossRef] [PubMed]

Blanken, J.; De Moor, R.J.; Meire, M.; Verdaasdonk, R. Laser induced explosive vapor and cavitation resulting in effective irrigation of the root canal. Part 1: A visualization study. *Laser Surg. Med.* **2009**, *41*, 514–519. [CrossRef] [PubMed]

Aranha, A.C.; Domıngues, F.B.; Franco, V.O.; Gutknecht, N.; Eduardo, C.P. Effects of Er:YAG and Nd:YAG lasers on dentin permeability in root surfaces: A preliminary in vitro study. *Photomed. Laser Ther.* **2005**, *23*, 504–508. [CrossRef]

Sahar-Helft, S.; Sarp, A.S.; Stabholtz, A.; Gutkin, V.; Redenski, I.; Steinberg, D. Comparison of positive-pressure, passive ultrasonic, and laser-activated irrigations on smear-layer removal from the root canal surface. *Photomed. Laser Surg.* **2015**, *33*, 129–135. [CrossRef]

21. de Groot, S.D.; Verhaagen, B.; Versluis, M.; Wu, M.K.; Wesselink, P.R.; van der Sluis, L.W. Laser-activated irrigation within r canals: Cleaning efficacy and flow visualization. *Int. Endod. J.* **2009**, *42*, 1077–1083. [CrossRef]
22. Eriksson, A.R.; Albrektsson, T. Temperature threshold levels for heat-induced hone tissue injury: A vital microscopic inj A vital microscope study in the rabbit. *J. Prosth. Dent.* **1983**, *50*, 101–107. [CrossRef] [PubMed]
23. Lan, W.H. Temperature elevation on the root surface during Nd:YAG laser irradiation in the root canal. *J. Endod.* **1999**, *25*, 155– [CrossRef] [PubMed]
24. Mazaheri, P. Temperaturentwicklung auf der Wurzeloberfläche bei einer endodonitischen Behandlung mit einem Diodenla Ph.D. Thesis, RWTH Aachen University, Aachen, Germany, 2001.
25. Gutknecht, N.; Van Gogswaardt, D.; Conrads, G.; Apel, C.; Schubert, C.; Lampert, F. Diode laser radiation and its bacterici effect in root canal wall dentin. *J. Clin. Laser Med. Surg.* **2000**, *18*, 57–60. [CrossRef]
26. Gutknecht, N. Irradiation of infected root canals with Nd:YAG lasers. A review. *LaserZahnheilkunde* **2004**, *4*, 219–226.
27. Sebbane, N.; Steinberg, D.; Keinan, D.; Vogt Sionov, R.; Farber, A.; Sahar-Helft, S. Antibacterial Effect of Er:YAG Laser Irradiat Applied by a New Side-Firing Spiral Tip on Enterococcus faecalis Biofilm in the Tooth Root Canal—An Ex Vivo Stud. *Appl.* **2022**, *12*, 12656. [CrossRef]
28. Matsumoto, H.; Yoshimine, Y.; Akamine, A. Visualization of irrigant flow and cavitation induced by Er:YAG laser within a canal model. *J. Endod.* **2011**, *37*, 839–843. [CrossRef]
29. Gregorcic, P.; Jamsek, M.; Lukač, M.; Jezersek, M. Synchronized delivery of Er:YAG laser-pulse energy during oscillation vapor bubbles. *J. Laser Health Acad.* **2014**, *1*, 6.
30. De Moor, R.J.; Blanken, J.; Meire, M.; Verdaasdonk, R. Laser induced explosive vapor and cavitation resulting in effective irrigat of the root canal. Part 2: Evaluation of the efficacy. *Lasers Surg. Med.* **2009**, *41*, 520–523. [CrossRef]
31. Zehnder, M. Root canal irrigants. *J. Endod.* **2006**, *32*, 389–398. [CrossRef]
32. Fraser, J.G. Chelating agents: Their softening effect on root canal dentin. *Oral Surg. Oral Med. Oral Pathol.* **1974**, *37*, 803– [CrossRef]
33. Bahcall, J.; Howard, P.; Miserendino, L.; Walia, H. Preliminary investigation of the histological effects of laser endodo treatment on the periradicular tissues in dogs. *J. Endod.* **1992**, *18*, 47–51. [CrossRef]
34. Kimura, Y.; Yonaga, K.; Yokoyama, K.; Kinoshita, J.; Ogata, Y.; Matsumoto, K. Root surface temperature increase during Er:Y laser irradiation of root canals. *J. Endod.* **2002**, *28*, 76–78. [CrossRef] [PubMed]
35. Jacobsen, T.; Norlund, A.; Englund, G.S.; Tranaeus, S. Application of laser technology for removal of caries: A systematic rev of controlled clinical trials. *Acta. Odont. Scandinav.* **2011**, *69*, 65–74. [CrossRef] [PubMed]
36. Rohde, M.; Mehari, F.; Klämpfl, F.; Adler, W. The differentiation of oral soft and hard tissues using laser induced breakdo spectroscopy—A prospect for tissue specific laser surgery. *J. Biophotonics* **2017**, *10*, 1250–1261. [CrossRef] [PubMed]

photonics

MDPI

nmunication

# dvantages of Using Hard X-ray Photons for Ultrafast iffraction Measurements

adimir Lipp [1,2,*], Ichiro Inoue [3] and Beata Ziaja [1,2,*]

1 Institute of Nuclear Physics, Polish Academy of Sciences, Radzikowskiego 152, 31-342 Kraków, Poland
2 Center for Free-Electron Laser Science CFEL , Deutsches Elektronen-Synchrotron DESY, Notkestr. 85, 22607 Hamburg, Germany
3 RIKEN SPring-8 Center, 1-1-1 Kouto, Sayo 679-5148, Hyogo, Japan; inoue@spring8.or.jp
* Correspondence: vladimir.lipp@desy.de (V.L.); ziaja@mail.desy.de (B.Z)

**Abstract:** We present a comparative theoretical study of silicon crystals irradiated with X-ray free-electron laser pulses, using hard X-ray photons of various energies. Simulations are performed with our in-house hybrid code XTANT based on Monte Carlo, Tight Binding and Molecular Dynamics simulation techniques. By comparing the strength of the coherently scattered signal and the corresponding electronic radiation damage for three X-ray photon energies available at the SACLA free-electron laser facility, we conclude that it would be beneficial to use higher photon energies for "diffraction-before-destruction" experiments.

**Keywords:** X-ray diffraction; X-ray; computer simulation; diffraction before destruction

check for updates

tion: Lipp, V.; Inoue I.; Ziaja, B. antages of Using Hard X-ray tons for Ultrafast Diffraction surements. *Photonics* **2023**, *10*, https://doi.org/10.3390/ tonics10080948

eived: 2 June 2023
ised: 31 July 2023
epted: 16 August 2023
lished: 18 August 2023



## 1. Introduction

X-ray free-electron lasers (XFELs) provide ultrashort X-ray pulses with unique characteristics, such as high spatial and temporal coherence, unprecedented brilliance, and MHz repetition rates [1–3]. Thanks to these characteristics, they enable applications that were not feasible with the synchrotron or high-harmonic generation sources, such as studies of ultrafast structural transformations with pump-and-probe experiments and "diffraction-before-destruction" applications [4–7]. The latter can provide opportunities for the structure determination of radiation-sensitive samples, but they also introduce many challenges [8], such as a potential target disordering during diffraction, which can deteriorate the quality of structure determination. In an earlier work, we showed that the X-ray-induced electron cascading can cause a delay in the structural changes at high photon energies [9,10], which could improve the reliability of "diffraction-before-destruction" applications for sufficiently short pulse durations.

Although the "diffraction-before-destruction" concept has been successfully applied to many protein crystals and has enabled the identification of their atomic structure [11,12], it is not clear whether one can realize advanced structure determination with this concept, going beyond the determination of atomic positions. Namely, the visualization of electron density distribution in chemical bonds and molecular orbitals still remains a challenge with ultrafast X-ray imaging. Such advanced structural studies are routinely performed at synchrotron X-ray diffraction beamlines [13]. However, applying XFEL pulses for such studies is not straightforward, as the strong excitation of valence electrons triggered by intense radiation on femtosecond timescales complicates the interpretation of the measurements. For this reason, theoretical predictions are essential to fully benefit from diffraction imaging opportunities at the XFEL facilities.

In this work, we study the effect of various hard X-ray photon energies on electronic damage and coherently scattered signals from X-ray irradiated silicon crystals. We perform computations with our hybrid code XTANT [14,15], based on the Monte Carlo (MC), Tight Binding (TB) and Molecular Dynamics (MD) simulation techniques. For realistic irradiation parameters, available at the SACLA XFEL facility in Japan [16], at the absorbed do typical for serial femtosecond crystallography of inorganic materials and small-molec systems with micrometer-size focused beams [17–19], we predict the average numbe X-ray coherent and incoherent scattering events as well as the number of ionization eve The results of our simulations point out the advantage of using harder X-ray photons ultrafast diffraction measurements.

## 2. Computational Approach

For the simulations, we use our in-house MC/TB/MD code XTANT [14,15]. It ta into account X-ray photon absorption by core and valence electrons, impact ionization the photoelectrons, Auger electrons and secondary high-energy electrons (i.e., with energ above ∼10 eV). These processes are described using atomistic approximation with a class Monte Carlo approach. The corresponding cross-sections, rates and energy levels are ta from the EPDL97 database [20], EADL97 database [21], complex dielectric function [ and from Ref. [22]. Low-energy electrons in the valence and in the conduction band ( latter with energies below 10 eV) obey the Fermi–Dirac distribution, which is updatec every time step, taking into account how many electrons were excited to or de-exci from the high-energy electron domain. Atomic trajectories are traced with a classi MD module, using potential energy surface calculated from a TB Hamiltonian. The n adiabatic electron–ion coupling, which leads to atomic heating on picosecond time sca is taken into account with rate equations, including the Boltzmann collision integral [ The code has been successfully validated in earlier studies (for a recent review, see [1 It provides comprehensive information about the evolution of X-ray irradiated sol including transient electron density and temperature, electronic and atomic configurat and optical parameters.

## 3. Results

Using the XTANT code, we performed simulations of silicon irradiated with an X- pulse at photon energies of 4, 15, and 20 keV, and at X-ray absorbed doses of 1.5, 5, a 10 eV/atom [23], comparable to those used in current serial femtosecond crystallograp of non-biological samples [17–19]. Here, we set the pulse duration of the X-ray pu to 6 fs (FWHM), which is a typical value in the standard machine operation mode SACLA [24]. For simplicity, we assumed that the sample thickness was comparable w the photon attenuation length (which is ∼10, 443 and 1038 μm for 4, 15, and 20 keV phot respectively [25]). This results in isochoric, almost homogeneous X-ray energy absorpt in the sample. Hence, we could perform constant-volume simulations for bulk silicon us a 512-atom supercell, with periodic boundary conditions.

The calculations were performed with a 5 fs time step. At each time step, the M results were averaged over 500,000 iterations.

Figure 1 shows the evolution of the number of high- and low-energy excited electr (the latter denoted as conduction band electrons), as well as the number of photons absorb by time $t$ for the case of an absorbed dose of 1.5 eV/atom. This is above the nontherm structural damage threshold in silicon (0.9 eV/atom [23,26]). Following [26], this thresh dose can be translated into the critical number of excited conduction band electro which lies in this case between 0.27 and 0.4 electrons per atom. For 4 keV, 15 keV a 20 keV photon energy (Figure 1a–c), the number of excited conduction band electr never exceeds the critical number. Note that absorbing the threshold dose yields a f disordering of the silicon structure on timescales much longer than the pulse duratior 6 fs considered here [23]. Still, some atomic dislocations may have possibly started on su a short timescale. However, at time zero, the number of excited electrons ranges from o 0.03 for 4 keV photons down to 0.006 for 20 keV photons. Because time zero correspo to the maximal intensity of the X-ray pulse, yielding the strongest diffraction signal, su a prediction indicates that during the 6 fs long X-ray pulse, the X-ray-induced sam damage should not significantly affect the quality of the structure determination in all the considered cases.

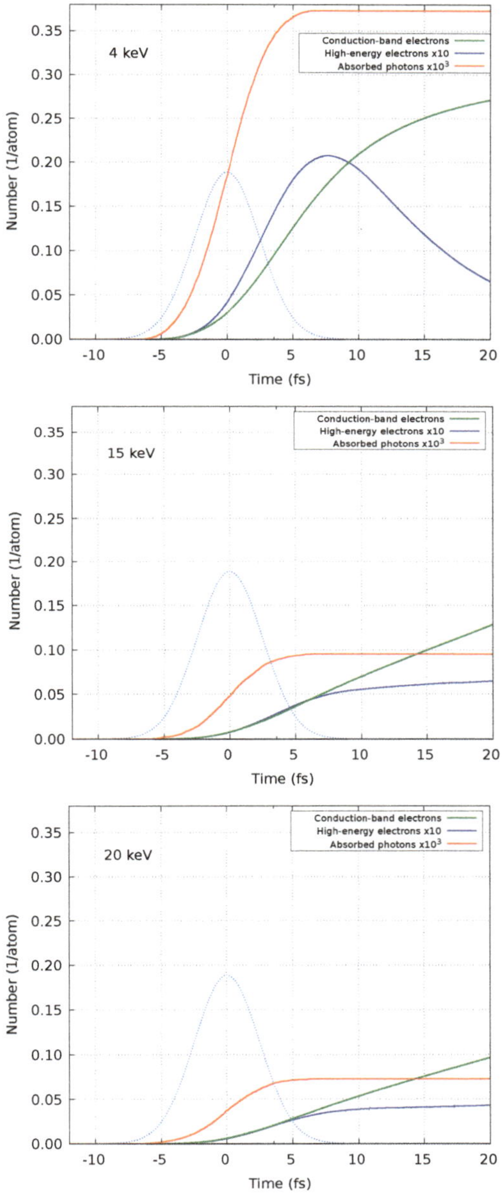


(a) (b) (c)

**Figure 1.** Number of conduction band (i.e., low-energy) electrons (green solid line) and high-energy electrons (blue line) in silicon as a function of time. This was calculated with the XTANT code, assuming irradiation of silicon by an X-ray pulse with 4 (**a**), 15 (**b**), and 20 keV (**c**) photons. The pulse duration was 6 fs, and the average absorbed dose was 1.5 eV/atom. The pulse shape is shown with a light-blue, dotted line in arbitrary units. The number of absorbed X-ray photons by time $t$ is also depicted for comparison (red line).

Similar observations are made for higher absorbed doses. Figures 2 and 3 pres the results regarding the number of excited electrons and absorbed photons for the X- absorbed doses of 5 eV/atom and 10 eV/atom, respectively. For the lowest photon ene 4 keV, the number of excited electrons exceeds the critical electron number at the enc the pulse. However, for higher photon energies, 15 keV and 20 keV, this number ne exceeds 0.20 electrons per atom. Also, at time zero, the number of excited electron atom ranges from 0.07 for 4 keV photons down to 0.02 for 20 keV photons at the absorl dose of 5 eV/atom, and from 0.13 for 4 keV photons down to 0.03 for 20 keV photon the absorbed dose of 10 eV/atom. Again, like in case of the 1.5 eV/atom dose (Figure the application of higher photon energies at the same absorbed dose yields a much sma average ionization degree in the irradiated sample than in the case of the lower pho energy applied.

This shows a clear advantage of using harder X-ray photons for high-resolut crystallography. Given that the typical crystallographic reliability factor in charge den studies is in the region of a few percent [27], the small electron excitation predicted by simulation for 20 keV photons implies the feasibility of the nearly damage-free visualizat of electron density distribution even when using an intense X-ray beam providing absorbed dose above the damage threshold.

Using the XTANT code with the electron–phonon coupling switched on (see [23]), performed test simulations for the case of 4 keV photon energy, and 1.5 and 5 eV/at absorbed doses. The obtained plots were indistinguishable from those obtained with the electron–phonon coupling taken into account (Figures 1a and 2a). We then concluc that the electron–ion coupling did not affect the presented results on the short time sca considered. Therefore, all the results were calculated without taking this into accou which decreased the overall computational costs.

Concerning the scattering signal, Figure 4 compares the number of coherent and ir herent photon scatterings with the number of photoabsorption events for the conside simulation conditions. The numbers were estimated with the respective cross secti taken from the NIST XCOM database [28]. This shows that the coherent scattering sig in the single-shot imaging increases with increasing photon energy, which is benefic The disadvantage is the increase in the incoherent signal contributing to the backgrou However, such an increase in the incoherent signal would not deteriorate the quality structural studies of crystalline materials because the contribution of incoherent scatter can be readily eliminated by estimating the background signal for each Bragg spot [29]

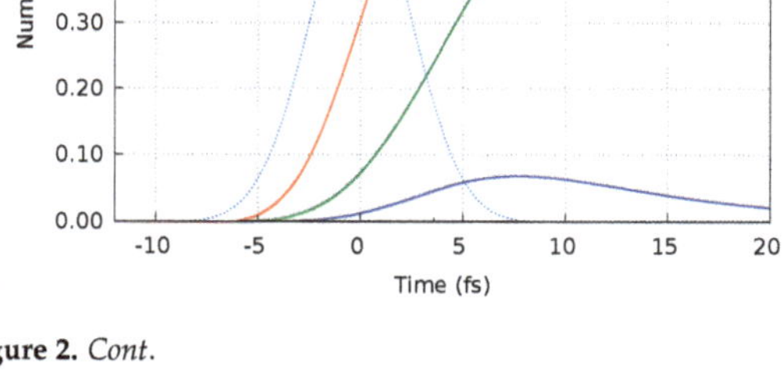


(a)

**Figure 2.** *Cont.*

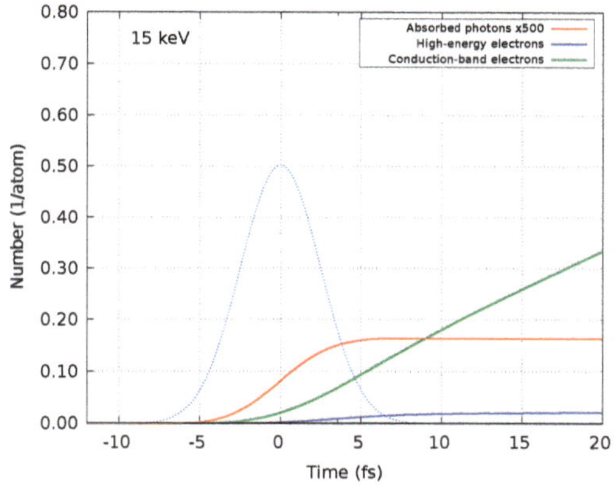


(**b**)

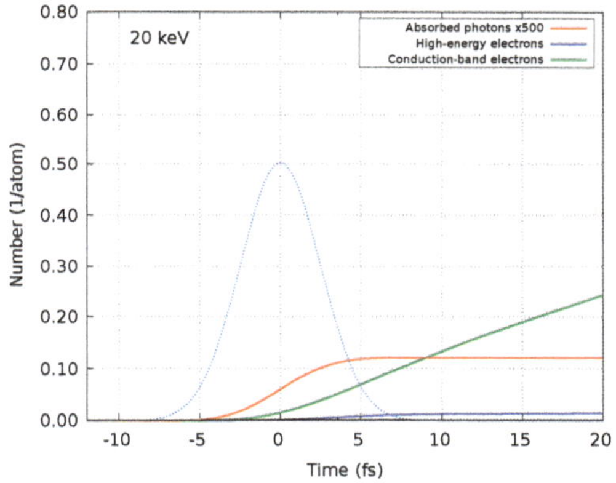


(**c**)

**Figure 2.** Number of conduction band (i.e., low-energy) electrons (green solid line) and high-energy electrons (blue line) in silicon as a function of time. This was calculated with the XTANT code, assuming irradiation of silicon by an X-ray pulse with 4 (**a**), 15 (**b**), and 20 keV (**c**) photons. The pulse duration was 6 fs, and the average absorbed dose was 5 eV/atom. The pulse shape is shown with a light-blue, dotted line in arbitrary units. The number of absorbed X-ray photons by time *t* is also depicted for comparison (red line).

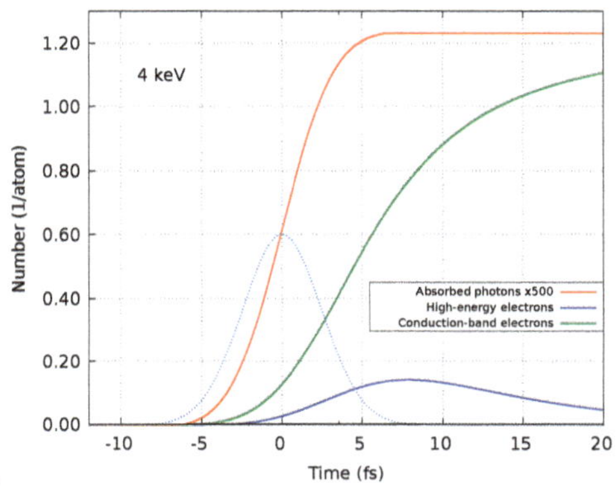


(**a**)

**Figure 3.** *Cont.*

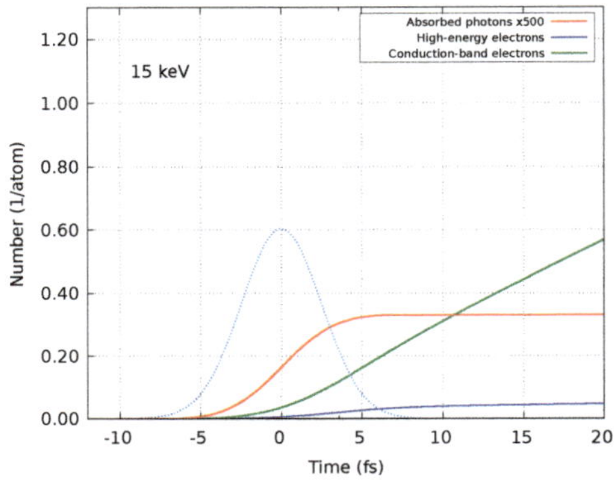


**(b)**

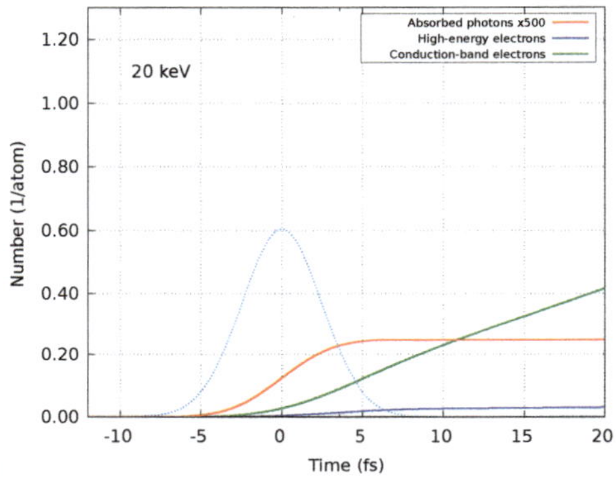


**(c)**

**Figure 3.** Number of conduction band (i.e., low-energy) electrons (green solid line) and high-ene electrons (blue line) in silicon as a function of time. This was calculated with the XTANT c assuming irradiation of silicon by an X-ray pulse with 4 (**a**), 15 (**b**), and 20 keV (**c**) photons. The p duration was 6 fs, and the average absorbed dose was 10 eV/atom. The pulse shape is shown w a light-blue, dotted line in arbitrary units. The number of absorbed X-ray photons by time *t* is a depicted for comparison (red line).

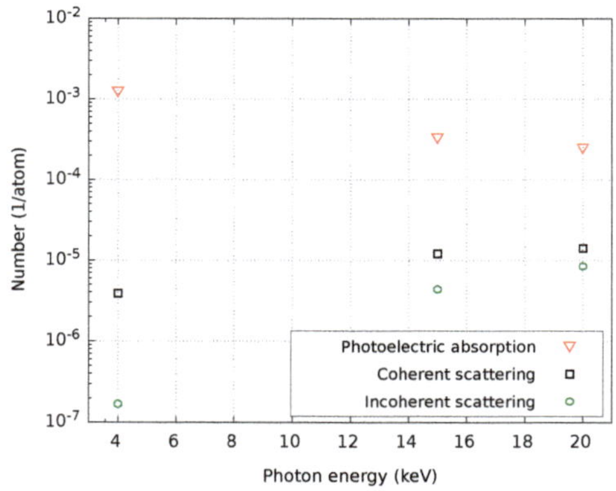


**Figure 4.** Number of coherent and incoherent photon scattering events and number of photoioniza events in silicon during a 6 fs long FEL pulse. This was obtained for photon energies of 4, and 20 keV at the absorbed dose of 5 eV/atom. The corresponding cross-sections were taken fr the NIST XCOM database [28] .

## 4. Discussion

As we mentioned above, we assume a homogeneous energy distribution across the whole simulation volume, because we do not perform quantitative studies to be compared to the results of an experiment, but aim to identify predominant physical processes and trends, following hard X-ray irradiation. In Ref. [30], we showed how one could take the spatial pulse profile into account and perform a rigorous volume integration of the simulated diffraction signal. Such computationally expensive research is not necessary for the scope of the current work; we aim to advertise here a general idea of using hard X-rays for imaging without attempting to perform a comparison with any concrete experimental result.

Concerning the temporal pulse profile, the XFEL pulses can have a spiky structure due to the finite bandwidth (for instance, see Figure 4 of Ref. [31] and Figure 21 in Ref. [3]). When analyzing experimental data on nonlinear X-ray effects, it is important to consider these spiky structures because the instantaneous intensity can be higher than the average intensity. However, we do not expect such effects in the cases studied here, because (i) the electron photoexcitation process is linear, and (ii) the timescale of the electron photoexcitation processes is much longer than the temporal spike width. Hence, we use a temporal Gaussian pulse profile in the simulations.

Considering an eventual effect of the finite spread of pulse energy on electron excitation, since the photoabsorption cross-section will be almost the same for the whole spectral range of the XFEL pulse (which is located far from any inner-shell ionization threshold, and with very small relative energy spread of $dE/E \sim 0.3\%$ or less), we can accurately describe the electron excitation processes using the monochromatic XFEL pulse.

Our simulations of X-ray-irradiated silicon crystals have shown that the electronic damage of the sample quantified by its average ionization degree (i.e., the number of excited conduction band electrons per atom) at a fixed absorbed dose decreases with the increasing X-ray photon energy. This also results in a much slower increase in the sample ionization degree with time for higher photon energies, increasing the onset for structural changes in the irradiated sample (see [10]). Therefore, it should be beneficial for the diffraction-before-destruction experiments to use higher-energy photons. They are already available at FEL facilities such as SACLA, for example [16].

Our study demonstrates that electronic damage can be to a large extent mitigated by the choice of proper photon energy. This opens a promising pathway towards advanced structure determination, i.e., the visualization of electron density distribution with ultrafast X-ray pulses.

## 5. Conclusions and Outlook

In summary, we studied electronic damage in silicon during and shortly after its irradiation with an XFEL pulse. The simulations were performed at the absorbed X-ray doses relevant for ultrafast diffraction experiments. The results demonstrate that electronic excitation becomes less significant as the photon energy increases, indicating a potential benefit of using hard X-ray photons for structure determination. In particular, the use of 20 keV XFEL pulses with a duration of 6 fs, as available at SACLA, allows the suppression of the electron excitation to the level where one can visualize electron density distribution in a nearly damage-free condition. This advantage of using hard X-ray photons also holds at higher absorbed doses, such as those typically used in diffraction experiments with silicon. Our simulations thus indicate the feasibility of charge density studies with intense XFEL pulses. This will provide new scientific opportunities that are challenging with conventional X-ray sources, such as the damage-free visualization of chemical bonding in inorganic–organic hybrid materials and the visualization of ultrafast electron transfer induced by femtosecond optical laser pulses.

**Author Contributions:** Conceptualization, B.Z. and I.I.; formal analysis, V.L. and B.Z.; investiga V.L. and B.Z.; methodology, V.L. and B.Z.; software, V.L. and B.Z.; supervision, B.Z.; validation, visualization, V.L.; writing—original draft, V.L.; writing—review and editing, V.L., B.Z. and I.I. authors have read and agreed to the published version of the manuscript.

**Funding:** This research received no funding.

**Data Availability Statement:** The data presented in this study are available on request from corresponding author, V.L.

**Acknowledgments:** We thank DESY Maxwell cluster for the computational resources.

**Conflicts of Interest:** The authors declare no conflict of interest.

## Abbreviations

The following abbreviations are used in this manuscript:

| | |
|---|---|
| XFEL | X-ray free-electron laser |
| MC | Monte Carlo |
| TB | Tight Binding |
| MD | Molecular Dynamics |

## References


1. Yabashi, M.; Tanaka, H. The next ten years of X-ray science. *Nat. Photonics* **2017**, *11*, 12–14. [CrossRef]
2. McNeil, B.W.; Thompson, N.R. X-ray free-electron lasers. *Nat. Photonics* **2010**, *4*, 814–821. [CrossRef]
3. Pellegrini, C.; Marinelli, A.; Reiche, S. The physics of X-ray free-electron lasers. *Rev. Mod. Phys.* **2016**, *88*, 015006. [CrossRef]
4. Neutze, R.; Wouts, R.; Van der Spoel, D.; Weckert, E.; Hajdu, J. Potential for biomolecular imaging with femtosecond X-ray pu *Nature* **2000**, *406*, 752–757. [CrossRef] [PubMed]
5. Gaffney, K.; Chapman, H.N. Imaging atomic structure and dynamics with ultrafast X-ray scattering. *Science* **2007**, *316*, 1444–1 [CrossRef]
6. Boutet, S.; Lomb, L.; Williams, G.J.; Barends, T.R.; Aquila, A.; Doak, R.B.; Weierstall, U.; DePonte, D.P.; Steinbrene Shoeman, R.L.; et al. High-resolution protein structure determination by serial femtosecond crystallography. *Science* **2** *337*, 362–364. [CrossRef] [PubMed]
7. Chapman, H.N.; Caleman, C.; Timneanu, N. Diffraction before destruction. *Philos. Trans. R. Soc. B* **2014**, *369*, 20130313. [Cross
8. Ziaja, B.; Jurek, Z.; Medvedev, N.; Saxena, V.; Son, S.K.; Santra, R. Towards Realistic Simulations of Macromolecules Irradia under the Conditions of Coherent Diffraction Imaging with an X-ray Free-Electron Laser. *Photonics* **2015**, *2*, 256–269. [Cross
9. Inoue, I.; Deguchi, Y.; Ziaja, B.; Osaka, T.; Abdullah, M.M.; Jurek, Z.; Medvedev, N.; Tkachenko, V.; Inubushi, Y.; Kasai, H.; e Atomic-scale visualization of ultrafast bond breaking in X-ray-excited diamond. *Phys. Rev. Lett.* **2021**, *126*, 117403. [CrossRe
10. Inoue, I.; Tkachenko, V.; Kapcia, K.J.; Lipp, V.; Ziaja, B.; Inubushi, Y.; Hara, T.; Yabashi, M.; Nishibori, E. Delayed Onset Directionality of X-Ray-Induced Atomic Displacements Observed on Subatomic Length Scales. *Phys. Rev. Lett.* **2022**, *128*, 223 [CrossRef]
11. Schlichting, I. Serial femtosecond crystallography: The first five years. *IUCrJ* **2015**, *2*, 246–255. [CrossRef] [PubMed]
12. Barends, T.R.; Stauch, B.; Cherezov, V.; Schlichting, I. Serial femtosecond crystallography. *Nat. Rev. Methods Prim.* **2022**, *2* [CrossRef] [PubMed]
13. Coppens, P. *X-ray Charge Densities and Chemical Bonding*; International Union of Crystallography: Chester, UK , 1997; Volum
14. Medvedev, N.; Jeschke, H.O.; Ziaja, B. Nonthermal phase transitions in semiconductors induced by a femtosecond extre ultraviolet laser pulse. *New J. Phys.* **2013**, *15*, 015016. [CrossRef]
15. Medvedev, N.; Tkachenko, V.; Lipp, V.; Li, Z.; Ziaja, B. Various damage mechanisms in carbon and silicon materials un femtosecond X-ray irradiation. *4Open* **2018**, *1*, 3. [CrossRef]
16. Yumoto, H.; Mimura, H.; Koyama, T.; Matsuyama, S.; Tono, K.; Togashi, T.; Inubushi, Y.; Sato, T.; Tanaka, T.; Kimura, T.; e Focusing of X-ray free-electron laser pulses with reflective optics. *Nat. Photonics* **2013**, *7*, 43–47. [CrossRef]
17. Schriber, E.A.; Paley, D.W.; Bolotovsky, R.; Rosenberg, D.J.; Sierra, R.G.; Aquila, A.; Mendez, D.; Poitevin, F.; Blaschke, Bhowmick, A.; et al. Chemical crystallography by serial femtosecond X-ray diffraction. *Nature* **2022**, *601*, 360–365. [Cross [PubMed]
18. Støckler, L.J.; Krause, L.; Svane, B.; Tolborg, K.; Richter, B.; Takahashi, S.; Fujita, T.; Kasai, H.; Sugahara, M.; Inoue, I.; e Towards pump–probe single-crystal XFEL refinements for small-unit-cell systems. *IUCrJ* **2023**, *10*, 103. [CrossRef] [PubMed
19. Takaba, K.; Maki-Yonekura, S.; Inoue, I.; Tono, K.; Hamaguchi, T.; Kawakami, K.; Naitow, H.; Ishikawa, T.; Yabashi, Yonekura, K. Structural resolution of a small organic molecule by serial X-ray free-electron laser and electron crystallogra *Nat. Chem.* **2023**, *15*, 491–497. [CrossRef]
20. Cullen, D.E.; Hubbell, J.H.; Kissel, L. *EPDL97: The Evaluated Photon Data Library*; 97 Version; Technical Report; Lawre Livermore National Lab. (LLNL): Livermore, CA, USA, 1997.

Perkins, S.; Cullen, D.; Chen, M.; Rathkopf, J.; Scofield, J.; Hubbell, J. *Tables and Graphs of Atomic Subshell and Relaxation Data Derived from the LLNL Evaluated Atomic Data Library (EADL). Z = 1–100*; Technical Report; Lawrence Livermore National Lab.: Livermore, CA, USA, 1991.

Palik, E.D. *Handbook of Optical Constants of Solids*; Academic Press: San Diego, CA, USA, 1985.

Medvedev, N.; Li, Z.; Ziaja, B. Thermal and nonthermal melting of silicon under femtosecond x-ray irradiation. *Phys. Rev. B* **2015**, *91*, 054113. [CrossRef]

Inoue, I.; Hara, T.; Inubushi, Y.; Tono, K.; Inagaki, T.; Katayama, T.; Amemiya, Y.; Tanaka, H.; Yabashi, M. X-ray Hanbury Brown-Twiss interferometry for determination of ultrashort electron-bunch duration. *Phys. Rev. Accel. Beams* **2018**, *21*, 080704. [CrossRef]

Henke, B.; Gullikson, E.; Davis, J. X-ray interactions: Photoabsorption, scattering, transmission, and reflection at E = 50–30.000 eV, Z = 1–92. *At. Data Nucl. Data Tables* **1993**, *54*, 181–342. [CrossRef]

Follath, R.; Koyama, T.; Lipp, V.; Medvedev, N.; Tono, K.; Ohashi, H.; Patthey, L.; Yabashi, M.; Ziaja, B. X-ray induced damage of $B_4C$-coated bilayer materials under various irradiation conditions. *Sci. Rep.* **2019**, *9*, 2029. [CrossRef]

Koritsanszky, T.S.; Coppens, P. Chemical applications of X-ray charge-density analysis. *Chem. Rev.* **2001**, *101*, 1583–1628. [CrossRef]

Berger, M.J.; Hubbell, J.H.; Seltzer, S.M.; Chang, J.; Coursey, J.S.; Sukumar, R.; Zucker, D.S.; Olsen, K. XCOM: Photon Cross Sections Database. **2010**, NIST, PML, Radiation Physics Division. https://dx.doi.org/10.18434/T48G6X (accessed on 17 August 2023).

White, T.A.; Kirian, R.A.; Martin, A.V.; Aquila, A.; Nass, K.; Barty, A.; Chapman, H.N. CrystFEL: A software suite for snapshot serial crystallography. *J. Appl. Crystallogr.* **2012**, *45*, 335–341. [CrossRef]

Tkachenko, V.; Abdullah, M.M.; Jurek, Z.; Medvedev, N.; Lipp, V.; Makita, M.; Ziaja, B. Limitations of Structural Insight into Ultrafast Melting of Solid Materials with X-ray Diffraction Imaging. *Appl. Sci.* **2021**, *11*, 5157. . [CrossRef]

Inubushi, Y.; Tono, K.; Togashi, T.; Sato, T.; Hatsui, T.; Kameshima, T.; Togawa, K.; Hara, T.; Tanaka, T.; Tanaka, H.; et al. Determination of the Pulse Duration of an X-Ray Free Electron Laser Using Highly Resolved Single-Shot Spectra. *Phys. Rev. Lett.* **2012**, *109*, 144801. [CrossRef] [PubMed]

photonics

Article

# Temperature-Dependent Optical Properties of Bismuth Triborate Crystal in the Terahertz Range: Simulation of Terahertz Generation by Collinear Three-Wave Mixing in the Main Crystal Planes

**Dmitry Ezhov [1,*], Nazar Nikolaev [2,3], Valery Antsygin [2], Sofia Bychkova [2], Yury Andreev [4] and Valery Svetlichnyi [1,*]**

1 Laboratory of Advanced Materials and Technology, Siberian Physical Technical Institute, Tomsk State University, Tomsk 634050, Russia
2 Laboratory of Terahertz Photonics, Institute of Automation and Electrometry, Siberian Branch of the Rus Academy of Sciences, Novosibirsk 630090, Russia; nazar@iae.nsk.su (N.N.); antsigin@iae.nsk.su (V.A.)
3 Laboratory of Laser Biophysics, Institute of Laser Physics, Siberian Branch of the Russian Academy of Sciences, Novosibirsk 630090, Russia
4 Laboratory of Geosphere-Biosphere Interactions, Institute of Monitoring of Climatic and Ecological Syst Siberian Branch of the Russian Academy of Sciences, Tomsk 634055, Russia; yuandreev@yandex.ru
* Correspondence: ezhov_dm@phys.tsu.ru (D.E.); v_svetlichnyi@bk.ru (V.S.)

**Abstract:** Terahertz (THz) frequency generation via nonlinear optical techniques is of partic interest due to the immense potential of this type of radiation in various scientific fields, rang from medicine to telecommunications. Selecting suitable nonlinear media for laser frequency do conversion presents a challenging task. Considering an approach that uses nonlinear crystals w high radiation resistance, pumped by intense laser pulses near their damage threshold, we suggest crystal of bismuth triborate ($BiB_3O_6$, BIBO). Compared to other borate-class crystals, BIBO exhi relatively high coefficients of quadratic susceptibility. In this paper, we have studied the opt properties of BIBO samples in a wide spectral range from 0.1 to 2.1 THz at temperatures of 473, 295, and 77 K using Terahertz Time-Domain Spectroscopy (THz-TDS). Furthermore, we simula collinear three-wave interactions with nonzero efficiency for difference frequency generation (D in the THz range. For the pump wavelengths of about 800 nm, we determined phase-matching ( conditions and compared the generation efficiency for different crystal cuts. The potential of utiliz BIBO crystal for terahertz frequency generation is discussed.

**Keywords:** bismuth triborate; terahertz frequency; refractive index; absorption coeffici temperature dependencies; three-wave interactions; THz frequency generation; phase matching



**Citation:** Ezhov, D.; Nikolaev, N.; Antsygin, V.; Bychkova, S.; Andreev, Y.; Svetlichnyi, V. Temperature-Dependent Optical Properties of Bismuth Triborate Crystal in the Terahertz Range: Simulation of Terahertz Generation by Collinear Three-Wave Mixing in the Main Crystal Planes. *Photonics* **2023**, *10*, 713. https://doi.org/10.3390/photonics10070713

Received: 5 May 2023
Revised: 16 June 2023
Accepted: 20 June 2023
Published: 22 June 2023



## 1. Introduction

The rapid development of terahertz (THz) technologies in recent decades in the fi of biomedicine [1,2], monitoring and security systems [3], remote sensing of the at sphere [4–6], wireless communications [7,8], and scientific research requires high-intens and efficient radiation sources with the necessary spectral properties. There are numer techniques for generating THz radiation, from classical radio-technical antenna devi to synchrotron sources. Devices based on photoconductive antennas or optical rectif tion in nonlinear crystals are widely used to solve many spectroscopic problems [9, The approach based on phase-matched (PM) three-wave mixing or difference freque generation (DFG) in nonlinear crystals has great potential for high-intensity, spectr bright THz generation that is tunable in frequency over a wide range [11,12]. This appro is common for optical parametric generation in the infrared (IR) range [13].

The efficiency of DFG depends not only on the corresponding nonlinear susceptibilities but also on the number of properties of a nonlinear crystal, such as the laser damage threshold, absorption coefficients, birefringence, and group-velocity mismatch in different spectral ranges, PM conditions, and geometric dimensions. All these properties depend in a complex way on the pump radiation parameters, including the pump wavelength, pulse duration, beam spatial parameters, and initial pump beam intensity [14,15]. For example, in an LBO crystal that does not possess high nonlinear susceptibility, a record, up to 90%, energy efficiency of second-harmonic generation (SHG) was achieved at a pump intensity of 775 MW/cm$^2$ [16].

The first works devoted to DFG in the THz range considered semiconductor crystals with large nonlinear coefficients, such as $ZnGeP_2$ and GaSe, popular for parametric generation in the mid-IR range [17,18]. In recent years, oxygen-containing nonlinear crystals, for example, β-BBO, LBO, BBO, etc., have been considered for THz frequency generation [19–22]. They have much lower nonlinear coefficients, but, at the same time, possess laser-induced damage thresholds that are several orders of magnitude higher and a much lower absorption coefficient in the main transparency window. This fact should allow for the frequency conversion of high-peak-intensity emission from short-pulsed laser sources at visible wavelengths. The lower effective nonlinear coefficient and, hence, lower figure-of-merit (FOM) values compared to semiconductor crystals do not prevent the higher conversion efficiency since it is proportional to the pump intensity [23].

Among the known nonlinear crystals of the borate family, special attention should be paid to bismuth triborate, $BiB_3O_6$ (BIBO). BIBO is a positive biaxial crystal that belongs to the monoclinic crystal system, space group C2 [24]. Due to its chemical composition, structure, and mature growing technology, it has a low absorption coefficient and a high laser damage threshold (>300 GW/cm$^2$ under 45 fs pump at 800 nm) in the visible and near-infrared spectral ranges occupied by powerful solid-state lasers (Nd:YAG, Ti:sapphire) [25,26]. Compared to other borate crystals, its nonlinear coefficients are approximately 4 and 1.7 times larger than those of the widely used LBO and β-BBO nonlinear crystals, respectively. These mark the BIBO crystal as an efficient device for an optical parametric generation of femtosecond pulses [27,28]. Furthermore, BIBO's linear absorption coefficient at the wavelength of 1064 nm is two times lower than that of the β-BBO crystal [29]. These properties make BIBO more attractive for high-intensity THz wave generation. In addition, the crystal is non-hygroscopic, thermally stable (has no phase transitions from 4 K and up to the melting point of 1019 K [30]), has sufficient hardness (5–6 on the Mohs scale), is good for optical polishing, and shows chemical stability, which is important for practical applications, especially in outdoor scenarios. To date, the terahertz optical properties of BIBO have not been studied in detail. Our preliminary study on absorption coefficients and refractive indices in the THz range for the principal optical axes of the crystal performed at room temperature allowed us to estimate possible PM conditions for DFG, revealing the prospects of BIBO [31,32]. To discover the full potential of the crystal for THz frequency generation, further verification is required.

A recently published article attempted to obtain THz generation from the BIBO crystal [33]. However, the authors used a sample that was cut for second-harmonic generation ($\theta$ = 28.9°). As a result, the registered THz generation output was likely inefficient, and the provided THz spectra were questionable and possibly misinterpreted.

In this work, the optical properties of the BIBO crystal were thoroughly studied in the THz range at different temperatures. The data obtained were used to simulate all possible PM conditions for collinear three-wave mixing, considering the THz radiation generation. We evaluated the efficiency and determined the optimal conditions for nonlinear conversion. This knowledge is crucial for the development of efficient and reliable nonlinear optical devices that employ BIBO crystals.

## 2. Materials and Methods

### *2.1. Properties of Samples under Test*

We studied the THz optical properties of four BIBO crystal samples with a cl aperture diameter of 1 cm. Two samples were cut perpendicular to the dielectric *y*-a and further referred to as y-cut samples with thicknesses of 526 and 5080 μm. The ot two samples were x-cut with thicknesses of 518 and 5430 μm. Using additional th samples allows us to measure absorption coefficients at a frequency range < 0.4 THz m accurately compared with the thin samples. The samples were obtained from Castec and their thicknesses were measured using a calibrated micrometer and refined us a measuring microscope. According to Castech's inspection report, the samples w cut with an angle tolerance not exceeding 5′ for thick samples and 10′ for thin samp Note that generally, for monoclinic point group 2 crystals, only one of the dielectric a coincides with the crystallographic one. Thereby, the *xyz* frame rotates around one of axes, and the angle depends on both wavelength and temperature [34]. Specifically, in case of the BIBO crystal, the *y*- and *z*-axes rotate about the crystallographic *b*-axis, but th wavelength dependence does not exceed $\Delta\Phi = \pm 1°$ in the main transparency window [ It is important to consider this fact when conducting polarimetry measurements to ens accurate results. The arrangement of the dielectric (*xyz*), crystallophysical (*XYZ*), a crystallographic (*abc*) coordinate frames for the y-cut sample is depicted in Figure 1.

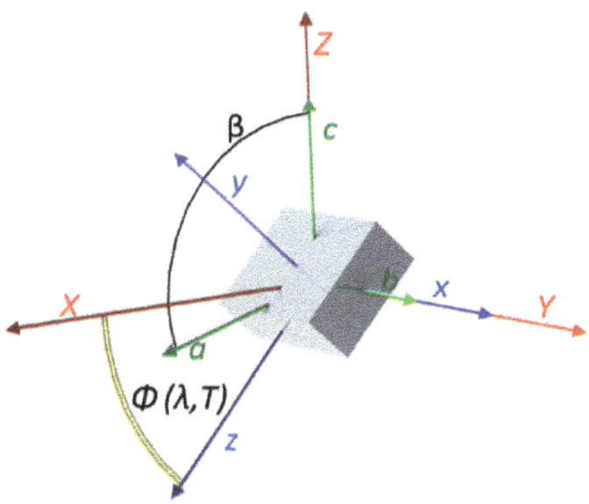


**Figure 1.** Mutual arrangement of the BIBO crystal coordinate frames in the y-cut sample.

### *2.2. Temperature-Dependent THz Polarimetry*

Terahertz optical properties of the BIBO crystals were measured in the transmiss geometry using manufactured in-house broadband THz-TDS, the scheme of which can found in [35,36]. As a source of the femtosecond pulses, we used the emission of the seco harmonic of the Er-doped fiber laser FFS-SHG (Toptica Photonics AG, Germany, Gräfelfi with a mean power of 80 mW, central wavelength of 775 nm, and pulse duration of 13 The THz waves were generated by an interdigitated large-area photoconductive anten iPCA-21-05-300-800-h (Batop, Germany, Jena), and detected utilizing the free space elec optic sampling technique in a (110)-cut 2 mm thick ZnTe crystal. Samples were placed i heating cell controlled by a PID regulator, TRM251 (Owen, Russia, Moscow), capable maintaining the temperature with an accuracy of 0.1 °C. To avoid a temperature gradien the crystal, we preliminary measured the time required to establish thermal equilibrium a heating cell using a calibrated thermocouple mounted on the surface of the sample un test before conducting THz measurements. The use of polytetrafluoroethylene windo covering the heating cell minimized temperature fluctuations caused by air drift. average, it took about 10 min to establish temperature equilibrium.

The optical properties of two principal dielectric axes of the BIBO samples were measured without rotating and replacing the sample, following the procedure described in [37]. According to this technique, the sample understudy was installed at an approximately 45° angle relative to the initial polarization of the THz field. The crystal's principal axis of interest was selected by simultaneously rotating in concert two high-performance grid polarizers at ±45°.

Terahertz absorption coefficients and refractive indices of the samples under study were extracted from the time-domain signals using a technique described in [38].

## 3. Results

### *3.1. Optical Properties of BIBO Crystal in the THz Range at Room Temperature*

The THz optical properties of the BIBO crystal were measured for both thick and thin samples at room temperature. The use of thin samples enabled the spectra to be acquired at frequencies higher than 1.5 THz, where the absorption coefficient exceeded the value of 10 $cm^{-1}$. Conversely, thick samples allowed us to obtain valid data in the frequency range below 0.5 THz, where the absorption coefficients were less than 0.1 $cm^{-1}$. By combining the spectra for both thicknesses of the samples, reliable data were obtained in the range of 0.1–2.1 THz. As a result, the extracted absorption coefficients and the refractive indices for the BIBO principal axes in a broad spectral range are shown in Figure 2.

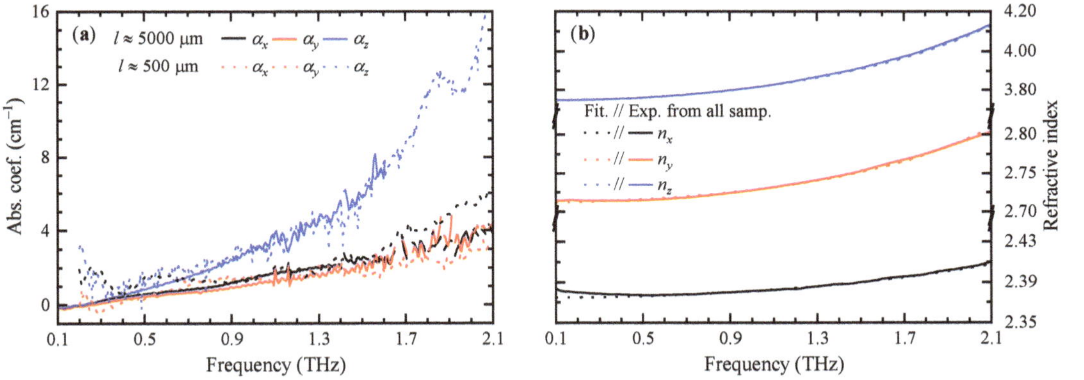


**Figure 2.** Optical properties of the BIBO crystal in the THz range at room temperature: (**a**) measured absorption coefficient of the crystal principal dielectric axes for two different sets of samples with thicknesses of 0.5 and 5 mm; (**b**) combined refractive indices of principle axes and their Sellmeier fitting.

The absorption coefficient along the dielectric *z*-axis varies from almost "0" $cm^{-1}$ (considering the measurement error) in the frequency region below 0.3 THz up to 16 $cm^{-1}$ at 2.1 THz. The absorption coefficients along the *x*- and *y*-axes were found to be practically similar to each other and increased slightly up to 4 $cm^{-1}$ at a frequency of 2.1 THz. The refractive index of the *x*-axis $n_x$ slightly deviates upwards from a straight line in the frequency range below 0.4 THz. This deviation is most probably due to the Gouy phase shift of the focused THz wave in the thick (5 mm) sample, which, in our case, is not compensated by the spectra-processing algorithms [39]. Therefore, the data points in this range were omitted while fitting the curve with the Sellmeier equation. The final dispersion expressions for the principal axes of the BIBO crystal at room temperature are shown below:

$$n_x^2 = 4.940 + \frac{0.700 \cdot \lambda^2}{\lambda^2 - 3777},$$
$$n_y^2 = 6.372 + \frac{0.992 \cdot \lambda^2}{\lambda^2 - 7100},$$
$$n_z^2 = 7.610 + \frac{6.456 \cdot \lambda^2}{\lambda^2 - 6515}.$$

### *3.2. Temperature-Dependent Optical Properties*

Temperature-dependent optical properties were only measured for the thick samp of the BIBO crystals. To obtain valid data on refractive indices and absorption coefficie we considered the linear thermal expansion of the crystal. Linear thermal expans coefficients were taken from [30]. Since the $y$ dielectric axis lies between the $a$ an crystallographic axes, which have different signs of thermal expansion coefficients, resulting thermal expansion of the y-cut sample is almost negligible. Cooling the sam down from room temperature to the liquid nitrogen (LN) temperature causes its len to increase by 6.3 μm, while heating from 22 to 200 °C leads to a length compressio 5.6 μm. As the $x$ dielectric axis coincides with the $b$ crystallographic axis, which has highest linear expansion coefficient, the resulting expansion of the x-cut sample is m noticeable. Cooling the x-cut sample down to the LN temperature shrinks its length 58 μm, while heating it to 200 °C leads to a 52 μm increase. Figure 3 shows the temperat dependencies of the absorption coefficients for all principal axes of the BIBO crystal.

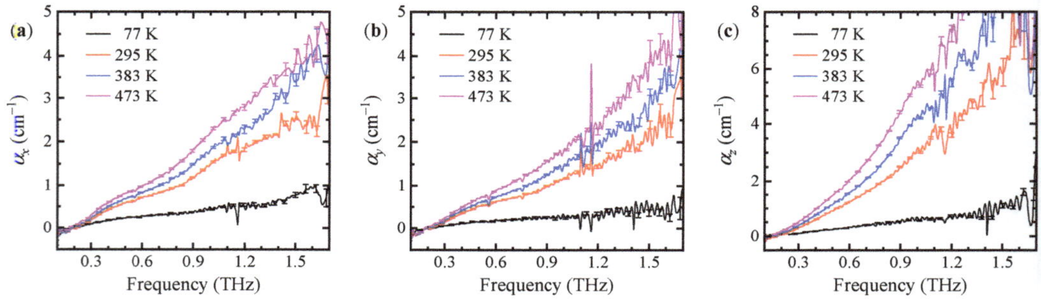


**Figure 3.** Absorption coefficient of the BIBO crystals in the THz range at different temperatu (77–473 K): (**a**) $x$-axis; (**b**) $y$-axis; (**c**) $z$-axis.

At longer wavelengths (>1500 μm), the absorption coefficients for all components less than 0.1 cm$^{-1}$ and practically do not change with the temperature within the lin of measurement accuracy. Heating the samples leads to a small linear increase in absorption coefficients in the short-wavelength region (<1500 μm or >0.2 THz). In range, the absorption coefficients along the $z$-axis of the crystal are the largest (8 cm$^{-}$ 300 μm), while the absorption along the $x$- and $y$-axes is about 1.6 times lower. Addition cooling the samples down to 77 K at these wavelengths leads to a drastic decrease absorption coefficients (from 8 to 10 times). This behavior can be described by an anal with data on the temperature-dependent THz absorption of α-BBO crystal by Prof. X Zhang et al. [40]. According to this work, the phonon absorption peak at 2.5 THz becor narrower and shifts to a higher frequency at a low temperature due to the Bose–Einst distribution of vibrational modes. And vice versa, at higher temperatures, the cer frequency of the phonon experiences a red shift while the peak width broadens. In case of the BIBO crystal, the absorption is more likely caused by the low-frequency tail phonon peak. According to Raman spectroscopy, the center frequency of this peak sho be located near 4.3 THz [41,42]. Unfortunately, the limited bandwidth of our THz-T does not allow us to reveal this phonon directly.

The temperature dependencies of the BIBO refractive indices are shown in Figure 4.

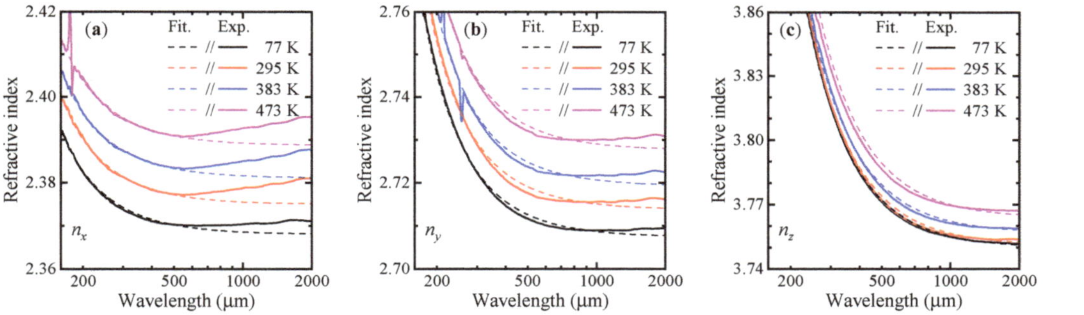

**Figure 4.** Experimentally measured and Sellmeier-fit refractive indices of the BIBO crystal in the THz range at different temperatures (77–473 K): (**a**) $x$-axis; (**b**) $y$-axis; (**c**) $z$-axis.

The long-wavelength tails of the refractive indices along the $x$- and $y$-axes decrease in proportion to the square of the temperature change upon cooling. The behavior of the refractive index along the $z$-axis is linear with respect to the temperature change. The upward deviation from the straight line of the long-wavelength tails of the refractive indices along the $x$- and $y$-axes may be related to diffraction in the thick samples, which is also confirmed by the presence of an interference pattern in the long-wavelength region of the absorption spectra in Figure 3a,b. Due to this fact, experimental points with a wavelength > 1000 μm were omitted during the fitting process. Temperature-dependent Sellmeier coefficients were found in the vicinity of the values corresponding to the room temperature (1).

The temperature dependence of the fitted Sellmeier coefficients for all principal axes is shown in Figure 5 and Table 1.

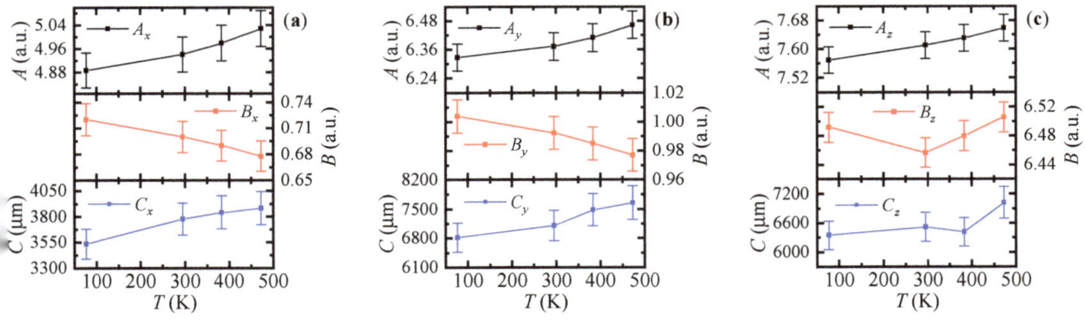

**Figure 5.** Temperature dependence of Sellmeier coefficients: (**a**) $x$-axis; (**b**) $y$-axis; (**c**) $z$-axis.

The temperature dependencies of the Sellmeier coefficients $A_x$ and $A_y$ were fitted with a quadratic polynomial, while for all other coefficients a linear fit was performed. The first derivatives of these coefficients with respect to the temperature are shown in Table 2.

**Table 1.** Sellmeier coefficients for different temperatures.

| Principle Axis ($n_i$) | $T$ (°C) | $T$ (K) | $A_i$ (a.u.) | $B_i$ (a.u.) | $C_i$ (μm$^2$) |
|---|---|---|---|---|---|
| $n_x$ | −196 | 77 | 4.888 | 0.720 | 3532 |
| | 22 | 295 | 4.940 | 0.700 | 3777 |
| | 110 | 383 | 4.979 | 0.690 | 3841 |
| | 200 | 473 | 5.028 | 0.678 | 3881 |
| $n_y$ | −196 | 77 | 6.326 | 1.004 | 6808 |
| | 22 | 295 | 6.372 | 0.992 | 7100 |
| | 110 | 383 | 6.409 | 0.985 | 7477 |
| | 200 | 473 | 6.463 | 0.977 | 7646 |
| $n_z$ | −196 | 77 | 7.569 | 6.492 | 6345 |
| | 22 | 295 | 7.610 | 6.456 | 6515 |
| | 110 | 383 | 7.630 | 6.480 | 6410 |
| | 200 | 473 | 7.658 | 6.506 | 7020 |

**Table 2.** Form of temperature derivatives of the fitted Sellmeier coefficients.

| Principle Axis ($n_i$) | $dA_i \times 10^5$ (K$^{-1}$) | $dB_i \times 10^4$ (K$^{-1}$) | $dC_i$ (μm$^2$·K$^{-1}$) |
|---|---|---|---|
| $n_x$ | $0.127 \cdot T + 0.69$ | −1.053 | 0.91 |
| $n_y$ | $0.155 \cdot T - 8.30$ | −0.667 | 2.16 |
| $n_z$ | 21.9 | 0.214 | 1.33 |

As a result, we propose a form of dispersion equations as an integral function c based on the corresponding temperature coefficients and their temperature derivative

$$n_i^2(T) = A_{i0} + \int_{T_0}^{T} dA_i dT + \frac{\left(B_{i0} + \int_{T_0}^{T} dB_i dT\right) \cdot \lambda^2}{\lambda^2 - \left(C_{i0} + \int_{T_0}^{T} dC_i dT\right)}.$$

Such modified equations allow one to determine the refractive indices in the wa length range of 150–3000 μm (0.1–2.0 THz) for any desired temperature in the range 77–473 K. We suggest using the coefficients from the equations obtained at room temp ture as the initial ones since they were derived from the combined experimental data two different thicknesses of the BIBO samples, which makes them more accurate.

To determine the refractive index value at the pump wavelength, we used ther optics dispersion formulas presented in [43]. Although these formulas were only valida for a temperature range of 20 °C to 120 °C, their linear dependence allows us to estim the refractive index at other temperatures as well.

The resulting equations enable the numerical calculation of the temperature-depend PM conditions for the down-conversion of femtosecond laser radiation into the THz ra Finding such conditions, also known as birefringent phase matching, for fixed waveleng is numerically equivalent to adjusting the refractive indices of the corresponding extrao nary waves according to the polar and azimuthal angles $\theta$ and $\varphi$ in such a way as to z the wavevector mismatch:

$$
\begin{aligned}
k_3 - k_2 - k_1 &= 0 \\
k_i(\lambda, T, \varphi, \theta) &= 2\pi \cdot n_i^{(e,o)}(\lambda, T, \varphi, \theta) \cdot \lambda_i^{-1}
\end{aligned} \quad (3)
$$

where $\lambda_i$ and $k_i$ are wavelengths in μm and wave vectors. In this case, $k_3$ and $k_2$ correspond to the pump and idler waves in the near-IR spectral range, and $k_1$ is a signal wave in the THz range ($\lambda_1 > \lambda_2 > \lambda_3$). Assuming the conversion of a broadband spectrum of the pump wave, for simplicity, we can fix $\lambda_3$. Then, for any predetermined signal wavelength $\lambda_1$ it is possible to uniquely choose an idler wavelength $\lambda_2$ in the vicinity of $\lambda_3$ to fulfill the energy conservation condition:

$$
\lambda_2 = \frac{\lambda_1 \lambda_3}{\lambda_3 + \lambda_1}. \quad (4)
$$

It is necessary to keep in mind that all types of three-wave mixing processes can be realized in the case of down-conversion into the THz range, i.e., beyond the "Reststrahlen" band of the crystal, if the corresponding effective nonlinear coefficient is not equal to zero. For example, it was found that the mixing of three extraordinary waves in GaSe crystal leads to effective THz generation [44]. To avoid any misconduct, for BIBO being a biaxial crystal, we have chosen "slow" and "fast" notation for interacting waves instead of "ordinary" and "extraordinary", generally used in the case of uniaxial crystals. For instance, in the case of three-wave mixing in the $xz$ plane of a BIBO crystal for which the condition $n_z > n_y > n_x$ is satisfied both in the optical and THz ranges, the $k_1^e = k_3^e - k_2^e$ type of interaction should be denoted as $k_1^f = k_3^f - k_2^f$ when $\theta_{int} < V_z$ and as $k_1^s = k_3^s - k_2^s$ when $\theta_{int} > V_z$ [45], where $V_z$ is an angle of the optical axes with the principal axis $z$.

The calculated wavevector mismatch for all three principal planes of BIBO is shown in Figure 6. Here, we choose $\lambda_1$ = 1000 μm as a signal wavelength in the THz range and $\lambda_3$ = 0.8 μm as a pump wavelength, corresponding to a Ti:sapphire femtosecond laser.

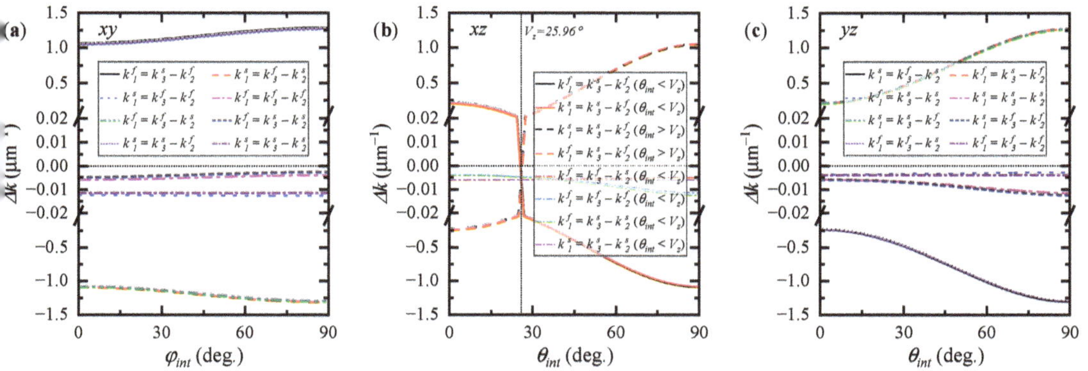


**Figure 6.** Wavevector mismatch for down-conversion into the THz range in the principal planes of BIBO crystal: (**a**) $xy$ plane; (**b**) $xz$ plane; (**c**) $yz$ plane.

All types of wave interactions in the $xy$ (Figure 6a) and $yz$ (Figure 6c) principal planes exhibit a nonzero phase mismatch for $\lambda_1$ = 1000 μm at any temperature considered. It indicates the non-fulfillment of the phase-matching conditions in these planes. On the other hand, calculations show that near the crystal optical axis in the $xz$ principal plane, the PM conditions are satisfied for $s - f = f$ and $s - f = s$ mixing when $\theta_{\text{int}} < V_z$ and for $s - f = s$ and $s - f = f$ interactions when $\theta_{\text{int}} > V_z$ (Figure 6b). However, it should be noted that these types of frequency conversion are highly sensitive to changes in angle and possess an angular bandwidth $\Delta\theta$ of only 6 mrad for a 1 mm sample. In other words, even a slight rotation of the sample by 1° results in a PM frequency offset of 1.2 THz due to the large birefringence of the BIBO crystal. These four interactions turn out to be pairwise symmetric with respect

to the angle $V_z$. The values of the effective nonlinear coefficient $d_{eff}$ correspondin
them are numerically identical in the same pairs [29]: $d_{eff} = -d_{14} \sin 2\theta$ for $s - f$
($\theta_{\text{int}} < V_z$) and $s - f = s$ ($\theta_{\text{int}} > V_z$) types, and $d_{eff} = d_{12} \cos \theta$ for $s - f = s$ ($\theta_{\text{int}} <$
and $s - f = f$ ($\theta_{\text{int}} > V_z$), assuming Kleinman symmetry due to close-to-zero absorpt
coefficients in the considered range.

## 4. Discussion

We have calculated temperature-tuning curves for the types of interactions involv
PM in the *xz* plane. The dependence of the PM angles on $\lambda_1$ is shown in Figure 7a.
semi-transparent curves correspond to PM at the LN temperature and 473 K (200 °C). Th
curves are again pairwise symmetrical with respect to the $V_z$ angle and follow it u
heating. The $V_z$ changes from 26.8° to 25.2° when the crystal is heated from 77 K to 47
Figure 7b represents these changes at the wavelength of $\lambda_3$ = 0.8 μm (dashed blue cu
and the corresponding temperature shift of the PM curves presented for the fixed sig
wavelength at $\lambda_1$ = 1000 μm.

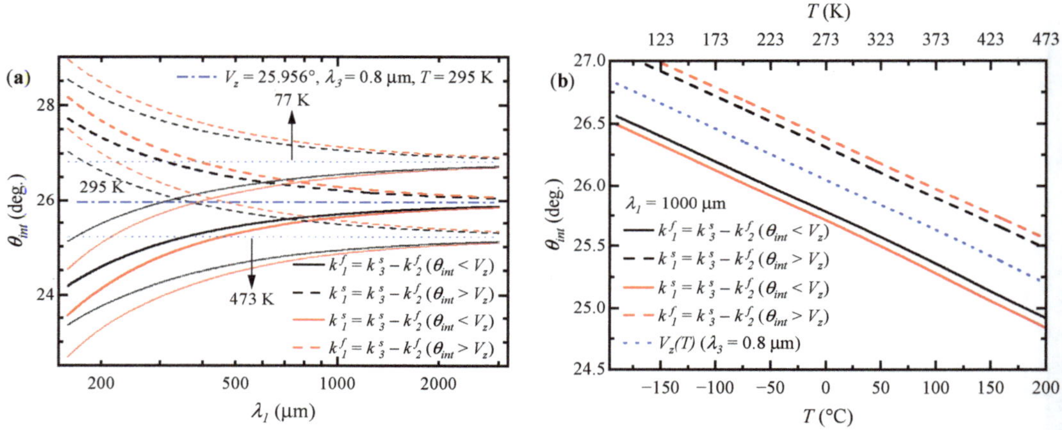


**Figure 7.** Temperature dependence of phase-matching conditions for the *xz* plane: (**a**) phase-matc
angles for different signal wavelengths $\lambda_1$; (**b**) phase-matching angles for $\lambda_1$ = 1000 μm.

Let us consider a crystal cut with $\theta_{int}$ = 25.68° and $\phi$ = 0° which corresponds to
PM wavelength of $\lambda_1$ = 1000 μm at room temperature. Temperature-tuning curves for
case are shown in Figure 8. Changing the crystal temperature makes it possible to tune
signal wavelength $\lambda_1$ over a wide spectral range. Varying the temperature from −50
to 50 °C, allows for access to the wavelength of 590–2165 μm (0.14–0.65 THz). Howe
for temperatures from 70 °C to 106 °C, the curves show quite a radical slope, wh
makes tuning almost unusable in real life since it requires the most precise temperat
control. Moreover, in this case, the signal wavelength $\lambda_1$ shifts to the millimeter rang
microwaves rather than the THz region.

In this case, the selected crystal cut has a "walk-off" angle $\rho$ that does not exc
60 mrad. Therefore, the effective aperture length $L_a$ for a pump beam with a diam
of 1 mm is found to be more than 16 mm. Additionally, the coherence length for
optical rectification of femtosecond pulses generating THz radiation should be ab
1 mm, according to [46]. Finally, these data allow us to simulate the resulting spectr
THz generation under a high-intensity pump. To simulate DFG THz spectra, we use
formula for the DFG conversion efficiency in the infinite plane wave under a non-deple
pump approximation, including the effects of the linear absorption from [23] and the val
of corresponding nonlinearity tensors in the dielectric frame from [29]. Simulated D

THz spectra for a 5 mm thick PM cut crystal, as well as for an x-cut (*yz* plane) crystal of the same thickness under a 1 TW·cm$^{-2}$ pump at 800 nm, at room temperature, are shown in Figure 9.

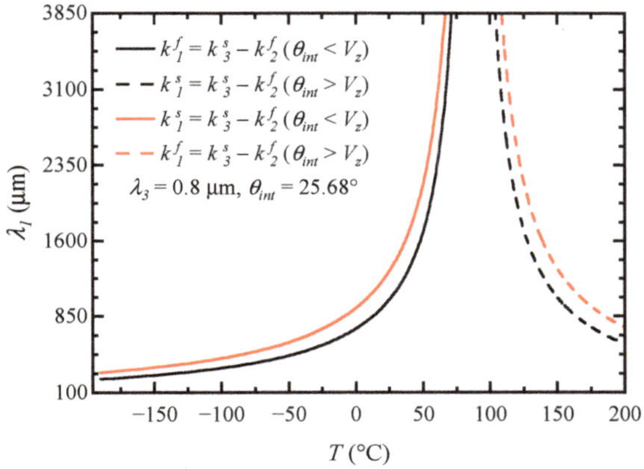

**Figure 8.** Temperature dependence of the phase-matching curves for the BIBO cut at $\theta_{int}$ = 25.68° in the *xz* plane.

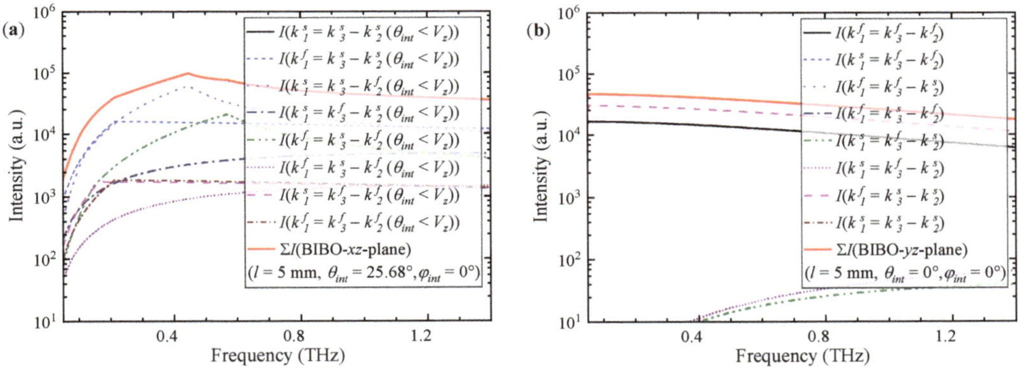

**Figure 9.** Simulation of terahertz generation by collinear three-wave mixing: (**a**) phase-matched cut crystal; (**b**) x-cut crystal.

The total intensity of the THz-generated spectra for the PM cut crystal (Figure 9a) shows two peaks corresponding to the PM interactions and a broad band substrate caused by optical rectification. The simulated intensities in these spectra are mostly limited by the effective lengths of the sample. In the case of PM interactions, the sample length limits the intensity, and in the case of optical rectification, the coherence length limits it. It is interesting to note that our calculations show that the total intensity of optical rectification in the x-cut sample is only one order of magnitude lower than the intensity of PM interaction in the PM cut crystal. Taking into account the high values of its nonlinear coefficients, it renders BIBO as one of the most prospective crystals of the borate family for phase- and non-phase-matched THz generation. The FOM for THz DFG in the BIBO crystal is estimated to be 1.5 times greater than that of the β-BBO crystal and almost 75 times greater than that of the LBO crystal.

## 5. Conclusions

In this research, we performed a thorough experimental study of the optical prope of the bismuth triborate nonlinear crystal in the range of 0.1–2.1 THz at temperatu from −196 °C to 200 °C. As a result, a new set of thermo-optic dispersion formulas v derived. We found that the BIBO crystal has nearly "zero" absorption coefficients be 0.33 THz; moreover, cooling it down to the LN temperature expands its transparency u 1 THz. Our calculations on the PM conditions for collinear three-wave mixing proces providing efficient THz radiation generation allowed us to identify the optimal crystal The PM conditions are fulfilled in the principal *xz* plane of the crystal for $\theta_{int}$ = 25.68 room temperature. Due to the narrow angular bandwidth of the DFG, we propose use of temperature tuning for the PM frequency instead of crystal rotation. Varying crystal temperature from −50 °C to 50 °C corresponds to the frequency tunning rang 0.14–0.65 THz. Taking into account the smaller absorption coefficients of BIBO, this rev its higher potential efficiency of THz nonlinear generation compared to the other crys of the borate family. Despite the absence of the PM interactions in the *xy* and *yz* pla the optical rectification coherence length for these planes is also in the range of 1 to 5 r making them viable for high-intensity frequency conversion into the THz range.

**Author Contributions:** Conceptualization, Y.A. and N.N.; methodology, N.N. and D.E.; softw D.E.; validation, V.A., Y.A. and V.S.; formal analysis, V.S.; investigation, D.E. and S.B.; resources, V data curation, D.E. and S.B.; writing—original draft preparation, D.E. and V.S.; writing—review editing, D.E., Y.A. and N.N.; visualization, D.E.; supervision, Y.A. and V.S.; project administrat V.A.; funding acquisition, N.N and Y.A. All authors have read and agreed to the published versio the manuscript.

**Funding:** The study was carried out within the framework of the State Assignment project Institute of Automation and Electrometry SB RAS (IA&E SB RAS), Institute of Laser Physics SB R and Institute of Monitoring of Climatic and Ecological Systems SB RAS (IMCES SB RAS).

**Institutional Review Board Statement:** The study did not require ethical approval.

**Informed Consent Statement:** The study did not involve humans.

**Data Availability Statement:** Not applicable.

**Acknowledgments:** The authors express their gratitude to the Shared Equipment Center "S troscopy and Optics" of the IA&E SB RAS and the Shared Research Center "VTAN" of the Novosib State University for the provided instrumentational support.

**Conflicts of Interest:** The authors declare no conflict of interest. Funding organization had no rol the design of the study; in the collection, analyses, and interpretation of data; in the writing of manuscript; and in the decision to publish the results.

## References

1. Son, J.-H. *Terahertz: Biomedical Science & Technology*; CRC Press: Boca Raton, FL, USA, 2014; ISBN 9780367576127.
2. Chen, X.; Lindley-Hatcher, H.; Stantchev, R.I.; Wang, J.; Li, K.; Hernandez Serrano, A.; Taylor, Z.D.; Castro-Camus, E.; Pickw MacPherson, E. Terahertz (THz) biophotonics technology: Instrumentation, techniques, and biomedical applications. *Chem. F Rev.* **2022**, *3*, 011311. [CrossRef]
3. Takida, Y.; Nawata, K.; Minamide, H. Security screening system based on terahertz-wave spectroscopic gas detection. *Opt. Exp* **2021**, *29*, 2529–2537. [CrossRef]
4. Brown, E.R. Fundamentals of terrestrial millimeter-wave and THz remote sensing. *Int. J. High Speed Electron. Syst.* **2003**, 995–1097. [CrossRef]
5. Kim, G.-R.; Moon, K.; Park, K.H.; O'Hara, J.F.; Grischkowsky, D.; Jeon, T.-I. Remote $N_2O$ gas sensing by enhanced 91 propagation of THz pulses. *Opt. Express* **2019**, *27*, 27514. [CrossRef] [PubMed]
6. Liu, L.; Weng, C.; Li, S.; Husi, L.; Hu, S.; Dong, P. Passive remote sensing of ice cloud properties at terahertz wavelengths ba on genetic algorithm. *Remote Sens.* **2021**, *13*, 735. [CrossRef]
7. Hwu, S.U.; Desilva, K.B.; Jih, C.T. Terahertz (THz) wireless systems for space applications. In Proceedings of the 2013 I Sensors Applications Symposium Proceedings, Galveston, TX, USA, 19–21 February 2013; pp. 171–175. [CrossRef]
8. Huang, Y.; Shen, Y.; Wang, J. From Terahertz Imaging to Terahertz Wireless Communications. *Engineering* **2023**, *22*, 106– [CrossRef]

Naftaly, M.; Vieweg, N.; Deninger, A. Industrial Applications of Terahertz Sensing: State of Play. *Sensors* **2019**, *19*, 4203. [CrossRef]

Naftaly, M. Characterisation of Emitters and Detectors. In *Next Generation Wireless Terahertz Communication Networks*; CRC Press: Boca Raton, FL, USA, 2021; pp. 89–111, ISBN 9781003001140.

Ahn, J.; Efimov, A.; Averitt, R.; Taylor, A. Terahertz waveform synthesis via optical rectification of shaped ultrafast laser pulses. *Opt. Express* **2003**, *11*, 2486. [CrossRef]

Bernerd, C.; Segonds, P.; Debray, J.; Roux, J.-F.; Hérault, E.; Coutaz, J.-L.; Shoji, I.; Minamide, H.; Ito, H.; Lupinski, D.; et al. Evaluation of eight nonlinear crystals for phase-matched Terahertz second-order difference-frequency generation at room temperature. *Opt. Mater. Express* **2020**, *10*, 561–576. [CrossRef]

Lanin, A.A.; Voronin, A.A.; Stepanov, E.A.; Fedotov, A.B.; Zheltikov, A.M. Frequency-tunable sub-two-cycle 60-MW-peak-power free-space waveforms in the mid-infrared. *Opt. Lett.* **2014**, *39*, 6430. [CrossRef]

Hafez, H.A.; Chai, X.; Ibrahim, A.; Mondal, S.; Férachou, D.; Ropagnol, X.; Ozaki, T. Intense terahertz radiation and their applications. *J. Opt.* **2016**, *18*, 093004. [CrossRef]

Dong, Y.Q.; Yin, Y.; Huang, J.J.; Zhang, X.L.; Andreev, Y.M.; Ezhov, D.M.; Grechin, S.G. Optimization on the frequency conversion of $LiGaS_2$ crystal. *Laser Phys.* **2019**, *29*, 095403. [CrossRef]

Kokh, A.; Kononova, N.; Mennerat, G.; Villeval, P.; Durst, S.; Lupinski, D.; Vlezko, V.; Kokh, K. Growth of high quality large size LBO crystals for high energy second harmonic generation. *J. Cryst. Growth* **2010**, *312*, 1774–1778. [CrossRef]

Shi, W.; Ding, Y.J.; Fernelius, N.; Vodopyanov, K. Efficient, tunable, and coherent 0.18–5.27-THz source based on GaSe crystal. *Opt. Lett.* **2002**, *27*, 1454–1456. [CrossRef]

Rowley, J.D.; Wahlstrand, J.K.; Zawilski, K.T.; Schunemann, P.G.; Giles, N.C.; Bristow, A.D. Terahertz generation by optical rectification in uniaxial birefringent crystals. *Opt. Express* **2012**, *20*, 16958–16965. [CrossRef]

Lubenko, D.M.; Losev, V.F.; Andreev, Y.M.; Lanskii, G.V. Model studies of THz-range generation via down conversion of the radiation of Ti:Sapphire lasers in LBO crystals. *Bull. Russ. Acad. Sci. Phys.* **2017**, *81*, 1239–1243. [CrossRef]

Mamrashev, A.; Nikolaev, N.; Antsygin, V.; Andreev, Y.; Lanskii, G.; Meshalkin, A. Optical properties of KTP crystals and their potential for Terahertz generation. *Crystals* **2018**, *8*, 310. [CrossRef]

Lubenko, D.M.; Lansky, G.V.; Nikolaev, N.A.; Sandabkin, E.A.; Losev, V.F.; Andreev, Y.M. Generation of THz emission in nonlinear BBO crystal at room temperature. In Proceedings of the XIV International Conference on Pulsed Lasers and Laser Applications, Tomsk, Russia, 15–20 September 2019; SPIE: Bellingham, Washington, USA, 2019; Volume 11322, pp. 498–503.

Rybak, A.; Antsygin, V.; Mamrashev, A.; Nikolaev, N. Terahertz optical properties of $KTiOPO_4$ crystal in the temperature range of (−192)–150 °C. *Crystals* **2021**, *11*, 125. [CrossRef]

Sutherland, R.L. *Handbook of Nonlinear Optics*; CRC Press: Boca Raton, FL, USA, 2003; ISBN 9781135541194.

Hellwig, H.; Liebertz, J.; Bohatý, L. Exceptional large nonlinear optical coefficients in the monoclinic bismuth borate $BiB_3O_6$ (BIBO). *Solid State Commun.* **1999**, *109*, 249–251. [CrossRef]

Kroupa, J.; Kasprowicz, D.; Majchrowski, A.; Michalski, E.; Drozdowski, M. Optical properties of Bismuth Triborate (BIBO) single crystals. *Ferroelectrics* **2005**, *318*, 77–82. [CrossRef]

Jang, J.H.; Yoon, I.H.; Yoon, C.S. Cause and repair of optical damage in nonlinear optical crystals of $BiB_3O_6$. *Opt. Mater.* **2009**, *31*, 781–783. [CrossRef]

Apurv Chaitanya, N.; Aadhi, A.; Singh, R.P.; Samanta, G.K. Type-I frequency-doubling characteristics of high-power, ultrafast fiber laser in thick BIBO crystal. *Opt. Lett.* **2014**, *39*, 5419–5422. [CrossRef]

Galletti, M.; Pires, H.; Hariton, V.; Alves, J.; Oliveira, P.; Galimberti, M.; Figueira, G. Ultra-broadband near-infrared NOPAs based on the nonlinear crystals BiBO and YCOB. *High Power Laser Sci. Eng.* **2020**, *8*, e29. [CrossRef]

Petrov, V.; Ghotbi, M.; Kokabee, O.; Esteban-Martin, A.; Noack, F.; Gaydardzhiev, A.; Nikolov, I.; Tzankov, P.; Buchvarov, I.; Miyata, K.; et al. Femtosecond nonlinear frequency conversion based on $BiB_3O_6$. *Laser Photon. Rev.* **2010**, *4*, 53–98. [CrossRef]

Bubnova, R.; Volkov, S.; Albert, B.; Filatov, S. Borates—Crystal Structures of Prospective Nonlinear Optical Materials: High Anisotropy of the Thermal Expansion Caused by Anharmonic Atomic Vibrations. *Crystals* **2017**, *7*, 93. [CrossRef]

Li, Y.; Huang, J.; Huang, Z.; Nikolaev, N.; Lanskii, G.; Mamrashev, A.; Andreev, Y. Optical properties of $BiB_3O_6$ in the terahertz range. *Results Phys.* **2020**, *16*, 102815. [CrossRef]

Nikolaev, N.A.; Mamrashev, A.A.; Antsygin, V.D.; Ezhov, D.M.; Lubenko, D.M.; Svetlichnyi, V.A.; Andreev, Y.M.; Losev, V.F. Millimetre-wave range optical properties of BIBO. *J. Phys. Conf. Ser.* **2021**, *2067*, 012011. [CrossRef]

Ghorui, C.; Rudra, A.M.; Chatterjee, U.; Chaudhary, A.K.; Ganesh, D. Efficient second-harmonic and terahertz generation from single $BiB_3O_6$ crystal using nanosecond and femtosecond lasers. *Appl. Opt.* **2021**, *60*, 5643–5651. [CrossRef] [PubMed]

Tzankov, P.; Petrov, V. Effective second-order nonlinearity in acentric optical crystals with low symmetry. *Appl. Opt.* **2005**, *44*, 6971–6985. [CrossRef]

Mamrashev, A.; Minakov, F.; Maximov, L.; Nikolaev, N.; Chapovsky, P. Correction of Optical Delay Line Errors in Terahertz Time-Domain Spectroscopy. *Electronics* **2019**, *8*, 1408. [CrossRef]

Mamrashev, A.A.; Nikolaev, N.A.; Kuznetsov, S.A.; Gelfand, A.V. Broadband metal-grid polarizers on polymeric films for terahertz applications. In *AIP Conference Proceedings, Proceedings of the 5th International Conference on Metamaterials and Nanophotonics Metanano 2020, St. Petersburg, Russia, 14–18 September 2020*; AIP Publishing LLC: Melville, NY, USA, 2020; Volume 2300, p. 020083.

Mamrashev, A.; Minakov, F.; Nikolaev, N.; Antsygin, V. Terahertz Time-Domain Polarimetry for Principal Optical Axes of Anisotropic Crystals. *Photonics* **2021**, *8*, 213. [CrossRef]

38. Duvillaret, L.; Garet, F.; Coutaz, J.-L. A reliable method for extraction of material parameters in terahertz time-domain spectrosc *IEEE J. Sel. Top. Quantum Electron.* **1996**, *2*, 739–746. [CrossRef]
39. Naftaly, M. *Terahertz Metrology*; Naftaly, M., Ed.; Artech House: Boston, MA, USA; London, UK, 2015; ISBN 9781608077779.
40. Liu, J.; Zhang, X.C. Birefringence and absorption coefficients of alpha barium borate in terahertz range. *J. Appl. Phys.* 2 *106*, 023107. [CrossRef]
41. Xia, H.R.; Li, L.X.; Teng, B.; Zheng, W.Q.; Lu, G.W.; Jiang, H.D.; Wang, J.Y. Raman scattering from bismuth triborate. *J. Ra Spectrosc.* **2002**, *33*, 278–282. [CrossRef]
42. Kasprowicz, D.; Runka, T.; Szybowicz, M.; Ziobrowski, P.; Majchrowski, A.; Michalski, E.; Drozdowski, M. Characterizatio bismuth triborate single crystal using Brillouin and Raman spectroscopy. *Cryst. Res. Technol.* **2005**, *40*, 459–465. [CrossRef]
43. Umemura, N.; Miyata, K.; Kato, K. New data on the optical properties of $BiB_3O_6$. *Opt. Mater.* **2007**, *30*, 532–534. [CrossRef]
44. Nazarov, M.M.; Sarkisov, S.Y.; Shkurinov, A.P.; Tolbanov, O.P. Efficient terahertz generation in GaSe via eee-interaction type Proceedings of the 2011 International Conference on Infrared, Millimeter, and Terahertz Waves, Houston, TX, USA, 2–7 Oct 2011; pp. 1–2.
45. Powers, P.E. *Field Guide to Nonlinear Optics*; SPIE: Bellingham, WA, USA, 2013; ISBN 9780819496362.
46. Schneider, A.; Neis, M.; Stillhart, M.; Ruiz, B.; Khan, R.U.A.; Günter, P. Generation of terahertz pulses through optical rectifica in organic DAST crystals: Theory and experiment. *J. Opt. Soc. Am. B* **2006**, *23*, 1822–1835. [CrossRef]

photonics

icle

# gmented Four-Element Photodiodes in a Three-Dimensional ıser Beam Angle Measurement

nislav Konov [1], Aleksander Frolov [1,2], Petr Shapovalov [1,2], Pavel Peretyagin [1,3,*] and Sergey Grigoriev [1]

1 Department of High-Efficiency Machining Technologies, Moscow State Technological University "STANKIN", 127055 Moscow, Russia; s.konov@stankin.ru (S.K.); a.frolov@stankin.ru (A.F.); p.shapovalov@stankin.ru (P.S.); s.grigoriev@stankin.ru (S.G.)
2 Department 1030, Central Research Institute of Automation and Hydraulics (CNIIAG), 127018 Moscow, Russia
3 Scientific Department, A.I. Evdokimov Moscow State University of Medicine and Dentistry, Delegatskaya St., 20, p. 1, 127473 Moscow, Russia
* Correspondence: p.peretyagin@stankin.ru; Tel.: +7-(49)-99732370

**Abstract:** Based on the registration of two laser beam projections, a method for measuring the angular deviation of a laser beam from its initial position in three-dimensional space is proposed and experimentally demonstrated. The laser ray is directed into the beam-splitting cube, which distributes the ray into two mutually orthogonal parts, on the path of which, two four-segment photodiodes are located at different distances from the beam-splitting cube. The value of the angular deviation from a certain initial position of the laser beam in three-dimensional space is obtained after the mathematical processing of the measured position of the laser beam projections' centers in the photodiodes coordinates systems. The innovation of this method is that the resulting angular deviation of the laser beam is in three-dimensional space due to the registration of a pair of projections. The experimental results also demonstrate that the proposed method has a high speed-measuring potential and can be used for solving a wide range of problems.

**Keywords:** quadrant slit photodiode; photogrammetry; angular deviation; motion tracking; non-contact measurement

check for updates

ion: Konov, S.; Frolov, A.; ovalov, P.; Peretyagin, P.; oriev, S. Segmented -Element Photodiodes in a e-Dimensional Laser Beam e Measurement. *Photonics* **2023**, )4. https://doi.org/10.3390/ onics10070704

ived: 11 April 2023
sed: 14 June 2023
pted: 16 June 2023
ished: 21 June 2023



## 1. Introduction

It is possible to conclude that most modern information-measuring systems are oriented to the solution of contact-coordinate-measuring tasks [1,2]. Usually, the problem statement for a coordinate-measuring system is formulated based on the design and technological limitations of the measured object. Thus, most parts of coordinate measuring, realized nowadays in mechanical engineering, are reduced to control the product elements for the observance of permissible deviations in the dimensions, forms, locations, and microreliefs of different surfaces, set in project documentation [3]. These measurements are supposed to be performed under normal conditions (temperature 20 °C, atmospheric pressure 101.3 kPa, relative humidity 60%, vibration frequency 0.1–30 Hz, vibration amplitude 0.075 mm, and so on).

It should be noted that there is another segment of measuring tasks related to the control of the above parameters in conditions close to the operating conditions of a finished product, i.e., with certain vibrations in different ranges of frequencies and amplitudes, temperature fluctuations, static loads, and other influencing physical factors. Essentially, it is about carrying out full-scale tests on special stands [4] with a realization of the possibility of measuring orientations in space and the linear dimensions of geometrical elements on controlled items.

The urgency of the realization of such measurements is extremely high due to the limited instrumental methods of control that often arise during investigation. For example, in the case of vibration or shock loading, developers have to carry out cycles of vibra loading, measuring only the parameters of the vibration itself, so they have to m judgements about its influence on the geometrical parameters of the investigated ob only using the results of the following static measurements of an object by mean coordinate-measuring systems. With such an approach, it is impossible to trace the dyna deformations of a measured object, which occur only under the action of alternating lo and stopping in an idle condition. More often, in the design of complex lightwe constructions, developers only use the modeling [5] of vibrations and shock loads with carrying out full-scale tests; however, obtaining a new measuring method will allow th to follow the change in the geometrical parameters of the surfaces of the researched obj in the process of their tests and to receive the distribution of the controlled values chang in the process of the research.

In this paper, the authors consider a solution to the problem of tracking, in real ti the spatial position of a certain plane (part of the studied product) under the conditi of specific influences on the product. In the process of measurement, the produc subjected to the effects of vibration on the vibration test bench, and the implementatio this measurement method allows for solutions to a number of design problems, such ensuring the required rigidity of structural elements under certain external influences. required minimum frequency of measuring an information update is about 10 kHz and measured object, in this case, makes linear-angular movements in all directions with approximate ranges of its linear fluctuations within 3 mm and angular deviations wi 2°. The specified limitations are determined by the requirements for the test condition the products. It is necessary to form a sequence of values for the angular deviations of plane at each moment of measurement (not less than 10 thousand samples per seco with the minimum possible error of measurement as the result of the measurements.

## 2. Materials and Methods

### *2.1. Measuring Method*

Based on the existing technical requirements of the measurement process, it is possible to use contact methods of measurement due to the inability to standardize measuring force under vibration conditions. The use of measuring systems based on C matrices (photo and video cameras) is not possible due to the high requirements for frequency of measuring data updates (not less than 10 kHz) and, in the case of us high-speed cameras, the resolution of these cameras decreases significantly and the cos the system increases extremely.

It is useful to convert the measurement information into the angular position of laser beam directed outward from the target item by placing a reflector or light source the monitored surface. The laser beam is distributed as a conditionally flat wave with extremely small divergence angle of its light beam, which makes it possible to examine laser beam projections at different distances. Interferometer-based systems (e.g., Michel scheme) or high-speed photodetectors [6,7] for determining the spatial distribution of li on a diode surface (quadrant split photodiodes—QPD) can be used as alternative varia of sensing elements to determine the linear-angle position of the beam in space.

The interferometer allows for a registration of the change in the geometric path of laser beam with a high speed and accuracy, but not the angular deviation of the beam order to perform the set task using the interferometer, three beams would be required to fo three reference points on the plane under investigation. This makes both the design of measurement system and the algorithm for processing the measurement information very ficult, which would also lead to high system implementation costs, since three interferome (one for each of the beams) with sophisticated high-precision optics are required.

The use of QPD to solve the problems of measuring the angular deviation of a la beam has been widely described in the literature [8–10], but there is one limitation t authors have not described. For example, in the implementation of the atomic fo microscope circuit (Figure 1), the nano-movements of the cantilever Δx are converted i the angular deviation of the laser beam $\alpha$ by projecting the beam onto the back mirror surface of the cantilever. The reflected beam falls on the QPD, which registers the power of the light flux coming to each of the four photosensitive segments and, using a calculation of the pair differences of the readings, information is received about the displacement of the spot center of the laser beam projection, relative to the center of the QPD in the photodiode's coordinate system. In this case, it is essential that the mirrored back surface of the cantilever is a console, one end of which is rigidly fixed in the device body, making linear movements of the mirrored surface essentially impossible. In the measurement problem described in the present article, linear movements of the surface under study are possible along all the axes, in connection with which, the use of the widespread measurement scheme with one QPD is not possible, because a linear displacement of the laser beam without changing its angular position will lead to a shift in the projection spot on the QPD surface, which will be falsely interpreted as an angular deviation.

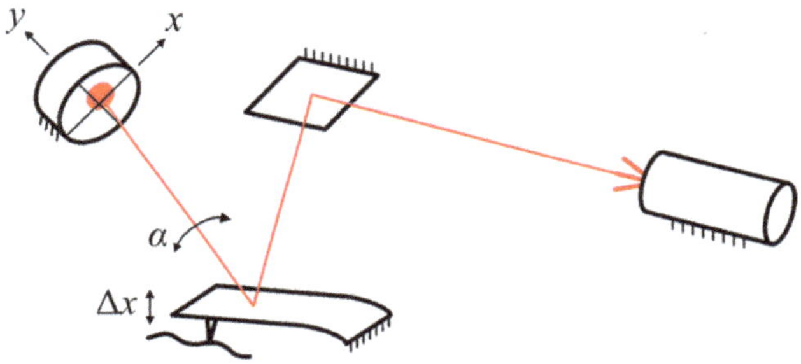


**Figure 1.** Scheme for measuring the angular deviation of a laser beam in an atomic force microscope.

To solve this problem, the authors propose the use of approaches close to photogrammetry [11,12]: using two beam projections (Figure 2) to determine the spatial location of the light source, including linear (x, y, z) and angular deviations ($\alpha_x$, $\alpha_y$, $\alpha_z$) in the global coordinate system (X, Y, Z), using the coordinates of two projections ($x'$, $y'$ and $x''$, $y''$) in the QPDs coordinate systems, ($X'$, $Y'$) and ($X''$, $Y''$), respectively.

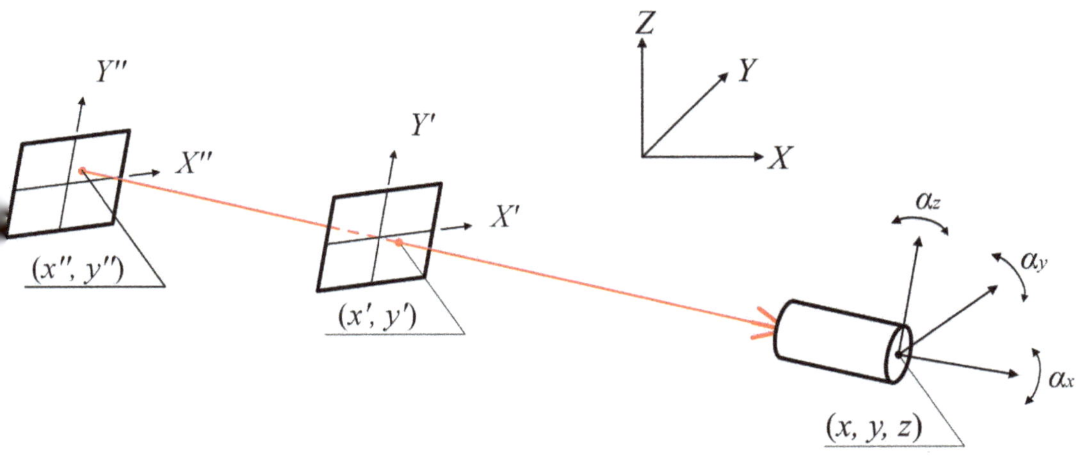


**Figure 2.** Laser beam trace in two projection planes.

### 2.2. Measurement Circuit Diagram

In the measurement scheme proposed by the authors (Figure 3), a laser diode, us a fixture, is attached to the studied plane of the part and the laser beam is oriented al the normal to the surface, which falls onto the beam-splitter cube, where it splits i two parts [13], each of which hits the QPD [14], located at a different distances from beam-splitter cube.

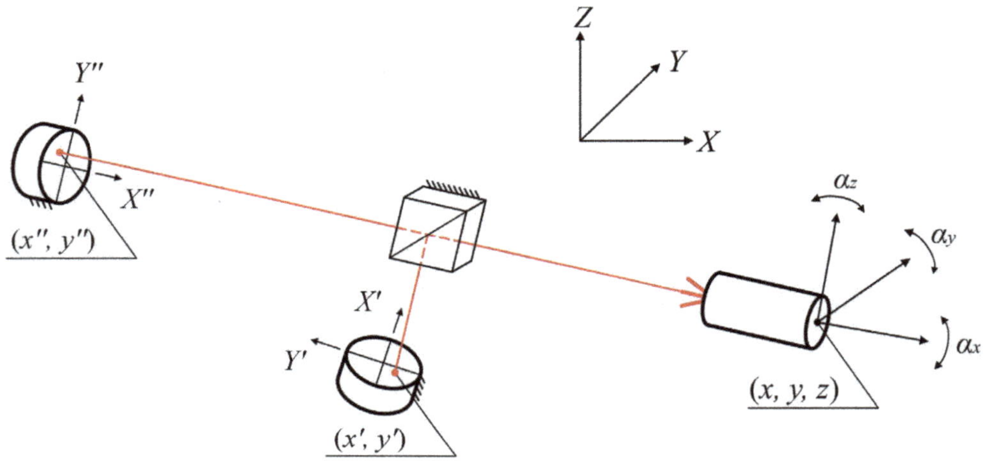


**Figure 3.** Schematic diagram of the proposed measuring system.

Thus, each of the photodiodes implements the projection plane of the laser be and the difference in the distances from each of the QPDs to the light-splitting cube is distance between the projection planes.

The presented approach is sensitive to both linear and angular displacements of laser source in all directions. To determine the deviation value, it is necessary to meas the deviation of the laser spot projections on the plane of each of the QPDs [15,16] in QPD coordinate systems (X′, Y′ and X″, Y″), and then recalculate them into the ang deviation value, relative to some position taken as the initial one, in accordance with mathematical model of the system [17].

### 2.3. Mathematical Model

To determine the effective distance from the measurement object to the sensi elements (QPD), it is necessary to determine the zone of the QPD's sensitivity and the va of possible angular deviations, as dictated by the set task. As a base model, it is reasona to proceed from the characteristics of the QPD (in the model chosen by the authors, diameter of the sensitive zone of the photodiode is 9 mm), and also from the approxim distance from the farthest photodiode to the object of measurement (about 150–200 m Based on the geometry [18,19] of the measurement scheme with zero initial deviatio the laser beam center projection, the following dependence between the deviation in laser beam projection center from the center of the QPD δ, the distance from the QPD to measurement object L, and the angle of deviation in the laser beam from the initial (z position α can be used:

$$\alpha(\delta, L) = \frac{180}{\pi} \cdot atan\left(\frac{\delta}{L}\right);$$

Taking into view the given relations, when using the laser beam with a beam diam of 4 mm (at values smaller than half of the QPD diameter, the range of limiting deviati will be limited by the diameter of the laser beam), the maximum deviation in the laser be projection center from the photodiode center δ from relation (1), provided that the laser beam is completely placed in the sensitive area of the QPD and the projection is present in each of the four quadrants of photodiode, will be −2 mm. Thus, given the value of the maximum linear deviation in the laser beam projection center from the center of the QPD, the dependence of the angular deviation in the beam on the distance between the farthest QPD and the object of measurement is given (Figure 4).

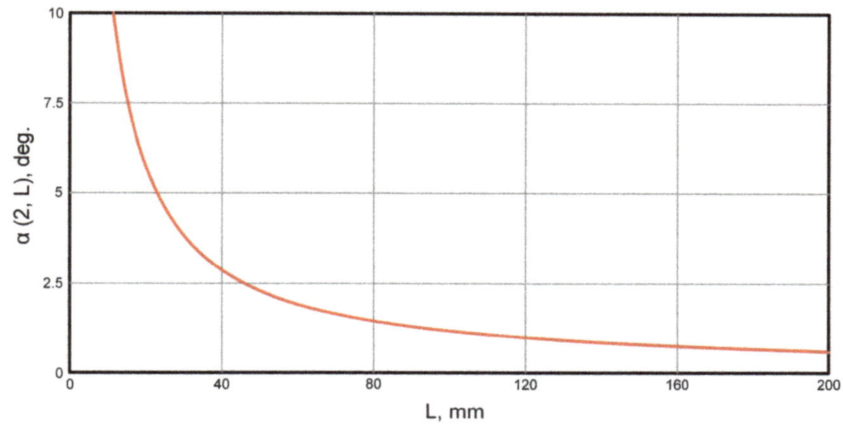


**Figure 4.** Angle of deviation α of the laser beam from the initial position depending on the distance between the QPD and the object of measurement.

The above calculations and dependencies are intended to illustrate the orders of values and dependencies between them.

Taking into account the physical principles of silicon photodiodes, it should be kept in mind that the output voltage of the photodiode segment is proportional to the surface area of the QPD segment illuminated by the light beam. Thus, the characteristics of the output voltage of the photodiode segments depend on the laser beam shape.

Taking into account the circular shape of the laser beam cross-section, it is possible to determine the expressions for the regions covered by the laser beam for each part of the QPD segments (Figure 5). For this purpose, it is first necessary to determine the equations for the upper and lower half-circle of the laser spot:

$$y_{Ru}(x, xc, yc, d) = yc + \sqrt{\left(\frac{d}{2}\right)^2 - (x - xc)^2};$$
$$y_{Rd}(x, xc, yc, d) = yc - \sqrt{\left(\frac{d}{2}\right)^2 - (x - xc)^2} \quad (2)$$

where $y_{Ru}(x, xc, yc, d)$ is the contour function of the upper half-circle of the laser beam projection spot, $y_{Rd}(x, xc, yc, d)$ is the contour function of the lower half-circle of the laser beam projection spot, $xc$ and $yc$ are coordinates of the center of the laser beam projection spot in the $X'$, $Y'$ coordinate system (Figure 5), and $d$ is the diameter of the laser beam projection spot.

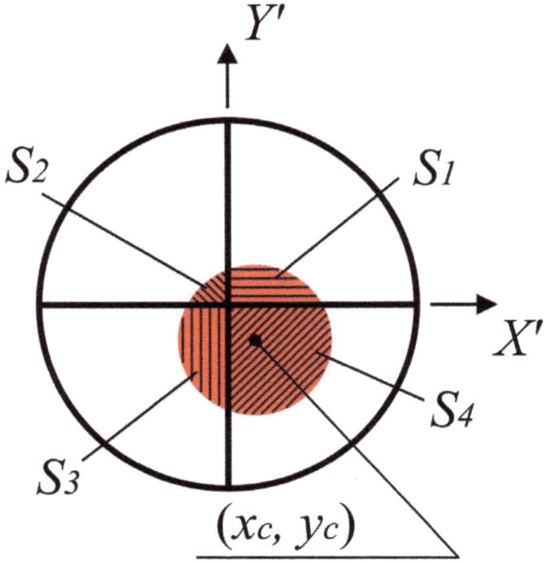


**Figure 5.** Projection of a circular laser beam onto photodiode segments.

Then, the surface areas of the QPD segments illuminated by the laser beam, depenc on the coordinates of the spot center of the laser beam projection and spot diameter, t the look:

$$
\begin{aligned}
S_1(xc, yc, d) &= \int_0^{xc+\frac{d}{2}} y_{Ru}(x, xc, yc, d)dx - \int_0^{xc+\frac{d}{2}} y_{Rd}(x, xc, yc, d)dx, \ when\ y_{Ru} \geq 0\ and\ y_{Rd} \geq 0; \\
S_2(xc, yc, d) &= \int_{xc-\frac{d}{2}}^{0} y_{Ru}(x, xc, yc, d)dx - \int_{xc-\frac{d}{2}}^{0} y_{Rd}(x, xc, yc, d)dx, \ when\ y_{Ru} \geq 0\ and\ y_{Rd} \geq 0; \\
S_3(xc, yc, d) &= \int_{xc-\frac{d}{2}}^{0} y_{Ru}(x, xc, yc, d)dx - \int_{xc-\frac{d}{2}}^{0} y_{Rd}(x, xc, yc, d)dx, \ when\ y_{Ru} \leq 0\ and\ y_{Rd} \leq 0; \\
S_4(xc, yc, d) &= \int_0^{xc+\frac{d}{2}} y_{Ru}(x, xc, yc, d)dx - \int_0^{xc+\frac{d}{2}} y_{Rd}(x, xc, yc, d)dx, \ when\ y_{Ru} \leq 0\ and\ y_{Rd} \leq 0;
\end{aligned}
$$

Figure 6 shows three-dimensional value graphs of the photodiode segments' sur areas, covered by a spot of laser beam projection, depending on the coordinates of spot center *xc, yc* at $d = 4$ mm. The signals at the output of the QPD segments wil proportional to the values of the areas covered by the spot of the laser beam projection as follows from relations (2) and (3), will have a non-linear character of change.

With voltage signals at the photodiodes output, which are proportional to the a of the QPD segments covered by the spot of the laser beam projection, it is possibl determine the estimated coordinates of the center of the laser beam projection on the C plane, using the well known relations:

$$
\begin{aligned}
xc' &= U_1 + U_4 - U_2 - U_3; \\
yc' &= U_1 + U_2 - U_4 - U_3;
\end{aligned}
$$

where $xc'$ and $yc'$ are the estimated coordinates of the center of the laser beam projectio the photodiode plane, and $U_1$, $U_2$, $U_3$, and $U_4$ are the measurement signals (digitized va of the voltages of the corresponding QPD segments, numbering according to Figure 5

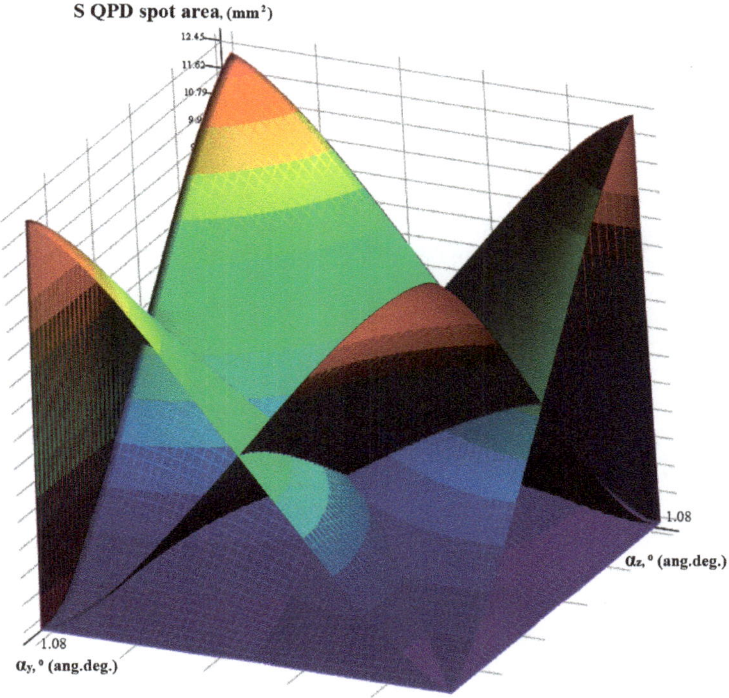


**Figure 6.** Diagrams of changes in the areas of the QPD segments covered by the laser beam projection spot depending on the coordinates of the center of the laser beam projection spot *xc*, *yc*.

In this case, $xc$ and $yc$ will differ from $xc'$ and $yc'$ due to the nonlinearity of Expressions (2) and (3), which can be further corrected using the total conversion function [20]. Given that, according to the measurement scheme (Figure 3), a pair of photodiodes located at different distances from the light source along the laser beam path is used to determine the total angular deviation $\alpha$ of the laser beam from its initial position, there are, in fact, two pairs of measuring signals (one each from QPD 1 and QPD 2), which can be recalculated according to Expression (4) into two pairs of estimated coordinates for the projection centers ($xc'$, $yc'$, $xc''$, $yc''$) of the laser beam on the planes of photodiode 1 and photodiode 2.

Thus, it is necessary to establish the relation between the estimated coordinates of the laser beam projection centers ($xc'$, $yc'$, $xc''$, $yc''$) and the total angle of deviation $\alpha$ of the laser beam from some initial (zero) position, nominally orthogonal to the planes of both QPDs, at which, the beam passes through the centers of both QPDs. Based on the results of empirical studies, the authors propose the following function for the total angle depending on the values of the two pairs of the projection centers' estimated coordinates:

$$\alpha(xc',yc',xc'',yc'') = k_0 \cdot \sqrt{(k_{11} \cdot xc'' - k_{12} \cdot xc' + b_1)^2 + (k_{21} \cdot yc'' - k_{22} \cdot yc' + b_2)^2}; \quad (5)$$

where $\alpha$ is the total angle of deviation in the laser beam initial position, $xc'$, $yc'$, $xc''$, and $yc''$ are coordinates of the laser beam projection centers in the projection planes of two slit photodiodes, $k_0$ is the scale factor conditioned by the laser beam diameter, $k_{11}$, $k_{12}$, $k_{21}$, and $k_{22}$ are scale factors, which are responsible for the linear distortions at the registration of the coordinates of the laser beam projections, such as a deviation in the photodiodes normals from the nominal mutually orthogonal location, and $b_1$ and $b_2$ are linear displacement coefficients along the QPD axes, which are responsible for the deviation in the photodiodes' coordinate centers from the nominal location in the normal plane. The coefficients $k_0$, $k_{12}$, $k_{21}$, $k_{22}$, $b_1$, and $b_2$ are determined via a calibration of the measuring system.

## 3. Results and Discussion

### 3.1. Experimental Stand

In order to verify the theoretical statements outlined in the second section of paper, the authors developed an experimental stand consisting of two quadrant segmer photodiodes (vendor OSI model SPOT-9DMI) and a light-splitting cube (model G 401112). A laser diode was used as a source of laser emission with a wavelength of 635 a transmitter power of 5 mW, and a spot diameter at a distance of 100–200 mm in range of 3.5–4 mm. The scheme included a current-limiting resistor of 88 ohms to red the intensity of the light radiation in connection with the limitations imposed by manufacturer of the QPD.

Quadrant segmented photodiodes with a sensing zone diameter of 9 mm an response time of less than 10 ns were use and the QPDs were included in a photovol circuit. To digitize the signals [21] of the photodiodes, an internal 12-bit ADC of an 8 microcontroller [22], with an internal eight-channel multiplexer [23] and the strappin the signal lines with capacitors with a 0.1 µF rating, was used. An external high-sp Flash memory circuit with a nominal volume of 64 Mbit was used to realize high-sp sensor polling with a rate above 10,000 samples per second.

The experimental stand is shown in Figure 7; the body of the experimental bench created using the SLS-printing method from polyamide. The simulator was connected personal computer using the USB 2.0 or a higher interface with a hardware USB-UART brid

**Figure 7.** Photo of the measuring system stand.

### 3.2. Software and Calibration

To collect and evaluate the measurement information [24], special software was de oped in C++ language and the software component of the system contained two parts

- The microprogram of the microcontroller [25];
- Software for the personal computer;

The algorithm of the microcontroller's work was built using the logic of the fi automata theory and the control was created by sending control commands from computer; thus, the system had several working modes:

- High-speed data acquisition (cyclic digitalization of sensor output signals from eight channels, which were recorded into external memory until it was full);
- Data transfer to computer via USB interface (complete transfer of all the recorded data from the external memory—more than 700 thousand records);
- System calibration (transfer of current sensor readings directly to the computer at a low speed for the calibration and calculation of the Equation (5) coefficients);
- Idle mode;

The interface of the computer program was realized in the format of a single-window application, implementing graphic images of the QPD's center crosses by drawing in them the laser beam projections and calculations of the coordinates of the laser beam projection centers. The mode of the data collection from the external memory of the stand, which was recorded in a text file for the subsequent analysis, was realized as an independent thread.

The main target of the calibration process was determining the internal parameters of the system, $k_0$, $k_{11}$, $k_{12}$, $k_{21}$, $k_{22}$, $b_1$, and $b_2$, from Expression (5). To calibrate the system, it was necessary to provide a laser beam falling with a number of known angles $\alpha$ and to fix the values of $xc'$, $yc'$, $xc''$, and $yc''$ for each $\alpha$ angle. With at least seven described data sets, it was possible to solve the obtained system of equations relative to the required coefficients $k_0$, $k_{11}$, $k_{12}$, $k_{21}$, $k_{22}$, $b_1$, and $b_2$, and increasing the number of data sets made possible to solve the system with control, reducing the calibration error via an application of LSM or other solution search methods.

An important characteristic of the calibration process was the necessity of maintaining a certain location of the laser beam projection centers in the QPDs' planes; the best solution was a uniform distribution over the entire surface or a distribution with increasing density as one approached the QPD centers. To set exemplary influences (known values of angular deviation $\alpha$), the authors developed a device to place the light source on the working surface of the coordinate-measuring machine. After committing the micro-movements of the beam [26], the angle was determined by the coordinate-measuring machine as a position of the precise cylinder based in a device rigidly connected to the laser transmitter.

### 3.3. Experimental Results

Based on the results of the simulation, the authors determined the optimal number of data sets for the calibration to be equal to nine. Their arrangement, with the total distribution of the theoretical measurement error within $\pm 1''$ (one angular second) and the reference angles of the beam direction deviation relative to the initial position (orthogonal to the photodiode planes), was $0°$; $\pm 0.345°$; and $\pm 0.53°$ when the laser beam rotated alternately in two mutually orthogonal directions ($\alpha_y$ pitch and $\alpha_z$ yaw).

The error map is shown in Figure 8, with the laser beam angular deviation limits being in each of the two mutually orthogonal directions ($\alpha_y$ pitch and $\alpha_z$ yaw) of $\pm 0.764°$ and about 75% of the error map falling within $\pm 1''$. The remaining 25% of the measurement range could also be used, but with a correction for a larger measurement error value. It is necessary to notice that the resulting values of the limits of the measurements of the angular deviations did not include all the possibilities of the obtained measuring system, so if the distance from a far-end QPD to the investigated object was reduced from the nominal value of 150 mm used by authors, it was possible to receive measuring ranges of angular deviations up to $10°$. It was also possible to increase the measuring range of the instrument using special wide-angle lenses that allowed for an increase in the range of the deviations in the input beams, but the measurement error would be increased and would most likely change the character of the distribution, which would require additional recalculations. In addition to using special optical lenses, it was possible to increase the sensitivity range of the sensors (QPD) by using scattering matte glasses, which made the intensity distribution of the laser beam incident be on the sensor smoother. These variants in the modifications of the measuring system are a subject of interest and will undoubtedly be carried out in the future.

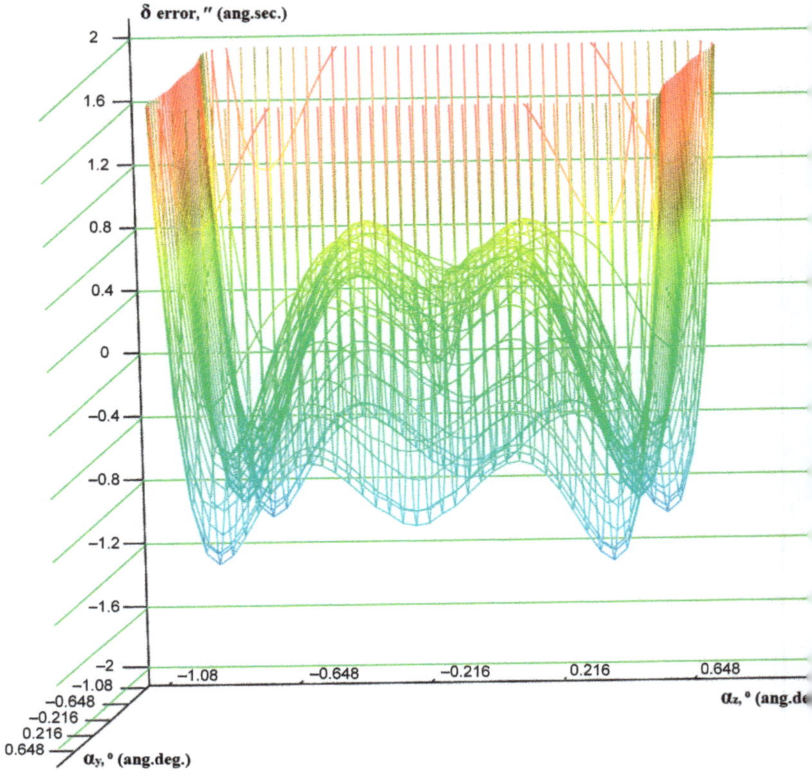


**Figure 8.** Measuring system error map.

According to the results of the experiments implemented by authors, on the cre and calibrated stand, it was possible to achieve a relative error of measurement over entire measurement range (which was around ±5° on an experimental stand mad authors) within 2%, and in the central part, up to 0.5%.

## 4. Conclusions

The authors proposed a scheme for measuring the angular deviation in critical chinery product surfaces in dynamic mode (including measurements performed du tests under conditions close to operating conditions). The described solution, in cont to similar and widely known methods, was not sensitive to linear displacements of light source along all the axes and allowed for controlling the rotation of the exami surface in all directions. Depending on the task, the measurement ranges could vary from fractions of a degree to tens of degrees, and the error of measurement [28] from units of the angular seconds to fractions of a degree. The most important advantag the proposed system was the use of a laser beam as a carrier of measuring informa and a pair of QPDs as a receiver, which made it possible to achieve a high frequenc interrogation in the measuring system (tens of thousands samples per second). One of important applications of the obtained results is the use of real-time measurement res as feedback for CNC systems [29,30]. The authors have outlined further directions for development of the research; they plan to develop new modifications for the measu system, allowing for the achievement of a larger measurement range and less sensitivit external measurement conditions, with a preservation of the orders of the measuren error range.

**Author Contributions:** Conceptualization, S.K.; methodology, S.G.; software, S.K.; validation, S.G.; formal analysis, P.S.; investigation, A.F.; resources, P.P.; data curation, A.F.; writing—original draft preparation, S.K.; writing—review and editing, P.P.; visualization, S.K.; supervision, S.K.; project administration, S.G.; funding acquisition, P.S. All authors have read and agreed to the published version of the manuscript.

**Funding:** This research was funded by Ministry of Science and Higher Education of the Russian Federation, project number FSFS-2021-0003.

**Institutional Review Board Statement:** Not applicable.

**Informed Consent Statement:** Not applicable.

**Data Availability Statement:** Data underlying the results presented in this paper are not publicly available at this time but may be obtained from the authors upon reasonable request.

**Acknowledgments:** The work was carried out on the equipment of the Collective Use Center of MSTU "STANKIN" (project No. 075-15-2021-695).

**Conflicts of Interest:** The authors declare no conflict of interest.

## Nomenclature

| Abbreviation | Description |
|---|---|
| CCD | Charge-coupled device |
| QPD | Quadrant segmented photodiode |
| ADC | Analog to digital converter |
| LSM | Least square method |
| CNC | Computer numeric control |

## erences


Gao, W.; Kim, S.W.; Bosse, H.; Haitjema, H.; Chen, Y.L.; Lu, X.D.; Knapp, W.; Weckenmann, A.; Estler, W.T.; Kunzmann, H. Measurement technologies for precision positioning. *CIRP Ann.* **2015**, *64*, 773–796. [CrossRef]

Su, X.; Zhang, Q. Dynamic 3-D shape measurement method: A review. *Opt. Lasers Eng.* **2010**, *48*, 191–204. [CrossRef]

Konov, S.G. Mobile contact-type coordinate-measurement system based on a photogrammetric system. *Meas. Tech.* **2010**, *53*, 149–152. [CrossRef]

Khuc, T.; Catbas, F.N. Computer vision-based displacement and vibration monitoring without using physical target on structures. *Struct. Infrastruct. Eng.* **2017**, *13*, 505–516. [CrossRef]

Fan, W.; Qiao, P. Vibration-based Damage Identification Methods: A Review and Comparative Study. *Struct. Health Monit.* **2011**, *10*, 83–111. [CrossRef]

Cincotta, S.; Neild, A.; He, C.; Armstrong, J. Visible Light Positioning Using an Aperture and a Quadrant Photodiode. In Proceedings of the IEEE Globecom Workshops (GC Wkshps), Singapore, 4–8 December 2017; pp. 1–6.

Chong, K.K.; Yew, T.K. Novel Optical Scanner Using Photodiodes Array for Two-Dimensional Measurement of Light Flux Distribution. *IEEE Trans. Instrum. Meas.* **2011**, *60*, 2918–2925. [CrossRef]

Song, Y.; Bhushan, B. Atomic-scale topographic and friction force imaging and cantilever dynamics in friction force microscopy. *Phys. Rev. B* **2006**, *74*, 165401. [CrossRef]

Klochkova, V.; Sheldakova, J.; Galaktionov, I.; Nikitin, A.; Kudryashov, A.; Belousov, V.; Rukosuev, A. Local Correction of the Light Position Implemented on an FPGA Platform for a 6 Meter Telescope. *Photonics* **2022**, *9*, 322. [CrossRef]

Kudryashov, A.; Alexandrov, A.; Rukosuev, A.; Samarkin, V.; Galarneau, P.; Turbide, S.; Châteauneuf, F. Extremely high-power CO2 laser beam correction. *Appl. Opt.* **2015**, *54*, 4352–4358. [CrossRef]

Konov, S.G.; Krutov, A.V. Development of a system for tracking the spatial movements of objects based on the technology of photogrammetry. *Meas. Tech.* **2014**, *56*, 1386–1389. [CrossRef]

Riveiro, B.; González-Jorge, H.; Varela, M.; Jauregui, D.V. Validation of terrestrial laser scanning and photogrammetry techniques for the measurement of vertical underclearance and beam geometry in structural inspection of bridges. *Measurement* **2013**, *46*, 784–794. [CrossRef]

Rayas, J.A.; León-Rodríguez, M.; Martínez-García, A.; Genovese, K.; Medina, O.M.; Cordero, R.R. Using a single-cube beam-splitter as a fringe pattern generator within a structured-light projection system for surface metrology. *Opt. Eng.* **2017**, *56*, 044103. [CrossRef]

Roa, J.; Oncins, G.; Diaz, J.; Sanz, F.; Segarra, M. Calculation of Young's Modulus Value by Means of AFM. *Recent Pat. Nanotechnol.* **2011**, *5*, 27–36. [CrossRef] [PubMed]

Ishola, F.; Cho, M. Experimental Study on Photodiode Array Sensor Aided MEMS Fine Steering Mirror Control for Laser Communication Platforms. *IEEE Access* **2021**, *9*, 100197–100207. [CrossRef]

16. Zečević, N.; Hofbauer, M.; Goll, B.; Zimmermann, H.; Tondini, S.; Chalyan, A.; Fontana, G.; Pavesi, L.; Testa, F.; Stracca, S.; et a 3D Photonic-Electronic Integrated Transponder Aggregator with 48 × 16 Heater Control Cells. *IEEE Photonics Technol. Lett.* 2 *30*, 681–684. [CrossRef]
17. Iosifidis, D.; Koivisto, T. Scale Transformations in Metric-Affine Geometry. *Universe* **2019**, *5*, 82. [CrossRef]
18. Du, Z.-G.; Pan, J.-S.; Chu, S.-C.; Luo, H.-J.; Hu, P. Quasi-Affine Transformation Evolutionary Algorithm with Communica Schemes for Application of RSSI in Wireless Sensor Networks. *IEEE Access* **2020**, *8*, 8583–8594. [CrossRef]
19. Meng, Z.; Pan, J.-S.; Xu, H. QUasi-affine transformation evolutionary (QUATRE) algorithm: A cooperative swarm based algori for global optimization. *Knowl-Based Syst.* **2016**, *109*, 104–121. [CrossRef]
20. Dong, H.-W.; Su, X.-X.; Wang, Y.-S.; Zhang, C. Topological optimization of two-dimensional phononic crystals based on the f element method and genetic algorithm. *Struct. Multidiscip. Optim.* **2014**, *50*, 593–604. [CrossRef]
21. Jaanus, M.; Udal, A.; Kukk, V.; Umbleja, K. Using Microcontrollers for High Accuracy Analogue Measurements. *Elektro Elektrotech.* **2013**, *19*, 51–54. [CrossRef]
22. Grigoriev, S.N.; Martinov, G.M. An ARM-based Multi-channel CNC Solution for Multi-tasking Turning and Milling Mach *Procedia CIRP* **2016**, *46*, 525–528. [CrossRef]
23. Sharma, M.; Agarwal, N.; Reddy, S.R.N. Design and development of daughter board for USB-UART communication betw Raspberry Pi and PC. In Proceedings of the International Conference on Computing, Communication & Automation, Gre Noida, India, 15–16 May 2015; pp. 944–948.
24. Grigoriev, S.N.; Emelianov, P.N.; Masterenko, D.A.; Ped', S.E. Sampling by Variables for Rayleigh Distributed Lots. *Meas.* **2022**, *65*, 417–425. [CrossRef]
25. Barrett, S.F.; Pack, D.J. Microchip AVR Operating Parameters and Interfacing. In *Microchip AVR Microcontroller Primer: Programm and Interfacing*; Synthesis Lectures on Digital Circuits & Systems; Springer: Berlin/Heidelberg, Germany, 2019.
26. Gao, W.; Shi, H.; Tang, Q. A Contactless Planar Inductive Sensor for Absolute Angular Displacement Measurement. *IEEE A* **2021**, *9*, 160878–160886. [CrossRef]
27. Goyal, D.; Pabla, B.S. The Vibration Monitoring Methods and Signal Processing Techniques for Structural Health Monitorin Review. *Arch. Comput. Methods Eng.* **2016**, *23*, 585–594. [CrossRef]
28. Grigoriev, S.N.; Masterenko, D.A.; Teleshevskii, V.I.; Emelyanov, P.N. Contemporary state and outlook for developmer metrological assurance in the machine-building industry. *Meas. Tech.* **2013**, *55*, 1311. [CrossRef]
29. Grigoriev, S.N.; Martinov, G.M. The Control Platform for Decomposition and Synthesis of Specialized CNC Systems. *Pro CIRP* **2016**, *41*, 858–863. [CrossRef]
30. Grigoriev, S.N.; Martinov, G.M. Research and development of a cross-platform CNC kernel for multi-axis machine tool. *Pro CIRP* **2014**, *14*, 517–522. [CrossRef]

photonics

MDPI

nmunication

# vestigation of the Process of Evolution of Traces of Explosives arried by Fingerprints Using Polarimetric Macrophotography ıd Remote LF/LIF Method

gei Bobrovnikov [1,2], Evgeny Gorlov [1,2,*], Viktor Zharkov [1], Sergei Murashko [1,2], Alexander Vorozhtsov [2] Alexander Stykon [2]

1 V.E. Zuev Institute of Atmospheric Optics SB RAS, Tomsk 634055, Russia
2 Faculty of Radiophysics, National Research Tomsk State University, Tomsk 634050, Russia
* Correspondence: gorlov_e@mail.ru

**Abstract:** The results of a study of the degradation of the cyclotrimethylenetrinitramine (RDX) traces carried in fingerprints depending on the fingerprint number are presented. The surface concentration of the trace was assessed using macrophotography in polarized light and by the method of laser fragmentation/laser-induced fluorescence. A technique for estimating the surface concentration of RDX traces in sweat-fat fingerprints based on pixel-by-pixel scanning of macrophotographs is described. The data of parallel experiments on remote laser detection of RDX particles in fingerprints are presented. A comparison shows that the results of the direct measurements of the total trace area are in good agreement with the LF/LIF response data.

**Keywords:** explosives; cyclotrimethylenetrinitramine; traces; macrophotography; laser fragmentation; laser-induced fluorescence



ion: Bobrovnikov, S.; Gorlov, E.; kov, V.; Murashko, S.; zhtsov, A.; Stykon, A. stigation of the Process of ıtion of Traces of Explosives ed by Fingerprints Using imetric Macrophotography and ɔte LF/LIF Method. *Photonics* 10, 740. https://doi.org/ ɔ0/photonics10070740

ved: 17 April 2023
sed: 18 May 2023
pted: 31 May 2023
shed: 28 June 2023



## 1. Introduction

Explosive devices carried covertly can be identified by a number of unmasking signs, which include trace amounts of explosive substances on the hands of a person and various surfaces that have come into contact with these substances (for example, clothing, the surface of hand luggage, covers, packaging elements, body lock handles, etc.). Objects that had gas-air or direct contact with an explosive device or explosive substance can adsorb explosive particles on their surface and retain them for quite a long time. Attempts to get rid of such traces do not always achieve their goal. Traces of explosives on a person's hands last up to several hours, and even a single washing of hands with soap cannot guarantee getting rid of their presence.

The aim of the study presented in the publication was to develop a method for visualizing traces of explosives in sweat-fat fingerprints and using macrophotography to quantify surface concentration. These studies complement a large cycle of works carried out by the authors in recent years on the remote detection of traces by the LF/LIF method. That is why the results of the macrophotographic polarization method are compared with the results of the LF/LIF method, which is well known to the authors from previous works.

Briefly, the essence of the LF/LIF method is to use the effect of photodissociation (photofragmentation, photolysis) of optically inactive molecules in order to form characteristic fragments that are highly efficient in the LIF process. The choice of a functional group (or part of it) of the initial compound as a characteristic fragment makes it possible to reliably determine whether it belongs to a particular class of compounds. For example, for all nitro compounds, the presence of nitro or nitroso groups in their structure is common. The nitric oxide molecules (NO-fragments) formed during the photofragmentation of nitro compounds can obviously serve as markers of the initial compounds [1–17].

## 2. Microstructure of RDX Traces in Fingerprints

To simulate contact with explosives, the index finger of the hand was pressed to a face with powdered cyclotrimethylenetrinitramine (RDX). The pressure force (2.2 ± 0. in this case corresponded to the minimum grip force $F$ required to hold an object weig 200 g with two fingers at a friction coefficient of 0.5.

After the explosive particles were transferred to the finger of the hand, it was rep edly applied to a clean surface, leaving sweat-fat prints with explosive particles on i glass slide was used as a trace-receiving surface. Glasses with the first, tenth, twent thirtieth, fortieth, fiftieth, sixtieth, seventieth, eightieth, ninetieth, and hundredth fin prints were used for the subsequent study of the microstructure of traces of explosives experimental study of the possibility of their remote detection by the LF/LIF method.

It is known that RDX crystalline particles possess natural optical activity [18]. W linearly polarized light passes through such substances, the polarization plane of l turns out to be rotated relative to its initial position. The phenomenon of the polariza plane rotation can be used to increase the contrast in the recognition of crystalline R particles against the background of a sweat-fat fingerprint consisting of an amorph substance. To do this, the preparation must be placed in transmitted light between crossed polarizers.

Figure 1 shows a layout of stand for macrophotography of explosive traces in tr mitted light using a Canon EOS 500D camera with an EF-S 18-55 f/3.5-5.6 IS lens. Dig images of a glass slide placed between two crossed polarizers (Edmund Optics, St #45-668) before and after the 50th fingerprint was applied are shown in Figure 2a and respectively. Figure 2a shows that the image of a "clean" glass slide contains br blotches—a sign of unintentional contamination of the glass surface with particles "foreign" substance that rotates the plane of polarization of the transmitted radiation expected, their presence persists even after the application of the target substance to same surface (Figure 2b). The exclusion of such artifacts was carried out by pixel-by-p "subtraction" of the image (a) from the image (b) during digital processing using the *Im* program (https://imagej.nih.gov/ij/ (accessed on14 November 2022)) and obtaini difference image as a result. The result of such image processing is shown in Figur Difference images of all samples are shown in Figure 3.

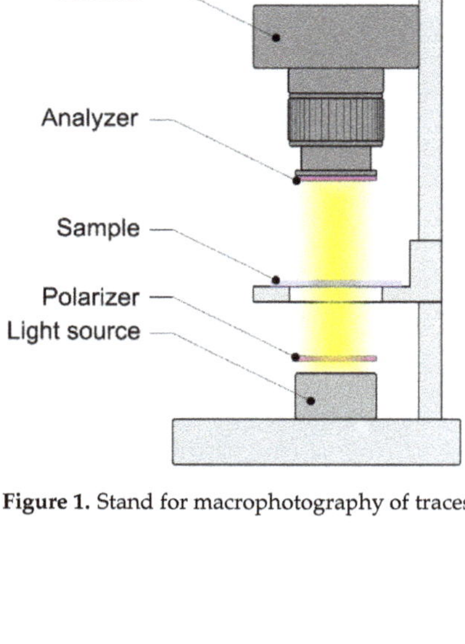


**Figure 1.** Stand for macrophotography of traces of explosives in transmitted light.

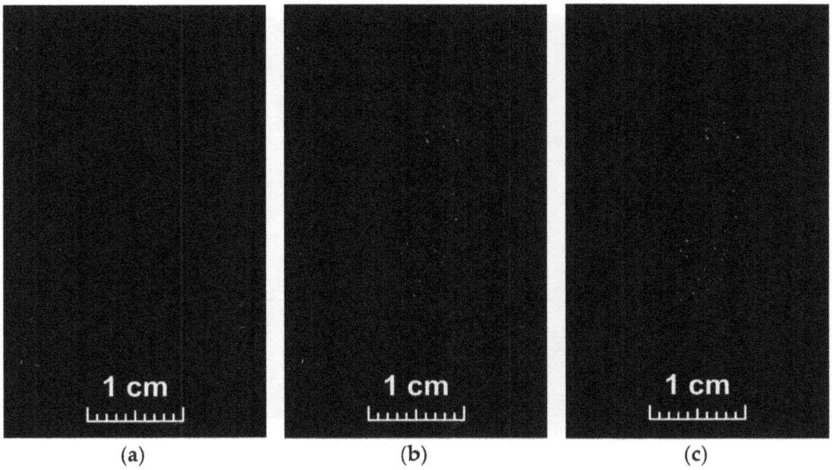


**Figure 2.** Digital images: (**a**) "clean" glass slide before fingerprinting; (**b**) the same glass slide with the 50th fingerprint containing traces of RDX; (**c**) difference image.

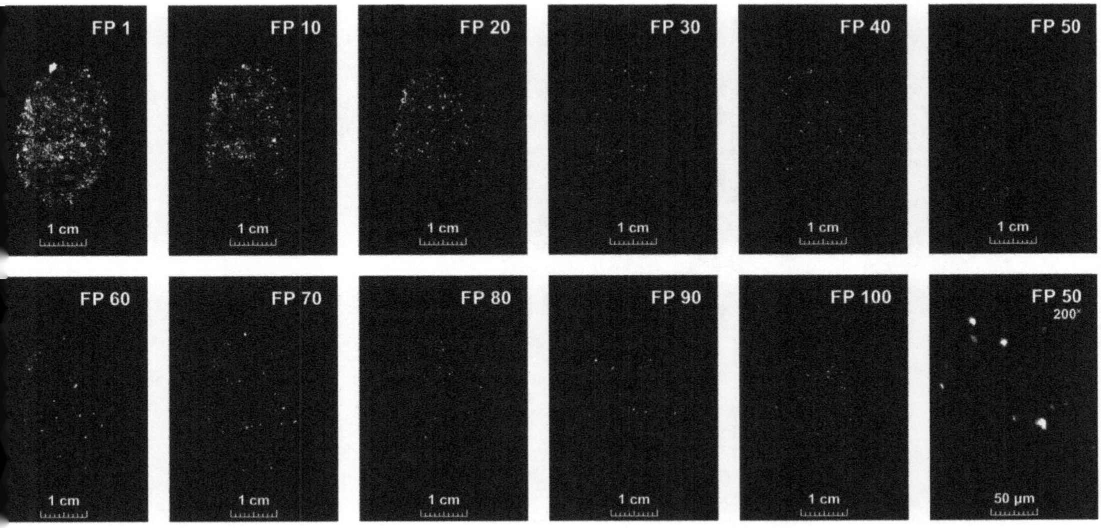


**Figure 3.** Difference images of sweat-fat fingerprints with RDX particles on a glass surface: FP-fingerprint; 1, 10, 20, ... 100—serial number of the fingerprint; $200^{\times}$—200-fold magnificationof the image fragment "OP 50" obtained using a polarizing microscope.

As can be seen from the figure, the trace area decreases significantly when moving from the 1st to the 100th fingerprint. Quantitative pixel-by-pixel analysis of macrophotographs showed that the decrease in the trace area occurs monotonically, and for the 100th fingerprint, this area is approximately two orders of magnitude smaller than the initial one (Figure 4).

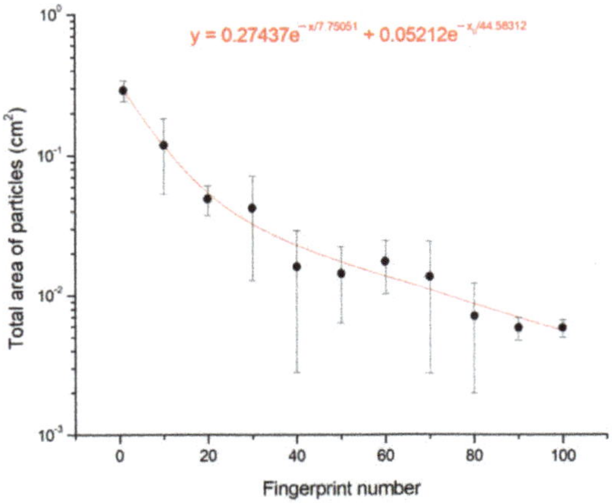


**Figure 4.** Dependence of the total area of trace particles on the serial number of the fingerprint.

Analysis of the trace macrophotographs shows that the projections of RDX parti onto the substrate plane have various shapes (Figure 5a–c). In this case, we are dealing w flat figures, the size of which can be characterized using some equivalent parameters. Th can be the maximum and minimum diameters (linear size), Feret and Martin diame etc., each of which determines the effective particle size.

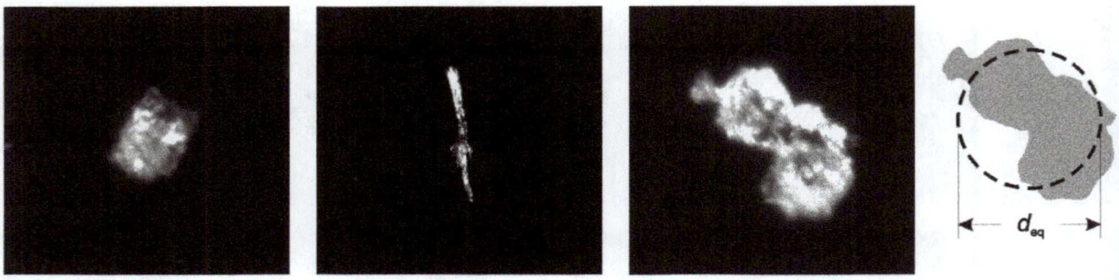


(a) (b) (c) (d)

**Figure 5.** Morphology of RDX particles. (**a–c**)—macrophotographs of RDX particles; (**d**)—sur equivalent diameter of particle (**c**).

To determine the effective volume of particles, one can use an approximation assumes the replacement of real particles by equivalent particles of a regular geom shape. In [18], to estimate the volume $V$ of RDX microscopic particles in fingerprints, it proposed to use the expression

$$V = \frac{1}{6}\pi d_{\text{eq}}^2 h,$$

where $d_{\text{eq}}$ is the surface-equivalent diameter—the diameter of a circle, the area of whic equal to the area of the surface projection of the original particle (Figure 5d) and $h \approx 0.31$ is the particle thickness.

Thus, knowing the volume of particles and the density of RDX (1.858 g/cm$^3$), can estimate the total mass of explosives in fingerprints. The results of such calculati showed that the mass of the transferred explosive at the first touch is ~3 mg and decre to ~30 μg in a hundredth fingerprint. Figure 6 shows the calculated dependence of surface concentration of RDX in fingerprints with an area of ~3 cm$^2$.

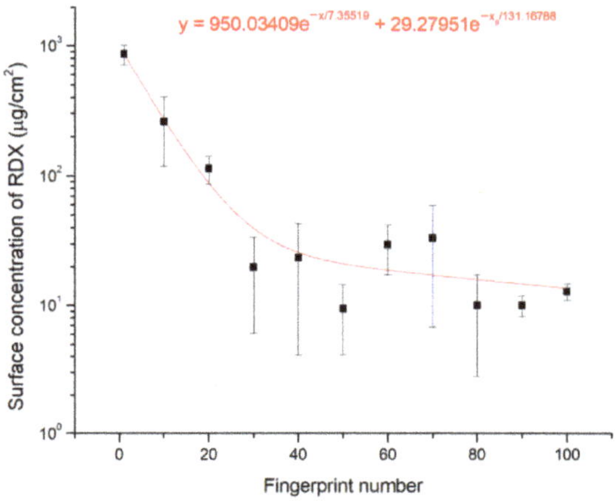


**Figure 6.** Dependence of the surface concentration of RDX on the serial number of the fingerprint.

## 3. Remote Detection of Traces of Explosives in Fingerprints Using the LF/LIF Method

The block diagram of the setup for experimental studies is shown in Figure 7.

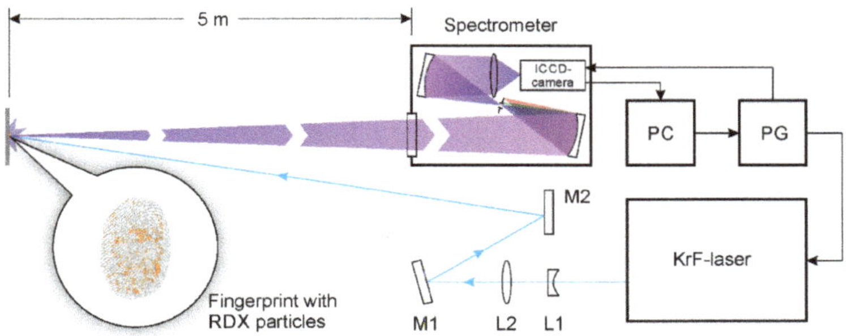


**Figure 7.** Block diagram of the setup for remote detection of RDX traces in fingerprints on a glass surface: M1 and M2 are the guiding mirrors, L1 and L2 are the lenses, PG is a pulse generator, and PC is the personal computer.

To perform the fragmentation of the RDX molecules and excite the fluorescence of their NO-fragments, we used a tunable KrF-laser with a narrow generation line of ~5 pm in the region of the bandhead of the $P_{12}$ branch of the absorption band $A^2\Sigma^+$ ($v' = 0$) – $X^2\Pi$ ($v'' = 2$) of the NO molecule (247.867 nm) [19].

With the help of rotary mirrors M1 and M2, the output laser beam was directed to the detection area. To provide the necessary energy density of laser radiation on the surface of the object, a two-lens L1–L2 system was used. Synchronous triggering of the laser and the photodetection system was carried out using a pulse generator PG controlled by a PC.

The optical response of the object surface to laser radiation was partially collected by the receiving optical system of a double diffraction spectrometer on concave holographic gratings. The spectrometer carried out spectral separation of the $\gamma(0, 0)$ fluorescence band of NO-fragments of nitro compounds in the wavelength range of 222–227 nm with a transmission of ~25% and an unshifted scattering line suppression level of 12 orders of magnitude. A time-gated intensified CCD camera (Andor iStar DH-712) was used to record the selected part of the spectrum in the full binning mode.

The experiments were carried out according to the following scheme (Figure 7). In the detection zone, at a distance of 5 m from the model of the remote detector of traces of

explosives, an object was installed with a sample with RDX traces fixed on its surface. N
a pulsed laser action was performed on the surface of the sample, followed by registra
of the optical response from it. The optical response was recorded in the photon coun
mode. At a signal accumulation volume of 50 laser pulses and a laser pulse repetition
of 20 Hz, the time of one measurement was 2.5 s. All measurements were carried o
a laboratory room at a temperature of 23 °C. The energy density in the laser pulse
~30 mJ/cm$^2$.

## 4. Experimental Results

The results of experiments on the detection of trace amounts of RDX in fingerprint
a glass slide are presented in Table 1. The developed experimental layout of the equipr
reliably recorded the response to the presence of RDX in the 100th fingerprint.

**Table 1.** The results of experiments on the detection of trace amounts of RDX in fingerprints
glass slide.

| Fingerprint Number | Fluorescence Intensity (Photocounts) | Signal-to-Background Ra |
|---|---|---|
| 1 | 138 | 137 |
| 10 | 36 | 35 |
| 20 | 17 | 16 |
| 30 | 13 | 12 |
| 40 | 11 | 10 |
| 50 | 14 | 13 |
| 60 | 15 | 14 |
| 70 | 4 | 3 |
| 80 | 4 | 3 |
| 90 | 6 | 5 |
| 100 | 3 | 2 |

The LIF intensity of NO-fragments of RDX, normalized to the intensity of those f
the first fingerprint, is shown in Figure 8. Here, for comparison, the dependence o
trace area on the number of the fingerprint is shown (Figure 4).

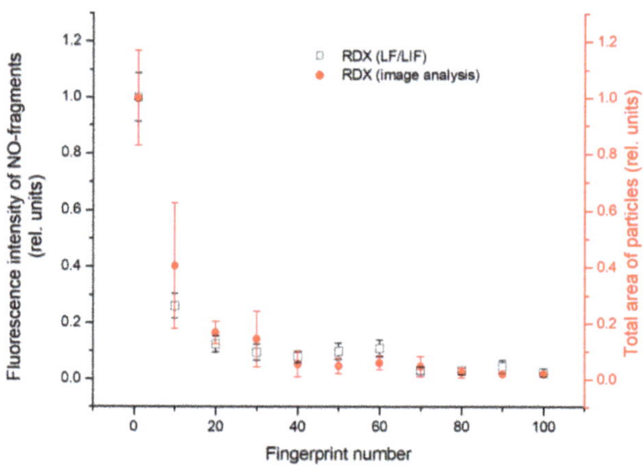


**Figure 8.** Dependences of the LIF intensity and the trace area on the number of the fingerprint.

It can be seen that both dependences are in good agreement with each other. Th
probably due to the fact that radiation with a wavelength of ~248 nm weakly penetr
into the volume of RDX particles (the penetration depth of the radiation pulse is ~55
For a series of 50 laser pulses, the layer of substance interacted with the radiation

be no thicker than 3 μm. The distribution of RDX particles over the parameter $h$ in the prepared samples (Figure 9) shows that the fraction of particles with $h \leq 10$ μm is about 9% of the total number of particles. Thus, when evaluating the effectiveness of using the LF/LIF method, the trace area, and not the mass of the substance in it, will be of decisive importance. Given this feature, it is obvious that the sensitivity of the method in terms of surface concentration (μg/cm$^2$) will depend on the trace structure.

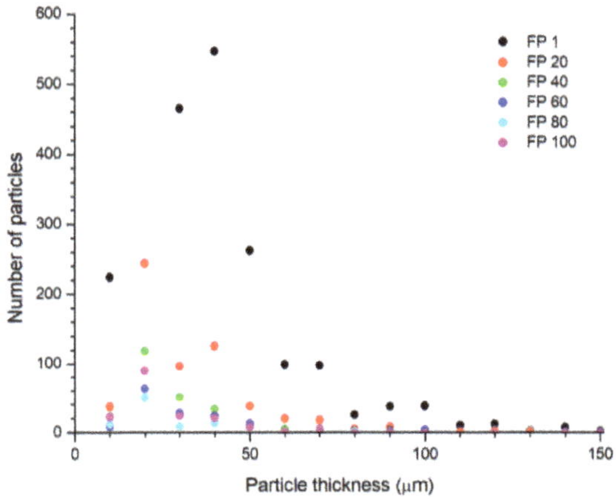


**Figure 9.** Distribution of RDX particles over the parameter $h$.

## 5. Conclusions

Thus, the use of high-resolution macrophotography in polarized light made it possible to study the evolution of the process of transferring the RDX microparticle trace from the 1st to 100th sweat-fat fingerprints. The pressure force during each act of fingerprint transfer was controlled and amounted to 2 N. This approach made it possible to stabilize the fingerprint transfer process and increase the repeatability of the result.

Pixel-by-pixel analysis of the trace macrophotographs made it possible to quantify the evolution of the total area of the particle contour projection onto the substrate plane and, within the framework of the model of flattened spheroids, to determine the mass of the trace in each fingerprint. The results of the experiments made it possible to plot the dependence of the trace mass on the fingerprint number. Analysis of the dependence shows that the decrease in the mass of the trace substance during transfer occurs according to an exponential law up to the 40th fingerprint, and then, the transfer process slows down significantly and, possibly, enters saturation. This phenomenon confirms the hypothesis of high trace preservation in the fingerprint and requires additional research.

A comparison shows that the data of direct measurements of the total trace area are in good agreement with the data of the LF/LIF responses.

**Author Contributions:** Conceptualization, S.B. and E.G.; methodology, E.G. and S.B.; validation, V.Z. and S.M.; formal analysis, E.G. and S.M.; resources, E.G., S.M., V.Z. and A.S.; data curation, E.G., V.Z. and S.M.; writing—original draft preparation, E.G.; writing—review and editing, S.B.; visualization, E.G. and S.M.; supervision, S.B.; project administration, A.V. All authors have read and agreed to the published version of the manuscript.

**Funding:** This study was supported by the Tomsk State University Development Programme (Priority-2030).

**Institutional Review Board Statement:** Not applicable.

**Informed Consent Statement:** Not applicable.

**Data Availability Statement:** https://cloud.mail.ru/public/LEGz/xFYbcSMCf.

**Conflicts of Interest:** The authors declare no conflict of interest.

## References


1. Rodgers, M.O.; Davis, D.D. A UV-Photofragmentation/Laser-Induced Fluorescence Sensor for the Atmospheric Detecti HONO. *Environ. Sci. Technol.* **1989**, *23*, 1106–1112. [CrossRef]
2. Galloway, D.B.; Bartz, J.A.; Huey, L.G.; Crim, F.F. Pathways and kinetic energy disposal in the photodissociation of nitroben *J. Chem. Phys.* **1993**, *98*, 2107–2114. [CrossRef]
3. Lemire, G.W.; Simeonsson, J.B.; Sausa, R.C. Monitoring of vapor-phase nitro compounds using 226-nm radiation: Fragment with subsequent NO resonance-enhanced multiphoton ionization detection. *Anal. Chem.* **1993**, *65*, 529–533. [CrossRef]
4. Galloway, D.B.; Glenewinkel-Meyer, T.; Bartz, J.A.; Huey, L.G.; Crim, F.F. The Kinetic and Internal Energy of NO fron Photodissociation of Nitrobenzene. *J. Chem. Phys.* **1994**, *100*, 1946–1952. [CrossRef]
5. Wu, D.D.; Singh, J.P.; Yueh, F.Y.; Monts, D.L. 2,4,6-Trinitrotoluene detection by laser-photofragmentation–laser-induced flu cence. *Appl. Opt.* **1996**, *35*, 3998–4003. [CrossRef] [PubMed]
6. Simeonsson, J.B.; Sausa, R.C. A critical review of laser photofragmentation/fragment detection techniques for gas phase che analysis. *Appl. Spectrosc. Rev.* **1996**, *31*, 1–72. [CrossRef]
7. Swayambunathan, V.; Singh, G.; Sausa, R.C. Laser photofragmentation–fragment detection and pyrolysis–laser-induced flu cence studies on energetic materials. *Appl. Opt.* **1999**, *38*, 6447–6454. [CrossRef] [PubMed]
8. Daugey, N.; Shu, J.; Bar, I.; Rosenwaks, S. Nitrobenzene detection by one-color laser photolysis/laser induced fluorescence o (v = 0–3). *Appl. Spectrosc.* **1999**, *53*, 57–64. [CrossRef]
9. Shu, J.; Bar, I.; Rosenwaks, S. Dinitrobenzene Detection by Use of One-color Laser Photolysis and Laser-Induced Fluorescer Vibrationally Excited NO. *Appl. Opt.* **1999**, *38*, 4705–4710. [CrossRef] [PubMed]
10. Shu, J.; Bar, I.; Rosenwaks, S. NO and PO photofragments as trace analyte indicators of nitrocompounds and organophospho *Appl. Phys. B* **2000**, *71*, 665–672. [CrossRef]
11. Shu, J.; Bar, I.; Rosenwaks, S. The use of rovibrationally excited NO photofragments as trace nitrocompounds indicators. *Phys. B* **2000**, *70*, 621–625. [CrossRef]
12. Arusi-Parpar, T.; Heflinger, D.; Lavi, R. Photodissociation Followed by Laser-Induced Fluorescence at Atmospheric Pressur 24 °C: A Unique Scheme for Remote Detection of Explosives. *J. Appl. Opt.* **2001**, *40*, 6677–6681. [CrossRef] [PubMed]
13. Heflinger, D.; Arusi-Parpar, T.; Ron, Y.; Lavi, R. Application of a unique scheme for remote detection of explosives. *Opt. Com* **2002**, *204*, 327–331. [CrossRef]
14. Wynn, C.M.; Palmacci, S.; Kunz, R.R.; Rothschild, M. Noncontact detection of homemade explosive constituents via photodi ation followed by laser-induced fluorescence. *Opt. Express.* **2010**, *18*, 5399–5406. [CrossRef] [PubMed]
15. Wynn, C.M.; Palmacci, S.; Kunz, R.R.; Aernecke, M. Noncontact optical detection of explosive particles via photodissoci followed by laser-induced fluorescence. *Opt. Express.* **2011**, *19*, 18671–18677. [CrossRef] [PubMed]
16. Bobrovnikov, S.M.; Gorlov, E.V. Lidar method for remote detection of vapors of explosives in the atmosphere. *Atmos. Ocean* **2011**, *24*, 235–241.
17. Bobrovnikov, S.M.; Vorozhtsov, A.B.; Gorlov, E.V.; Zharkov, V.I.; Maksimov, E.M.; Panchenko, Y.N.; Sakovich, G.V. Lidar dete of explosive vapors in the atmosphere. *Russ. Phys. J.* **2016**, *58*, 1217–1225. [CrossRef]
18. Verkouteren, J.R.; Coleman, J.L.; Cho, I. Automated Mapping of Explosives Particles in Composition C-4 Fingerprints. *J. For Sci.* **2010**, *55*, 334–340. [CrossRef] [PubMed]
19. Panchenko, Y.; Puchikin, A.; Yampolskaya, S.; Bobrovnikov, S.; Gorlov, E.; Zharkov, V. Narrowband KrF Laser for Lidar Sys *IEEE J. Quantum Electron.* **2021**, *57*, 1500105. [CrossRef]

photonics

MDPI

*munication*

# e Effect of Electron Escape Rate on the Nonlinear Dynamics Quantum Dot Lasers under Optical Feedback

gqing Wang [1,2], Zhengmao Wu [1,2], Yanfei Zheng [1,2] and Guangqiong Xia [1,2,*]

1 School of Physical Science and Technology, Southwest University, Chongqing 400715, China; dw1610020123@email.swu.edu.cn (Q.W.); zmwu@swu.edu.cn (Z.W.); yfzhengswu@email.swu.edu.cn (Y.Z.)

2 Chongqing Key Laboratory of Micro & Nano Structure Optoelectronics, Southwest University, Chongqing 400715, China

* Correspondence: gqxia@swu.edu.cn

**Abstract:** When theoretically investigating the nonlinear dynamics of quantum dot lasers (QDLs), the parameter value of the electron escape rate ($C^e$) is sometimes approximated to zero to simplify the calculation. However, the value of $C^e$ is dependent on the energy interval between the ground state (GS) and the excited state (ES) in the conduction band and is affected by the operation temperature. As a result, such simplified approximation treatments may lead to inaccurate results. In this study, after considering the effect of $C^e$, we investigate the nonlinear dynamics of QDLs with and without optical feedback based on the asymmetric electron-hole carrier rate equation model. The simulation results show that without optical feedback, the lasing conditions for ES and GS in free-running QDLs are dependent on the value of $C^e$. A larger $C^e$ is more helpful for the ES emission, and the GS emission will stop lasing if $C^e$ is large enough. Through analyzing the dynamical characteristics of GS and ES in QDLs with optical feedback under different $C^e$ values, it can be found that the dynamical characteristics are strongly correlative with $C^e$.

**Keywords:** quantum dot lasers (QDLs); nonlinear dynamics; optical feedback; electron escape rate ($C^e$)



on: Wang, Q.; Wu, Z.; Zheng, Y.; . The Effect of Electron Escape on the Nonlinear Dynamics of um Dot Lasers under Optical ack. *Photonics* **2023**, *10*, 878. //doi.org/10.3390/ nics10080878

ved: 9 July 2023
ed: 25 July 2023
ted: 27 July 2023
shed: 28 July 2023



## 1. Introduction

Semiconductor lasers (SLs) can exhibit diverse nonlinear dynamical behaviors under external perturbations, such as optical injection [1], optical feedback [2–4], and optoelectronic feedback [5]. These dynamical behaviors include: steady (S) state; period one (P1) state; period two (P2) state; multi-period (MP) state; chaotic (C) state; low frequency fluctuation (LFF) [6,7] and bistable state [8], which can be applied in photonic microwave generation [9,10]; all-optical logic gates [11]; reservoir computing [12,13]; chaotic secure communications [14]; random number generation [15,16]; and so on.

With the technical development of SLs and photonic integration [17], self-assembled nanostructured semiconductor quantum dot lasers (QDLs) have attracted the attention of many researchers [18,19]. The active region of QDLs is a zero-dimensional quantum dots (QD) nanostructure, which induces QDLs with discrete energy levels and state densities. Compared with traditional quantum well SLs, QDLs possess some advantages, such as low threshold current [20,21], temperature insensitivity [22,23], large modulation bandwidth [24], and low chirp [25]. Such unique properties make QDLs excellent candidate light sources in many fields, such as optical communications [26], all-optical logic gates [27], silicon photonic integrated circuits [28], photonic microwave generation [29,30], and so on. Related studies have demonstrated that QDLs can emit at the ground state (GS) and excited state (ES), either solely or simultaneously, by doping the active region of QDLs and changing the shape and size of QDs [31]. Since the GS and ES emission possesses a different wavelength and threshold current, it can be divided into three types: ground state quantum dot lasers (GS-QDLs), excited state quantum dot lasers (ES-QDLs), and dual-state quantum dot lasers (DSQDLs) [32,33]. For DSQDLs, there are two threshold currents. When the bias current arrives at the first threshold current, the GS starts to lase. With the further inc in the bias current, the number of carriers in the ES rapidly increases. Once the bias cu exceeds the second threshold, the GS and ES can simultaneously emit and the wavele difference between GS and ES emission is close to 100 nm [34]. For GS-QDLs and ES-Q there is always only one threshold current as the bias current increases. GS-QDLs po a lower threshold current compared with ES-QDLs owing to relatively low energy strong damping of relaxation oscillation of GS-QDLs. In recent years, the nonlinea namics of QDLs under external perturbations have received extensive attention [3 Under optical injection, hysteresis bistability has been observed in QDLs [37]. U optical feedback, mode competition of QDLs [38], energy exchange between longitu modes [39], two-color oscillations [40,41], and low-frequency fluctuations (LFF) [4 have been experimentally and theoretically reported successively. Theoretically, ther two models for analyzing the nonlinear dynamics of QDLs under external perturbat One is the symmetric exciton model [44], in which excitons are particles formed by coulomb interactions between electrons and holes in different energy levels of QD sem ductor materials. As a result, the energy level of GS and ES of QDLs presents electri neutral. This model can be applied to investigate the modulation characteristics [45], op noise characteristics [46], and photonic microwave generation [29], etc. The other is asymmetric electron-hole model [47], in which electrons and holes are heavily aggreg in different energy levels of QD semiconductor materials, respectively. Under this case coulomb interactions between electrons and holes are much smaller than the interact between electrons and electrons, and holes and holes, and then the symmetry betv electron-hole pairs is broken. The asymmetric electron-hole model can be adopte explain some behaviors of QDLs, such as the hysteresis bistability in QDLs under op injection [37], the external cavity mode competition [38], and all-optical switching [4 QDLs under optical feedback, etc. We have noted that the parameter value of elec escape rate ($C^e$) is sometimes estimated to be zero in order to simplify calculations du adopting the asymmetric electron-hole model. In fact, as pointed out in [39], the diffe $C^e$ values would affect the change in the GS-ES carrier number.

Based on above considerations, in this study, the effect of $C^e$ on the nonlinear dyna of QDLs under optical feedback is analyzed by adopting the asymmetric electron- model. The structure of the article is as follows: First, we analyze the factors influen the parameter values of $C^e$. Second, we inspect the effect of different $C^e$ on the la characteristics of QDLs without optical feedback. Next, via bifurcation diagrams, nonlinear dynamics of QDLs under optical feedback are analyzed for different valu Finally, through mapping the dynamical state distribution of QDLs in the space of p offset and optical feedback intensity, the effect of $C^e$ on the dynamical state distributi revealed.

## 2. Asymmetric Electron-Hole Carrier Rate Equation Model

Figure 1 shows a schematic of asymmetric electron-hole carrier dynamics and en level structure of QDLs, where the carriers include electrons with negative charge in conduction band and holes with positive charge in the valence band. In this model, only relatively low energy levels including the ground state (GS) and the first excited state energy level are considered. The active region of QDLs is assumed to contain only one of QD ensemble; meanwhile, the effect of non-uniform broadening is ignored. The car can be injected directly into the ES level through electrodes. Parts of the carriers are capt into the GS energy level by carrier scattering and Auger recombination processes [49] w capture rate of $B^{e,h}$, where the superscripts $e$ and $h$ represent electrons and holes, respecti Due to thermal excitation process [49], some carriers in the GS energy level escape into ES energy level, with an escape rate of $C^{e,h}$. Based on the asymmetric electron-hole ca dynamics and energy level structure model of QDLs, dimensionless rate equations [48 the nonlinear dynamics of QDLs under optical feedback can be described as:

$$\frac{dE_{GS}}{dt} = \frac{1}{2}[g_{GS}(n_{GS}^e + n_{GS}^h - 1) - 1]E_{GS} - i\frac{a_{GS}(n_{GS}^e + n_{GS}^h)}{2}E_{GS} + k_c e^{i\omega_{GS}\tau}E_{GS}(t-\tau) \tag{1}$$

$$\frac{dE_{ES}}{dt} = \frac{1}{2}[g_{ES}(n_{ES}^e + n_{ES}^h - 1) - 1]E_{ES} - i\frac{a_{ES}(n_{ES}^e + n_{ES}^h)}{2}E_{ES} + k_c e^{i\omega_{ES}\tau}E_{ES}(t-\tau) \tag{2}$$

$$\frac{dn_{GS}^e}{dt} = \eta[2B^e n_{ES}^e(1-n_{GS}^e) - 2C^e n_{GS}^e(1-n_{ES}^e) - n_{GS}^e - (n_{GS}^e + n_{GS}^h - 1)|E_{GS}|^2] \tag{3}$$

$$\frac{dn_{GS}^h}{dt} = \eta[2B^h n_{ES}^h(1-n_{GS}^h) - 2C^h n_{GS}^h(1-n_{ES}^h) - n_{GS}^h - (n_{GS}^e + n_{GS}^h - 1)|E_{GS}|^2] \tag{4}$$

$$\frac{dn_{ES}^e}{dt} = \eta[J^e(1-n_{ES}^e) - B^e n_{ES}^e(1-n_{GS}^e) + C^e n_{GS}^e(1-n_{ES}^e) - n_{ES}^e - (n_{ES}^e + n_{ES}^h - 1)|E_{ES}|^2] \tag{5}$$

$$\frac{dn_{ES}^h}{dt} = \eta[J^h(1-n_{ES}^h) - B^h n_{ES}^h(1-n_{GS}^h) + C^h n_{GS}^h(1-n_{ES}^h) - n_{ES}^h - (n_{ES}^e + n_{ES}^h - 1)|E_{ES}|^2] \tag{6}$$

where the subscripts GS and ES represent the ground state and excited state, respectively. $E$ and $n$ represent the complex electric field and carrier number, respectively. $J$ is the normalized injected current ($J^e = J^h = J$). $t$ represents the dimensionless time normalized to the photon lifetime $\tau_p$, and $\tau$ is the external cavity round trip time. $k_c$ stands for the normalized optical feedback intensity. $\omega$ represents the angular frequency normalized to $1/\tau_p$. $\eta$ represents the ratio of the photon lifetime to the non-radiative carrier lifetime. Considering that the GS and ES have two-fold degeneration and four-fold degeneration, respectively, we set $g_{GS} = 2g_0$ and $g_{ES} = 4g_0$ ($g_0$ is the effective gain factor). $1 - n$ describes Pauli blocking and $a(n^h + n^e)$ represents the frequency change induced by the resonant carrier factor $a$ [42]. The symmetric distribution of electron-hole pairs is broken due to the difference between $C^e$ and $C^h$. To determine the escape rates $C^{e,h}$, here we use the Kramers relation [46]:

$$C^{e,h} = B^{e,h}exp(-\frac{\Delta E^{e,h}}{k_b T}) \tag{7}$$

where $\Delta E^e$ and $\Delta E^h$ are the energy level interval between GS and ES in the conduction band and in the valence band, respectively, and $k_b$ represents the Boltzmann constant. $T$ represents the operation temperature of lasers. Generally, $\Delta E^h$ is near to 0, and therefore $C^h \approx B^h$. However, $\Delta E^e$ is usually tens of meV. Referring to [37], we set $\Delta E^e$ at 50 meV. In this case, the dependence of $C^e$ on the operation temperature of QDLs is given in Figure 2, where $B^h = B^e = C^h = 150$ [48]. Obviously, $C^e$ is increased with the increase in the operation temperature of QDLs. At room temperature, $k_b T$ = 25 meV, and then $C^e$ = 20.3.

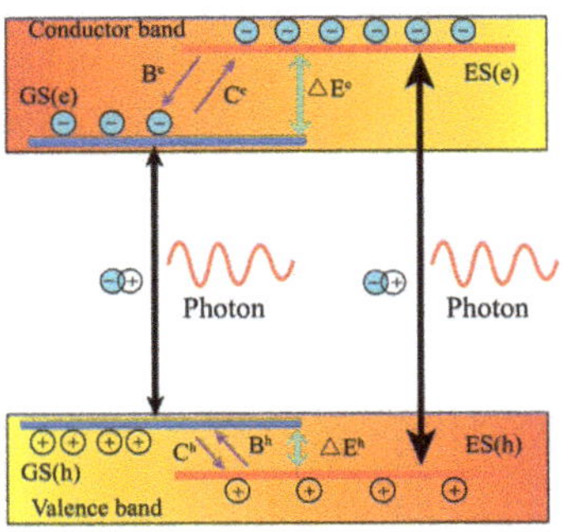


**Figure 1.** Schematic diagram of asymmetric electron-hole carrier dynamics and energy level structure of QDLs. ES(e): electrons in excited state; ES(h): holes in excited state; GS(e): electrons in ground state; GS(h): holes in ground state.

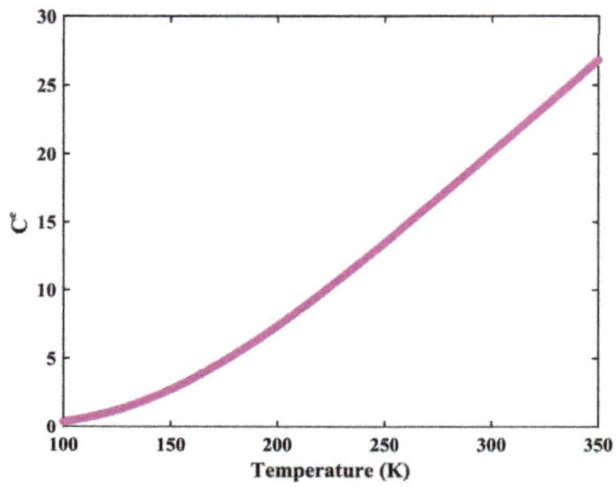


**Figure 2.** Electron escape rate $C^e$ as a function of the operation temperature of QDLs.

## 3. Results and Discussion

Equations (1)–(6) can be numerically solved by the fourth-order Runge–Kutta me via MATLAB software, with the parameters [48]: $\eta = 0.02$, $B^e = B^h = C^h = 150$, $g_0 =$ $a_{GS} = a_{ES} = 5$, $\omega_{GS} \approx 3015.9$, $\omega_{ES} \approx 3222.1$.

Figure 3 shows the normalized output power of GS (blue) and ES (red) as a functic normalized injection current $J$ for a free-running QDL under different $C^e$, and the thres currents of ES and GS as a function of $C^e$ for a free-running QDL. For $C^e = 0$ (Figure the QDL starts to lase when $J$ arrives at 3.0. When $J$ is within the range of 3.0–4.5 free-running QDL is operating at the GS emission. Once $J$ exceeds 4.5, the ES of starts to lase. With a further increase in $J$, the output power of GS decreases while output power of ES increases. When $J$ is within the range of 4.5–10.0, the free-running is operating at the coexistence of GS and ES. When $J$ is more than 10.0, the GS emis is completely suppressed. As shown in Figure 3b, for $C^e = 20.3$ (corresponding to r temperature), the threshold current of QDL is 3.0, and only the ES emits. Obviously QDL operating at room temperature, adopting the approximate treatment of $C^e = 0$ lead to inaccurate results. From Figure 3c, it can be seen that the threshold currents o and ES are depended on $C^e$. With the increase in $C^e$, the threshold current of ES decre firstly and then maintains a constant level, while the threshold current of GS remai a constant firstly and then the GS stops lasing. As a result, a relatively larger value (meaning a higher temperature) is more helpful for ES emission, which is consistent the experimental observation in [50]. In the following, taking $J = 3.5$ and $J = 5.7$ as exam we will analyze the effect of different $C^e$ values on the nonlinear dynamics of QDL optical feedback.

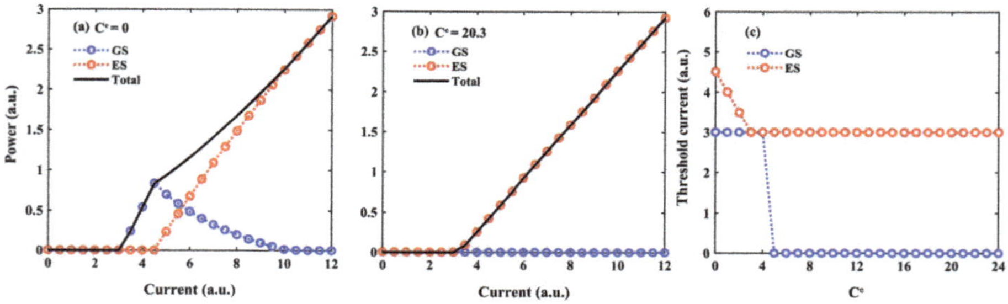


**Figure 3.** Normalized output power as a function of normalized the injection current for a without optical feedback under $C^e = 0$ (**a**) and $C^e = 20.3$ (**b**), respectively, and threshold currents and GS as a function of $C^e$ (**c**).

A bifurcation diagram is usually adopted to analyze the nonlinear dynamics evolution of QDLs with a certain parameter. The bifurcation diagram is obtained from a time series by sampling and plotting local peaks and valleys of the output waveform for different parameter values [51,52]. Figure 4 gives the bifurcation diagrams of power extreme and mean power of the GS (blue) and ES (red) as a function of optical feedback intensity ($k_c$) under $J = 3.5$ and delayed time $\tau = 100$. For $C^e = 0$ (Figure 4(a1)), the GS is stable (S) under $k_c < 0.0092$. For $k_c$ is within the range of (0.0092, 0.015), the output waveform of GS emission has two extreme values corresponding to the peaks and valleys, and the GS can be judged to operate at period one (P1) oscillation. When $k_c$ is located within the range of (0.015, 0.016), the output waveform of GS possesses four extreme values corresponding to the peaks and valleys, and the GS can be determined to be period two (P2) oscillation. When $k_c$ is within the range of (0.028, 0.031), the output waveform of GS emission has multiple extreme values corresponding to the peaks and valleys, then the GS operates at the multi-period (MP) oscillation. When $k_c$ exceeds the value of 0.031, the output waveform shows irregular oscillation, and the GS evolves into chaotic (C) oscillation. For ES emission, one can observe only for $0.02 < k_c < 0.028$, in which the ES behaves C oscillation. When $C^e = 5$ (Figure 4(b1)), only the ES emits under $k_c < 0.012$. When $k_c$ is located within the range of (0.012, 0.04), the GS undergoes an evolution from P1, S, P1, P2, MP to C oscillation. Once $k_c$ exceeds the values of 0.04, the GS appears as a new cascade bifurcation from S, MP into C oscillation. When $k_c$ is within the range of (0.012, 0.014), the ES enters P1 oscillation. When $k_c$ exceeds the values of 0.044, the ES evolves into C oscillation. When $C^e = 10$ (Figure 4(c1)), the ES has multiple dynamical states, including S, P1, P2, and MP oscillation when $k_c$ is in the range of (0, 0.035). When $k_c$ is located within the range of (0.057, 0.062) and (0.08, 0.083), the ES behaves as C oscillation. When $k_c$ is within the range of (0.036, 0.09), the GS exhibits from S, P2, MP to C oscillation. When $C^e = 15$ (Figure 4(d1)), the ES shows various dynamical states, such as S, P1, P2, MP, and C oscillation within the range of (0, 0.051). For GS emission, it only starts emission when the $k_c$ exceeds 0.053. When $k_c$ is located within the range of (0.053, 0.09), the GS exhibits multiply dynamical states such as S, P1, P2, and C oscillation. As shown in Figure 4(e1), when $C^e = 20.3$ (corresponding to the room temperature), only the ES emits for $k_c < 0.081$, and S, P1 and C oscillation can be observed. For GS, it starts to emit for $k_c > 0.081$, C and S oscillation can be observed. By comparing the bifurcation diagram of power extreme and mean power under different $C^e$, one can find that strongly optical feedback is favorable for GS emission under a given $C^e$; meanwhile, the dynamical evolution of GS and ES is significantly related to the value of $C^e$. Moreover, for a larger value of $C^e$ (corresponding to a higher temperature), the region of feedback strength for realizing chaotic output is narrower.

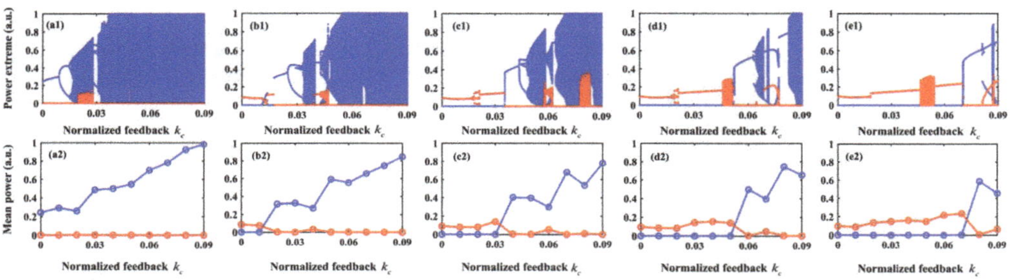


**Figure 4.** Bifurcation diagrams of power extreme and mean power as a function of feedback intensity of GS (blue) and ES (red) in QDL biased at 3.5 under optical feedback with $\tau = 100$ and (**a**) $C^e = 0$, (**b**) $C^e = 5$, (**c**) $C^e = 10$, (**d**) $C^e = 15$, (**e**) $C^e = 20.3$, respectively.

Figure 5 shows the bifurcation diagrams of power extreme and mean power of the GS (blue) and ES (red) as a function of $k_c$ under injection current $J = 5.7$ and delay time $\tau = 100$. For $C^e = 0$ (Figure 5(a1)), the GS and ES are stable under $k_c < 0.023$. When $k_c$

is within the range of (0.023, 0.025), the GS and ES show various dynamical states s as P1, P2, and MP oscillation. When $k_c$ exceeds the values 0.034, the GS and ES ent oscillation. When $C^e$ = 5 (Figure 5(b1)), only the ES emits under $k_c$ < 0.018. When located within the range of (0.018, 0.04), the GS starts emission and the GS and ES are S. GS and ES exhibit multiple dynamical states, including P1, P2, and MP, under the ra of (0.04, 0.042). When $k_c$ is within the range of (0.042, 0.1), the GS and ES evolve from C oscillation. When $C^e$ = 10 (Figure 5(c1)), only the ES emits under $k_c$ < 0.037. With increase in $k_c$ from 0.037 to 0.065, the GS and ES behave as S. Once $k_c$ exceeds 0.07, the and ES evolve into C oscillation. When $C^e$ = 15 (Figure 5(d1)), the ES is S under the ra of $k_c$ (0, 0.055). When $k_c$ is located within the range of (0.055, 0.075), the GS and ES sl rich dynamical states, including P1, P2, and MP oscillation. When $k_c$ exceeds 0.075, GS and ES enter C oscillation. As shown in Figure 5(e1), when $C^e$ = 20.3 (correspond to the room temperature), only the ES emits for $k_c$ < 0.065. When $k_c$ is within the rang (0.065, 0.098), the GS and ES emit simultaneously and multiple dynamical states suc P1, P2, and MP oscillations can be observed. When $k_c$ is located within the range of (0. 0.1), the GS emission operates at S, while the ES emission is suppressed. By comparing bifurcation diagram of power extreme and mean power under different $C^e$, it can be fo that the GS and ES always have the same dynamic evolution with the increase in $k_c$. a relatively large $C^e$, a lower $k_c$ is more favorable for the ES emission, and a stronger more favorable for the GS emission. Similarly, in the case of QDL biased at 3.5, for a la value of $C^e$ (corresponding to a higher temperature), the chaos state can be observed wi a narrower region of the feedback strength.

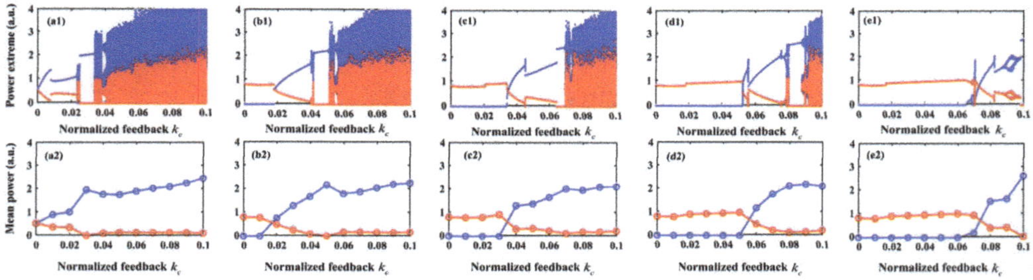


**Figure 5.** Bifurcation diagrams of power extreme and mean power as a function of feedback inte of GS (blue) and ES (red) in QDL biased at 5.7 under optical feedback with $\tau$ = 100 and (**a**) $C^e$ (**b**) $C^e$ = 5, (**c**) $C^e$ = 10, (**d**) $C^e$ = 15, (**e**) $C^e$ = 20.3, respectively.

Besides the optical feedback intensity ($k_c$), the phase offset (defined as $\Delta\tau w_{GS}$) ca by the slight variation in $\tau$ is another crucial factor to affect the nonlinear dynamics of Q Finally, we simulate the dynamical evolution of QDL in the parameter space compose such two crucial parameters; the corresponding results under $J$ = 3.5 are given in Figu For $C^e$ = 0 (Figure 6(a1,a2)), while the dynamical states of GS include S, P1, P2, MP, C, and the C state occupies the widest area. The dynamical states of ES include S P2, and MP, and most of the region is S state. When $C^e$ = 5 (Figure 6(b1)), there are various dynamical states, including S, P1, P2, MP, and C, but the C region is shrur compared with that of $C^e$ = 0. As shown in Figure 6(b2), although most areas still bel to the S region for ES, the regions for other dynamical states are expanded. For $C^e$ (Figure 6(c1,c2)), multiple dynamical states including S, P1, P2, MP and C can be obse for the GS and ES. For $C^e$ = 15 (Figure 6(d1,d2)) and $C^e$ = 20.3 (Figure 6(e1,e2)), the S re for GS becomes larger and other dynamic states become smaller, and more dynamical st including S, P1, P2, MP and C oscillation can still be observed for ES emission.

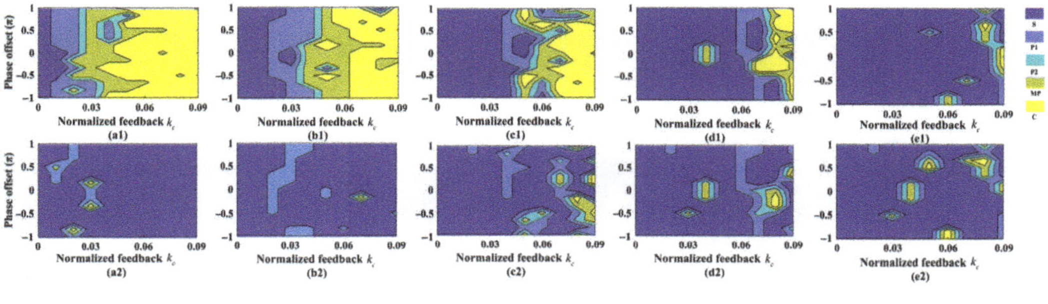


**Figure 6.** Mapping of the dynamical states of GS emission (the first row) and ES emission (the second row) in the parameter space of feedback intensity and phase offset for QDL biased at 3.5 under different $C^e$ where (**a**) $C^e = 0$, (**b**) $C^e = 5$, (**c**) $C^e = 10$, (**d**) $C^e = 15$, (**e**) $C^e = 20.3$. S: stable, P1: period one, P2: period two, MP: multi-period, and C: chaos.

For $J = 5.7$, the corresponding results are given in Figure 7. For $C^e = 0$ (Figure 7(a1,a2)), the dynamical states of GS and ES include S, P1, P2, MP, and C oscillation. In the dwhole parameter space, a small number of regions are located at S. For $C^e = 5$ (Figure 7(b1,b2)) and $C^e = 10$ (Figure 7(c1,c2)); there are also multiple dynamical states, such as S, P1, P2, MP, and C oscillation, and the C region is shrunken compared with that under $C^e = 0$. For $C^e = 15$ (Figure 7(d1,d2)) and $C^e = 20.3$ (Figure 7(e1,e2)), the S region of GS and ES widens in the whole space and other dynamic states of the region become smaller. From Figures 6 and 7, it can be concluded that the dynamical states distribution of GS and ES in the parameter space of optical feedback intensity and phase offset is strongly dependent on the parameter value of $C^e$.

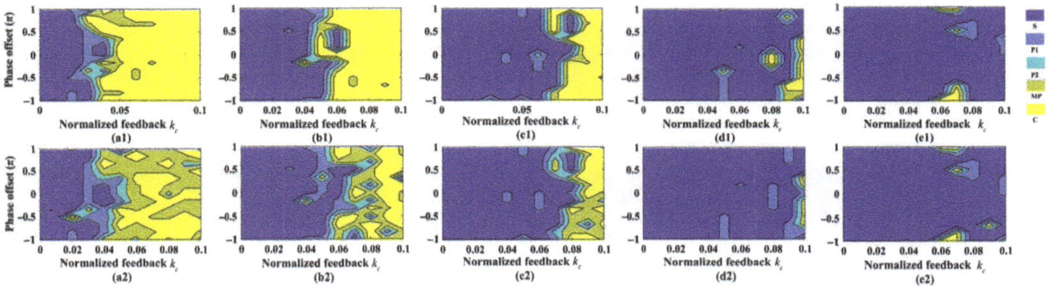


**Figure 7.** Mapping of the dynamical states of GS emission (the first row) and ES emission (the second row) in the parameter space of feedback intensity and phase offset for QDL biased at 5.7 under different $C^e$ where (**a**) $C^e = 0$, (**b**) $C^e = 5$, (**c**) $C^e = 10$, (**d**) $C^e = 15$, (**e**) $C^e = 20.3$. S: stable, P1: period one, P2: period two, MP: multi-period, and C: chaos.

## 4. Conclusions

In summary, via an asymmetric electron-hole carrier rate equation model, the effects of the different parameter values of electron escape rate ($C^e$) on the nonlinear dynamics of quantum dot lasers (QDLs), with or without optical feedback, are theoretically investigated. The simulation results show that, for QDLs without optical feedback, the emission threshold of ground state (GS) and excited state (ES) strongly depends on the parameter value of $C^e$. A relatively larger value of $C^e$ is more helpful for ES emission, and GS stops lasing when $C^e$ is large enough. For QDLs with optical feedback, various dynamical states, such as stable (S), period one (P1) oscillation, period two (P2) oscillation, multi-period (MP) oscillation, and chaotic (C) oscillation of GS and ES for QDLs, can be observed. By comparing the dynamical state distributions for GS and ES in the parameter space of feedback intensity and phase offset under different $C^e$, it can be concluded that the dynamic state of QI with optical feedback is strongly correlative with the parameter value of $C^e$. As a res when theoretically investigating the nonlinear dynamics of QDLs with or without opt feedback, the effect of $C^e$ should be taken into account for obtaining accurate results.

**Author Contributions:** Conceptualization, Q.W. and G.X.; methodology, Q.W.; validation, Q.W., and Z.W.; formal analysis, Q.W. and Y.Z.; investigation, Q.W. and Y.Z.; resources, G.X. and Z.W.; curation, Q.W.; writing—original draft preparation, Q.W.; writing—review and editing, G.X. Z.W.; visualization, Q.W.; supervision, Z.W.; project administration, G.X.; funding acquisition, and Z.W. All authors have read and agreed to the published version of the manuscript.

**Funding:** This research was funded by the National Natural Science Foundation of China (61875 the Chongqing Natural Science Foundation (CSTB2022NSCQ-MSX0313) and the Postgradu Research and Innovation Project of Chongqing (CYB22111).

**Institutional Review Board Statement:** Not applicable.

**Informed Consent Statement:** Not applicable.

**Data Availability Statement:** Not applicable.

**Conflicts of Interest:** The authors declare no conflict of interest.

## References


1. Lin, F.Y.; Tu, S.Y.; Huang, C.C.; Chang, S.M. Nonlinear dynamics of semiconductor lasers under repetitive optical pulse injec *IEEE J. Sel. Top. Quantum Electron.* **2009**, *15*, 604–611. [CrossRef]
2. Zhong, Z.Q.; Chang, D.; Jin, W.; Lee, M.W.; Wang, A.B.; Jiang, S.; He, J.X.; Tang, J.M.; Hong, Y.H. Intermittent dynamical s switching in discrete-mode semiconductor lasers subject to optical feedback. *Photonics Res.* **2021**, *9*, 1336–1342. [CrossRef]
3. Asghar, H.; Wei, W.; Kumar, P.; Sooudi, E.; McInerney, J.G. Stabilization of self-mode-locked quantum dash lasers by symm dual-loop optical feedback. *Opt. Express* **2018**, *26*, 4581–4592. [CrossRef] [PubMed]
4. Alrebdi, T.A.; Asghar, M.; Asghar, H. All Optical stabilizations of nano-structure-based QDash semiconductor mode-locked la based on asymmetric dual-loop optical feedback configurations. *Photonics* **2022**, *9*, 376. [CrossRef]
5. Chengui, G.R.G.; Jacques, K.; Woafo, P.; Chembo, Y.K. Nonlinear dynamics in an optoelectronic feedback delay oscillator piecewise linear transfer functions from the laser diode and photodiode. *Phys. Rev. E* **2020**, *102*, 042217. [CrossRef] [PubMe
6. Dou, X.Y.; Qiu, S.M.; Wu, W.Q. Numerical demonstration of the transmission of low frequency fluctuation dynamics gener by a semiconductor laser with optical feedback. *Photonics* **2022**, *9*, 483. [CrossRef]
7. Tiana-Alsina, J.; Masoller, C. Experimental and numerical study of locking of low-frequency fluctuations of a semiconductor with optical feedback. *Photonics* **2022**, *9*, 103. [CrossRef]
8. Chen, J.J.; Zhong, Z.Q.; Li, L.F. Characteristics of polarization switching and bistability in vertical-cavity surface-emitting subject to continuous variable-polarization optical injection. *Acta Opt. Sin.* **2022**, *42*, 0714003.
9. Chang, D.; Zhong, Z.Q.; Valle, A.; Jin, W.; Jiang, S.; Tang, J.M.; Hong, Y.H. Microwave photonic signal generation in an opti injected discrete mode semiconductor laser. *Photonics* **2022**, *9*, 171. [CrossRef]
10. Ji, S.K.; Xue, C.P.; Valle, A.; Spencer, P.S.; Li, H.Q.; Hong, Y.H. Stabilization of photonic microwave generation in vertical-ca surface-emitting lasers with optical injection and feedback. *J. Light. Technol.* **2018**, *36*, 4347–4353. [CrossRef]
11. Qureshi, K.K. Realization of all-optical logic gates using injection locking in a distributed feedback laser diode. *Microw. Technol. Lett.* **2016**, *58*, 940–944. [CrossRef]
12. Jiang, L.; Liang, W.Y.; Song, W.J.; Jia, X.H.; Yang, Y.L.; Liu, L.M.; Deng, Q.X.; Mou, X.Y.; Zhang, X. Characteristic study on par reservoir computation based on VCSEL dynamics with optoelectronic feedback. *IEEE J. Quantum Electron.* **2022**, *58*, 240 [CrossRef]
13. Lin, B.D.; Shen, Y.W.; Tang, J.Y.; Yu, J.Y.; He, X.M.; Wang, C. Deep time-delay reservoir computing with cascading injection-lo lasers. *IEEE J. Sel. Top. Quantum Electron.* **2023**, *29*, 7600408. [CrossRef]
14. Zhong, D.Z.; Zhao, K.K.; Hu, Y.L.; Zhang, J.B.; Deng, W.A.; Hou, P. Four-channels optical chaos secure communications wit rate of 400 Gb/s using optical reservoir computing based on two quantum dot spin-VCSELs. *Opt. Commun.* **2023**, *529*, 12 [CrossRef]
15. Cai, Q.; Li, P.; Shi, Y.C.; Jia, Z.W.; Ma, L.; Xu, B.J.; Chen, X.F.; Shore, K.A.; Wang, Y.C. Tbps parallel random number genera based on a single quarter-wavelength-shifted DFB laser. *Opt. Laser Technol.* **2023**, *162*, 109273. [CrossRef]
16. Kawaguchi, Y.; Okuma, T.; Kanno, K.; Uchida, A. Entropy rate of chaos in an optically injected semiconductor laser for phy random number generation. *Opt. Express* **2021**, *29*, 2442–2457. [CrossRef]
17. Yang, J.J.; Tang, M.C.; Chen, S.M.; Liu, H.Y. From past to future: On-chip laser sources for photonic integrated circuits. *Ligh Appl.* **2023**, *12*, 16. [CrossRef]

Moos, R.; Konieczniak, I.; Dos Santos, G.E.; Gobbi, Â.L.; Bernussi, A.A.; Carvalho, W.; Medeiros-Ribeiro, G.; Ribeiro, E. Assessing electronic states of InAsP/GaAs self-assembled quantum dots by photoluminescence and modulation spectroscopy. *J. Lumin.* **2019**, *206*, 639–644. [CrossRef]
Mitsuru, S.; Kohki, M.; Yoshiaki, N.; Koji, O.; Hiroshi, I. Performance and physics of quantum-dot lasers with self-assembled columnar-shaped and 1.3-*um* emitting InGaAs quantum dots. *IEEE J. Sel. Top. Quantum Electron.* **2000**, *6*, 462–474.
Al-Marhaby, F.A.; Al-Ghamdi, M.S. Experimental investigation of stripe cavity length effect on threshold current density for InP/AlGaInP QD laser diode. *Opt. Mater.* **2022**, *127*, 112191. [CrossRef]
Jung, D.; Norman, J.; Kennedy, M.J.; Shang, C.; Shin, B.; Wan, Y.T.; Gossard, A.C.; Bowers, J.E. High efficiency low threshold current 1.3 mu m InAs quantum dot lasers on on-axis (001) GaP/Si. *Appl. Phys. Lett.* **2017**, *111*, 122107. [CrossRef]
Mikhrin, S.S.; Kovsh, A.R.; Krestnikov, I.L.; Kozhukhov, A.V.; Livshits, D.A.; Ledentsov, N.N.; Shernyakov, Y.M.; Novikov, I.I.; Maximov, M.V.; Ustinov, V.M.; et al. High power temperature-insensitive 1.3 µm InAs/InGaAs/GaAs quantum dot lasers. *Semicond. Sci. Technol.* **2005**, *20*, 340–342. [CrossRef]
Cong, D.Y.; Martinez, A.; Merghem, K.; Ramdane, A.; Provost, J.G.; Fischer, M.; Krestnikov, I.; Kovsh, A. Temperature insensitive linewidth enhancement factor of P-type doped InAs/GaAs quantum-dot lasers emitting at 1.3 µm. *Appl. Phys. Lett.* **2008**, *92*, 191109. [CrossRef]
Qasaimeh, O. Modulation bandwidth of inhomogeneously broadened InAs/GaAs quantum dot lasers. *Opt. Commun.* **2004**, *236*, 387–394. [CrossRef]
Saito, H.; Nishi, K.; Kamei, A.; Sugou, S. Low chirp observed in directly modulated quantum dot lasers. *IEEE Photonics Technol. Lett.* **2000**, *12*, 1298–1300. [CrossRef]
Zhukov, A.E.; Kovsh, A.R. Quantum dot diode lasers for optical communication systems. *Quantum Electron.* **2008**, *38*, 409–423. [CrossRef]
Kotb, A. Simulation of high quality factor all-optical logic gates based on quantum-dot semiconductor optical amplifier at 1 Tb/s. *Optik* **2016**, *127*, 320–325. [CrossRef]
Dong, B.Z.; Chen, J.D.; Lin, F.Y.; Norman, J.C.; Bowers, J.E.; Grillot, F. Dynamic and nonlinear properties of epitaxial quantum-dot lasers on silicon operating under long- and short-cavity feedback conditions for photonic integrated circuits. *Phys. Rev. A* **2021**, *103*, 033509. [CrossRef]
Wang, C.; Raghunathan, R.; Schires, K.; Chan, S.C.; Lester, L.F.; Grillot, F. Optically injected InAs/GaAs quantum dot laser for tunable photonic microwave generation. *Opt. Lett.* **2016**, *41*, 1153–1156. [CrossRef]
Jiang, Z.F.; Wu, Z.M.; Yang, W.Y.; Hu, C.X.; Lin, X.D.; Jin, Y.H.; Dai, M.; Cui, B.; Yue, D.Z.; Xia, G.Q. Numerical simulations on narrow-linewidth photonic microwave generation based on a QD laser simultaneously subject to optical injection and optical feedback. *Appl. Opt.* **2020**, *59*, 2935–2941. [CrossRef]
Li, Q.Z.; Wang, X.; Zhang, Z.Y.; Chen, H.M.; Huang, Y.Q.; Hou, C.C.; Wang, J.; Zhang, R.Y.; Ning, J.Q.; Min, J.H.; et al. Development of modulation p-doped 1310 nm InAs/GaAs quantum dot laser materials and ultrashort cavity fabry–perot and distributed-feedback laser diodes. *ACS Photonics* **2018**, *5*, 1084–1093. [CrossRef]
Kelleher, B.; Dillane, M.; Viktorov, E.A. Optical information processing using dual state quantum dot lasers: Complexity through simplicity. *Light-Sci. Appl.* **2021**, *10*, 238. [CrossRef]
Lin, L.C.; Chen, C.Y.; Huang, H.; Arsenijevic, D.; Bimberg, D.; Grillot, F.; Lin, F.Y. Comparison of optical feedback dynamics of InAs/GaAs quantum-dot lasers emitting solely on ground or excited states. *Opt. Lett.* **2018**, *43*, 210–213. [CrossRef]
Xu, P.F.; Tao, Y.; Ji, H.M.; Cao, Y.L.; Gu, Y.X.; Liu, Y.; Ma, W.Q.; Wang, Z.G. Temperature-dependent modulation characteristics for 1.3 mu m InAs/GaAs quantum dot lasers. *J. Appl. Phys.* **2010**, *107*, 013102. [CrossRef]
Jiang, Z.F.; Wu, Z.M.; Jayaprasath, E.; Yang, W.Y.; Xia, G.Q. Nonlinear dynamics of exclusive excited-state emission quantum dot lasers under optical injection. *Photonics* **2019**, *6*, 58. [CrossRef]
Wang, X.H.; Wu, Z.M.; Jiang, Z.F.; Xia, G.Q. Nonlinear dynamics of two-state quantum dot lasers under optical feedback. *Photonics* **2021**, *8*, 300. [CrossRef]
Tykalewicz, B.; Goulding, D.; Hegarty, S.P.; Huyet, G.; Dubinkin, I.; Fedorov, N.; Erneux, T.; Viktorov, E.A.; Kelleher, B. Optically induced hysteresis in a two-state quantum dot laser. *Opt. Lett.* **2016**, *41*, 1034–1037. [CrossRef] [PubMed]
Virte, M.; Panajotov, K.; Sciamanna, M. Mode competition induced by optical feedback in two-color quantum dot lasers. *IEEE J. Quantum Electron.* **2013**, *49*, 578–585. [CrossRef]
Virte, M.; Pawlus, R.; Sciamanna, M.; Panajotov, K.; Breuer, S. Energy exchange between modes in a multimode two-color quantum dot laser with optical feedback. *Opt. Lett.* **2016**, *41*, 3205–3208. [CrossRef]
Kelleher, B.; Tykalewicz, B.; Goulding, D.; Fedorov, N.; Dubinkin, I.; Erneux, T.; Viktorov, E.A. Two-color bursting oscillations. *Sci. Rep.* **2017**, *7*, 8414. [CrossRef]
Meinecke, S.; Kluge, L.; Hausen, J.; Lingnau, B.; Ludge, K. Optical feedback induced oscillation bursts in two-state quantum-dot lasers. *Opt. Express* **2020**, *28*, 3361–3377. [CrossRef] [PubMed]
Viktorov, E.A.; Mandel, P.; O'Driscoll, I.; Carroll, O.; Huyet, G.; Houlihan, J.; Tanguy, Y. Low-frequency fluctuations in two-state quantum dot lasers. *Opt. Lett.* **2006**, *31*, 2302–2304. [CrossRef] [PubMed]
Yamasaki, K.; Kanno, K.; Matsumoto, A.; Akahane, K.; Yamamoto, N.; Naruse, M.; Uchida, A. Fast dynamics of low-frequency fluctuations in a quantum-dot laser with optical feedback. *Opt. Express* **2021**, *29*, 17962–17975. [CrossRef] [PubMed]

44. Wang, C.; Grillot, F.; Even, J. Impacts of wetting layer and excited state on the modulation response of quantum-dot lasers. *IE Quantum Electron.* **2012**, *48*, 1144–1150. [CrossRef]
45. Kayhani, K.; Rajaei, E. Investigation of dynamical characteristics and modulation response function of InAs/InP (311) B quan dot lasers with different QD size. *Photonics Nanostruct. Fundam. Appl.* **2017**, *25*, 1–8. [CrossRef]
46. Zhou, Y.G.; Duan, J.N.; Grillot, F.; Wang, C. Optical noise of dual-state lasing quantum dot lasers. *IEEE J. Quantum Electron.* *56*, 2001207. [CrossRef]
47. Viktorov, E.A.; Mandel, P.; Tanguy, Y.; Houlihan, J.; Huyet, G. Electron-hole asymmetry and two-state lasing in quantum lasers. *Appl. Phys. Lett.* **2005**, *87*, 053113. [CrossRef]
48. Virte, M.; Breuer, S.; Sciamanna, M.; Panajotov, K. Switching between ground and excited states by optical feedback in a quan dot laser diode. *Appl. Phys. Lett.* **2014**, *105*, 121109. [CrossRef]
49. Abusaa, M.; Danckaert, J.; Viktorov, E.A.; Erneux, T. Intradot time scales strongly affect the relaxation dynamics in quantum lasers. *Phys. Rev. A* **2013**, *87*, 063827. [CrossRef]
50. Drzewietzki, L.; The, G.A.P.; Gioannini, M.; Breuer, S.; Montrosset, I.; Elsasser, W.; Hopkinson, M.; Krakowski, M. Theoretical experimental investigations of the temperature dependent continuous wave lasing characteristics and the switch-on dynami an InAs/InGaAs quantum-dot semiconductor laser. *Opt. Commun.* **2010**, *283*, 5092–5098. [CrossRef]
51. Kim, B.; Locquet, A.; Li, N.Q.; Choi, D.; Citrin, D.S. Bifurcation-cascade diagrams of an external-cavity semiconductor l Experiment and theory. *IEEE J. Quantum Electron.* **2014**, *50*, 965–972. [CrossRef] [PubMed]
52. Mengue, A.D.; Essimbi, B.Z. Complex chaos and bifurcations of semiconductor lasers subjected to optical injection. *Opt. Quan Electron.* **2011**, *42*, 389–407. [CrossRef]

photonics MDPI

cle

# ven-Grooved-Rod, Side-Pumping Concept for Highly Efficient EM$_{00}$-Mode Solar Laser Emission through Fresnel Lenses

go Costa, Dawei Liang *, Joana Almeida, Miguel Catela, Dário Garcia, Bruno D. Tibúrcio Cláudia R. Vistas

Centre of Physics and Technological Research, Physics Department, NOVA School of Science and Technology, 2829-516 Caparica, Portugal
* Correspondence: dl@fct.unl.pt

**Abstract:** Low-cost, lightweight, and easily available Fresnel lenses are a more alluring choice for solar laser power production, when compared to the costly and complex heliostat-parabolic mirror systems. Therefore, a seven-rod solar laser head was designed and numerically studied to enhance the efficiency in TEM$_{00}$-mode laser power production, employing six Fresnel lenses with 10 m$^2$ total collection area for collection and concentration of sunlight. Six folding mirrors redirected the solar rays towards the laser head, composed of six fused silica aspheric lenses and rectangular compound parabolic concentrators paired together for further concentration, and a cylindrical cavity, in which seven Nd:YAG rods were mounted and side-pumped. With conventional rods, total TEM$_{00}$-mode laser power reached 139.89 W, which is equivalent to 13.99 W/m$^2$ collection efficiency and 1.47% solar-to-TEM$_{00}$-mode laser power conversion efficiency. More importantly, by implementing rods with grooved sidewalls, the total laser power was increased to 153.29 W, corresponding to 15.33 W/m$^2$ collection and 1.61% conversion efficiencies. The side-pumping configuration and the good thermal performance may ensure that the seven-grooved-rod system has better scalability than other previously proposed schemes.

**Keywords:** Fresnel lens; grooved rod; Nd:YAG; solar laser; TEM$_{00}$-mode



ion: Costa, H.; Liang, D.;
ida, J.; Catela, M.; Garcia, D.;
cio, B.D.; Vistas, C.R.
-Grooved-Rod, Side-Pumping
ept for Highly Efficient
-Mode Solar Laser Emission
gh Fresnel Lenses. *Photonics*
10, 620. https://doi.org/
0/photonics10060620

ved: 27 March 2023
ed: 22 May 2023
ted: 24 May 2023
shed: 26 May 2023



## 1. Introduction

The global energy crisis triggered an increasing interest in the adoption of renewable power, potentially making it surpass coal as the predominant source of electricity production in the near future, while helping to limit global warming to 1.5 °C [1]. Amid a multitude of renewable energy resources, solar is one of the most exploitable ones, with its field experiencing extensive development in recent years, mainly in photovoltaics [2]. Since the 1960s [3,4], natural sunlight has also been used to produce narrowband, collimated laser radiation in an efficient, simple, and reliable way, obviating the need for artificial pumping sources and their electrical power consumption and conditioning equipment. Solar laser is an emerging technology for both renewable energy and laser-based research and has been investigated as an instrument for materials processing [5] and optical communications [6], while showing great promise for space-based applications, such as wireless power transmission [7], deep-space optical data transmission and networking [8], laser propulsion [9], and asteroid deflection [10].

Even though solar laser emission is possible without any focusing optics [11], collectors and concentrators are normally employed to excite the active medium and obtain high efficiency. However, when using these components, the laser crystal is subjected to high power radiation, running the risk of experiencing significant thermal lens and thermal stress effects [12]. This has sparked researchers to look for ways to ameliorate the conditions for solar laser power production and reduce the probability of rod fracture, while being mindful of the laser efficiency. For example:

1. Nd:YAG proved to be a great rod material for highly intense solar pumping to its availability, reasonably low cost, good thermomechanical properties [13], spectroscopic properties of the dopant [14]. Sensitizers of the $Nd^{3+}$ ion emission, s as $Cr^{3+}$ and $Ce^{3+}$ ions, can also be added as co-dopants in the doped YAG hos increase the laser efficiency through the higher sunlight absorption and subseq transfer of the excitation energy to the $Nd^{3+}$ ions [15,16].
2. Better heat dissipation and lessening of the thermal lens effect can be achieved thro grooved rods, rather than the conventional rods with smooth sidewalls, since the of contact between the laser medium and the cooling water is increased due to shape of its sidewall, improving the laser efficiency and beam quality [17].
3. Multi-rod solar laser systems can facilitate an equal distribution of the total amc of concentrated radiation among the several rods [18–20]. This way, substar improvement of the thermal performance can be achieved, in comparison with single, thicker rod systems. The coordination of multiple laser beams, with each being optimized to perform a part of the overall procedure, has also produced res that were unobtainable with a single beam in various laser-based applications [21-
4. Designing a solar laser system in side-pumping configuration can be constit as a solution to provide alleviation of the thermal loading issue owed to the r uniformity of the pump light distribution along the rod axis that is ubiquitous in end-side-pumping approach, despite the latter being the configuration that has led to record-breaking efficiencies [15,20,26–32]. Furthermore, the free access to b rod ends enables the optimization of more laser resonator parameters, permitting efficient extraction of laser at low-order modes with improved beam quality [33- This is of great significance for many laser-based applications in which the opera in $TEM_{00}$-mode is desired due to the low divergence of the laser beam, allowing a small, focused spot [38]. Damage in the resonator optics could also be prever as a result of its smooth profile, avoiding the appearance of hot sites inside laser medium [39].

The most significant results were achieved through heliostat-parabolic mirror syst In 2017, 9.30 W continuous-wave (cw) $TEM_{00}$-mode laser power was obtained by u a 1.18 $m^2$ effective collection area and a solar laser head composed of a fused silica pheric lens and a conical pump cavity, in which a single Nd:YAG rod was mounted end-side-pumped [30]. This corresponded to 7.88 W/$m^2$ collection and 0.79% sola $TEM_{00}$-mode-laser power conversion efficiencies. For multimode operation, the cur record values were established in 2022 with the production of 16.50 W cw laser power f a 0.40 $m^2$ effective collection area and a similar laser head, this time with three Ce:Nd: rods within the single pump cavity, with collection and conversion efficiencies reacl 41.25 W/$m^2$ and 4.64% [20], respectively. Nevertheless, using heliostat-parabolic mi systems comes with a few drawbacks: namely, the need for complex and expensive in lations to accommodate heavy components, and the blockage of sunlight redirected f the heliostat through the placement of the laser head and its supporting mechanics in f of the primary concentrator.

Albeit having inherent chromatic aberration properties, the collection and conce tion of solar rays can be done through low-cost, lightweight, and easily available Fre lenses. In 2013, 2.30 W cw $TEM_{00}$-mode laser power was obtained with a 1.00 m diam Fresnel lens to side pump a single Nd:YAG rod, fixed inside a laser head comprise a fused silica aspheric lens, a two-dimensional compound parabolic concentrator (C and a V-shaped pump cavity [40]. The collection and conversion efficiencies were 2.93 W/$m^2$ and 0.33%, respectively. In total, 26.93 W cw multimode laser power produced in 2023 by end-side pumping a grooved Ce:Nd:YAG/YAG bonded crystal in a system that takes advantage of a 0.94 m diameter Fresnel lens and a laser head consists of a conical cavity and a quartz tube, within which the rod was fixed [41]. was equivalent to a 38.81 W/$m^2$ collection efficiency and a 3.88% conversion efficienc

Due to all the reasons mentioned above, the lack of experimental works regarding $TEM_{00}$-mode solar laser power production through Fresnel lenses (the most recent one being the aforementioned from 2013 [40], to the best of our knowledge) and the importance of studying the scalability of these systems, a seven-rod, side-pumping concept is presented here. Six Fresnel lenses with 10 $m^2$ total collection area, with the help of six folding mirrors, collected and concentrated sunlight towards a laser head composed of six aspheric lenses, six rectangular CPCs and a cylindrical pump cavity, within which seven grooved Nd:YAG rods were mounted and pumped. The design parameters were optimized using Zemax® (Kirkland, EA, USA) ray tracing in non-sequential mode and LASCAD™ software (Munich, Germany). Efficient production of 153.29 W total $TEM_{00}$-mode laser power could be achieved, corresponding to a 15.33 W/$m^2$ collection efficiency and a 1.61% solar-to-$TEM_{00}$-mode-laser power conversion efficiency.

## 2. Description of the Seven-Rod $TEM_{00}$-Mode Solar Laser Concept

The collection and first-stage concentration of solar rays were done by six Fresnel lenses, each with 0.73 m diameter and approximately 3.5 m focal length, totaling 10 $m^2$ of collection area. They were made of polymethylmethacrylate, which is a lightweight, clear, and stable polymer that has sunlight resistance, thermal stability up to at least 80 °C, and spectral transmissivity that matches the solar spectrum [42]. Six elliptical folding mirrors, with 1.5 m height, 1.0 m width, and 95% reflectivity, were tilted 45° to redirect the concentrated solar rays towards the laser head. Figure 1 depicts how the Fresnel lenses ($FL_1$–$FL_6$) and folding mirrors ($M_1$–$M_6$) were positioned in relation to it. The six pairs were divided into two groups: one was responsible for the concentration of solar rays towards the upper section of the laser head, while the other did the same for the lower section. In the upper-section group, the Fresnel lenses $FL_1$, $FL_3$, and $FL_5$ and folding mirrors $M_1$, $M_3$, and $M_5$ were positioned 1799.25 mm away from the optical axis of the laser head, with $FL_1$, $FL_3$, and $FL_5$ being 1637.50 mm above the middle of the laser head while $M_1$, $M_3$, and $M_5$ were only 6.75 mm above (=1630.75 mm below the corresponding Fresnel lens). In the lower-section group, the Fresnel lenses $FL_2$, $FL_4$, and $FL_6$ and folding mirrors $M_2$, $M_4$, and $M_6$ were positioned 1785.75 mm away from the optical axis of the laser head, with $FL_2$, $FL_4$, and $FL_6$ also being 1637.50 mm above the middle of the laser head, whereas $M_2$, $M_4$, and $M_6$ were 6.75 mm below (=1644.25 mm below the corresponding Fresnel lens).

Both the upper and lower sections of the laser head had three aspheric lenses and three rectangular CPCs that further concentrated the solar rays towards seven grooved Nd:YAG rods mounted inside a cylindrical pump cavity (Figure 2). The six aspheric lenses, each positioned on the same plane as the center of the corresponding folding mirror and 22 mm away from the optical axis of the laser head, had an input face of 110.0 mm diameter and −59 mm radius of curvature, and a plane output face of 25.4 mm diameter and 111 mm thickness. They were designed with fused silica material, which is transparent over the absorption spectrum of Nd:YAG and acts as a filter of undesired radiation, aside from having low thermal expansion coefficient and high resistance to scratching and thermal shock [43]. A rectangular CPC was placed below each aspheric lens, with its $21.4 \times 16.2$ $mm^2$ input face being 1.8 mm away from the lens's output face. The CPCs had an output face with $12.2 \times 7.8$ $mm^2$ area, 13.4 mm length, and inner mirrored walls with 95% reflectivity. The walls connecting the output edges of 12.2 mm with the input edges of 21.4 mm had an acceptance angle of 30°, whereas the walls perpendicular to those had a 26° acceptance angle.

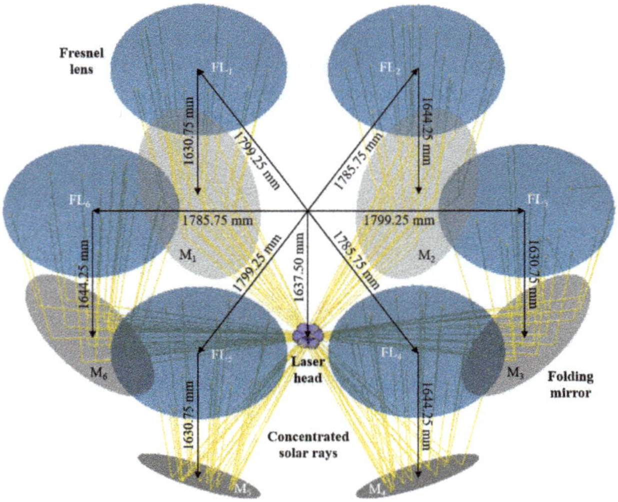


**Figure 1.** Collection and first-stage concentration system composed of six Fresnel lenses ($FL_1$–$FL_6$) with collection area totaling 10 $m^2$, and six folding mirrors ($M_1$–$M_6$) that redirected the concentr solar rays towards the laser head. $FL_1$, $FL_3$, and $FL_5$, and $M_1$, $M_3$, and $M_5$ were responsible fo concentration of solar rays towards the upper section of the laser head; $FL_2$, $FL_4$, and $FL_6$, and $M_4$, and $M_6$ had the same role for the lower section.

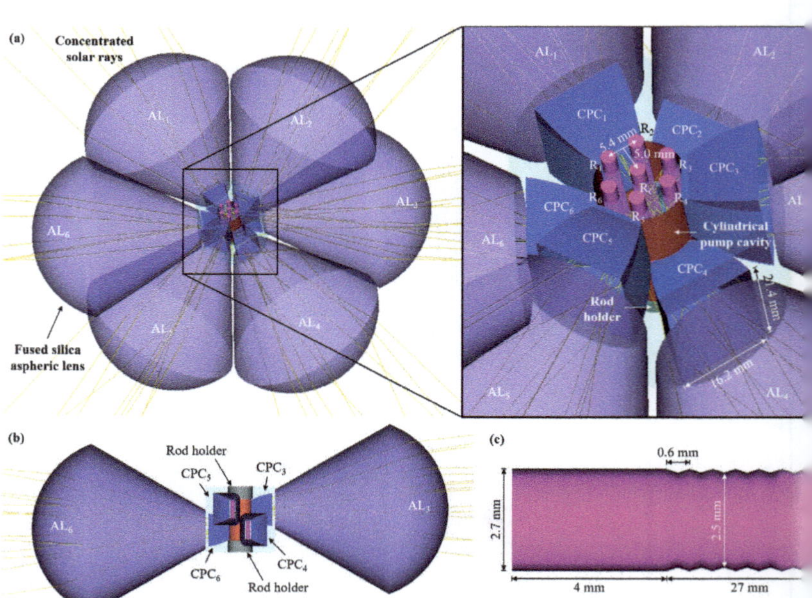


**Figure 2.** (**a**) Three-dimensional and (**b**) side views of the solar laser head. The rod holder fro upper section in (**a**) and fused silica aspheric lenses $AL_1$, $AL_2$, $AL_4$, and $AL_5$ in (**b**) were hi for better visualization of the pump cavity. $AL_1$, $AL_3$, and $AL_5$, and $CPC_1$, $CPC_3$, and $CPC_5$ responsible for the further concentration of solar rays in the upper section of the pump cavity; $AL_4$, and $AL_6$, and $CPC_2$, $CPC_4$, and $CPC_6$ had the same role in the lower section. (**c**) End of o the grooved rods designed in Zemax®.

The cylindrical pump cavity was a mirrored pipe with 15.7 mm diameter, 27.0 mm length, and 95% reflectivity. It had six 13.5 × 7.8 $mm^2$ area openings for the entrance of solar rays, with the 7.8 mm edges being connected with those of the CPCs. Each end of the cavity was closed with a rod holder of the same diameter and reflectivity for the mechanical fixation of 4 mm of the corresponding ends of the seven 2.7 mm diameter, 35 mm length grooved Nd:YAG rods (Figure 2b). The rods, apart from the 4 mm at each end reserved for the holder, had grooves along their length (Figure 2c) akin to those on the active media utilized in previous experimental works performed by our research group [36,44]: with 0.6 mm pitch and 0.1 mm depth. Additionally, the end faces of each rod presented a coating that is antireflective (AR) to the 1064 nm laser emission wavelength. As shown in Figure 2, one rod was positioned at the center of the pump cavity ($R_C$), while the other six were grouped into three pairs ($R_1$ and $R_2$, $R_3$ and $R_4$, and $R_5$ and $R_6$). Each pair was placed 5 mm away from the center of the cavity, while the optical axes of the grooved rods were 5.4 mm away from each other.

The plane output face of the aspheric lenses, the CPCs, the cylindrical pump cavity and the seven grooved rods were in direct contact with cooling water, entering from the CPCs of the upper section ($CPC_1$, $CPC_3$, and $CPC_5$) and exiting from the ones at the lower section of the pump cavity ($CPC_2$, $CPC_4$, and $CPC_6$) to help diminish the heat generation within the rods and the optical components.

Each of the seven laser resonators was comprised of two opposing mirrors aligned with the optical axis of the grooved rod, both with different coatings: highly reflective (HR) and partially reflective (PR) to the 1064 nm laser emission wavelength, through which the solar laser was emitted (Figure 3).

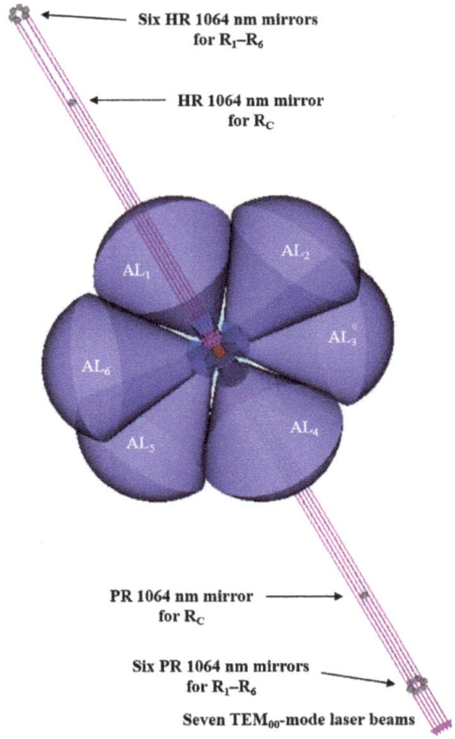


**Figure 3.** Three-dimensional view of the laser head with the laser resonators for rods $R_C$ and $R_1$–$R_6$.

## 3. Numerical Modeling

### 3.1. Modeling of the Design Parameters of the $TEM_{00}$-Mode Solar Laser System through Zem

Preparations to the Zemax® file prior to the design parameter optimization w done. A 950 $W/m^2$ terrestrial solar irradiance and the 16% overlap that the Nd:Y absorption spectrum has with the solar spectrum [45] were considered when calcula the effective pump power for each of the six light sources. After consultation of direct standard solar spectrum for one-and-a-half air mass [46], 22 spectral irradia values at the peak absorption wavelength for the 1.0 at.% Nd:YAG material, as well as absorption coefficient for each of those wavelengths [47], were used as reference data the light sources and in the glass catalog data for Nd:YAG, respectively. Both the absorp spectra and the wavelength-dependent refractive indices of fused silica and water were also added to the glass catalog data to account for the absorption losses that occu those media.

The numerical information from each rod was obtained using a detector volume. absorbed pump power in each rod was calculated by summing up the power in each of 87,500 voxels that the detector volume was comprised of. This number of voxels, along $6 \times 10^6$ analysis rays, enabled the acquisition of accurate results and good image resolu from each detector. The absorbed pump flux distribution in five transversal cross-secti of the seven grooved Nd:YAG rods and in the longitudinal cross-section of rods $R_C$ an is presented in Figure 4. Longitudinal profiles for rods $R_2$–$R_6$ were not included, as t are similar to that of rod $R_1$. The maximum pump flux is represented in red, whereas blue color identifies the areas of the rods where there was little or no absorption.

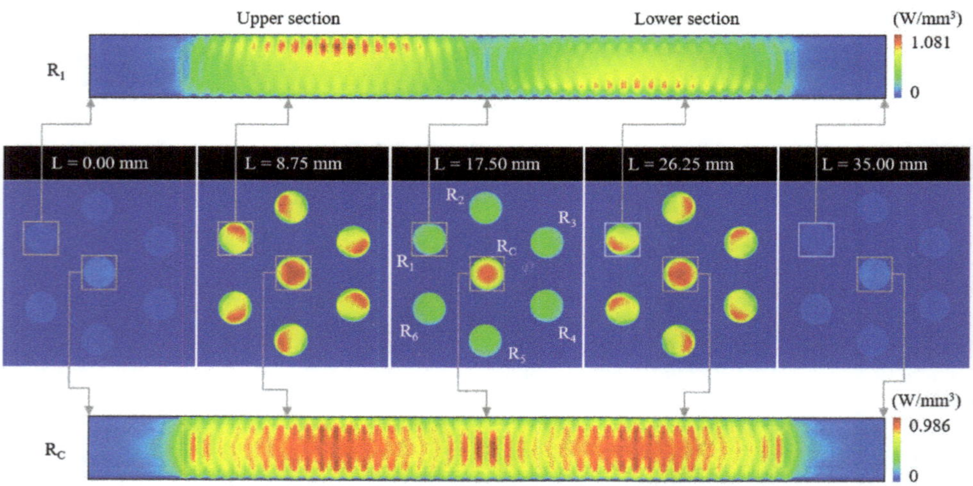


**Figure 4.** Absorbed pump flux distribution in five transversal cross-sections of the seven 2.7 diameter, 35 mm length, grooved Nd:YAG rods, as well as in the longitudinal cross-section o rods $R_C$ and $R_1$. L represents the position where each transversal cross-section profile was obtai

Each time the ray tracing finished performing, the absorbed pump flux data wer ported from Zemax® to be used in the LASCAD™ software afterwards. The quantifica of the thermal effects applied to the active media and the highest solar laser output po provided by the optimal resonator beam parameters in each case determined what mo cations were to be made to the design parameters to possibly lead to further enhancem in $TEM_{00}$-mode laser power and beam quality.

### *3.2. Modeling of the Laser Resonator Parameters through LASCAD™ for TEM$_{00}$-Mode Laser Beam Extraction*

In LASCAD™, the data analysis was done while considering a temperature-dependent stimulated emission cross-section given by:

$$\sigma(T) = 2.8 \times 10^{-19} - 3.9 \times 10^{-22}(T - 300), \quad (1)$$

where $2.8 \times 10^{-19}$ cm$^2$ is the stimulated emission cross-section at 300 K [39], $-3.9 \times 10^{-22}$ cm$^2$/K is the slope of the linear relationship between stimulated emission cross-section and temperature for the 1.0 at.% Nd:YAG medium [48], and $T$ the maximum temperature that the rod achieved. The 230 µs fluorescence lifetime [39], the 0.002 cm$^{-1}$ absorption and scattering loss for the 1.0 at.% Nd:YAG medium, and the 660 nm mean-absorbed and intensity weighted solar pump wavelength [13] were also considered.

Each laser resonator was comprised of the two mirrors with the HR and PR 1064 nm coatings (reflectivity of 99.98% and 90–99%, respectively), whose optical axes were aligned with that of the rod. The resonator length plays a crucial role in the extraction of the TEM$_{00}$-mode laser beam. In short resonators, the TEM$_{00}$-mode poorly matches the active region, and the laser oscillates in multiple modes, leading to high beam quality factors ($M_x^2$, $M_y^2$). By increasing its length, the TEM$_{00}$-mode beam size within the rod also increases, along with the diffraction losses at the rod edges, leading to the elimination of higher order modes. Long resonators were chosen so that only the TEM$_{00}$-mode could oscillate, improving the beam quality [39]. To facilitate the oscillation of the TEM$_{00}$-mode, small diameter laser rods were used due to their behavior as apertures. The separation length and radius of curvature (RoC) of the resonator mirrors were also optimized.

Figures 5 and 6 show the laser resonator design for the extraction of a TEM$_{00}$-mode laser beam from the rods R$_C$ and R$_1$, respectively. It is worth pointing out that the laser resonators for rods R$_2$–R$_6$ are similar to that of R$_1$. To extract the most efficient TEM$_{00}$-mode laser beam from rod R$_C$, HR and PR 1064 nm mirrors with an 8000 mm RoC and a 305.6 mm separation length on each side were used; this enabled the emission of a 27.11 W laser beam with good quality factors of $M_x^2$ = 2.25, $M_y^2$ = 1.00. On the other hand, the laser resonators for rods R$_1$–R$_6$ were longer, with 850 mm RoC mirrors and 423.8 mm separation lengths, resulting in six 21.03 W laser beams with even better quality factors ($M_x^2$ = 1.09, $M_y^2$ = 1.00) than those from rod R$_C$.

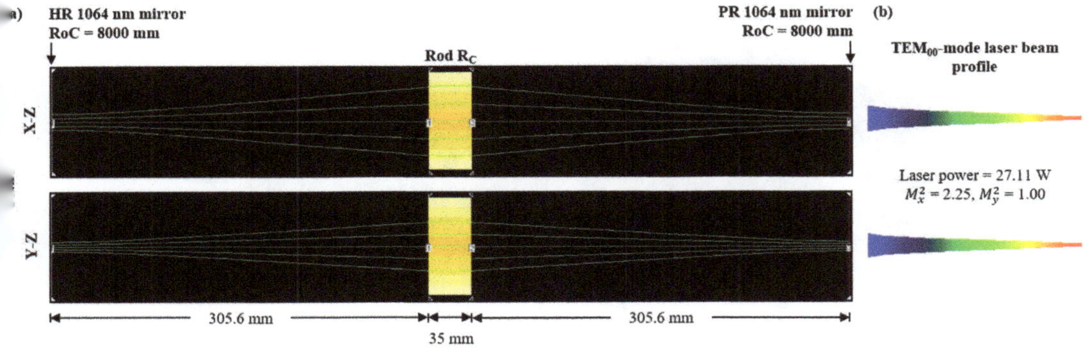


**Figure 5.** *Cont.*

(c)

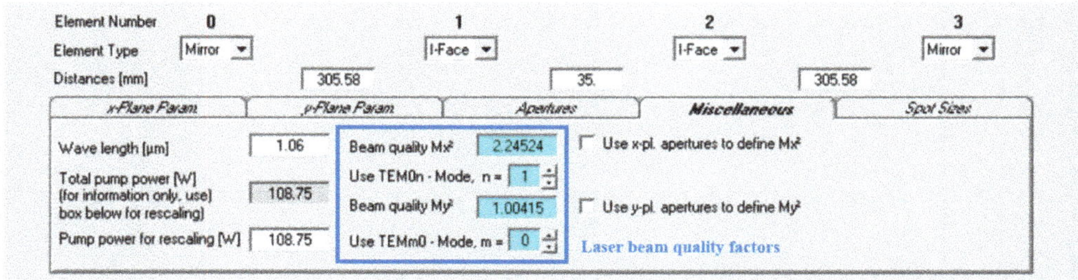


(d)

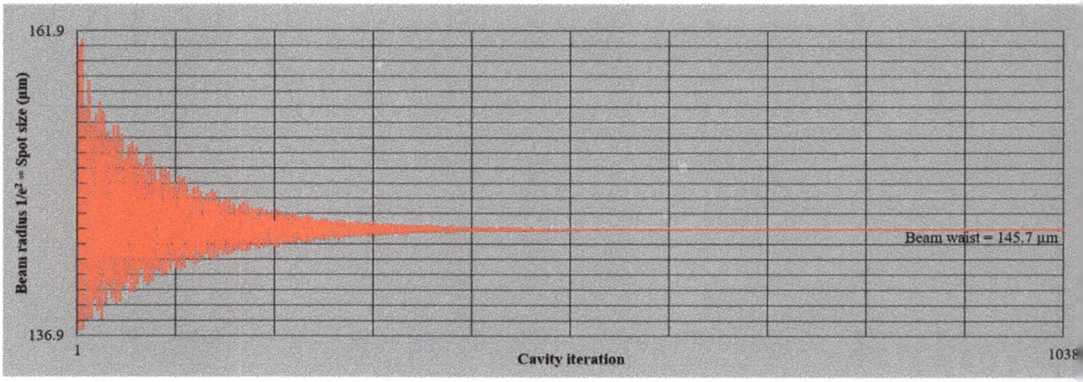


**Figure 5.** (**a**) Laser resonator design in LASCAD™ for the extraction of a $TEM_{00}$-mode laser b from rod $R_C$, with (**b**) the respective beam profile. (**c**) LASCAD™ window in which the laser b quality factors $M_x^2$, $M_y^2$ are presented. (**d**) Laser beam waist radius obtained at the PR 1064 nm ou mirror, through the LASCAD™ beam propagation method.

(a)

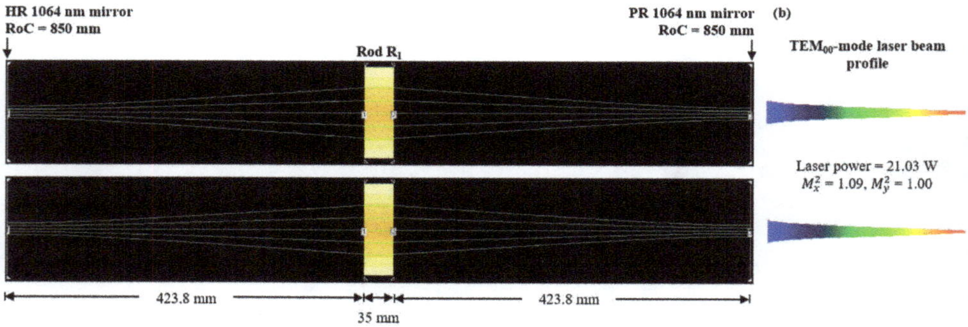


(c)

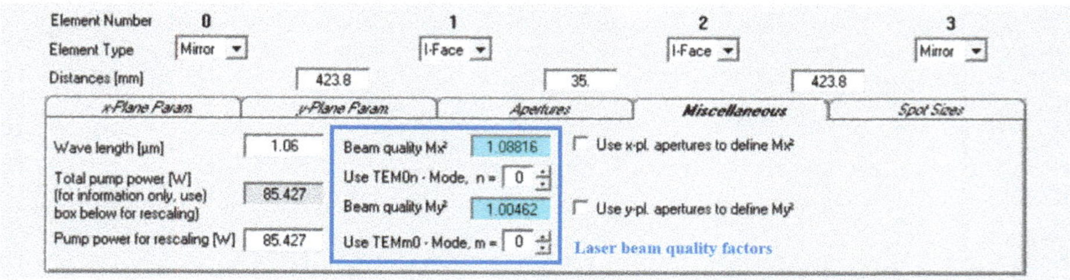


**Figure 6.** *Cont.*

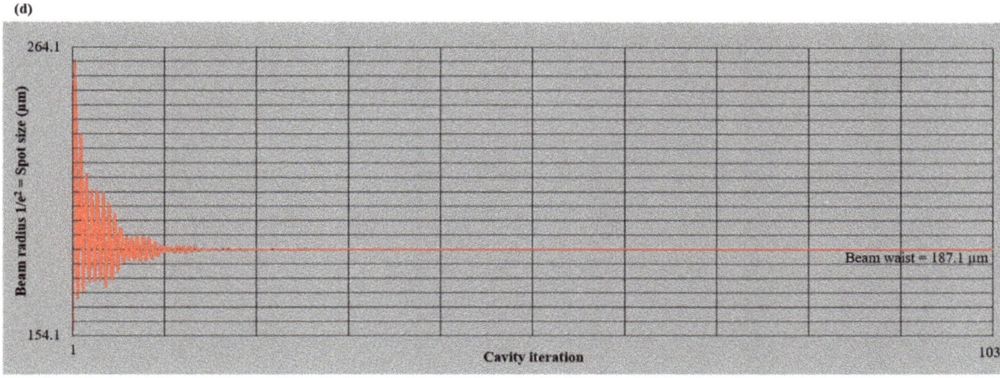


Figure 6. (a) Laser resonator design in LASCAD™ for the extraction of a $TEM_{00}$-mode laser beam from rod $R_1$, with (b) the respective beam profile. (c) LASCAD™ window in which the laser beam quality factors $M_x^2$, $M_y^2$ are presented. (d) Laser beam waist radius obtained at the PR 1064 nm output mirror, through the LASCAD™ beam propagation method.

## 4. Numerical Analysis of the Seven-Rod $TEM_{00}$-Mode Solar Laser System

### 4.1. Laser Power Analysis

The optimization process was firstly conducted for a scheme in which conventional rods were used. After finding the parameters with which the maximum $TEM_{00}$-mode laser power could be achieved, the rods were substituted by ones with grooved sidewalls. Figure 7 presents the total $TEM_{00}$-mode laser power as a function of the rod length using rods with the optimal diameter of 2.7 mm for both the conventional and the grooved rod schemes. In the conventional rod scheme, the maximum total laser power was 139.89 W (=25.53 + 6 × 19.06), attained with 2.7 mm diameter and 35 mm length rods, corresponding to collection and solar-to-$TEM_{00}$-mode-laser power conversion efficiencies of 13.99 W/m$^2$ and 1.47%, respectively. However, the adoption of grooved rods of the same dimensions increased the laser power that each of the seven rods would produce, totaling 153.29 W (=27.11 + 6 × 21.03). Collection efficiency, in this case, was 15.33 W/m$^2$, whereas the conversion efficiency reached 1.61%.

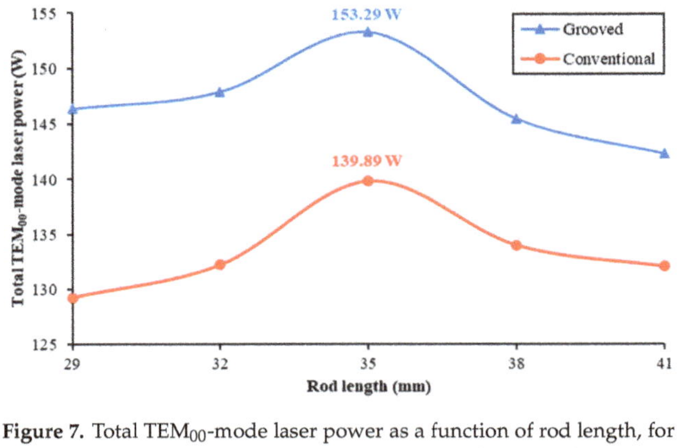


Figure 7. Total $TEM_{00}$-mode laser power as a function of rod length, for the conventional and the grooved Nd:YAG rod schemes while using rods with 2.7 mm diameter.

### *4.2. Thermal Analysis*

The thermal-induced effects—heat load, temperature, and stress intensity—on r $R_C$ and $R_1$ from the conventional and the grooved rod schemes were analyzed thro LASCAD™ and are presented in Figure 8.

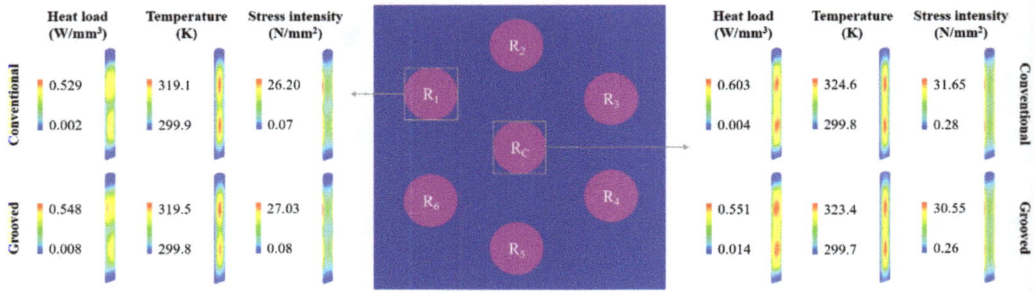


**Figure 8.** Heat load, temperature, and stress intensity for rods $R_C$ and $R_1$ from both the conventi and the grooved Nd:YAG rod schemes.

Due to opting for the side-pumping configuration and how small their diam was, the grooved rods ended up not playing a very relevant role in heat dissipation. rod $R_C$, the grooved sidewall provided a modicum of alleviation when compared to conventional one, registering decreases in the maximum heat load, temperature, and st intensity of only 8.62%, 0.37%, and 3.48%, respectively. For rod $R_1$, the opposite scen transpired, with the difference between values being even lower: heat load was 3.5 temperature 0.13%, and stress intensity 3.17% higher with the grooved rod. Neverthe in both case studies, the stress intensity was always substantially lower than the st fracture limit of 200 N/mm$^2$ for the Nd:YAG material [49], reflecting a potential g performance of the rods under highly intense solar pumping.

## 5. Discussion

In the proposed seven-rod TEM$_{00}$-mode solar laser scheme, opting for rods w grooved sidewalls instead of the conventional ones proved to be an added value to incr its efficiency, despite not being sufficiently pertinent in providing better heat dissipa This constituted improvements in comparison to other numerical works in which Fre lenses were used as the collectors and primary concentrators. Table 1 summarizes results of the present work and those from two other numerical studies of highly effic TEM$_{00}$-mode laser power production schemes by Liang et al. [50,51]:

1. A single-rod scheme, in which four Fresnel lenses with 4 m$^2$ total collection area, w the help of folding mirrors, collected and concentrated the solar rays towards a l head composed of four fused silica aspheric lenses with long rectangular light gu for solar flux homogenization, four hollow rectangular CPCs, and four V-sha pump cavities, where a 3 mm diameter and 76 mm length conventional Nd:YAG was mounted [50].
2. A seven-rod scheme, where a single 4 m$^2$ collection area Fresnel lens collected concentrated the solar rays, being then received by a laser head comprised of a fu silica aspheric lens, a conical pump cavity, and seven 2.5 mm diameter and 15 length conventional Nd:YAG rods [51].

The proposed seven-grooved-rod scheme is 1.04 times more efficient than the sin rod scheme by Liang et al. [50], as well as being able to produce six laser beams of be quality. Moreover, thermal lens and thermal stress effects can be more pronounced in single-rod scheme than in the one proposed here. On the other hand, the seven-rod sch by Liang et al. produced seven laser beams of overall better quality [51]; however, not is the proposed scheme 1.12 times more efficient, but a 4 $m^2$ collection area Fresnel lens is also not easily available on the market, which is a major deterrent for experimental testing, as opposed to six smaller Fresnel lenses with a 10 $m^2$ total collection area.

**Table 1.** Comparison between the results of the present work and those from [50,51].

| Parameter | Liang et al. (2014) [50] | Liang et al. (2021) [51] | Present Work |
|---|---|---|---|
| Type of rod | Conventional | Conventional | Grooved |
| Number of rods | 1 | 7 | 7 |
| Configuration | Side-pumping | End-side-pumping | Side-pumping |
| Number of Fresnel lenses | 4 | 1 | 6 |
| Collection area ($m^2$) | 4 | 4 | 10 |
| Solar irradiance ($W/m^2$) | 950 | 950 | 950 |
| Total $TEM_{00}$-mode laser power (W) | 59.10 | 54.64 (=7.85 + 6 × 7.80) | 153.29 (=27.11 + 6 × 21.03) |
| ɔllection efficiency ($W/m^2$) | 14.78 | 13.66 | 15.33 |
| Ɔonversion efficiency (%) | 1.56 | 1.44 | 1.61 |
| $M_x^2$, $M_y^2$ | 1.08, 1.43 | 1 × (1.36, 1.00)<br>6 × (1.00, 1.04) | 1 × (2.25, 1.00)<br>6 × (1.09, 1.00) |

Due to the disposition of the rods and the side-pumping configuration, a laser beam merging could be performed in the proposed scheme to combine the six individual $TEM_{00}$-mode laser beams from the external rods, with the help of folding mirrors [52], enabling the extraction of a single laser beam with higher power and better quality. However, this laser beam merging technique cannot be employed in the seven-rod scheme by Liang et al. [51], since the rod tilting and the end-side-pumping configuration do not make it possible for the folding mirrors to be mechanically fixed.

## 6. Conclusions

A seven-rod solar laser concept was proposed here to improve the efficiency in $TEM_{00}$-mode laser power production, employing six Fresnel lenses with 10 $m^2$ total collection area for collection and concentration of sunlight. Six folding mirrors redirected the solar rays towards the laser head comprised of six aspheric lenses, six rectangular CPCs, and a cylindrical cavity, in which seven Nd:YAG rods were mounted and side-pumped. With 2.7 mm diameter and 35 mm length conventional rods, 139.89 W total $TEM_{00}$-mode laser power was numerically achieved, which is equivalent to a 13.99 $W/m^2$ collection efficiency and a 1.47% solar-to-$TEM_{00}$-mode-laser power conversion efficiency. By adopting grooved rods of the same dimensions, the total laser power was increased to 153.29 W, corresponding to 15.33 $W/m^2$ collection and 1.61% conversion efficiencies. Regarding $TEM_{00}$-mode laser power production through the usage of Fresnel lenses, the proposed seven-grooved-rod scheme is 1.04 times more efficient than the one that held the previous numerical record, in which a single conventional rod was side-pumped [50]. It is also 1.12 times more efficient than another seven-conventional-rod concept, designed in the end-side-pumping configuration [51]. Furthermore, the side-pumping configuration paired with the good thermal performance may ensure the seven-grooved-rod system has better scalability than these previously proposed schemes, as well as opening the possibility for the application of a laser beam merging technique.

The implementation of Cr:Nd:YAG [15] or Ce:Nd:YAG [16] rods could lead to further improvements in efficiency, depending on how well the absorption spectra of these materials match with the solar spectrum. The use of easily available Fresnel lenses and the possibility of producing multiple $TEM_{00}$-mode laser beams of high efficiency reinforce the importance that solar lasers could have in many laser-based applications.

**Author Contributions:** Conceptualization, D.L., H.C. and J.A.; methodology, H.C., D.L., J.A. C.R.V.; software, H.C., D.L., J.A. and D.G.; validation, H.C., D.L. and J.A.; formal analysis, I D.L., J.A., M.C. and B.D.T.; investigation, H.C., D.L. and J.A.; resources, D.L. and J.A.; data cura H.C., D.L. and J.A.; writing—original draft preparation, H.C.; writing—review and editing, I D.L., J.A., M.C., D.G., B.D.T. and C.R.V.; visualization, H.C.; supervision, D.L. and J.A.; pr administration, D.L.; funding acquisition, D.L. All authors have read and agreed to the publis version of the manuscript.

**Funding:** This research was funded by Fundação para a Ciência e a Tecnologia—Ministéri Ciência, Tecnologia e Ensino Superior (FCT—MCTES), grant UIDB/00068/2020.

**Institutional Review Board Statement:** Not applicable.

**Informed Consent Statement:** Not applicable.

**Data Availability Statement:** Not applicable.

**Acknowledgments:** The FCT-MCTES fellowship grants 2021.06172.BD, CEECIND/03081/ SFRH/BD/145322/2019, PD/BD/142827/2018, and SFRH/BPD/125116/2016 of Hugo Costa, J Almeida, Miguel Catela, Dário Garcia, and Cláudia R. Vistas, respectively, are acknowledged.

**Conflicts of Interest:** The authors declare no conflict of interest. The funders had no role in the de of the study; in the collection, analyses, or interpretation of data; in the writing of the manuscrip in the decision to publish the results.

## References

1. IEA Renewable Power's Growth Is Being Turbocharged as Countries Seek to Strengthen Energy Security. Available online: h //www.iea.org/news/renewable-power-s-growth-is-being-turbocharged-as-countries-seek-to-strengthen-energy-secu (accessed on 2 February 2023).
2. McKinsey Renewable-Energy Development in a Net-Zero World. Available online: https://www.mckinsey.com/indust electric-power-and-natural-gas/our-insights/renewable-energy-development-in-a-net-zero-world (accessed on 2 February 2
3. Kiss, Z.J.; Lewis, H.R.; Duncan, R.C., Jr. Sun Pumped Continuous Optical Maser. *Appl. Phys. Lett.* **1963**, *2*, 93–94. [CrossRef
4. Young, C.G. A Sun-Pumped Cw One-Watt Laser. *Appl. Opt.* **1966**, *5*, 993. [CrossRef]
5. Yabe, T.; Bagheri, B.; Ohkubo, T.; Uchida, S.; Yoshida, K.; Funatsu, T.; Oishi, T.; Daito, K.; Ishioka, M.; Yasunaga, N.; 100 W-Class Solar Pumped Laser for Sustainable Magnesium-Hydrogen Energy Cycle. *J. Appl. Phys.* **2008**, *104*, 083104. [Cros
6. Guan, Z.; Zhao, C.M.; Yang, S.H.; Wang, Y.; Ke, J.Y.; Zhang, H.Y. Demonstration of a Free-Space Optical Communication Sy Using a Solar-Pumped Laser as Signal Transmitter. *Laser Phys. Lett.* **2017**, *14*, 055804. [CrossRef]
7. Lando, M.; Kagan, J.A.; Shimony, Y.; Kalisky, Y.Y.; Noter, Y.; Yogev, A.; Rotman, S.R.; Rosenwaks, S. Solar-Pumped Solid Laser Program. In Proceedings of the 10th Meeting on Optical Engineering in Israel, Jerusalem, Israel, 22 September Shladov, I., Rotman, S.R., Eds.; SPIE: Bellingham, WA, USA; Volume 3110, p. 196.
8. Hemmati, H.; Biswas, A.; Djordjevic, I.B. Deep-Space Optical Communications: Future Perspectives and Application *Proceedings of the IEEE*; IEEE: Piscataway, NJ, USA, 2011; Volume 99, pp. 2020–2039.
9. Rather, J.D.G.; Gerry, E.T.; Zeiders, G.W. Investigation of Possibilities for Solar Powered High Energy Lasers in Space. *NASA Rep. Serv.* **1977**. Available online: https://ntrs.nasa.gov/citations/19820010707 (accessed on 2 February 2023).
10. Vasile, M.; Maddock, C.A. Design of a Formation of Solar Pumped Lasers for Asteroid Deflection. *Adv. Space Res.* **201** 891–905. [CrossRef]
11. Masuda, T.; Iyoda, M.; Yasumatsu, Y.; Dottermusch, S.; Howard, I.A.; Richards, B.S.; Bisson, J.-F.; Endo, M. A Fully Planar Pumped Laser Based on a Luminescent Solar Collector. *Commun. Phys.* **2020**, *3*, 60. [CrossRef]
12. Hwang, I.H.; Lee, J.H. Efficiency and Threshold Pump Intensity of CW Solar-Pumped Solid-State Lasers. *IEEE J. Quantum Ele* **1991**, *27*, 2129–2134. [CrossRef]
13. Weksler, M.; Shwartz, J. Solar-Pumped Solid-State Lasers. *IEEE J. Quantum Electron.* **1988**, *24*, 1222–1228. [CrossRef]
14. Lupei, V.; Lupei, A.; Gheorghe, C.; Ikesue, A. Emission Sensitization Processes Involving $Nd^{3+}$ in YAG. *J. Lumin.* **2016** 594–601. [CrossRef]
15. Liang, D.; Vistas, C.R.; Tibúrcio, B.D.; Almeida, J. Solar-Pumped Cr:Nd:YAG Ceramic Laser with 6.7% Slope Efficiency. *Sol. E Mater. Sol. Cells* **2018**, *185*, 75–79. [CrossRef]
16. Vistas, C.R.; Liang, D.; Almeida, J.; Tibúrcio, B.D.; Garcia, D.; Catela, M.; Costa, H.; Guillot, E. Ce:Nd:YAG Side-Pumped Laser. *J. Photonics Energy* **2021**, *11*, 018001. [CrossRef]
17. Guan, Z.; Zhao, C.; Yang, S.; Wang, Y.; Ke, J.; Gao, F.; Zhang, H. Low Threshold and High Efficiency Solar-Pumped Laser Fresnel Lens and a Grooved Nd:YAG Rod. In Proceedings of the High-Power Lasers and Applications VIII, Beijing, C 9 November 2016; Li, R., Singh, U.N., Walter, R.F., Eds.; SPIE: Bellingham, WA, USA; Volume 10016, p. 1001609.
18. Liang, D.; Almeida, J.; Garcia, D.; Tibúrcio, B.D.; Guillot, E.; Vistas, C.R. Simultaneous Solar Laser Emissions from Three Nd: Rods within a Single Pump Cavity. *Sol. Energy* **2020**, *199*, 192–197. [CrossRef]

Tibúrcio, B.D.; Liang, D.; Almeida, J.; Garcia, D.; Catela, M.; Costa, H.; Vistas, C.R. Tracking Error Compensation Capacity Measurement of a Dual-Rod Side-Pumping Solar Laser. *Renew. Energy* **2022**, *195*, 1253–1261. [CrossRef]

Liang, D.; Vistas, C.R.; Garcia, D.; Tibúrcio, B.D.; Catela, M.; Costa, H.; Guillot, E.; Almeida, J. Most Efficient Simultaneous Solar Laser Emissions from Three Ce:Nd:YAG Rods within a Single Pump Cavity. *Sol. Energy Mater. Sol. Cells* **2022**, *246*, 111921. [CrossRef]

Olsen, F.O.; Hansen, K.S.; Nielsen, J.S. Multibeam Fiber Laser Cutting. *J. Laser Appl.* **2009**, *21*, 133–138. [CrossRef]

Eifel, S.; Holtkamp, J. Multi-Beam Technology Boosts Cost Efficiency. Available online: https://www.laserfocusworld.com/industrial-laser-solutions/article/14216434/multibeam-technology-boosts-cost-efficiency (accessed on 10 February 2023).

Strite, T.; Gusenko, A.; Grupp, M.; Hoult, T. Multiple Laser Beam Material Processing. *Biul. Inst. Spaw.* **2016**, *60*, 30–33. [CrossRef]

Gillner, A.; Finger, J.; Gretzki, P.; Niessen, M.; Bartels, T.; Reininghaus, M. High Power Laser Processing with Ultrafast and Multi-Parallel Beams. *J. Laser Micro Nanoeng.* **2019**, *14*, 129–137. [CrossRef]

Thibault, F. Multibeam Laser Marking: The Individual Traceability Enabler. Available online: https://www.laserfocusworld.com/laser-processing/article/14270572/laser-marking-the-individual-traceability-enabler (accessed on 10 February 2023).

Yabe, T.; Ohkubo, T.; Uchida, S.; Yoshida, K.; Nakatsuka, M.; Funatsu, T.; Mabuti, A.; Oyama, A.; Nakagawa, K.; Oishi, T.; et al. High-Efficiency and Economical Solar-Energy-Pumped Laser with Fresnel Lens and Chromium Codoped Laser Medium. *Appl. Phys. Lett.* **2007**, *90*, 261120. [CrossRef]

Liang, D.; Almeida, J. Highly Efficient Solar-Pumped Nd:YAG Laser. *Opt. Express* **2011**, *19*, 26399. [CrossRef]

Dinh, T.H.; Ohkubo, T.; Yabe, T.; Kuboyama, H. 120 Watt Continuous Wave Solar-Pumped Laser with a Liquid Light-Guide Lens and an Nd:YAG Rod. *Opt. Lett.* **2012**, *37*, 2670. [CrossRef]

Xu, P.; Yang, S.; Zhao, C.; Guan, Z.; Wang, H.; Zhang, Y.; Zhang, H.; He, T. High-Efficiency Solar-Pumped Laser with a Grooved Nd:YAG Rod. *Appl. Opt.* **2014**, *53*, 3941. [CrossRef] [PubMed]

Liang, D.; Almeida, J.; Vistas, C.R.; Guillot, E. Solar-Pumped Nd:YAG Laser with 31.5 W/m$^2$ Multimode and 7.9 W/m$^2$ TEM$_{00}$-Mode Collection Efficiencies. *Sol. Energy Mater. Sol. Cells* **2017**, *159*, 435–439. [CrossRef]

Guan, Z.; Zhao, C.; Li, J.; He, D.; Zhang, H. 32.1 W/m$^2$ Continuous Wave Solar-Pumped Laser with a Bonding Nd:YAG/YAG Rod and a Fresnel Lens. *Opt. Laser Technol.* **2018**, *107*, 158–161. [CrossRef]

Garcia, D.; Liang, D.; Vistas, C.R.; Costa, H.; Catela, M.; Tibúrcio, B.D.; Almeida, J. Ce:Nd:YAG Solar Laser with 4.5% Solar-to-Laser Conversion Efficiency. *Energies* **2022**, *15*, 5292. [CrossRef]

Lando, M.; Kagan, J.; Linyekin, B.; Dobrusin, V. A Solar-Pumped Nd:YAG Laser in the High Collection Efficiency Regime. *Opt. Commun.* **2003**, *222*, 371–381. [CrossRef]

Liang, D.; Almeida, J.; Vistas, C.R.; Guillot, E. Solar-Pumped TEM$_{00}$ Mode Nd:YAG Laser by a Heliostat-Parabolic Mirror System. *Sol. Energy Mater. Sol. Cells* **2015**, *134*, 305–308. [CrossRef]

Almeida, J.; Liang, D.; Vistas, C.R.; Bouadjemine, R.; Guillot, E. 5.5 W Continuous-Wave TEM$_{00}$-Mode Nd:YAG Solar Laser by a Light-Guide/2V-Shaped Pump Cavity. *Appl. Phys. B* **2015**, *121*, 473–482. [CrossRef]

Vistas, C.R.; Liang, D.; Almeida, J.; Guillot, E. TEM$_{00}$ Mode Nd:YAG Solar Laser by Side-Pumping a Grooved Rod. *Opt. Commun.* **2016**, *366*, 50–56. [CrossRef]

Liang, D.; Almeida, J.; Vistas, C.R.; Oliveira, M.; Gonçalves, F.; Guillot, E. High-Efficiency Solar-Pumped TEM$_{00}$-Mode Nd:YAG Laser. *Sol. Energy Mater. Sol. Cells* **2016**, *145*, 397–402. [CrossRef]

Duocastella, M.; Arnold, C.B. Bessel and Annular Beams for Materials Processing. *Laser Photonics Rev.* **2012**, *6*, 607–621. [CrossRef]

Koechner, W. *Solid-State Laser Engineering*, 5th ed.; Springer Series in Optical Sciences; Springer: Berlin/Heidelberg, Germany, 1999; Volume 1, ISBN 978-3-662-14221-9.

Liang, D.; Almeida, J. Solar-Pumped TEM$_{00}$ Mode Nd:YAG Laser. *Opt. Express* **2013**, *21*, 25107. [CrossRef] [PubMed]

Cai, Z.; Zhao, C.; Zhao, Z.; Zhang, J.; Zhang, Z.; Zhang, H. Efficient 38.8 W/m$^2$ Solar Pumped Laser with a Ce:Nd:YAG Crystal and a Fresnel Lens. *Opt. Express* **2023**, *31*, 1340. [CrossRef] [PubMed]

Xie, W.T.; Dai, Y.J.; Wang, R.Z.; Sumathy, K. Concentrated Solar Energy Applications Using Fresnel Lenses: A Review. *Renew. Sustain. Energy Rev.* **2011**, *15*, 2588–2606. [CrossRef]

Bernardes, P.H.; Liang, D. Solid-State Laser Pumping by Light Guides. *Appl. Opt.* **2006**, *45*, 3811. [CrossRef]

Vistas, C.R.; Liang, D.; Almeida, J. Solar-Pumped TEM$_{00}$ Mode Laser Simple Design with a Grooved Nd:YAG Rod. *Sol. Energy* **2015**, *122*, 1325–1333. [CrossRef]

Zhao, B.; Changming, Z.; He, J.; Yang, S. The Study of Active Medium for Solar-Pumped Solid-State Lasers. *Acta Opt. Sin.* **2007**, *27*, 1797–1801.

*ASTM G173-03(2012)*; Standard Tables for Reference Solar Spectral Irradiances: Direct Normal and Hemispherical on 37° Tilted Surface. ASTM International: West Conshohocken, PA, USA, 2020.

Prahl, S. Nd:YAG—Nd:Y3Al5O12. Available online: https://omlc.org/spectra/lasermedia/html/052.html (accessed on 14 March 2023).

Dong, J.; Rapaport, A.; Bass, M.; Szipocs, F.; Ueda, K. Temperature-Dependent Stimulated Emission Cross Section and Concentration Quenching in Highly Doped Nd3+:YAG Crystals. *Phys. Status Solidi* **2005**, *202*, 2565–2573. [CrossRef]

Almeida, J.; Liang, D.; Vistas, C.R.; Guillot, E. Highly Efficient End-Side-Pumped Nd:YAG Solar Laser by a Heliostat–Parabolic Mirror System. *Appl. Opt.* **2015**, *54*, 1970. [CrossRef]

50. Liang, D.; Almeida, J.; Vistas, C.R. Scalable Pumping Approach for Extracting the Maximum $TEM_{00}$ Solar Laser Power. *Appl.* **2014**, *53*, 7129. [CrossRef] [PubMed]
51. Liang, D.; Almeida, J.; Tibúrcio, B.D.; Catela, M.; Garcia, D.; Costa, H.; Vistas, C.R. Seven-Rod Pumping Approach for the M Efficient Production of $TEM_{00}$ Mode Solar Laser Power by a Fresnel Lens. *J. Sol. Energy Eng.* **2021**, *143*, 061004. [CrossRef]
52. Tibúrcio, B.D.; Liang, D.; Almeida, J.; Garcia, D.; Catela, M.; Costa, H.; Vistas, C.R. Enhancing $TEM_{00}$-Mode Solar Laser v Beam Merging and Ring-Array Concentrator. *J. Sol. Energy Eng.* **2022**, *144*, 061005. [CrossRef]

photonics

*le*

# gh Efficiency and High Stability for SHG in an $Nd{:}YVO_4$ ser with a KTP Intracavity and Q-Switching through ırmonic Modulation

uel Mardoqueo Afanador Delgado [1], Juan Hugo García López [1], Rider Jaimes Reátegui [1], nte Aboites [2], José Luis Echenausía Monroy [1,3] and Guillermo Huerta Cuellar [1,*]

[1] Dynamical Systems Laboratory, CULagos, Universidad de Guadalajara, Centro Universitario de los Lagos, Enrique Díaz de León 1144, Paseos de la Montaña, Lagos de Moreno 47460, Jalisco, Mexico; samuel.adelgado@academicos.udg.mx (S.M.A.D.); jhugo.garcia@academicos.udg.mx (J.H.G.L.); rider.jaimes@academicos.udg.mx (R.J.R.); echenausia@cicese.mx (J.L.E.M.)

[2] Lasers Laboratory, Optical Research Center, Loma del Bosque 115, Col. Lomas del Campestre, León 37150, Guanajuato, Mexico; aboites@cio.mx

[3] Applied Physics Division, Center for Scientific Research and Higher Education at Ensenada (CICESE), Carr. Ensenada-Tijuana 3918, Zona Playitas, Ensenada 22860, Baja California, Mexico

* Correspondence: guillermo.huerta@academicos.udg.mx

**Abstract:** In this paper, the stabilization and high efficiency of an unstable Second Harmonic Generation (SHG) of an $Nd{:}YVO_4$ laser with a KTP intracavity is demonstrated. By using a passive Q-switching crystal ($Cr^{4+}{:}YAG$) and a parametric modulation method (harmonic modulation), the stabilization of the laser is reached. An harmonic modulation was applied to the pumping of the $Nd{:}YVO_4$-KTP laser to control the amplitude and frequency of the laser emission. The results were characterized by using power spectra analysis, optical spectrum, bifurcation diagrams, and temporal series of the laser intensity. The promising application of this green light source is materialized when such light is necessary for high-density optics, such as in the treatment materials industry or in some aesthetic applications.

**Keywords:** $Nd{:}YVO_4$ laser; second harmonic generation; Q-switched lasers; frequency doubled

check for updates

ɪn: Afanador Delgado, S.M.; López, J.H.; Reátegui, R.J.; s, V.; Echenausía Monroy, J.L.; ı Cuellar, G. High Efficiency gh Stability for SHG in an $O_4$ Laser with a KTP ıvity and Q-Switching through onic Modulation. *Photonics* 0, 454. https://doi.org/ /photonics10040454

ed: 11 March 2023
d: 8 April 2023
ted: 11 April 2023
hed: 14 April 2023



## 1. Introduction

Currently, one of the most widely used solid-state lasers in industrial applications, such as materials processing, cutting, and welding, is the $Nd{:}YVO_4$ laser (yttrium orthovanadate crystal doped with neodymium) [1]. The $Nd{:}YVO_4$ is used due to its high absorption coefficient and its wide section of stimulated emission; in addition, it has a better performance as compared to an $Nd{:}YAG$ ($Y_2Al_5O_{12}$ aluminum oxide and itrium oxide doped with neodymium $Nd^{3+}$) laser. However, thermal effects (thermal lens) are more significant in the $Nd{:}YVO_4$; so, it is necessary to monitor the glass temperature to keep it low, as in [2] by using both acousto-optic modulator and GaAs saturable absorber. Mukhopadhyay et al. presented an analysis by incorporating a nonlinear loss term due to the intracavity second-harmonic generation to the general recurrence relation for the mode-locked pulses under the Q-switched envelope at the fundamental wavelength [3]. The use of a suturable absorber has also been demonstrated in a compact diode-pumped with Q-switched extracavity for frequency-doubled Nd:YVO4/KTP green-pulse laser generation [4,5]. Other researchers demonstrated that the diffusion bonding crystal method is an effective method to relieve the thermal lens for the end-pumping laser crystal, and opens the possibility to increase the optical power of the system [6,7].

Lasers doped with $Nd^{3+}$, such as Nd:YAG and $Nd{:}YVO_4$, usually emit radiation at a wavelength of 1064 nm, supported by a $KTiOPO_4$ crystal (potassium titanyl phosphate crystal or KTP). Inside the laser cavity, it is possible to double the frequency by conve the infrared light of 1064 nm into green light of 532 nm; this phenomenon is called Se Harmonic Generation (SHG) [8]. Due to the nonlinear coupling between modes in the the crystal causes irregular fluctuations in the laser output, which is amplified by a Q within the laser cavity due to the presence of the amplifying laser [9], also known a green problem. This irregular behavior has been studied in detail and attributed t destabilization of the relaxation of the oscillations, which has always been present i type of lasers due to the nonlinear coupling of the longitudinal modes [10,11].

There are several approaches in order to stabilize the green light generated by in a solid-state laser. One of them is the use of external modulation, where low emi power is achieved by using nonlinear waves for SHG. The stabilization of $Nd{:}YVO_4$ pulses with passive Q-switching has been studied [12,13]. This solution provides conversion efficiency, and high power can be achieved by using flat waves [14]. $Nd{:}YVO_4$ laser, placement of the KTP crystal is fundamental to obtain green emi by folding the frequency. When the crystal is placed inside the cavity green radiati emitted at high power, since the energy density of the radiation is higher inside the o cavity. The green light intensity produced by the KTP crystal is proportional to the sc of the fundamental wavelength intensity, but tends to be unstable. This instability is ca by the competition of modes in the cavity; if the KTP is placed outside the cavity, it is possible to generate the second harmonic, and the emission, in this case, is stable, but low power [15].

An approach to suppressing unwanted fluctuations is to control the injection curre the pump diode by means of electronic feedback. In 1992 [10], Rajarshi Roy et al. succe in stabilizing the output waveforms of the laser using a modification of the chaos co method proposed by Ott–Grebogy–Yorke (OGY) [16]; this control technique is know Occasional Proportional Feedback (OPF), and in other systems, such as $CO_2$ lasers method has been used to stabilize unstable periodic orbits [17]. A parametric modul technique has been implemented by adding a sinusoidal signal to the system param in lasers made of erbium-doped fibers [18,19] to induce multistability and to stab the laser. A fast saturable absorber in a single-mode infrared laser induces a sequ of periodic orbits (P3, P4, P5, etc), which are stable. In the context of infrared lasers has been observed in simulations and studied via numerical continuation in bifurc analysis in the $CO_2$ laser with a saturable absorber [20,21]. The harmonic modul technique consists in adding a periodic signal to the input of the system; in this ca is a sinusoidal periodic signal to an accessible parameter of the laser system to obt modulated parameter.

There is a method to increase the output power of a laser by modulating losses. method is known as Q-switching, and it is divided into two techniques: active and pa Q-switching. In active Q-switching, losses are modulated by active elements, suc acousto-optic or electro-optic modulators; in passive Q-switching, losses are automati modulated by a saturable absorber, and the pulse is generated when the energy stor the medium reaches the appropriate level [22]. Normally, the average power of lase Q-switch operation is lower than the optical power of a continuous wave laser an instabilities are reduced by placing a Q-switch crystal inside the cavity [23–26]. Peri modulation of the pump leads to different stable states in the experiment, for inst another infrared laser ($CO_2$ laser with modulated losses) where changing the periodic l modulation depth "m" or frequency modulation leads to different chaotic and peri attractors. This dependence on the losses modulation depth "m" and frequency has studied for two different models in the $CO_2$ laser: the two-level model and the four- model [27,28].

In this paper, an experimental study of an $Nd{:}YVO_4$ laser with a KTP crystal intrac emitting at 1064 nm is shown. In such case, the KTP crystal produces a laser emissi 532 nm, which tends to be unstable and gives rise to the so-called "green problem" tre in recent years [10,11]. By applying controlled external modulation and using passi switching, it is possible to stabilize the emission and obtain a high conversion efficiency. The emission of the laser is analyzed based on the response in terms of optical power, optical spectra, frequency spectra, and time series by using the local maxima and Inter-Spike-Interval (ISI). This work focuses on stabilizing the green emission in frequency and pulse amplitude by adding harmonic modulation to the pumping current, where this modulation is applied by changing the amplitude and frequency. In Section 2, the experimental setup of an $Nd{:}YVO_4$ laser with a KTP crystal intracavity for second harmonic generation with external modulation is presented. Then, in Section 3, some results on stabilizing the frequency and amplitude of the SHG of the $Nd{:}YVO_4$ laser are shown. Finally, in Section 4, the results obtained and the conclusions are presented.

## 2. Preliminaries

There are a few methods to generate green light, but the best known is probably the second harmonic generation. *The green problem* arises when the cavity generates not only the second harmonic but also other frequencies due to the competition of longitudinal modes, which produce chaotic behavior [9]. The generation of those frequencies, as well as the generation of the second harmonic, are related to the polarization of the medium. This polarization occurs when an electric field is applied to the material as follows [29]:

$$\vec{P} = \epsilon_0 \chi \vec{E}, \tag{1}$$

where $\epsilon_0$ is the permittivity in the free space and $\chi$ is the linear susceptibility. If the electric field is small, the approximation is linear, but if the electric field is large, the best approximation is nonlinear; then, we can expand the equation of polarization in power series as:

$$\begin{aligned} \vec{P} &= \epsilon_0 \chi \vec{E} + \epsilon_0 \chi^{(2)} \vec{E}^2 + \epsilon_0 \chi^{(3)} \vec{E}^3 + \cdots + \epsilon_0 \chi^{(n)} \vec{E}^n, \\ \vec{P} &= \vec{P}^{(1)} + \vec{P}^{(2)} + \vec{P}^{(3)} + \cdots + \vec{P}^{(n)}, \end{aligned} \tag{2}$$

note that $\chi^{(2)}, \chi^{(3)}, \cdots, \chi^{(n)}$ are the nonlinear susceptibilities of second, third, and $n$th-order, respectively, and $P^{(n)}$ is the polarization of order $n$. Next, suppose an electromagnetic wave

$$\vec{E} = E_1 e^{-\omega_1 t} + E_2 e^{-\omega_2 t} + c.c.\,, \tag{3}$$

where $\omega_1$ and $\omega_2$ are frequencies, $E_1$ and $E_2$ are the amplitudes of the electromagnetic waves, while $c.c.$ represents a complex conjugate term.

Substituting Equation (3) into the polarization of second order of Equation (2), the following equation is obtained:

$$\begin{aligned} \vec{P}^{(2)} &= \epsilon_0 \chi^{(2)} (E_1 e^{-i\omega_1 t} + E_2 e^{-i\omega_2 t} + c.c.)^2, \\ \vec{P}^{(2)} &= \epsilon_0 \chi^{(2)} (E_1 e^{-i\omega_1 t} + E_2 e^{-i\omega_2 t} + E_1^* e^{i\omega_1 t} + E_2^* e^{i\omega_2 t})^2, \\ \vec{P}^{(2)} &= 2\epsilon_0 \chi^{(2)} (E_1 E_1^* + E_2 E_2^*) + \epsilon_0 \chi^{(2)} (E_1 E_1^* e^{-i2\omega_1 t} + E_2 E_2^* e^{-i2\omega_2 t}) + \\ & \quad 2\epsilon_0 \chi^{(2)} E_1^* E_2 e^{-i(\omega_1+\omega_2)t} + 2\epsilon_0 \chi^{(2)} E_1 E_2^* e^{-i(\omega_1-\omega_2)t}. \end{aligned} \tag{4}$$

By analyzing Equation (4), terms with different frequencies corresponding to the four non-linear effects of second order (Equations (5)–(8)) can be identified: Optical Rectification (OR), Second Harmonic Generation (SHG), Generation of Frequency Sum (GFS), and Generation of Difference Frequency (GDF):

$$\vec{P}^{(2)}_{OR} = 2\epsilon_0 \chi^{(2)} (E_1 E_1^* + E_2 E_2^*), \tag{5}$$

$$\vec{P}^{(2)}_{SHG} = \epsilon_0 \chi^{(2)} (E_1 E_1^* e^{-i2\omega_1 t} + E_2 E_2^* e^{-i2\omega_2 t}), \tag{6}$$

$$\vec{P}^{(2)}_{GFS} = 2\epsilon_0 \chi^{(2)} E_1^* E_2 e^{-i(\omega_1+\omega_2)t}, \tag{7}$$

$$\vec{P}^{(2)}_{GDF} = 2\epsilon_0 \chi^{(2)} E_1 E_2^* e^{-i(\omega_1-\omega_2)t}. \tag{8}$$

In the cavity of a laser, it is possible to generate other longitudinal modes, cre
more frequency combinations and making the behavior more unstable [30]. The
time this phenomenon was observed in a laser was in 1961, when it was reported th
addition to the laser's natural wavelength (694.3 nm), a second harmonic (347.2 nm)
been generated [8]. With this finding, the study of nonlinear optics and the use of SH
different areas of science began [31,32].

## 3. Experimental Setup

In Figure 1, the scheme of the experimental arrangement of an $Nd{:}YVO_4$ laser a KTP crystal intracavity for second harmonic generation is shown. As can be seen arrangement is the simplest in order to find the easiest way to generate the SHG. To ach Q-switching, a $Cr^{4+}:YAG$ crystal is used, which generates short pulses from 10 to 5 This experiment was carried out with a laser diode pumping at 808 nm (LD, Modu ML2020), a current source (Bk-Precision 1696), and a lens (L) with focal length $f = 5$ which is used to focus the beam onto the $Nd{:}YVO_4$. The cavity consists of the $Nd{:}Y$ crystal and a KTP intracavity glass, which have special coatings that work as refle surfaces at the wavelengths used (Figure 1), and at the output there are two beam spli BS1 (BS010), designed for wavelengths from 400 to 700 nm, and BS2 (BS011), is desi for wavelengths of 700 to 1100 nm. The resulting beam of BS1 is filtered by a bandpass (FL532-10, FG) at 532 nm, and the BS2 beam is filtered by a bandpass filter (FL1064-10, at 1064 nm. Such filtering affects photo-detectors $PD_1$ and $PD_2$, respectively, which co the optical signal to an electrical voltage to be analyzed by the oscilloscope (DPO7 the output of $PD_2$ is also analyzed by a frequency spectrum analyzer (FSA Aeroflex). unfiltered beam is introduced into an optical fiber to be analyzed in an optical spec analyzer (OSA MS9030A). All the technical specifications of the crystals and other op components can be consulted in the Alphalas web page [33].

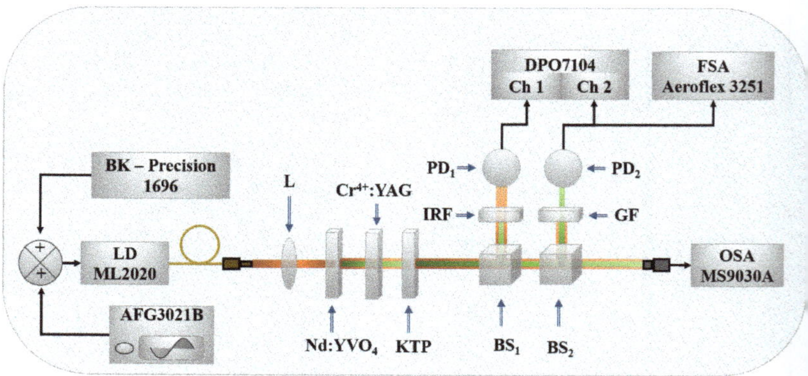


**Figure 1.** Schematic arrangement of the $Nd{:}YVO_4$-KTP laser with a KTP crystal intracavity for se harmonic generation and modulated by an external signal.

By carrying out the implementation of the experimental setup shown in Figure will be possible to know the unstable behavior that characterizes this type of laser its output as a second harmonic. With this, we can analyze the original response o $Nd{:}YVO_4$-KTP laser which will serve as a starting point before applying the harm modulation $A_m sin(2\pi f_m t)$ to the system.

In order to analyze the emission of the $Nd{:}YVO_4$-KTP laser with external modula in the experimental setup shown in Figure 1, a sinusoidal modulation signal given b arbitrary function generator (AFG) is added to the pump current. With the AFG device pump current $I_m$ is given as

$$I_m = I_0 + u'_m(t),$$

where $I_0$ is the initial current parameter, and $u'_m$ corresponds to the current obtained by the use of an Operational Amplifier (OA) that sums the signal $u_m(t) = A_m sin(2\pi f_m t)$ that comes from the AFG as external harmonic modulation. In order to reach the desired control, the modulation signal, with a suitable gain $A_m$ and modulation frequency $f_m$ able to eliminate coexisting states, must be defined [11,12].

Using the same arrangement shown in Figure 1, it is possible to separate the green (532 nm) and infrared (1064 nm) emissions through the use of beam splitters (BS1 and BS2) at the output of the laser; then, we characterize the behavior of the laser when applying the modulation. The acquisition of data can be achieved through the use of $PD_1$ and $PD_2$, that convert the optical signal to voltage amplitude; then, the signal is received by the oscilloscope (temporal series), the optical spectrum analyzer (optical spectra), and the frequency spectrum analyzer (frequency spectrum), to construct the figures of temporal series, optical spectrum, frequency spectrum, and other related with this measurements.

In order to obtain the laser emission from the $Nd{:}YVO_4$ it is necessary to feed the pump laser diode (LD) with an adequate voltage and current (Figure 1), these were fixed as $V = 7$ V and $I = 2.1$ A, respectively. The applied harmonic modulation was varied in frequency as 10 kHz < $f_m$ < 90 kHz in increments of 1 kHz, while modulation amplitude was increased in steps of 1 V within a range of 2 V < $A_m$ < 20 V. Figure 2a–e shows the emission of LD at 808 nm for $f_m = 60$ kHz and $A_m$ variable, and Figure 3a–e shows the emission of the laser diode with $A_m = 15$ V and variable $f_m$.

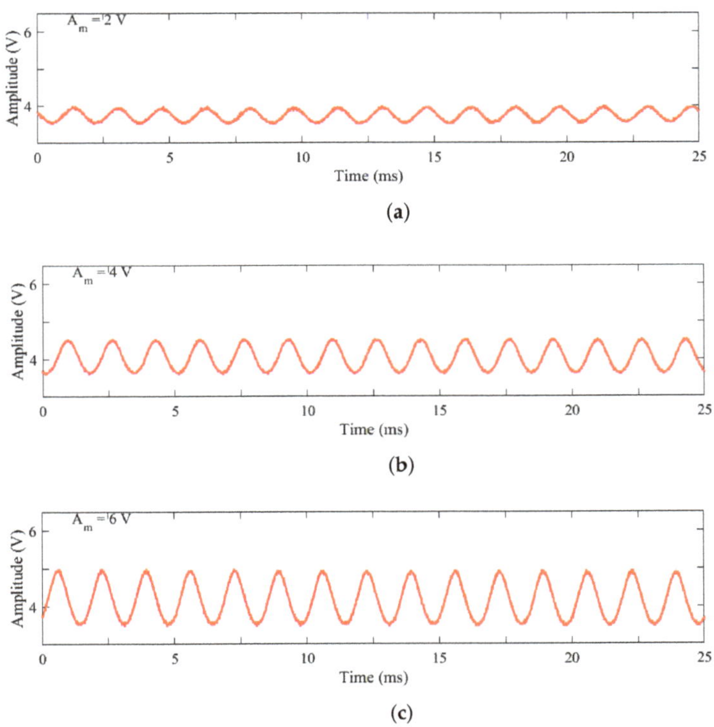


**Figure 2.** *Cont.*

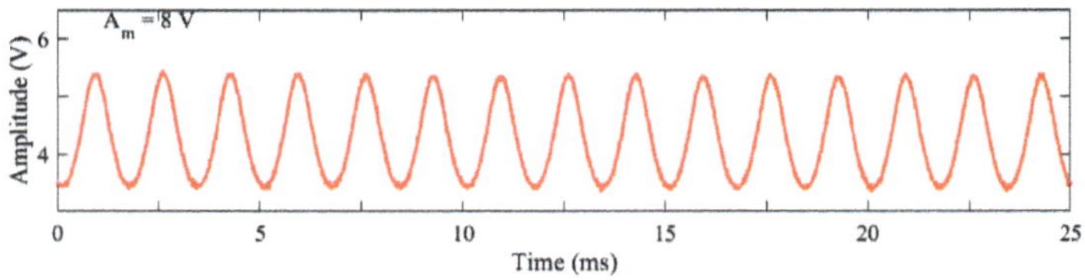


(**d**)

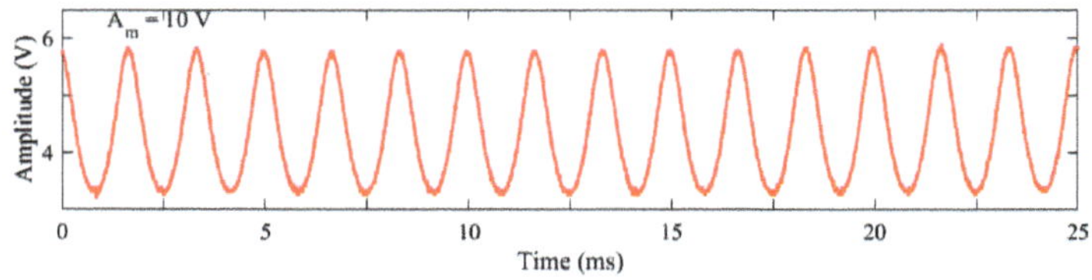


(**e**)

**Figure 2.** Pump laser diode emission by varying the harmonic modulation amplitude for a frequency $f_m$ = 60 kHz and modulation amplitude $A_m$: (**a**) 2 V, (**b**) 4 V, (**c**) 6 V, (**d**) 8 V, (**e**) 10 V.

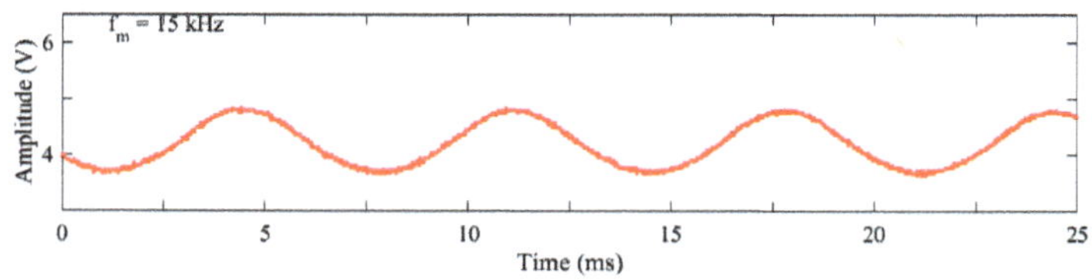


(**a**)

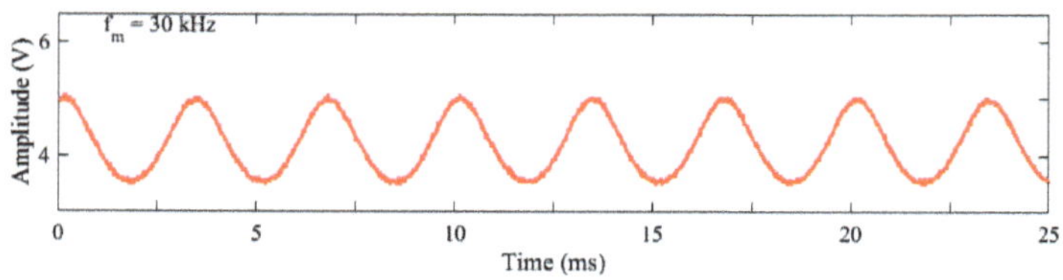


(**b**)

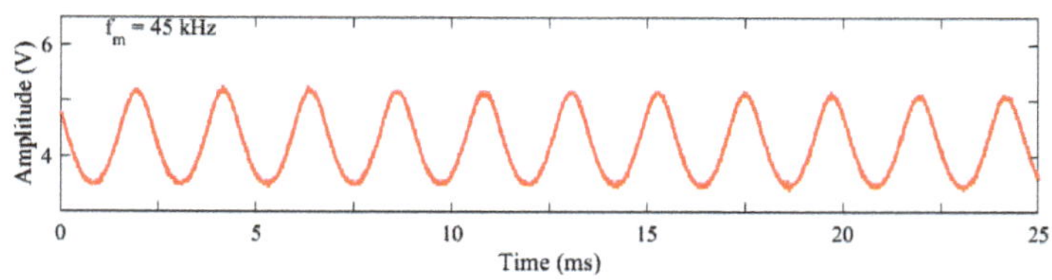


(**c**)

**Figure 3.** *Cont.*

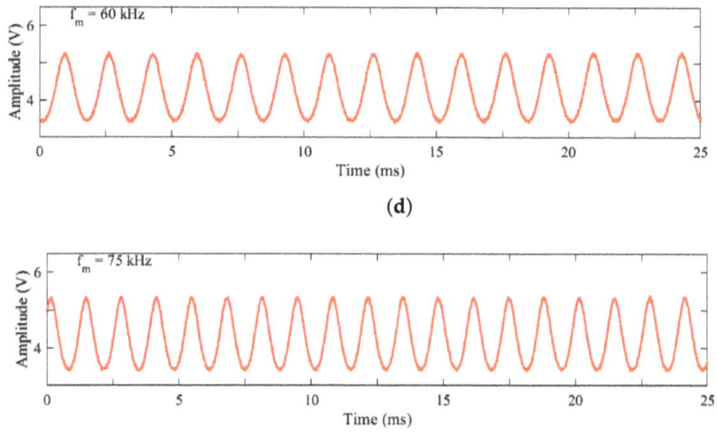


**Figure 3.** Laser diode emission by varying the harmonic modulation frequency and with fixed amplitude $A_m$ = 7.5 V and modulation frequency $f_m$: (**a**) 15 kHz, (**b**) 30 kHz, (**c**) 45 kHz, (**d**) 60 kHz, (**e**) 75 kHz.

Figure 4 shows the pump power necessary for the laser to emit at the 1064 nm wavelength and at 532 nm in continuous wave and with Q-switch. It is observed that the emission power in a continuous wave is higher than using Q-switch, also the emission threshold is lower in a continuous wave.

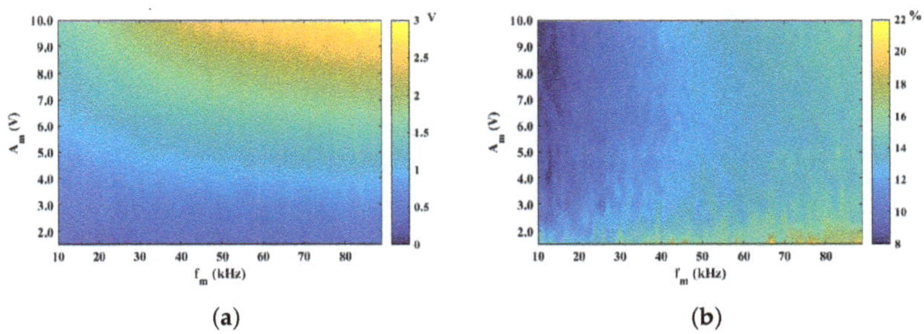


**Figure 4.** (**a**) voltage map of the DL emission when modulated with signal $u_m(t) = A_m \sin(2\pi f_m t)$. (**b**) percentage relationship of the voltage that presents the intensity of the emission of the LD with respect to the modulation amplitude $A_m$.

In Figure 4, we can observe that the voltage of the emission increases by increasing $f_m$ or $A_m$, but the intensity of the emission (Figure 4a vertical color bar, given in volts (V)), does not have the same amplitude to $A_m$, and only a percentage is observed Figure 4b; this relationship between the amplitude of the modulation and the amplitude of the emission is shown in Figure 5. The peak-to-peak voltage of the LD emission was calculated by varying modulation amplitude (Figure 5a) and modulation frequency (Figure 5b).

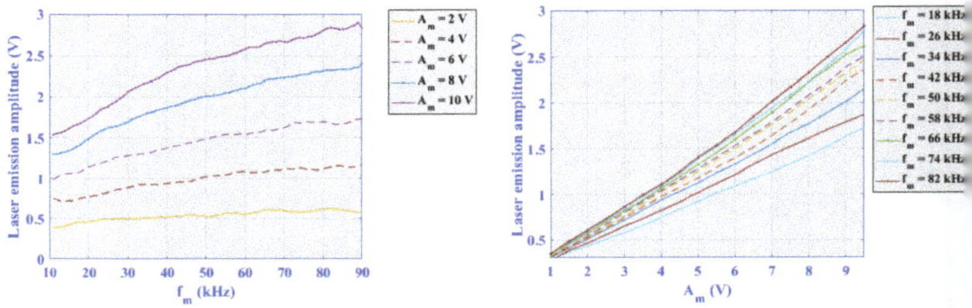


(**a**) (**b**)

**Figure 5.** Curves of the voltage of the emission of the LD varying: (**a**) amplitude ($A_m$), and (
quency ($f_m$).

## 4. Results of the Stabilization in Frequency and Amplitude of SHG of the *Nd:YV*(

The stability of the intensity of the emission of the second harmonic in *Nd:Y* KTP lasers is known to be normally irregular; so, in this experiment, it was work stabilize the intensity of the emission and also to achieve greater conversion effici The wavelength of the second harmonic (532 nm) with respect to the emission a fundamental wavelength (1064 nm). Passive Q-switching has been used to obtai emission intensity with less stability, but it may present greater conversion efficiency c laser power. It will be shown that, with the harmonic modulation $u_m(t) = A_m \sin(2\pi$ there is an emission intensity that has stability and greater conversion efficiency. Ir section, the modulation parameters for amplitude ($A_m$) and frequency ($f_m$) are define which the emission of the second harmonic has greater stability.

In order to know the behavior of the output power as a function of the pum power, from the experimental arrangement shown in Figure 1, a sweep of 0 to 4.5 optical pumping power was made, and the results can be seen in Figure 6, which sh the $Nd:YVO_4$ laser output, as a continuous wave and in Q-Switching mode at 1064 n well as the power output as a continuous wave and in Q-Switching mode at 532 nm.

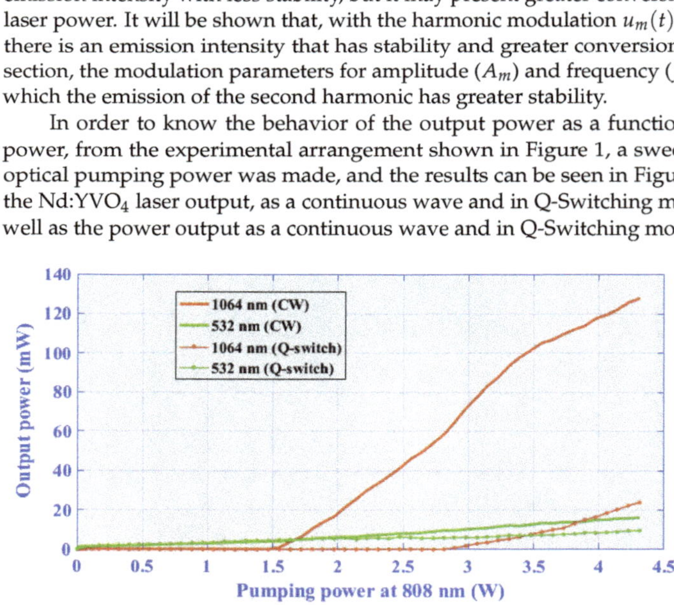


**Figure 6.** Power of the laser with KTP intra-cavity.

The intracavity conversion efficiency was found by relating the optical pow 532 nm to the optical power at 1064 nm. In Table 1, we can see that if the laser set-up w the Q-switching mode, efficiency was 40%, higher than the results in previous works whereas if the laser set-up was not in Q-switching mode, efficiency was only 12%.

In order to analyze the laser emission, an optical spectrum analyzer was used, in w modulated and unmodulated laser measurements were made, and it was verified tha the modulation ranges used, the laser emission lengths stay the same. Figure 7. show optical spectra amplitude at wavelength range 350 nm $< \lambda <$ 1300 nm of the laser emission $Nd{:}YVO_4$, with a modulation amplitude range 7 V < $A_m$ < 10 V and modulation frequency fixed to $f_m$ = 50 kHz to know the behavior of current output when modulating the diode current. The characteristic wavelengths of 808 nm of the pump laser diode, 1064 nm of the emission of $Nd{:}YVO_4$ and 532 nm of the second harmonic when using the KTP are observed. The wavelength of 350 nm corresponds to the frequency sums phenomenon of the fundamental wavelengths 1064 nm and second harmonic (532 nm).

**Table 1.** Optical efficiency conversion with intra-cavity configuration without modulation at $V = 7$ V and $I = 2.1$ A.

| | Q-Switch | Efficiency (%) |
|---|---|---|
| Intracavity | X | 12.7 |
| | ✓ | 40.0 |

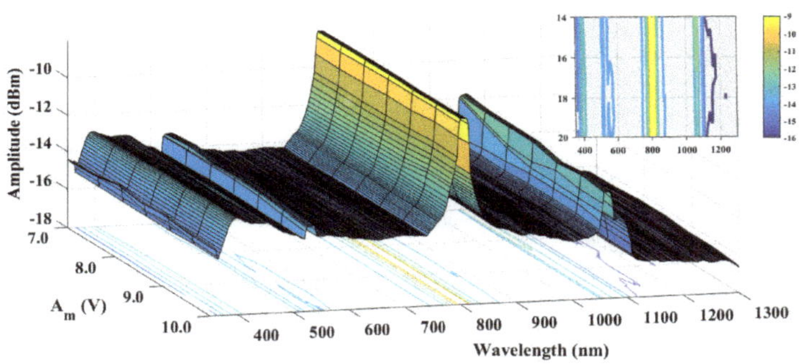


**Figure 7.** Surface of optical spectra.

Using the experimental arrangement for the laser configuration shown in Figure 1, it was found that the pulse frequency without modulation is $F_Q \approx 57$ kHz; this is observed in Figure 8, where the parameters of the modulation signal were kept constant with voltage $V = 7$ V and current $I = 2.1$ A.

When a harmonic modulation is added to the pump, it is expected that there will be a harmonic resonance between the modulation signal and the relaxation frequency of the laser ($F_Q$). Figure 9 shows the emission at 1064 nm, 532 nm, and the frequency spectrum of the emission at 532 nm for the modulation parameters $A_m = 7.5$ V and $f_m = 57$ kHz, showing a conversion efficiency of 25% (Table 2). Table 2 shows the conversion efficiency obtained for a modulation sweep $2V < A_m < 10$ V with a fixed frequency $f_m = 57$ kHz in the intracavity emission with Q-switching.

Here we can see that the frequency peak at 57 kHz is narrower, this can confirm harmonic resonance. An analysis of the frequency emission was carried out. As observed in the experimental setup, the emission at 532 nm was analyzed in frequency with a frequency spectrum analyzer, for each parameters variation ($A_m$, $f_m$) ten spectra were taken and the average of these spectra was calculated. With the averaged spectra, frequency maps were built, one for each modulation amplitude. Figure 10 shows the frequency spectrum maps of the $Nd{:}YVO_4$ laser emission that were obtained by varying the modulation frequency 1 kHz < $f_m$ < 100 kHz while using a range of amplitude variation modulation of 2 V < $A_m$ < 10 V. In the maps, each row represents a frequency spectrum taken from 1 to 100 kHz, the y-axis represents the modulation frequency variation.

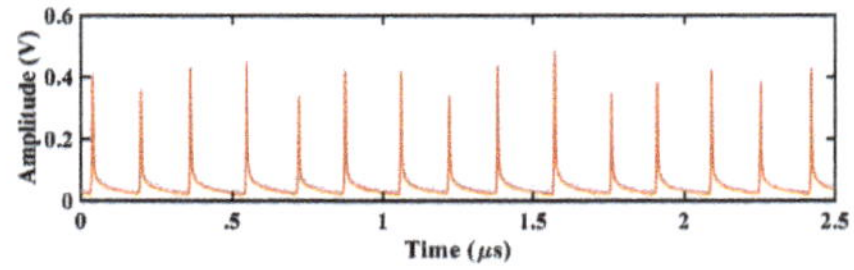


(**a**)

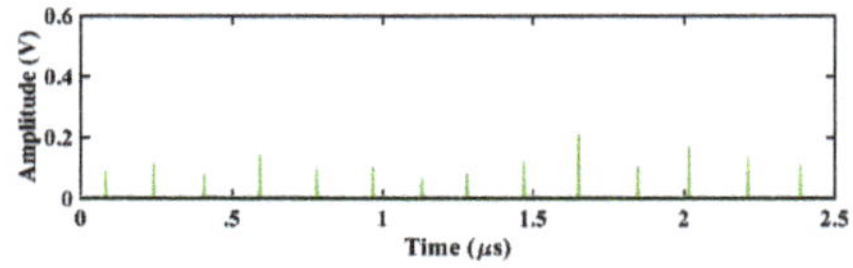


(**b**)

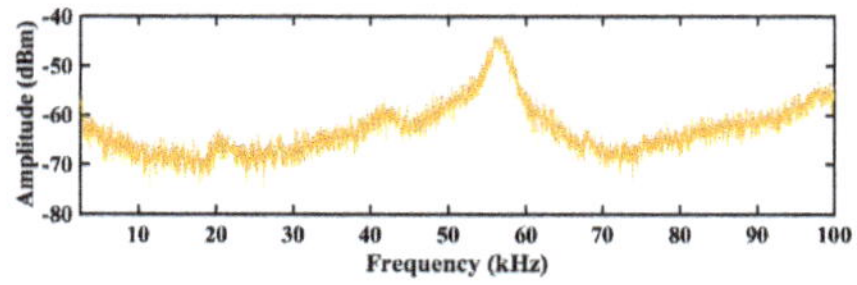


(**c**)

**Figure 8.** Time series and frequency spectra of the emission laser without modulation. (**a**) series of the 1064 nm wavelength emission. (**b**) Time series of the wavelength radiation of 53 (**c**) Spectrum frequency of the emission of 532 nm.

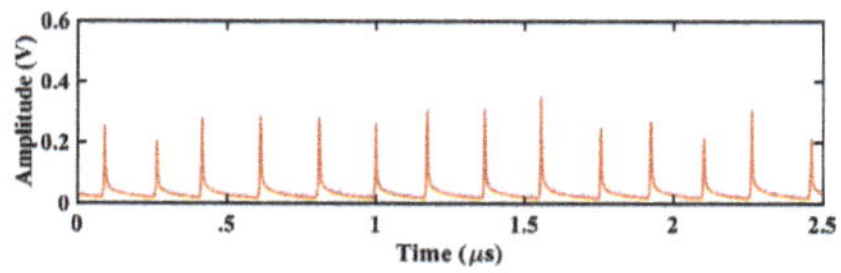


(**a**)

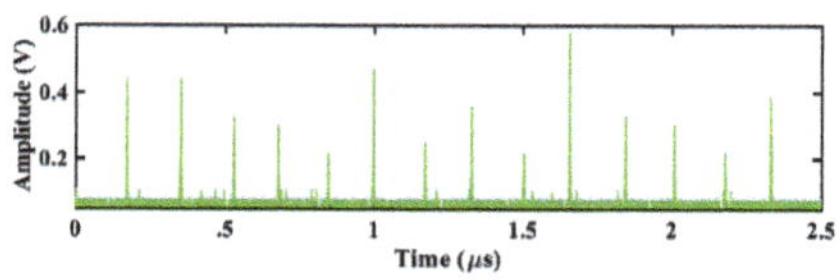


(**b**)

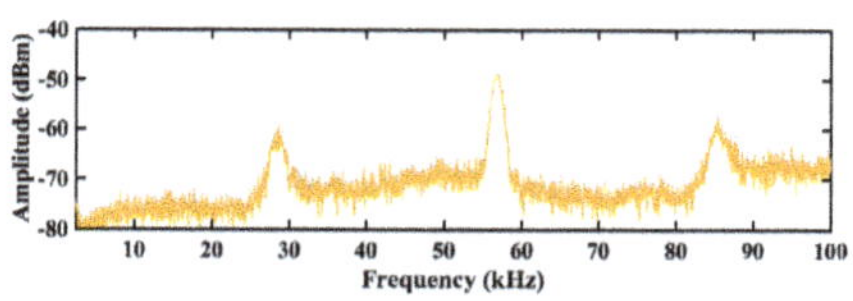


(**c**)

**Figure 9.** Time series of the laser emission with the modulation parameters $A_m = 7.5$ V $f_m = 57$ kHz. (**a**) Time series of the 1064 nm wavelength emission. (**b**) Time series of the wavel radiation of 532 nm. (**c**) Spectrum frequency of the emission of 532 nm.

**Table 2.** Optical efficiency conversion with intra-cavity configuration maintaining the pump of $V = 7$ V and $I = 2.1$ A, with modulation parameters 2 V $< A_m <$ 10 V and $f_m = 57$ kHz.

| | Limits | Efficiency (%) |
|---|---|---|
| Intracavity | Min | 14.0 |
| | Max | 25.0 |

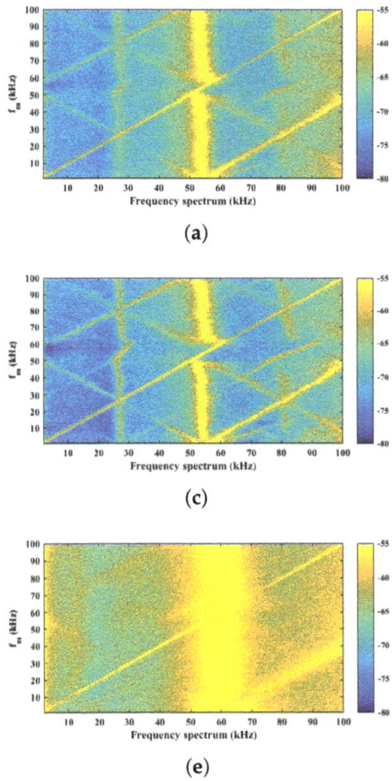


**Figure 10.** Frequency spectrum maps obtained by varying harmonic modulation frequency 1 kHz < $f_m$ < 100 kHz and modulation amplitude: (**a**) $A_m = 2$ V; (**b**) $A_m = 4$ V; (**c**) $A_m = 6$ V; (**d**) $A_m = 8$ V; (**e**) $A_m = 10$ V.

The local maxima of the time series obtained for each parameter set ($A_m$, $f_m$) were determined and with this the standard deviation was calculated using an algorithm in Matlab. In Figure 11, we can see the standard deviation for each set of parameters. The black areas represent the minimums of the standard deviation: Figure 11a represents the standard deviation of the local maxima, Figure 11b represents the standard deviation of the interspike-interval (ISI), and Figure 11c represents the values of the parameters for which there are minimums of standard deviation of the ISI and of the local maxima of the emission at 532 nm.

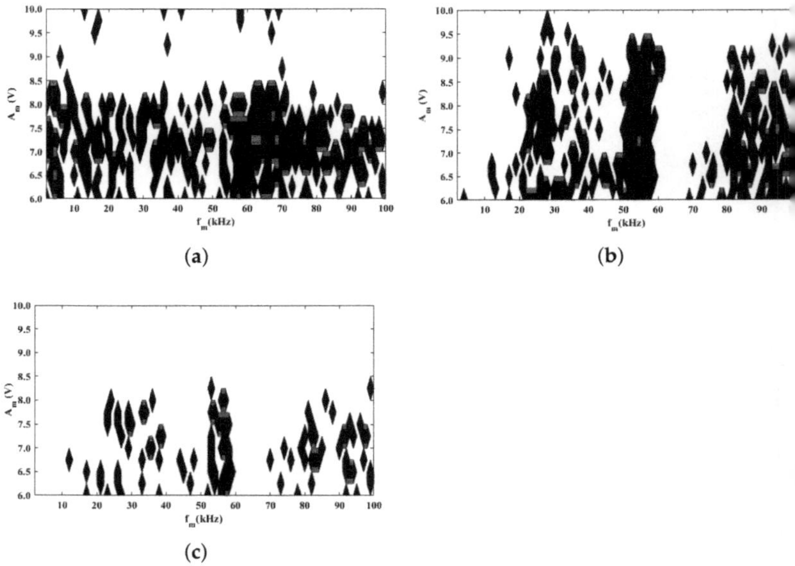


**Figure 11.** Stability zones for different amplitudes $A_m$ and frequencies $f_m$. (**a**) represents the sta deviation of the local maxima, (**b**) represents the standard deviation of the interspike-interva and (**c**) represents the values of the parameters for which there are minimums of standard dev of the ISI and of the local maxima of the emission at 532 nm.

## 5. Conclusions

The implementation of an intracavity $Nd{:}YVO_4$-KTP laser with passive Q-swit ($Cr^{4+}{:}YAG$) was presented, using a harmonic modulation technique to stabilize the sion in amplitude and frequency of the laser.

For the optical emission without modulation, the relaxation frequency coincides the peak of the frequency spectrum $F_Q = 57$ kHz, we observe in the Figure 8 that this is broad, which means that the frequency can have small variations with respect to the peak. With respect to the optical power, it was found that the conversion efficiency bet the fundamental wavelength 1064 nm and the wavelength of 532 nm, in pulsed emi (crystal for Q-switching) and using a saturable absorber ($Cr^{4+}{:}YAG$) intracavity, result 40% conversion efficiency. Analyzing the time series of the laser emission when modul signal $u_m(t) = A_m \sin(2\pi f_m t)$ is added to the pumping, the behavior of the laser emi is modified. In the frequency spectra, the relaxation frequency $F_Q$ and the modul frequency $f_m$, in the maps Figure 10, the frequency locking or resonance can be obse This resonance is achieved at the dominant frequency range 45 kHz < $f_m$ < 60 kH the amplitude 5 V < $A_m$ < 10 V ; under these conditions, the laser emission is stabi in amplitude and frequency. The $F_Q$ relaxation frequency of the $Nd{:}\ YVO_4$ laser found using a modulation amplitude of 7 V < $A_m$ < 8 V and a modulation frequ $f_m = FQ = 57$ KHz. The best result appeared with the modulation parameters $A_m =$ and $f_m = 57$ kHz, with this set of parameters, the frequency spectrum presents a narr peak, with which it was possible to stabilize the emission laser in frequency. The stan deviation was calculated from local maxima of time series obtained for each paran variation ($A_m$, $f_m$), i.e., for each pair of fixed values of $f_m$ and $A_m$, we obtain all the series and take all the local maxima to construct the standard deviation maps. It can be that, in order to stabilize the emission of the $Nd{:}YVO_4$-KTP intracavity laser with passi switching, a modulation frequency that is approximately equal to the oscillation frequ of the laser ($f_m = F_Q$), must be sought a harmonic resonance between the modul frequency and the relaxation frequency of the laser. In addition, the conversion efficiency from 1064 nm to 532 nm in the laser with modulation was calculated, achieving an efficiency between 14% to 25%. Compared to the emission without modulation, the efficiency is lower, however, stable zones with greater frequency and amplitude are obtained.

**Author Contributions:** S.M.A.D.: Writing—Original Draft, Writing—Review and Editing, Methodology, Software, Validation, Data Curation, Visualization. J.H.G.L.: Writing—Review and Editing, Resources, Project administration. R.J.R.: Supervision, Funding acquisition, Writing—Review and Editing, Resources. V.A.: Writing—Review and Editing, Resources. J.L.E.M.: Writing—Review and Editing, Visualization, Conceptualization. G.H.C.: Writing—Original Draft, Writing—Review and Editing, Methodology, Software, Validation, Data Curation, Visualization, Project administration. All authors have read and agreed to the published version of the manuscript.

**Funding:** This project was supported by CONACYT under project number 320597.

**Acknowledgments:** S.M.A.D. acknowledges support from National Council of Science and Technology -CONACyT-, Mexico, CONACYT. J.L.E.M. thanks CONACYT for financial support (CVU-706850, project: A1-S-26123, and project: 320597). R.J.R. thanks CONACYT for financial support, project No. 320597. G.H.C. acknowledges the State Council of Science and Technology of Jalisco (COECYT-JAL), project number 10304-2022, with title "Control difuso aplicado a sistemas de tecnología solar térmica y fotovoltaica para implementación de planta piloto para pasteurización de leche" de la Convocatoriadel Fondo de Desarrollo Científico de Jalisco para Atender Retos Sociales "FODECIJAL 2022".

**Conflicts of Interest:** The authors certify that they have NO affiliations with or involvement in any organization or entity with any financial interest (such as honoraria; educational grants; participation in speakers' bureaus; membership, employment, consultancies, stock ownership, or other equity interest; and expert testimony or patent-licensing arrangements), or non-financial interest (such as personal or professional relationships, affiliations, knowledge, or beliefs) in the subject matter or materials discussed in this manuscript.

## rences


Eichhorn, M. *Laser Physics: From Principles to Practical Work in the Lab*; Springer Science & Business Media: Berlin/Heidelberg, Germany, 2014.

Zhang, Y.; Zhao, S.; Li, D.; Yang, K.; Li, G.; Zhang, G.; Cheng, K. Diode-pumped doubly Q-switched mode-locked YVO4/Nd: YVO4/KTP green laser with AO and GaAs saturable absorber. *Opt. Mater.* **2011**, *33*, 303–307. [CrossRef]

Mukhopadhyay, P.; Alsous, M.; Ranganathan, K.; Sharma, S.; Gupta, P.; George, J.; Nathan, T. Analysis of laser-diode end-pumped intracavity frequency-doubled, passively Q-switched and mode-locked Nd: YVO4 laser. *Appl. Phys. B* **2004**, *79*, 713–720. [CrossRef]

Wang, X.; Li, M. Continuous-wave passively mode-locked Nd:YVO4/KTP green laser with a semiconductor saturable absorber mirror. *Laser Phys.* **2010**, *20*, 733–736. [CrossRef]

Liu, J.; Yang, J.; Fan, X.; Wang, G. Symmetric and short green-pulse generation by doubly Q-switched Nd:YVO4 laser. *Laser Phys.* **2010**, *20*, 222–225. [CrossRef]

Zhuo, Z.; Li, T.; Li, X.; Yang, H. Investigation of Nd:YVO4/YVO4 composite crystal and its laser performance pumped by a fiber coupled diode laser. *Opt. Commun.* **2007**, *274*, 176–181. [CrossRef]

Fields, R.; Birnbaum, M.; Fincher, C. Highly efficient Nd:YVO4 diode-laser end-pumped laser. *Appl. Phys. Lett.* **1987**, *51*, 1885–1886. [CrossRef]

Franken, P.; Hill, A.E.; Peters, C.E.; Weinreich, G. Generation of optical harmonics. *Phys. Rev. Lett.* **1961**, *7*, 118. [CrossRef]

McDonagh, L.; Wallenstein, R. Low-noise 62 W CW intracavity-doubled TEM 00 Nd: YVO 4 green laser pumped at 888 nm. *Opt. Lett.* **2007**, *32*, 802–804. [CrossRef]

Roy, R.; Murphy, T., Jr.; Maier, T.; Gills, Z.; Hunt, E. Dynamical control of a chaotic laser: Experimental stabilization of a globally coupled system. *Phys. Rev. Lett.* **1992**, *68*, 1259. [CrossRef]

Campos-Mejía, A.; Pisarchik, A.; Pinto-Robledo, V.; Sevilla-Escoboza, R.; Jaimes-Reátegui, R.; Huerta-Cuellar, G.; Vera-Avila, V. Synchronization of infrared and green components in a loss-modulated dual-cavity Nd:YAG laser with second harmonic generation. *Eur. Phys. J. Spec. Top.* **2014**, *223*, 2799–2806. [CrossRef]

Villafana-Rauda, E.; Chiu, R.; Mora-Gonzalez, M.; Casillas-Rodriguez, F.; Medel-Ruiz, C.; Sevilla-Escoboza, R. Dynamics of a Q-switched Nd:YVO4/Cr:YAG laser under periodic modulation. *Results Phys.* **2019**, *12*, 908–913. [CrossRef]

Tang, J.; Bai, Z.; Zhang, D.; Qi, Y.; Ding, J.; Wang, Y.; Lu, Z. Advances in All-Solid-State Passively Q-Switched Lasers Based on Cr4+:YAG Saturable Absorber. *Photonics* **2021**, *8*, 93. [CrossRef]

Jensen, O.B.; Andersen, P.E.; Sumpf, B.; Hasler, K.H.; Erbert, G.; Petersen, P.M. 1.5 W green light generation by single-pass second harmonic generation of a single-frequency tapered diode laser. *Opt. Express* **2009**, *17*, 6532–6539. [CrossRef] [PubMed]

15. Hitz, B. Generating High Green Power Without the Green Problem. *Photonics Spectra* **2007**, *41*, 92–93.
16. Ott, E.; Grebogi, C.; Yorke, J.A. Controlling chaos. *Phys. Rev. Lett.* **1990**, *64*, 1196. [CrossRef]
17. Pisarchik, A.; Corbalán, R.; Chizhevsky, V.; Vilaseca, R.; Kuntsevich, B. Dynamic stabilization of unstable periodic orbi $CO_2$ laser by slow modulation of cavity detuning. *Int. J. Bifurc. Chaos* **1998**, *8*, 1783–1789. [CrossRef]
18. Pisarchik, A.N.; Kir'yanov, A.V.; Barmenkov, Y.O.; Jaimes-Reátegui, R. Dynamics of an erbium-doped fiber laser with modulation: Theory and experiment. *JOSA B* **2005**, *22*, 2107–2114. [CrossRef]
19. Sevilla-Escoboza, R.; Pisarchik, A.N.; Jaimes-Reátegui, R.; Huerta-Cuellar, G. Selective monostability in multi-stable sy *Proc. R. Soc. Math. Phys. Eng. Sci.* **2015**, *471*, 20150005. [CrossRef]
20. Doedel, E.J.; Pando L., C.L. Isolas of periodic passive Q-switching self-pulsations in the three-level:two-level model for a with a saturable absorber. *Phys. Rev. E* **2011**, *84*, 056207. [CrossRef]
21. Doedel, E.J.; Pando, C.L. Multiparameter bifurcations and mixed-mode oscillations in *Q*-switched $CO_2$ lasers. *Phys.* **2014**, *89*, 052904. [CrossRef]
22. Khanin, Y.I. *Fundamentals of Laser Dynamics*; Cambridge International Science Publishing: Cambridge, UK, 2008.
23. Kee, H.H.; Lee, G.; Newson, T.P. Narrow linewidth CW and Q-switched erbium-doped fibre loop laser. *Electron. Lett.* *34*, 1318–1319. [CrossRef]
24. Yang, X.Q.; Wang, H.X.; He, J.L.; Zhang, B.T.; Huang, H.T. A compact passively Q-switched intra-cavity frequency double YAG/Cr4+: YAG composite crystal green laser. *Laser Phys.* **2009**, *19*, 1964. [CrossRef]
25. Tsai, T.Y.; Fang, Y.C. A saturable absorber Q-switched all-fiber ring laser. *Opt. Express* **2009**, *17*, 1429–1434. [CrossRef]
26. Degnan, J.J. Theory of the optimally coupled Q-switched laser. *IEEE J. Quantum Electron.* **1989**, *25*, 214–220. [CrossRef]
27. Pando L, C.; Meucci, R.; Ciofini, M.; Arecchi, F. $CO_2$ laser with modulated losses: Theoretical models and experiments chaotic regime. *Chaos Interdiscip. J. Nonlinear Sci.* **1993**, *3*, 279–285. [CrossRef]
28. Pando, C.; Acosta, G.L.; Meucci, R.; Ciofini, M. Highly dissipative Hénon map behavior in the four-level model of the $CO_2$ with modulated losses. *Phys. Lett. A* **1995**, *199*, 191–198. [CrossRef]
29. Boyd, R.W. *Nonlinear Optics*; Academic Press: Cambridge, MA, USA, 2020.
30. Boyd, R.W. The nonlinear optical susceptibility. *Nonlinear Opt.* **2008**, *3*, 1–67.
31. Alam, M.; Voravutinon, N.; Warycha, M.; Whiting, D.; Nodzenski, M.; Yoo, S.; West, D.P.; Veledar, E.; Poon, E. Compa effectiveness of nonpurpuragenic 595-nm pulsed dye laser and microsecond 1064-nm neodymium: Yttrium-aluminum-g laser for treatment of diffuse facial erythema: A double-blind randomized controlled trial. *J. Am. Acad. Dermatol.* **20** 438–443. [CrossRef]
32. Shen, Y. Surface properties probed by second-harmonic and sum-frequency generation. *Nature* **1989**, *337*, 519–525. [Cro
33. Alphalas-Inc. DPSS Laser Kit for Education and Research: LASKIT®-500. Available online: https://www.alphalas products/lasers/dpss-laser-kit-for-education-and-research-laskit-500.html?etcc_med=SEA&etcc_par=Google&etcc_ Laser%20Kit&etcc_grp=193189973&etcc_bky=laser%20kit&etcc_mty=p&etcc_plc=&etcc_ctv=304748847243&etcc_bd etcc_var=CjwKCAjw0N6hBhAUEiwAXab-Talimf0tJ_9sk93ikW0uWykv7-2Aprbr6ObiFyslIHYJ9kw3AXmXIRoCbqoQ BwE&gclid=CjwKCAjw0N6hBhAUEiwAXab-Talimf0tJ_9sk93ikW0uWykv7-2Aprbr6ObiFyslIHYJ9kw3AXmXIRoCbqoQ BwE (accessed on 13 April 2023).
34. Matía-Hernando, P.; Guerra, J.M.; Weigand, R. An Nd:YLF laser Q-switched by a monolayer-graphene saturable-abs mirror. *Laser Phys.* **2013**, *23*, 025003. [CrossRef]

MDPI
St. Alban-Anlage 66
4052 Basel
Switzerland
www.mdpi.com

*Photonics* Editorial Office
E-mail: photonics@mdpi.com
www.mdpi.com/journal/photonics

www.ingramcontent.com/pod-product-compliance
Lightning Source LLC
Chambersburg PA
CBHW041752220926
51847CB00014B/178
* 9 7 8 3 7 2 5 8 0 9 1 8 9 *